Shock Wave Science and Technology Reference Library

Collection Editors

Hans Grönig

Hans Grönig is Professor emeritus at the Shock Wave Laboratory of RWTH Aachen University, Germany. He obtained his Dr. rer. nat. degree in Mechanical Engineering and then worked as postdoctoral fellow at GALCIT, Pasadena, for one year. For more than 50 years he has been engaged in many aspects of mainly experimental shock wave research including hypersonics, gaseous and dust detonations. For about 10 years he was Editorin-Chief of the journal Shock Waves.

Yasuyuki Horie

Professor Yasuyuki (Yuki) Horie is internationally recognized for his contributions in high-pressure shock compression of solids and energetic materials modeling. He is a co-chief editor of the Springer series on Shock Wave and High Pressure Phenomena and the Shock Wave Science and Technology Reference Library, and a Liaison editor of the journal Shock Waves. He is a Fellow of the American Physical Society, and Secretary of the International Institute of Shock Wave Research. His current interests include fundamental understanding of (a) the impact sensitivity of energetic solids and its relation to microstructure attributes such as particle size distribution and interface morphology, and (b) heterogeneous and nonequilibrium effects in shock compression of solids at the mesoscale.

Kazuyoshi Takayama

Professor Kazuyoshi Takayama obtained his doctoral degree from Tohoku University in 1970 and was then appointed lecturer at the Institute of High Speed Mechanics, Tohoku University, promoted to associate professor in 1975 and to professor in 1986. He was appointed director of the Shock Wave Research Center at the Institute of High Speed Mechanics in 1988. The Institute of High Speed Mechanics was restructured as the Institute of Fluid Science in 1989. He retired in 2004 and became emeritus professor of Tohoku University. In 1990 he launched Shock Waves, an international journal, taking on the role of managing editor and in 2002 became editorin-chief. He was elected president of the Japan Society for Aeronautical and Space Sciences for one year in 2000 and was chairman of the Japanese Society of Shock Wave Research in 2000. He was appointed president of the International Shock Wave Institute in 2005. His research interests range from fundamental shock wave studies to the interdisciplinary application of shock wave research.

More information about this series at http://www.springer.com/series/8774

Shock Wave Science and Technology Reference Library 9

Collection Editors

Hans Grönig
Yasuyuki Horie
Kazuyoshi Takayama

The new Springer collection, Shock Wave Science and Technology Reference Library, conceived in the style of the famous Handbuch der Physik has as its principal motivation to assemble authoritative, state-of-the-art, archival reference articles by leading scientists and engineers in the field of shock wave research and its applications. A numbered and bounded collection, this reference library will consist of specifically commissioned volumes with internationally renowned experts as editors and contributing authors. Each volume consists of a small collection of extensive, topical and independent surveys and reviews. Typical articles start at an elementary level that is accessible to non-specialists and beginners. The main part of the articles deals with the most recent advances in the field with focus on experiment, instrumentation, theory, and modeling. Finally, prospects and opportunities for new developments are examined. Last but not least, the authors offer expert advice and cautions that are valuable for both the novice and the well-seasoned specialist.

Ozer Igra · Friedrich Seiler
Editors

Experimental Methods of Shock Wave Research

Springer

Editors
Ozer Igra
Department of Mechanical Engineering
Ben Gurion University of the Negev
Beer Sheva
Israel

Friedrich Seiler
Institute for Fluid Mechanics (ISTM)
Karlsruhe Institute of Technology (KIT)
Karlsruhe
Germany

Shock Wave Science and Technology Reference Library
ISBN 978-3-319-23744-2 ISBN 978-3-319-23745-9 (eBook)
DOI 10.1007/978-3-319-23745-9

Library of Congress Control Number: 2015949234

Springer Cham Heidelberg New York Dordrecht London

Printed on acid-free paper

Springer International Publishing AG Switzerland is part of Springer Science+Business Media (www.springer.com)

Preface

For a layman the term 'shock wave' may sound bizarre and completely superfluous for everyday life. However, this term is relevant in many daily events, for example the fast energy released during every lightning generates a shock wave that eventually decays to a sound wave. A shock wave accompanies every volcanic eruption, and it can be a hazardous event in coal mines; once an unexpected explosion occurs, a shock is transmitted through the mine's shafts, enhancing its strength due to burning of the coal dust entrained behind the shock front. Attenuating such a shock/blast wave is therefore an important issue. Shock waves accompany every supersonic flight and every missile/spacecraft entering into the earth's atmosphere. Recently, weak shock waves have been used for treating health problems. Nowadays it is normal practice to shatter kidney stones using focused weak shock waves. Shock waves are also used for returning a broken bone to its original location and there is an ongoing research on using shock waves for eliminating cancer growths. It is clear from this brief foreword that studying the physical behavior of shock and blast waves is essential for developing reliable ways for attenuating them, in the case of destructive shocks, and for proper design of supersonic airplanes and/or missiles. It is not surprising therefore that much effort has been devoted to developing laboratory facilities in which shock and blast waves can be generated and studied in a safe way. The facilities proposed, built, and used for studying shock and blast waves include shock tubes, shock tunnels, expansion tubes, ram accelerators, light gas guns, and ballistic ranges. In the present volume a variety of experimental methods which are used in shock tubes, shock tunnels, and expansion tubes facilities is presented. Details regarding ram accelerators, light gas gun, and ballistic range facilities will appear in Volume 10. When possible, in addition to the technical description of the facility, some typical results obtained using such facilities are described. In addition to descriptions of facilities mentioned above, this book includes techniques for measuring physical properties of blast waves and electrically generated shock waves.

Information about active shock wave laboratories at different locations around the world that are not described in the following chapters is given in the appendix.

This list is far from being complete. It includes only laboratories that responded favorably to our request for information. Additional information can be obtained from the research laboratories mentioned there.

The chapters in this book were written by different authors, each an expert in the described field/technique. We would like to thank all of them for their contributions to this book. Also, special thanks are due to our wives, Heidrun and Irene, for their patience and support exhibited while we spent much time bringing the book to publication.

Beer Sheva, Israel
Karlsruhe, Germany
May 2015

Ozer Igra
Friedrich Seiler

Contents

About the Editors

Ozer Igra has devoted most of his research activities to studying various aspects of shock and blast waves phenomena. His studies include both experimental and numerical investigations of strong (ionizing shocks in argon) and moderate to weak shock/blast wave in gases and in suspensions. Results of his investigations can be found in his many publications available in leading professional journals dealing with gas-dynamic flows and shock waves. He received his B.Sc. and M.Sc. degrees from the Department of Aeronautical Engineering of the Technion, Israel Institute of Technology and his Ph.D. from the Institute for Aerospace Engineering, University of Toronto, Canada. He Joined the Ben Gurion University of the Negev in 1971. There he established the Shock Waves Laboratory, supervised many masters, doctorate, and postdoc students and served as the chairman of the Department of Mechanical Engineering and thereafter as the Dean of the Faculty of Engineering. He is on the International Advisory Committee of the International Symposium on Shock Waves (ISSW) and the International Symposium on Shock Interactions (ISIS), and on the editorial board of the Shock Waves Journal.

Friedrich Seiler is well-known for his studies in shock tube technology at the French-German Research Institute of Saint-Louis in France. As scientist at the ISL, from 1980 on, his research mainly deals with high-velocity flight aerodynamics, hypersonics, optical measuring techniques and fundamental research in the field of ram acceleration. Until his retirement in the year 2011, he became head of the Aerothermodynamics & Shock Tube Laboratory at ISL. He contributes in the editorial board to the progress of the "International Journal on Shock Waves" and is also a member of the International Advisory committee of the "International Symposium on Shock Waves (ISSW)" and the "International Symposium on Flow Visualization (ISFV)". From the University of Karlsruhe in Germany he received his "Dipl.-Phys." degree in physics and his "Dr.-Ing." degree from the same University. Also the University of Karlsruhe, now Karlsruhe Institute of Technology (KIT), has awarded him in 1992 the rank "Professor" and since then he is Lecturing Professor for Fluid Mechanics at KIT. His current interest is focused on the dynamics of Mach waves in supersonic jets. He contributed to a theory which describes the Mach wave behavior using a new approach.

Part I
Shock Wave Generation in Gases

Shock Tubes

Ozer Igra and Lazhar Houas

1 Introduction

A shock tube is a facility that can produce transmitted shock waves in controlled laboratory conditions. The facility is a duct divided into two sections as shown schematically in Fig. 1. One part, the driver, accommodates high pressure gas (P_4) while the other section (the driven) contains the low pressure test gas (P_1). The two sections are separated by a barrier (diaphragm). Once the barrier is suddenly removed a shock wave propagates into the driven section and a rarefaction wave propagates into the driver section. By appropriate choice of the two pressures, P_4 and P_1, and gas species a desired strength of the resulting/generated shock wave (its Mach number, M_s) can be obtained. As shown subsequently, the initial thermodynamic conditions of the gases used dictate the post-shock flow i.e., the conditions behind the shock wave.

The first to describe an experimental investigation conducted in a shock tube was Vielle [30]. Thereafter shock tubes were frequently used for studying various types of compressible flows and for simulating reentry flows. For a historical description of shock wave investigations see Chap. 1 in the *Handbook of Shock Waves*, Ben-Dor et al. [4]. In the following sections a general description of the flow evolved in a shock tube is given; this is followed by descriptions of different geometries and techniques used for generating desired shock-tube-flows and the different diagnostics used for recording these flows.

O. Igra
Ben Gurion University, 84105 Beer Sheva, Israel
e-mail: ozer@bgu.ac.il

L. Houas (✉)
CNRS, Aix-Marseille Université, IUSTI UMR 7343, 13013 Marseille, France
e-mail: Lazhar.houas@univ-amu.fr

O. Igra and F. Seiler (eds.), *Experimental Methods of Shock Wave Research*,
Shock Wave Science and Technology Reference Library 9,
DOI 10.1007/978-3-319-23745-9_1

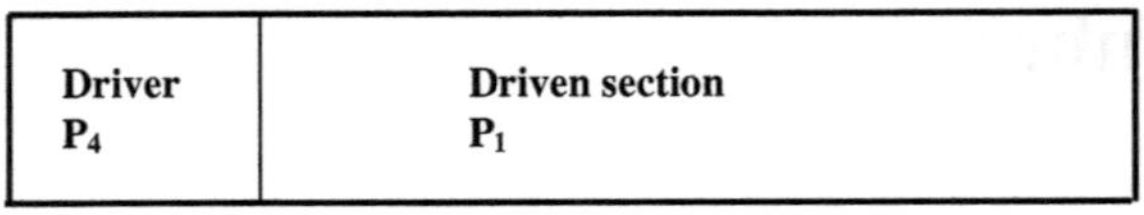

Fig. 1 Schematic description of a shock tube

2 Shock Tube Flow

The flow generated in a straight, horizontal shock tube with uniform cross-section upon removal of the barrier separating between the driver and driven sections of the tube is one-dimensional non-steady flow. It consists of a shock wave propagating into the driven section and an expansion wave, centered at the membrane position, propagating into the driver section; see schematic presentation in Fig. 2a. A contact surface, propagating into the driven section, behind the transmitted shock wave separates between the driven and the driver gases. It should be mentioned here that Fig. 2a represents the ideal case in which the incident shock wave is generated instantly after removal of the separating barrier (diaphragm) and the rarefaction wave is a centered wave. In reality this is not necessarily the case as will be shown subsequently. The prevailing pressure and temperature variations in an ideal shock tube flow are shown schematically in Fig. 2b. Symbols (1) and (4) represent the initial states in the driven and the driver sections, respectively; (2) and (3) represent the flow fields prevailing behind the transmitted shock wave and behind the contact surface, respectively, and (5) and (6) indicate flow field prevailing behind the reflected shock wave and the reflected rarefaction wave, respectively. In the following, equations relating the flow fields across these waves are presented. Indexes numbered from 1 to 6 refer to flow fields (1) to (6). A detailed derivation of these equations is available in Glass and Hall [8] and in many gasdynamic textbooks, e.g., see Sect. 5.12 in Aksel and Eralp [1].

2.1 Flow Relations Across the Shock Front

For a perfect gas, which is a reasonable assumption for many gases in which the pressure and temperature jump across the shock front is moderate, the following relations can be derived from the one-dimensional conservation equations of mass, momentum and energy,

$$\frac{\rho_2}{\rho_1} = \frac{u_1}{u_2} = \frac{(\gamma+1)M_s^2}{2+(\gamma-1)M_s^2},$$

where M_s is the shock wave Mach number, ρ and u are the gas density and velocity, respectively, and index 1 refers to pre-shock conditions while 2 refers to post-shock conditions. γ is the gas specific heat capacity ratio.

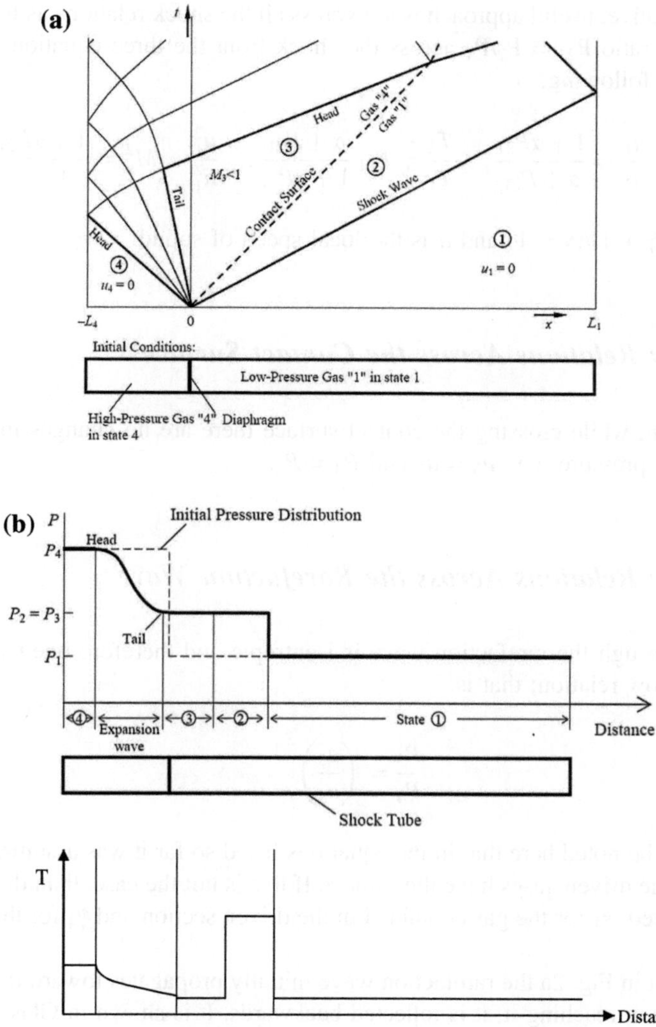

Fig. 2 **a** Schematic description of waves generated in a shock tube. **b** Schematic description showing pressure and temperature variations inside a shock tube

The pressure and temperature relations across the shock front are given in the following two equations:

$$\frac{P_2}{P_1} = 1 + \frac{2\gamma}{(\gamma+1)}\left(M_s^2 - 1\right)$$

$$\frac{T_2}{T_1} = 1 + \frac{2(\gamma-1)}{(\gamma+1)^2}\frac{\gamma M_s^2 + 1}{M_s^2}\left(M_s^2 - 1\right)$$

An alternative, useful approach is to express all the shock relations as function of the pressure ratio $P_{21} \equiv P_2/P_1$ across the shock front the three equations then are given in the following:

$$\frac{\rho_2}{\rho_1} = \frac{u_1}{u_2} = \frac{1+\alpha P_{21}}{\alpha + P_{21}}; \quad \frac{T_2}{T_1} = P_{21}\frac{\alpha + P_{21}}{1+\alpha P_{21}}; \quad \frac{u_1^2}{a_1^2} = M_s^2 = \frac{1+\alpha P_{21}}{1+\alpha},$$

where $\alpha \equiv (\gamma + 1)/(\gamma - 1)$ and a is the local speed of sound.

2.2 Flow Relations Across the Contact Surface

By definition, while crossing the contact surface there are no changes in the flow velocity and pressure, i.e., $u_3 = u_2$ and $P_3 = P_2$.

2.3 Flow Relations Across the Rarefaction Wave

The flow through the rarefaction wave is isentropic and therefore one can use the isentropic flow relation; that is:

$$\frac{P_3}{P_4} = \left(\frac{a_3}{a_4}\right)^{\frac{2\gamma}{\gamma-1}}.$$

It should be noted here that in the equations listed so far it was assumed that the driver and the driven gases have the same γ. If this is not the case, than different γ's should be used. γ_1 for the gas contained in the driven section and γ_4 for the driver's gas.

As shown in Fig. 2a the rarefaction wave initially propagates toward the driver's end-wall; upon reaching it, it is reflected backwards. It is shown in Glass and Hall [8] that along the Mach waves composing the rarefaction wave the following relation holds: $u + \frac{2}{\gamma-1}a = Const$. Therefore, we can relate state 3 to state 4 by the following equation,

$$u_4 + \frac{2}{\gamma - 1}a_4 = u_3 + \frac{2}{\gamma - 1}a_3.$$

Based on equations listed so far one can obtain a relation between the initial pressure ratio P_{41} ($\equiv P_4/P_1$) and the resulting shock wave Mach number, M_s. It is:

$$P_{41} = \left[1 + \frac{2\gamma}{\gamma+1}\left(M_s^2 - 1\right)\right]\left[1 - \frac{\gamma-1}{\gamma+1}\frac{a_1}{a_4}\left(M_s - \frac{1}{M_s}\right)\right]^{-\frac{2\gamma}{\gamma-1}}.$$

It should be noted that the last equation assumes equal γ's in the driver and the driven gases.

The pressure ratio ($P_{52} \equiv P_5/P_2$) across the reflected shock wave from the driven section end-wall is: $P_{52} = \frac{\alpha+2-P_{12}}{1+\alpha P_{12}}$; the density ratio across the reflected shock wave from the driven section end-wall is: $\rho_{52} = \frac{1+\alpha P_{52}}{\alpha+P_{52}}$ and the temperature ratio is, $T_{52} = \frac{P_{52}(\alpha+P_{52})}{1+\alpha P_{52}}$. For detailed derivation of the above listed relations see Glass and Hall [8]. In this book one also finds these relations in a graphical form as well as additional details and relations concerning deviation from ideal conditions and interactions between various waves occurring at later time after the barrier removal. An internet site where the ideal shock tube flow equations listed here were numerically solved and graphically presented is available at http://www.bgu.ac.il/me/laboratories/shockwave/. By providing the initial geometrical and physical conditions one can easily obtain all the resulting flow properties.

Deviations from ideal shock tube operation result from ignoring the following facts: In reality the gases used are not perfect and therefore ignoring viscous losses should be justified. In many cases, because of the very short flow duration in the shock tube energy losses due to friction are negligibly small. Furthermore, in developing the shock tube relations it was assumed that the gas is a perfect gas. This could be reasonable when the incident shock wave has a moderate strength; for air $M_s < 5$. When this is not the case the above listed equations have to be adapted to include energy losses and real gas effects; for details see Anderson [2]. It is also assumed in developing the flow equations for the flow-field generated inside the shock tube that the incident shock wave is generated instantly upon removal of the barrier separating between the driver and the driven gases, and also that a centred rarefaction wave propagates into the driver; as shown schematically in Fig. 2a. However, in reality the barrier used cannot be opened/removed instantly. As discussed subsequently, when different types of suitable barriers for shock tube operation are presented, it takes a finite time for either the diaphragm breaking or the opening of a quick-opening-valve that is used for a barrier. As a result of these non-ideal processes, initially compression waves propagate into the driven section, and downstream of the diaphragm they coalesce to a well defined normal shock wave, and a non-centred rarefaction wave propagates into the high pressure driver gas.

3 Review of Different Shock Tubes

A variety of shock tubes have been and are used worldwide during the past century; a chronological description of studies regarding shock and blast wave phenomena can be found in Chap. 1 of Ben-Dor et al. [4] and in Krehl [23]. In laboratories the most convenient and safe way to generate shock/blast waves in a controllable way is a shock tube. The geometrical shape/size and the range of initial conditions to be used (pressure and temperature) depend on the type of flow to be generated inside the driven section (test section) of the tube. In the following a brief description is given of only a few typical designs along with a brief description of studies conducted in these shock tubes. The selection of described tubes was made based on information available to the writers.

3.1 UTIAS Hypervelocity Shock Tube, Canada

The UTIAS combustion driven hypersonic shock tube at University of Toronto, the Institute for Aerospace Studies was designed to simulate high enthalpy, real gas flow typical of hypersonic flight. The combustion driver is capable of generating shock speeds in excess of 7 km per second. It was built in 1964 at the Institute for Aerospace Studies, University of Toronto, Canada. A brief description of this tube is outlined subsequently; a detailed description is given in Boyer [5]. A schematic drawing of the tube is given in Fig. 3. Although designed for generating hypervelocity flows it is suitable for studying a wide range of incident shock wave Mach numbers, from a bit higher than one and up to about 20. When flows generated behind weak/moderate incident shock wave Mach number are investigated

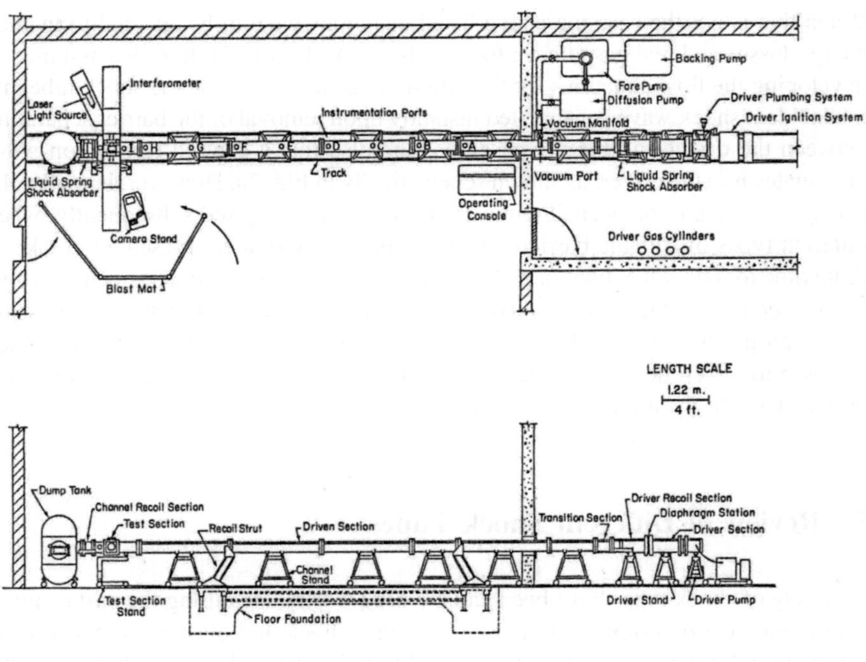

Fig. 3 An overview (*on top*) and a side view (*at the bottom*) of the UTIAS 4 by 7 in. shock tube

($M_s < 5$), there is no need for a combustible gas mixture and normal high pressure gases, released from compressed gas bottles are used.

A photographic view of this shock tube is given in Fig. 4. The driven section of the shock tube has a rectangular cross-section of 4 in. by 7 in. (10 cm by 18 cm) and the driver (56 in. long) consists of a 6 in., inner-diameter chrome-molybdenum steel

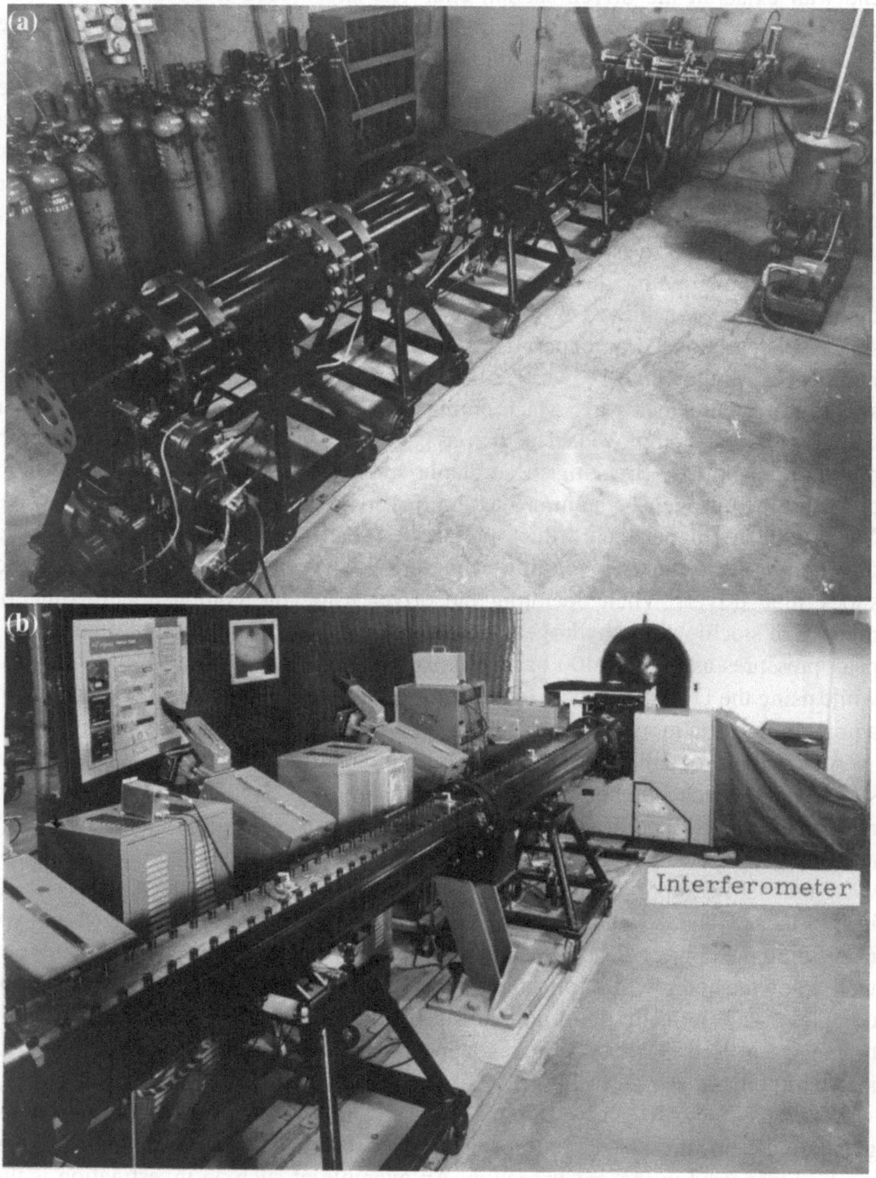

Fig. 4 General view of the 4 × 7 in. UTIAS shock tube. **a** Driver section. **b** Driven section

cylinder with a wall thickness of 4 in. Such a thick wall enabled holding very high pressures, up to 11,000 psi with a safety factor of 6. The connection between the two sections, the driver and the driven, is done via a transition section having a varying cross-section from circular to rectangular. The driven section of the shock tube is about 50 ft. long and is terminated by a vertical, cylindrical dump tank. The dump tank prevents existence of a reflected shock, unavoidable in cases when an end-wall exists at the driven section end. The shock tube test-section is located approximately 46 ft. downstream from the diaphragm separating the driver from the driven sections. The test section had high quality glass windows having a 7×10 in. field of view and a 3.75 in. thickness. The windows were made from borosilicate crown optical glass (BK-7), with surfaces ground and polished flat to within a wedge angle less than 2 s of arc, and specified to have a maximum peak to peak variation in transmitted wave front over the entire diameter of less than one-quarter of a wavelength at 6328 Å. These extreme requirements were needed as the principal diagnostic tool was a Mach-Zehnder interferometer.

The wide range of incident shock wave Mach numbers that could be generated in this shock tube is achieved by using a variety of diaphragms separating the two shock tube sections. For experiments in which the incident shock wave Mach number was in the range $1 < M_S < 2$ thin mylar diaphragms are used; for a medium strength incident shock wave Mach number, i.e., $2 < M_S < 8$ a few layers of mylar diaphragms were employed. The driver gas was compressed gas fed from high pressure gas-bottles. Breaking these diaphragms was reached by introducing high pressure air, nitrogen or helium into the driver section. For incident shock wave Mach numbers higher than 8 thick pre-scribed steel diaphragms were used. These diaphragms were burst by igniting a mixture of hydrogen and oxygen inside the shock tube driver. When operating the driver using the combustion mixture (a mixture of stochiometric hydrogen and oxygen with 77.5 % dilution of other gases) peak pressures as high as 400 bar were reached. In the following, results obtained while using the UTIAS 4 by 7 in. shock tube driver in its two different modes, first in a non-combustion mode and thereafter as a combustion driver, are briefly presented.

An example of an experimental investigation in which the generated incident shock wave Mach number was within the range of $2 \le M_s \le 8$ is the study on regions and transitions of nonstationary oblique shock wave diffractions in perfect and imperfect gas conducted by Ben Dor [3]. In this study the diffraction of a shock wave by sharp compressive corners ($2° \le \theta_w \le 60°$) was investigated. It was shown in Ben Dor [3] that seven shock wave diffraction domains exist in nitrogen and six in argon in the range $1 \le M_s \le 8$ and $0° \le \theta_w \le 90°$. The domains consist of the four well known shock wave reflections: regular reflection, single-Mach reflection, complex-Mach reflection and double-Mach reflection. The range of initial conditions used in Ben Dor's experiments appears in Table 1 and a sample of obtained interferograms is shown in Fig. 5.

When using the 10 cm by 18 cm (4 by 7 in.) UTIAS hypersonic shock tube for generating very strong incident shock waves ($10 \le M_s \le 20$) a combustible gas mixture was used in the driver section. An example of such an investigation is the work of Igra [18] where the recombination rate constant of ionized argon was

Table 1 The range of initial conditions used in Ben Dor's experiments

Initial conditions for obtaining various shock wave mach numbers in nitrogen			Initial conditions for obtaining various shock wave mach numbers in argon		
Ms	P_{41}	Driving gas	Ms	P_{41}	Driving gas
2.0	110	CO_2	2.0	60	CO_2
3.7	220	He	3.0	70	He
4.8	690	He	4.4	350	He
6.2	800	He	5.2	800	He
7.0	1550	H_2	6.1	520	He
7.8	3320	H_2	6.9	1000	H_2
			7.9	1900	H_2

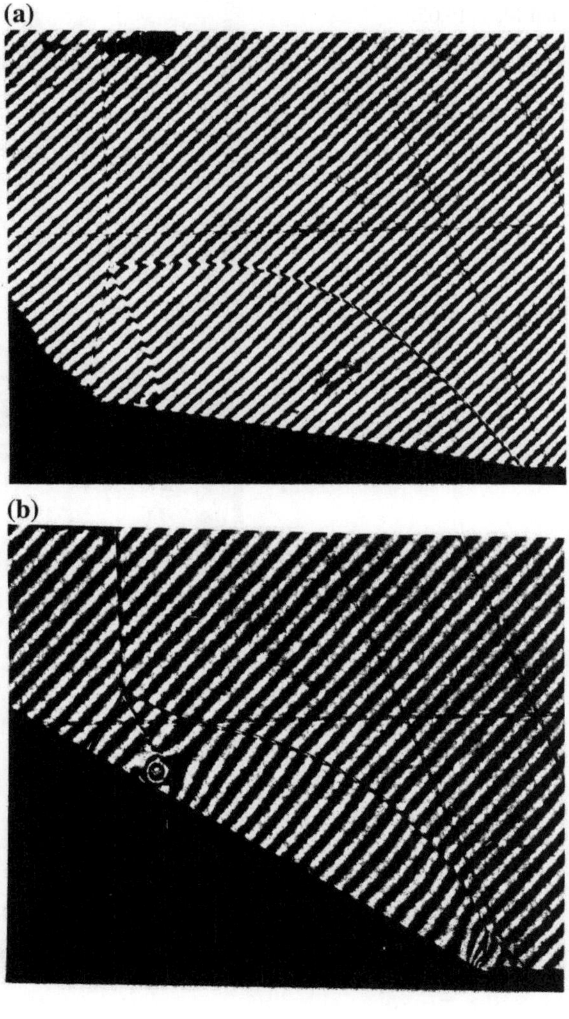

Fig. 5 Interferograms of different oblique shock wave diffractions in nitrogen in nonstationary flows; taken from Ben Dor [3]. **a** $M_S = 4.72$, $\Theta_W = 10.00°$, $P_O = 15.00$ torr, $T_O = 295.0$ K. **b** $M_S = 3.74$, $\Theta_W = 30.00°$, $P_O = 15.27$ torr, $T_O = 297.3$ K

investigated. Experiments using a 15° corner expansion model placed in flow of ionized argon, generated by a strong shock wave provided values of the recombination rate constant as a function of temperature. It also provided information on variations of the plasma flow quantities with distance from the corner. The data were obtained by using mainly a Mach-Zehnder interferometer which had a nine inch diameter field of view. The interferometer had a dual frequency laser light source and was driven by either a ruby or neodymium rod. The range of the incident shock Mach number, electron number density, temperature and initial channel (driven section) pressure and temperature were as follows: $13 < M_s < 19$, $10^{16} < n_e < 1.5 \times 10^{17}$ cm^{-3}, $9000 < T < 13{,}000$ K, $2.2 < P_1 < 10$ torr, $T_1 \approx 300$ K.

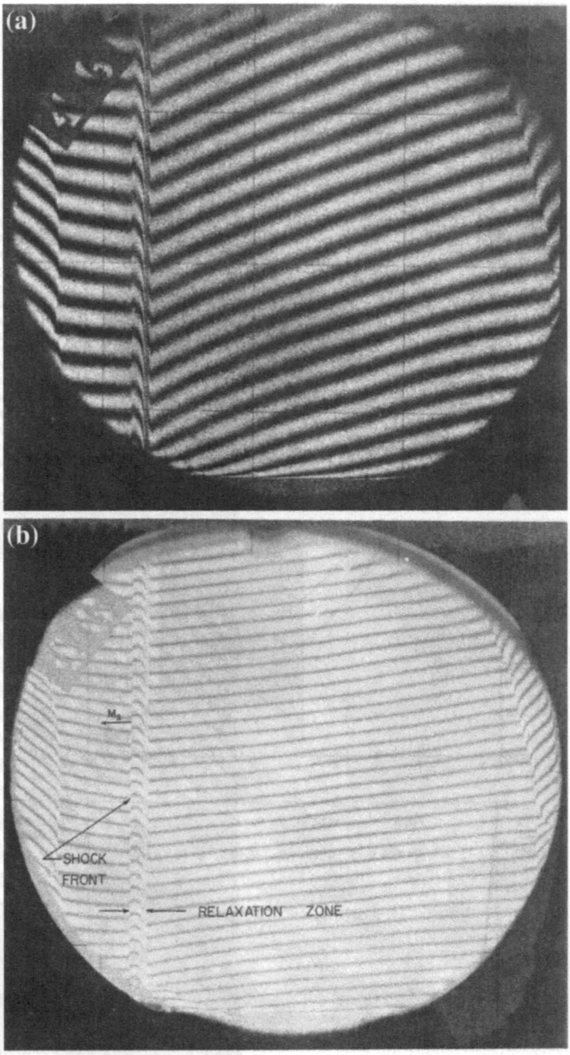

Fig. 6 Relaxation zone behind strong shock wave. $P_1 = 1.83$ torr, $T_1 - 296.1$ K and $M_s = 19.05$. **a** $\lambda = 10{,}600$ Å, **b** $\lambda = 5300$ Å

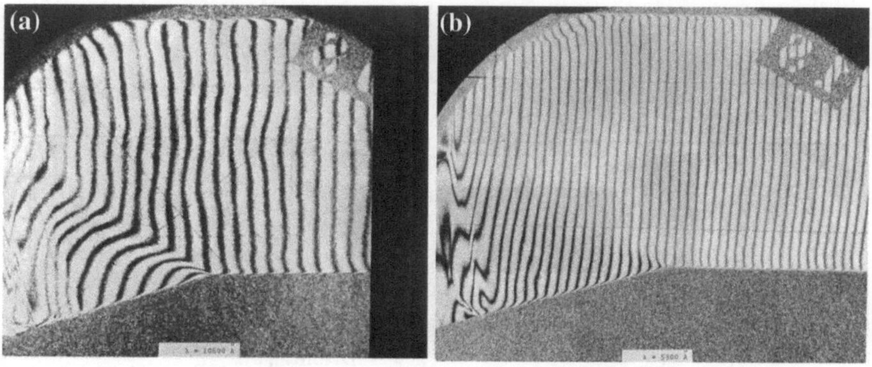

Fig. 7 Interferograms of expansion argon plasma. $P_1 = 8.60$ torr, $T_1 - 298.2$ K and $M_s = 13.3$. **a** $\lambda = 10{,}600$ Å, **b** $\lambda = 5300$ Å

Typical interferograms from Igra's [18] study are shown in Figs. 6 and 7. In Fig. 6 the relaxation zone prevailing behind the strong incident shock wave is shown and in Fig. 7 the expansion of the ionized argon plasma is exhibited.

3.2 Refurbishing the UTIAS Hypervelocity Shock Tube

In 1994, the 10 cm by 18 cm UTIAS hypersonic shock tube was transported to the Shock Wave Research Center of the Institute of Fluid Science, Tohoku University, Japan. A large diameter shock tube for generating strong shock waves requires a thick diaphragm that separates between the high pressure driver gas and the low-pressure test gas and hence its rupture is not easily controllable. As a result the shock tube Mach number reproducibility is difficult to achieve. Furthermore, in the original version diaphragm changing was laborious and time consuming. In order to overcome these two limitations a modification altering this shock tube to a diaphragm-less one was implemented. A quick opening valve, whose operational concept and design is described in Ikui et al. [19] and in Oguchi et al. [26], is shown in Fig. 8a. This valve was added replacing the original diaphragm section. In addition, 12 pressure transducers were added along the driven section and the original dump-tank was abolished. Instead of the dump-tank a 290 mm i.d. and 2.5 m long steel tube was added. For attenuating the reflected shock wave an alloy, 25 mm diameter and 350 mm long with coarse threads was installed at the end of the modified dump tank.

When the back-up pressure was reduced, the piston started to recede smoothly sliding inside the co-axial teardrop shaped section; see Fig. 8a. For quickly reducing the high-pressure gas, an auxiliary section was connected to a double diaphragm section shown in Fig. 8a. The diaphragm rupture initiates the piston's recessive motion. It may be a demerit of this diaphragm-less operation system that the shock

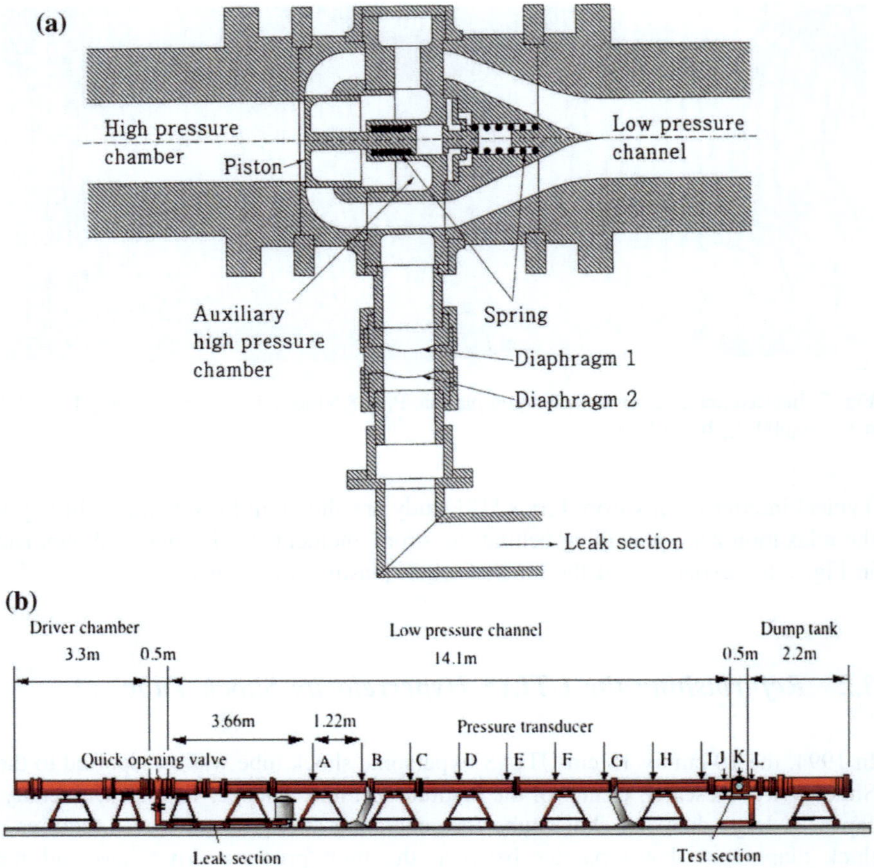

Fig. 8 **a** The quick opened piston valve replacing the diaphragm in the refurbished shock tube. **b** The modified UTIAS, diaphragm-less, shock tube

formation length was slightly longer than that observed when using a diaphragm rupturing system. However, the scatter in repeatability of the generated incident shock wave is much less when using the diaphragm-less operation system; for details see Ikui et al. [19].

Additional refurbishment made in the original UTIAS shock tube is the replacement of the original driver-section with a single piece of 155 mm diameter and 4 m long high tensile strength steel tube shown in Fig. 9.

Elongating the high-pressure chamber postponed the arrival of the reflected expansion wave from the driver end-wall, thereby increasing the shock tube test time. This extension of the high-pressure chamber is beneficial when a detonation driven system would be adopted with the combination of diaphragm-less operation.

Also enlarged is the original test section; it was extended to a field of view of 180 mm × 1,100 mm. This wide field of view enables quantitative visualization of the entire slug length prevailing behind the incident shock front. Such a long test

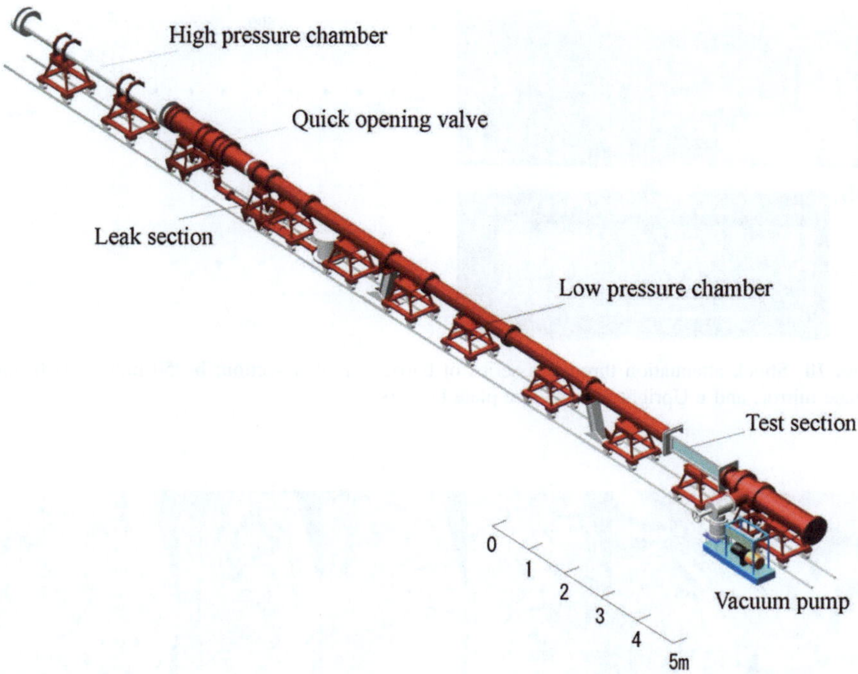

Fig. 9 Refurbished shock tube

section is needed, for example, in studying the relaxation zone behind strong shock waves or for detailed observation of shock attenuation while passing through a series of plate barriers. As an example, experiments regarding shock wave propagation through a series of plate barriers, conducted using the modified UTIAS shock tube, are briefly shown. It was possible to conduct such experiments due to the enlarged test-sect. (1.1 m long); observations were performed with double exposure holographic interferometry equipped with a pair of 1 m diameter schlieren mirrors; for details see Ohtomo et al. [27]. Description of the test section and two different sets of barriers are shown in Fig. 10. A sample of holographic interferometry photos obtained for these barriers along with appropriate pressure histories, recorded at stations 1, 2, and “a” to “h” are shown in Figs. 11 and 12.

It is clear from Figs. 11 and 12 that while propagating through the sequence of barriers the shock wave experiences significant attenuation; it exits the barriers as a sound pulse (a Mach wave). A detailed discussion of such a complex flow field is available in Ohtomo et al. [27] and in Igra and Igra [17].

It should be mentioned that in light of the modifications made, significantly increasing the test section length and using improved flow visualization (holographic interferometry), it was possible to accurately record a large part of the flow field using plain Perspex windows as is evident from Figs. 11 and 12.

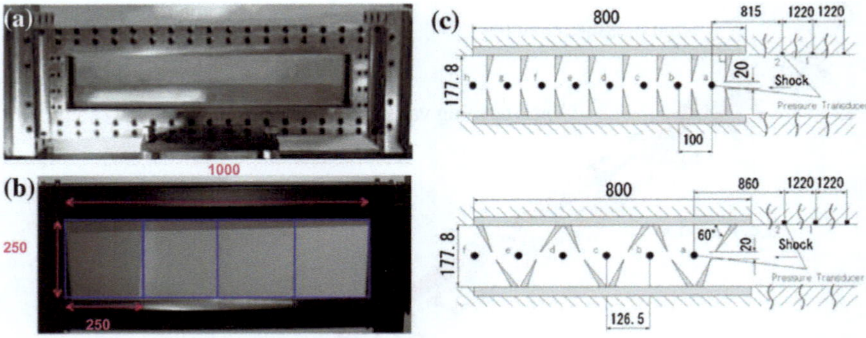

Fig. 10 Shock attenuation through a series of barriers: **a** Test section; **b** 250 mm × 1000 mm plane mirror; and **c** Upright and oblique plate barriers

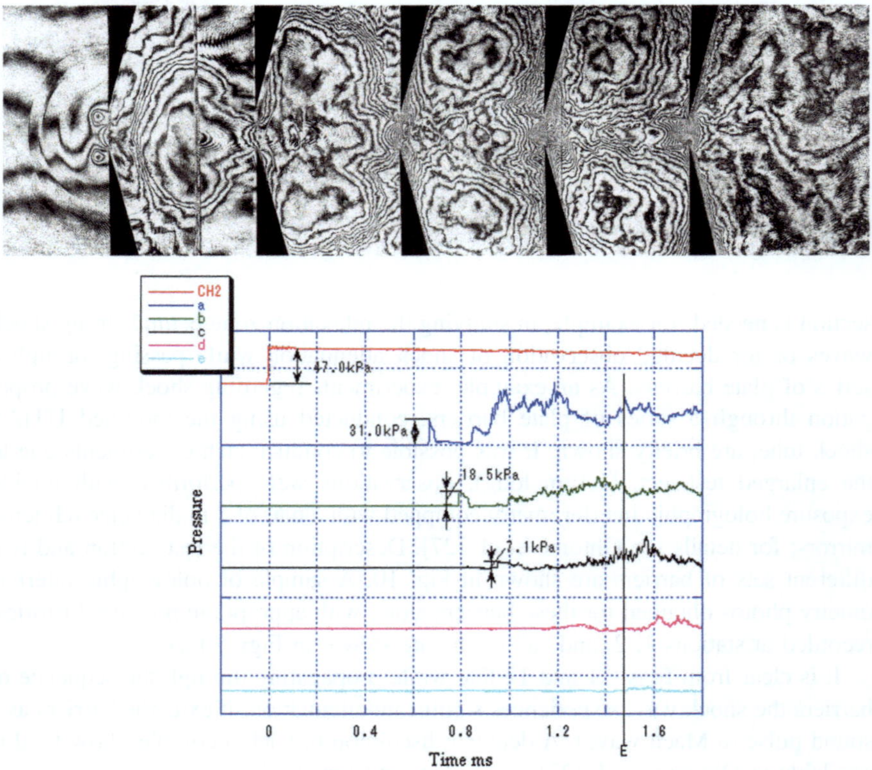

Fig. 11 Shock attenuation through upright series of barriers, M_s = 3.0; taken from Ohtomo et al. [27]

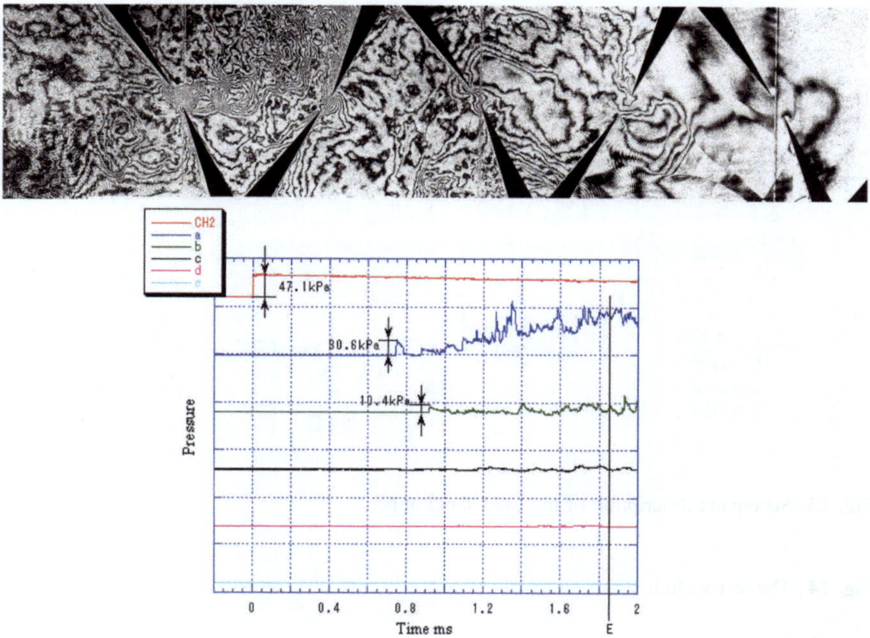

Fig. 12 Shock attenuation through oblique series of barriers, $M_s = 3.0$; taken from Ohtomo et al. [27]

3.3 The Shock Tube in the Prefectural University, Toyama, Japan

For evaluating the motion of a solid object in a gaseous medium, one has to know the drag coefficient of the object. In experimental investigations aimed at finding the drag coefficient of a solid sphere moving through a fluid various techniques have been employed, among others, shock tubes have been used. In these experiments the movement of a sphere (or spheres) exposed to a traveling shock wave was recorded. From the recorded spheres trajectories the relevant drag coefficient was deduced. In such shock tube experiments a technical problem arises, how to keep the investigated sphere(s) away from the tube walls. One solution, offered by Suzuki et al. [29], is described in the following.

To obtain the drag coefficient of a sphere in a nonstationary flow experiments were conducted in a horizontal shock tube shown schematically in Fig. 13. The driver section is 0.75 m long and 110 mm inner diameter; the driven section is 3.3 m long and has a square cross section of 75 mm by 75 mm.

The shock tube test section is shown in Fig. 14; its sidewalls are made from acrylic resin boards enabling observation of the sphere's motion. As shown in Fig. 14 the sphere is injected into the test section centre using a miniature shock

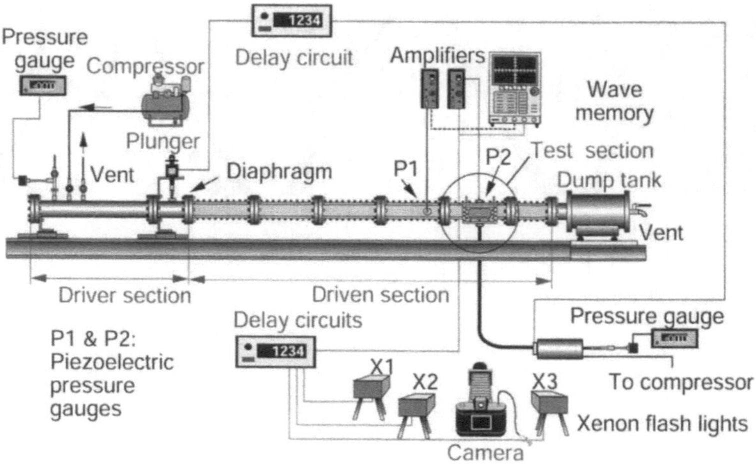

Fig. 13 Schematic description of the used shock tube

Fig. 14 The test-section

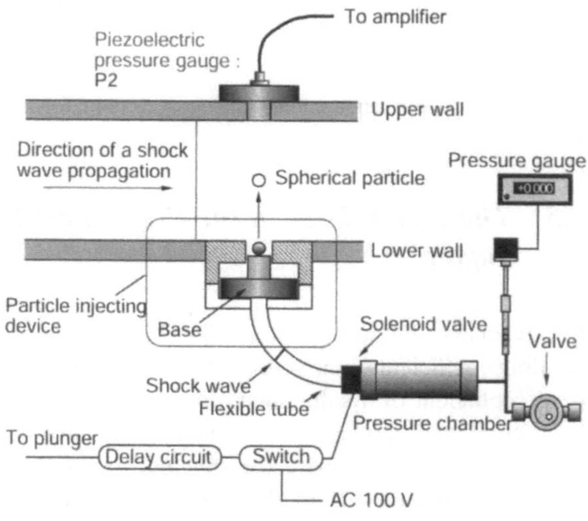

tube mounted on the lower wall of the test section. The experimental procedure was as follows:

The pressure chamber of the particle injecting device is filled with compressed air. A shock wave propagates through the flexible tube upon opening of the solenoid valve. Upon the shock reflection from the base the sphere is launched upwards; by proper selection of the chamber pressure the sphere is launched up to the test section centre. Using triggering from pressure gauges P1 and P2 (see Fig. 13) photos are taken during the particle movement induced by the flow behind the incident shock wave. A typical visualization is shown in Fig. 15.

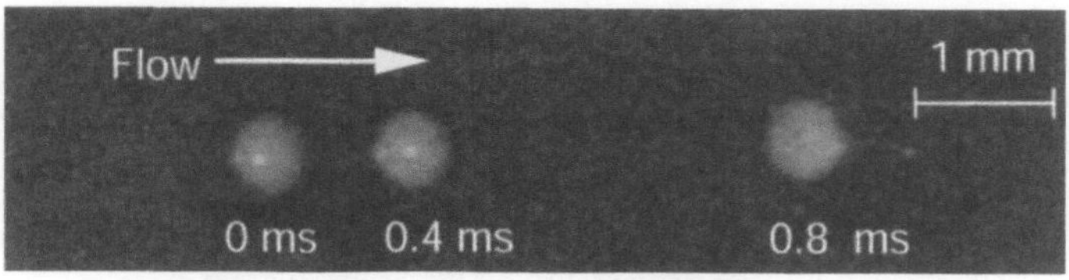

Fig. 15 Typical triple-exposure photograph of a moving sphere behind a planar shock wave. $M_s = 1.25$, the sphere dia. 0.65 mm and its density is 1300 kg/m^3

3.4 The Shock Tube at the Ben Gurion University, Beer Sheva, Israel

In the Shock Tube Laboratory of the Ben Gurion University there are seven different shock tube systems built for conducting different types of experiments. In Table 2 the tubes specifications are summarized.

These shock tubes make it possible to conduct investigations in the following topics:

1. Richtmeyer-Meskov instability.
2. Shock wave mitigation by geometrical means.
3. Mach reflection phenomena.
4. Shock wave mitigation by foams.
5. Shock—bubbles interaction.
6. Shock wave—obstacles interaction.
7. Blast wave loads on structures.
8. Shock wave propagation in granular materials.

The vertical, 4.5-m long shock tube (ST-02sv) shown in Fig. 16a, was upgraded and has become fully automated. The shock tube is fitted with a 2-m long driver section. The driven and driver sections are separated by a KB-40A fast opening valve (FOV) manufactured by ISTA Inc. Based on the manufacturer, its opening

Table 2 Shock tube specifications

System	M_{max}	Orientation	Cross section	Cross section dimension	Driver length (m)	Driven length (m)
ST-01sh	1.7	Horizontal	Square	80 mm × 80 mm	2	3.5
ST-02sv	5	Vertical	Square	80 mm × 80 mm	2	4.5
ST-03ch	2	Horizontal	Circular	D = 20 cm	2	8
ST-04sv	1.8	Vertical	Square	32 mm × 32 mm	1	1.5
ST-05sv	2	Vertical	Square	36 mm × 36 mm	1.5	3.5
ST-06sv	2.5	Vertical	Square	56 mm × 56 mm	1.5	3
ST-07sv	2	Vertical	Square	250 mm × 125 mm	2	6

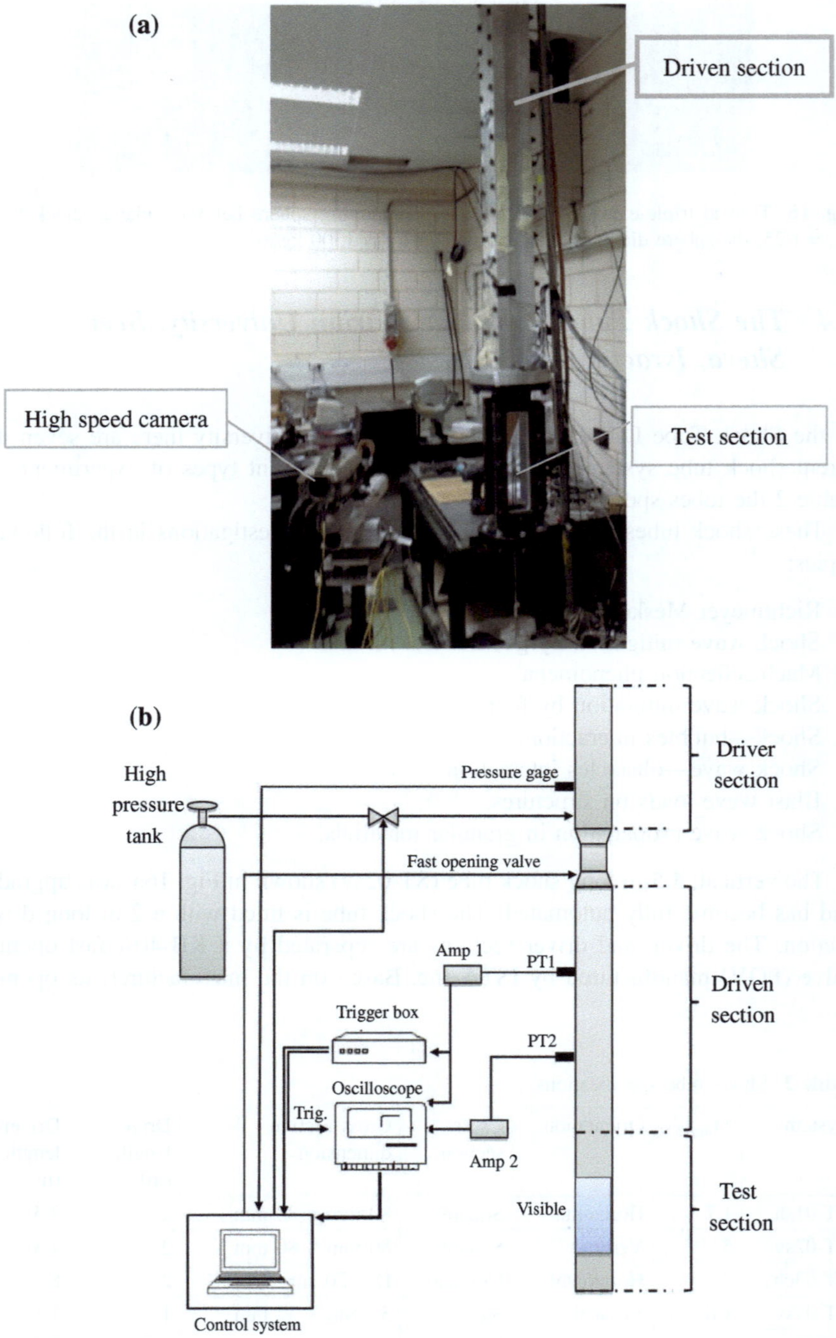

Fig. 16 **a** Shock Tube ST-02sv. **b** Schematic illustration of the experimental facility and the data acquisition system

time is less than 2 ms. After the pressurization of the driver section, the FOV is activated using a remote controlled servo motor. A schematic drawing of the experimental facility and the data acquisition system is shown in Fig. 16b. The shock wave and the data acquisition system are synchronized using an external, in-house designed, trigger box. The operation of the entire system is computer controlled through a self-developed LabView application. The control system is based on PCI-6602 and PCI-6035E National instruments I/O electronic cards. The driver gas is either air or helium and it is filled automatically to the desired pressure. The fast opening valve and the driver pressure automated control provided very good repeatability of the incident shock wave Mach number; to within 0.03 %. The incident shock wave Mach number is measured using two pressure transducers that are mounted 500-mm apart upstream of the test section. The operation of the facility and the data acquisition system are fully automatic. A set of any number of experiments could be conducted without any manual intervention.

3.5 *The Shock Tube Laboratory of the Ernst Mach Institute, Freiburg, Germany*

From the early 60s and up to the early 90s the Ernst-Mach-Institute in Freiburg, Germany was among the leading laboratories in shock wave research. Recognizing their reputation they were chosen to host the International Symposium on Shock Waves in 1967, the Mach Reflection Symposium in 1983 and in 1990, the International Symposium on Military Applications of Blast Simulation (MABS 10) in 1987 and the 3rd International Meeting on Random Choice Methods in Gas Dynamics and on Riemann Problems, in 1988. Intensive experimental shock wave research was conducted at the Ernst Mach Institute using their few available shock tubes. One of the shock tubes had a cross-section of 4 cm by 11 cm and was equipped with a Cranz-Schardin multiple-spark camera capable of taking a sequence of 24 alternating shadowgraph/schlieren photos during each experiment. A schematic drawing of this shock tube is shown in Fig. 17 and a photograph in Fig. 18.

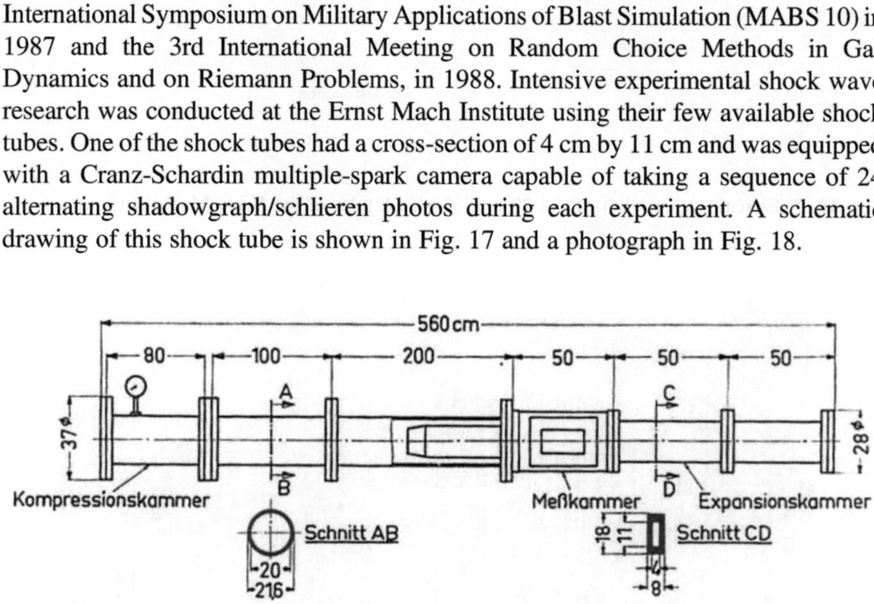

Fig. 17 Schematic view of the Ernst Mach Institute shock tube having a test-section of 4 cm × 11 cm

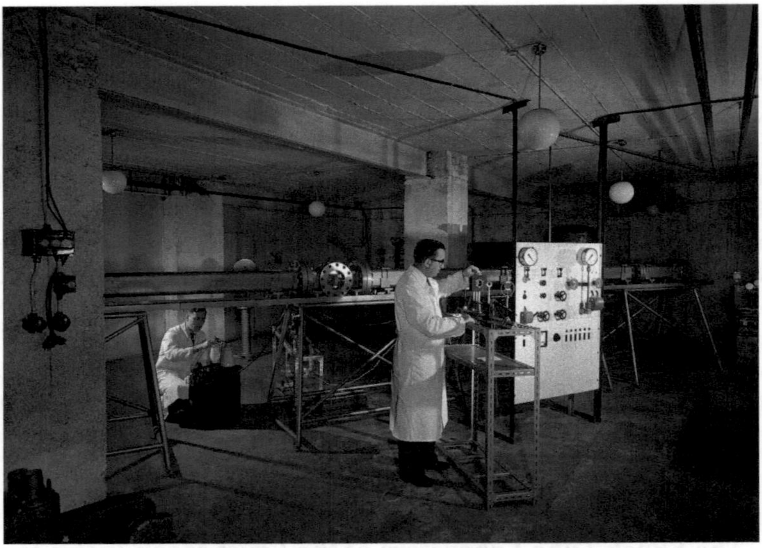

Fig. 18 Photograph of the shock tube of the Ernst Mach Institute having a test-section of 4 cm × 11 cm

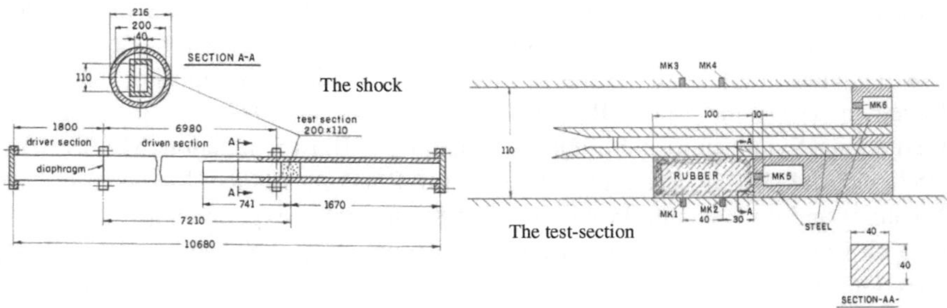

Fig. 19 Schematic description of the 20 cm i.d. shock tube of the Ernst Mach Institute

Fig. 20 Photograph of the 20 cm i.d. shock tube of the Ernst Mach Institute

Another shock tube was composed of a stainless steel pipe having an internal diameter of 20 cm; its driver section is 180 cm long and the driven section 888 cm long. The test section was located 721 cm downstream from the diaphragm; see schematic drawing in Fig. 19. As shown, for having a two-dimensional flow inside the test-section a cocky-cutter was introduced ahead of the test-section location. In Fig. 19 the configuration used for studying shock wave reflection from a non-rigid wall is shown. A photograph of this shock tube is shown in Fig. 20.

This shock tube generated incident shock waves within the range 1 < Ms < 5 and was used for studying a variety of shock wave flows. Among the not classified studies that were conducted in this shock tube were the following papers: Shock wave propagation over a cavity, Igra et al. [13], Starting flow in a nozzle, Igra et al. [15], Attenuating shock waves propagating in tunnels by introducing an expansion chamber in the tunnel, Igra et al. [16], Reflection of shock waves from a non-rigid walls, Mazor et al. [25] and Blast wave reflection from a wedge, Igra et al. [14]. For blast wave generation the tube's driver was significantly shortened; details are available in Igra et al. [14]. A sample from the optical diagnostic used in these investigations is presented in Figs. 21 and 22.

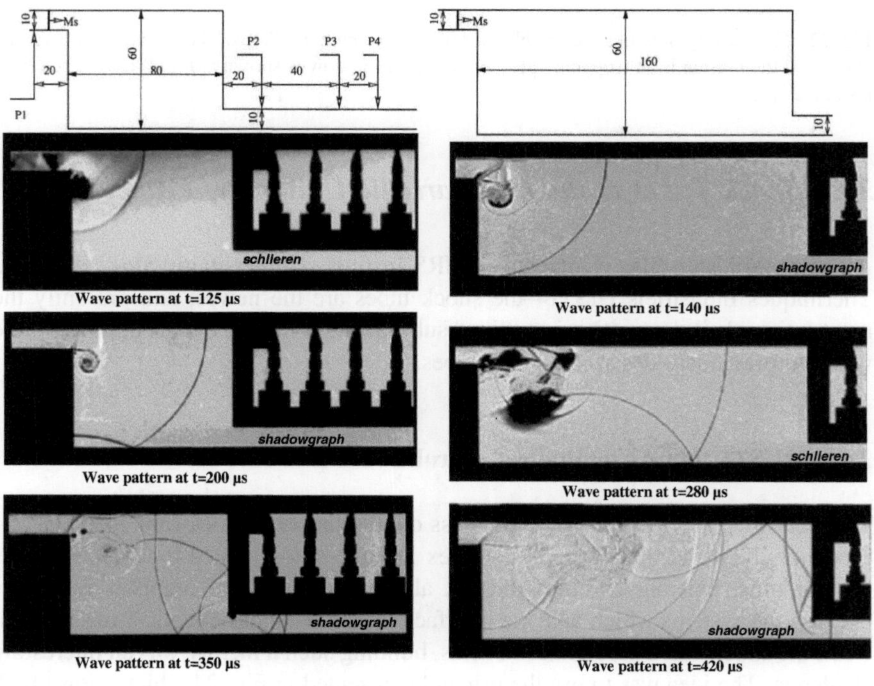

Fig. 21 Sample of schlieren/shadowgraph photos, taken from Igra et al. [16], showing shock wave propagation inside a duct having an expansion chamber

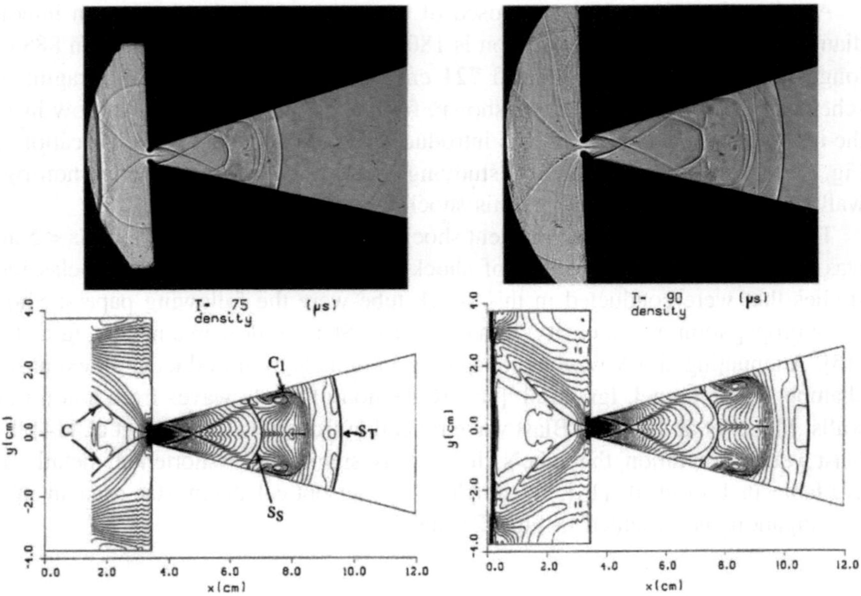

Fig. 22 The starting flow in a wedge-like nozzle, *on the left* at $t = 75$ μs. The *top* is a shadowgraph photo and the *bottom* is an isopycnic plot; *on the right* the flow is shown at $t = 90$ μs. (Taken from Igra et al. [15])

3.6 Shock Tubes at the Aix-Marseille University, France

At the Aix-Marseille University-CNRS-Institut Universitaire des Systemes Thermiques Industriels UISTI—the shock tubes are the newest and presently the most active, their description is outlined subsequently in more details than was done with the previously described shock tubes.

3.6.1 IUSTI Initial Conditions Controlled Shock Tube

The T200 IUSTI shock tube is a stainless double junction shock tube (see Fig. 23) especially built in 2000 to study complex interfaces accelerated by a shock wave. At that time, existing facilities did not allow observation of complex interfaces because the systems which take the interface in sandwich did not allow observing it when the shock wave accelerates it. Thus, building such a facility was an interesting challenge. The idea was to use the principle presented in Fig. 24 which allows both excellent viewing and control of the shape of the interface when it is hit by the shock wave.

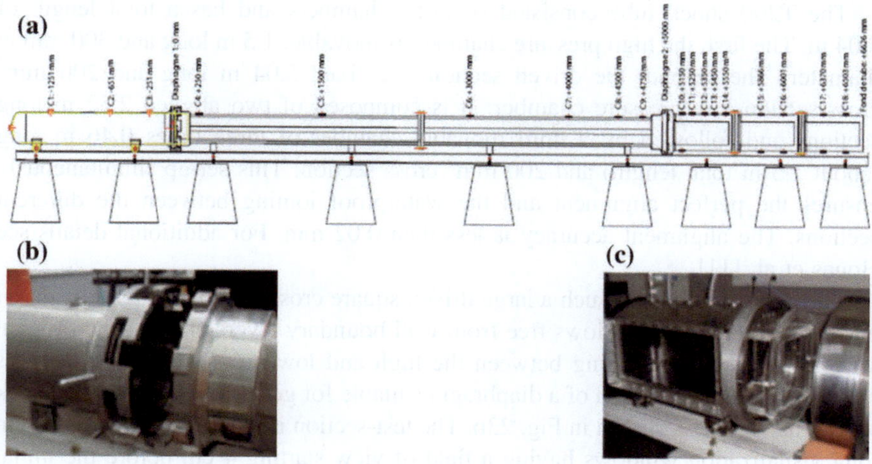

Fig. 23 **a** Overview of T200 IUSTI shock tube; **b** photographs of its rapid sixth turn device connecting the high pressure and the low pressure chambers; **c** and its original device allowing control of the initial conditions between the low pressure chamber and the test-section junction with the interface compartment

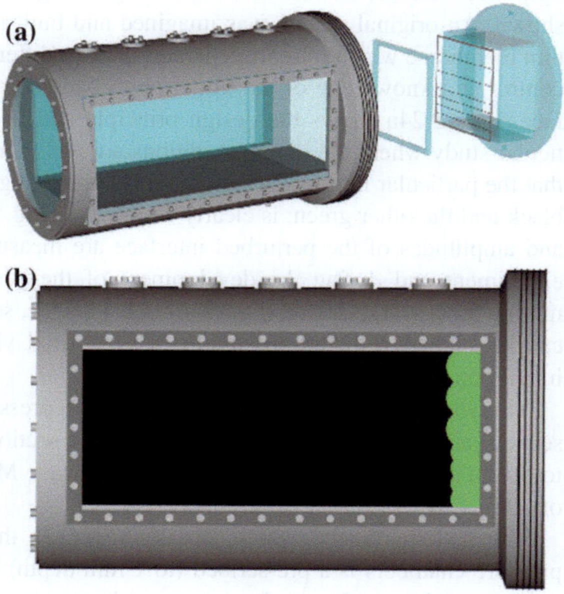

Fig. 24 **a** Schematic description of the T200 shock tube test-section and its original device allowing control of the initial conditions; **b** a view of the initial interface (before the shock arrival) (Richtmyer-Meshkov instability). Taken from Houas et al. [11]

The first junction allows placing the diaphragm to create the shock wave and the second the phenomenon to be studied in order to control the initial conditions (in the example shown it is an interface between two different gases expected to mix together after being accelerated by the incident shock wave).

The T200 shock tube consisted of three chambers and has a total length of 7.04 m. The first, the high pressure chamber, is movable, 1.5 m long and 300 mm in diameter. The second, the driven section is a fixed 5.04 m long and 200 mm^2 cross-section low pressure chamber. It is composed of two aligned 2.52 m long sections and followed by a third movable chamber of three times 0.46 m long (about 1.5 m total length) and 200 mm^2 cross section. This set-up simultaneously ensures the perfect alignment and the waterproof joining between the different sections. The alignment accuracy is less than 0.02 mm. For additional details see Houas et al. [11].

The interest in having such a large driven square cross-section chamber is that it allows studying uniform flows free from wall boundary layers effects. The original sixth-turn device connecting between the high and low pressure sections permits rapid and safe introduction of a diaphragm suitable for generating the desired shock wave; this device is shown in Fig. 23b. The test-section is composed of three 46 cm long visualization windows having a field of view starting 4 cm before the initial position of the interface and up to 1.5 m thereafter. At the test section end-wall a Plexiglas window is placed enabling possible use of two-dimensional diagnostic technique by beaming light through this window. The second junction, Fig. 23c, connecting the driven section and the test-sections, includes a tape that supports the needed initial conditions. In the scheme shown in Fig. 24 an example of a configuration studied, an interface separating two gases having different densities is shown. An original system was imagined and built in order to introduce a 0.5 μm thin membrane which initially separates the two different gases and permits both the control and knowledge of the initial perturbations imposed on that material interface. Figure 24a shows the design principle and in Fig. 24b the result for a particular study where the initial conditions are well controlled is shown. We observe that the particular interface which separates the two gases of different densities, one black and the other green, is clearly visible (see Fig. 24b). The initial wave lengths and amplitudes of the perturbed interface are measurable both before starting the experiment and during the development of the perturbations and the following mixing process. Of course, thanks to such a device, several other types of interfaces can be studied with the possibility of very good viewing and knowledge of the initial conditions.

This shock tube was designed to withstand a pressure of 20 bars (even in the test section) and a minimum pressure, in the driven section, as low as 0.5 mbar, in order to obtain a pressure ratio adequate for shock wave Mach numbers within the range of $1.05 \leq M_s \leq 4$ in air or any inert other gas.

The first diaphragm which initially separates the high pressure and the low pressure chambers is a pre-scribed (to 1 mm depth) aluminum disc. A rotary vane pump (an Alcatel 63 m^3/h Pascal series) is connected to the three tube sections. It enable pumping air from one or more sections in order to fill them with required test gas and/or to decrease the initial pressure for reaching higher Mach number of the incident shock wave. The driven section of the shock tube is equipped with ten ports for pressure gauges. Four are flush mounted high frequency PCB pressure transducers (113A21 type). They are connected, via a PCB amplifier (48A04 type), to a multichannel digital oscilloscope (Tektronix TDS) and allow, besides the shock wave velocity determination, triggering of the diagnostic system from the oscilloscope delayed triggering channel.

An example of results obtained with T200 shock tube is shown in Fig. 25. The investigation was related to the Richtmyer-Meshkov instability, where a particular interface separating two gases of different densities was accelerated by a shock wave. The aim was to follow and study the development of the initial perturbations of the interface, as to their wavelength and amplitude, in order to obtain information on the mixing of the two gases. The experimental diagnostic was laser sheet visualizations and two types of interfaces are shown in Fig. 25a, b where one clearly observes the initial interface, at t = 0 as well as its evolvement with time. The images allow plotting the evolution of the interface perturbations amplitude and to compare it with different theories and models. For details see Mariani et al. [24].

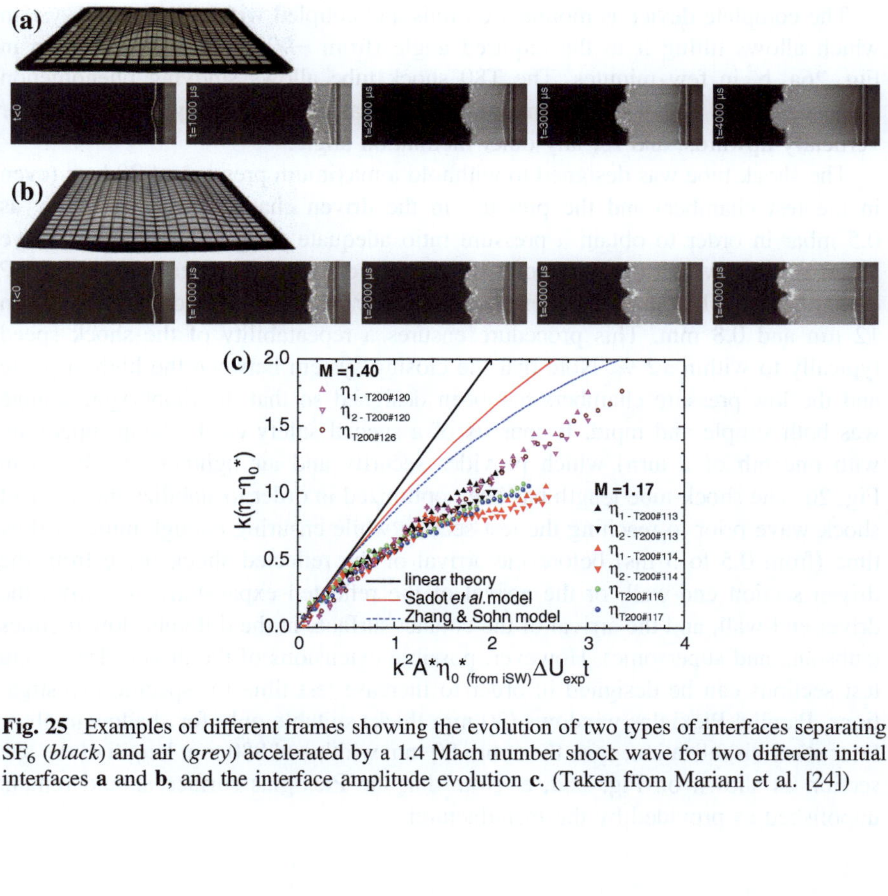

Fig. 25 Examples of different frames showing the evolution of two types of interfaces separating SF_6 (*black*) and air (*grey*) accelerated by a 1.4 Mach number shock wave for two different initial interfaces **a** and **b**, and the interface amplitude evolution **c**. (Taken from Mariani et al. [24])

Such an experiment shows that even with well controlled initial conditions, the diaphragm fragments slow the development of the interface perturbation and the experimental results remain below both the theoretical and numerical ones.

3.6.2 IUSTI Variable Inclination Shock Tube

The driving concept in designing the T80 IUSTI shock tube was to build a very easy to operate shock tube, easily transportable to another laboratory and capable of generating shock waves vertically upward or downward, in a variety of inclinations and for pure-gas and multi-phase flow investigations.

The T80 stainless steel 80 mm^2 cross section shock tube has a total length of 3.75 m. It is composed of a movable 0.75 m long high pressure chamber followed by a fixed 2.02 m long low pressure chamber and ending with a movable 0.98 m long test-section. All the tube internal wall surfaces are polished to within ±2 μm. The driver gases are nitrogen or carbon-dioxide or helium according to the desired strength of the incident shock wave. Additional details can be found in Jourdan et al. [22].

The complete device is mounted on rails and coupled with a pneumatic system which allows tilting it to the required angle (from −90° to +90°), as shown in Fig. 26a, b, in few minutes. The T80 shock tube allows studying phenomenon associated with a shock wave moving horizontally, or vertically downwards, or vertically upwards, and for any other inclination angle.

The shock tube was designed to withhold a maximum pressure of 20 bars (even in the test chamber) and the pressure in the driven chamber can be as low as 0.5 mbar in order to obtain a pressure ratio adequate for a required shock wave Mach number within the range of 1.05–4 in air or in any inert gases. The HP/LP diaphragm is a 150 mm diameter aluminum disc, pre-scribed to depth of between 12 μm and 0.8 mm. This procedure ensures a repeatability of the shock speed typically to within ±2 %. Note that the closing system between the high-pressure and the low-pressure chambers has been designed so that the diaphragm change was both simple and rapid. It consists of a special safety catch clamp (operating with one 6th of a turn) which provides security and air tightness as shown in Fig. 26. The shock-tube length has been optimized in order to stabilize the incident shock wave prior to reaching the test-section while ensuring enough uniform flow time (from 0.5 to 5 ms) before the arrival of the reflected shock wave from the driven section end-wall, or the arrival of the reflected expansion wave from the driver end-wall, and the arrival of the contact surface, in the different flow regimes (subsonic and supersonic). However, possible extensions of the driver, driven and test sections can be designed in order to increase test time for specific investigations. Parallel Plexiglas windows (30 mm thick) suitable only for shadowgraph or laser sheet visualization, span the entire 80 mm width and 880 mm length of the test section as shown on Fig. 26d, e. Note that the Plexiglas surface should remain unpolished as provided by the manufacturer.

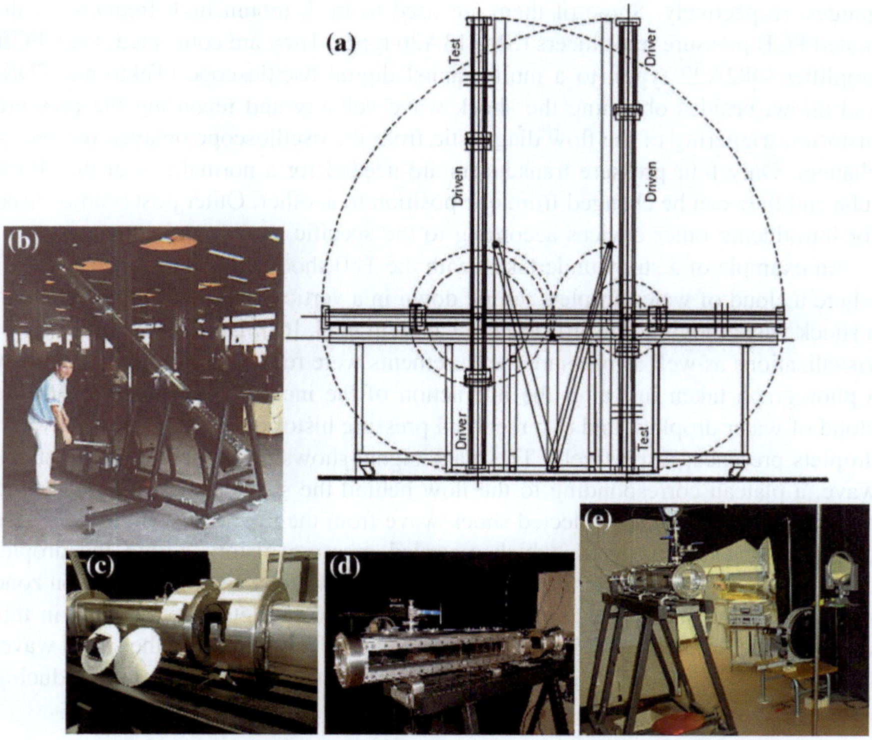

Fig. 26 **a** Schematic description of the T80 variable inclination shock tube; **b** with the inclination pneumatic system principle; **c** the original quick HP/LP chambers closing device; the experimental visualization chamber **d** and **e**. (Taken from Jourdan et al. [22])

The test section of the shock tube ends with a rigid Plexiglas plate which allows the introduction of a laser sheet. However, Plexiglas can easily be replaced with glass, Carbon or Lexan for windows suitable for other optical diagnostics (schlieren, interferometry or X-ray absorption).

Finally, the whole system is set up on rollers in order to move it easily into, or out of the laboratory and to optimize the available work space.

This installation has been mechanically well conceived so as to facilitate the preparation and the execution of a simple run (less than 10 min) including diaphragm insertion, pressure data reduction and camera loading, even by an inexperienced operator. The driver section is equipped with a compression pressure gauge to measure the pressure in the driver gas just before the shock wave release.

A rotary vane pump (an Alcatel 20 m^3/h SD series) is connected both to the driver and driven sections. It permits removing air from these sections in order to fill them with a desired gas or to decrease the pressure of the driven gas (down to 0.5 mbar) for high shock wave Mach number experiments. The driver and the driven sections of the shock tube are equipped with 4 and 11 posts for pressure gauges, respectively. Some of them are used to flush mount high frequency calibrated PCB pressure transducers (SM113A26 type). They are connected, via a PCB amplifier (482A22 type), to a multichannel digital oscilloscope (Tektronix TDS) and allow, besides obtaining the shock wave velocity and recording the pressure histories, triggering of the flow diagnostic from the oscilloscope delayed triggering channel. Only four pressure transducers are needed for a normal run of the shock tube and they can be changed from one position to another. Other posts can be used for introducing other devices according to the specific study requirements.

An example of a study undertaken with the T80 shock tube is shown in Fig. 27, where a cloud of water droplets falling down in a vertical orientation interacts with a shock wave moving upwards (See in Chauvin et al. [6]). During this study direct visualizations as well as pressure measurements were recorded. Figure 27b, c show a photograph taken just after the interaction of the incident shock wave with the cloud of water droplets, and two recorded pressure histories with and without water droplets presence, respectively. The black signal shows the jump across the shock wave, a plateau corresponding to the flow behind the shock wave and the second pressure jump across the reflected shock wave from the tube end-wall; all in a case of pure air. The red signal, which recorded pressure history within the droplet cloud, shows a strongly attenuated shock wave followed by a strong relaxation zone attributed to the secondary atomization of the water droplets. It was found in this work that the droplets burst into smaller droplets and are moved by the shock wave. Both phenomena absorb part of the shock wave kinetic energy; thereby reducing about 50 % of the shock energy.

3.6.3 IUSTI High Mach Number Shock Tube

The T85 shock tube is a double diaphragm shock tube, separating the driver from the driven section is made of either steel or aluminum, and the second is a thin membrane separating the two investigated gases. Its total length is about 9 m (see Fig. 28). It is composed of a steel high pressure chamber, followed by four identical aluminum sections and the test-section. Further details are available in Houas and Chemouni [10].

The driver section is composed of steel pipe having a 6 cm inner diameter, 1.3 m long and 4 cm thick. The low pressure section and the test-section have a quadratic cross-section of 8.5 × 8.5 cm. The connection between the driver and driven sections is via a transition section having varying cross-section from 6 cm diameter to 8.5 cm square. The length of the test-section can be changed from 0.8 to 1.52 m. The high pressure chamber is equipped with 13 spark plugs; when introducing a stoichiometric mixture of H_2 and O_2 and igniting it, a shock wave propagating at Mach 9 can be achieved. A photography of the high pressure chamber is presented in Fig. 28b.

The gases in the low pressure section are initially separated by a thin plastic or nitrocellulose membrane. Before introducing the test gas into the test-section, a vacuum of about 10^{-2} Torr was produced on both sides of the thin membrane using

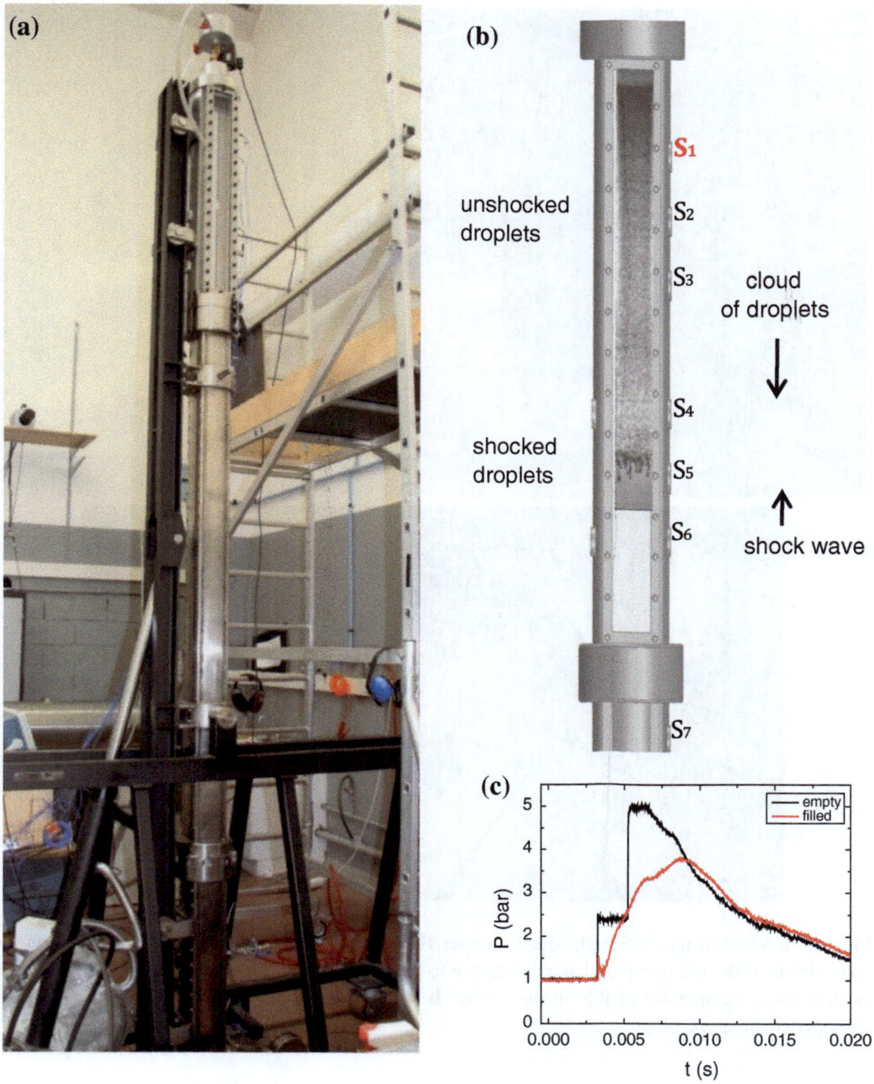

Fig. 27 **a** Example of experimental study conducted with T80 shock tube in a vertical orientation where a water cloud of calibrated droplets, 500 μm in diameter; **b** moving downward interacts with a 1.5 Mach number shock wave moving upward; **c** and two recorded pressure signals with (*red*) and without (*black*) droplet cloud. (Taken from Chauvin et al. [6]

a rotary pump followed by a diffusion pump. During the phases of pumping and gas filling, the maximum amplitude of the thin membrane deformation is controlled with a U tube capable of measuring pressure difference of less that 1 mm.

For obtaining moderate Mach number shock waves ranging from 3 to 6, a vacuum is induced in the low pressure chamber and helium is used as driver gas.

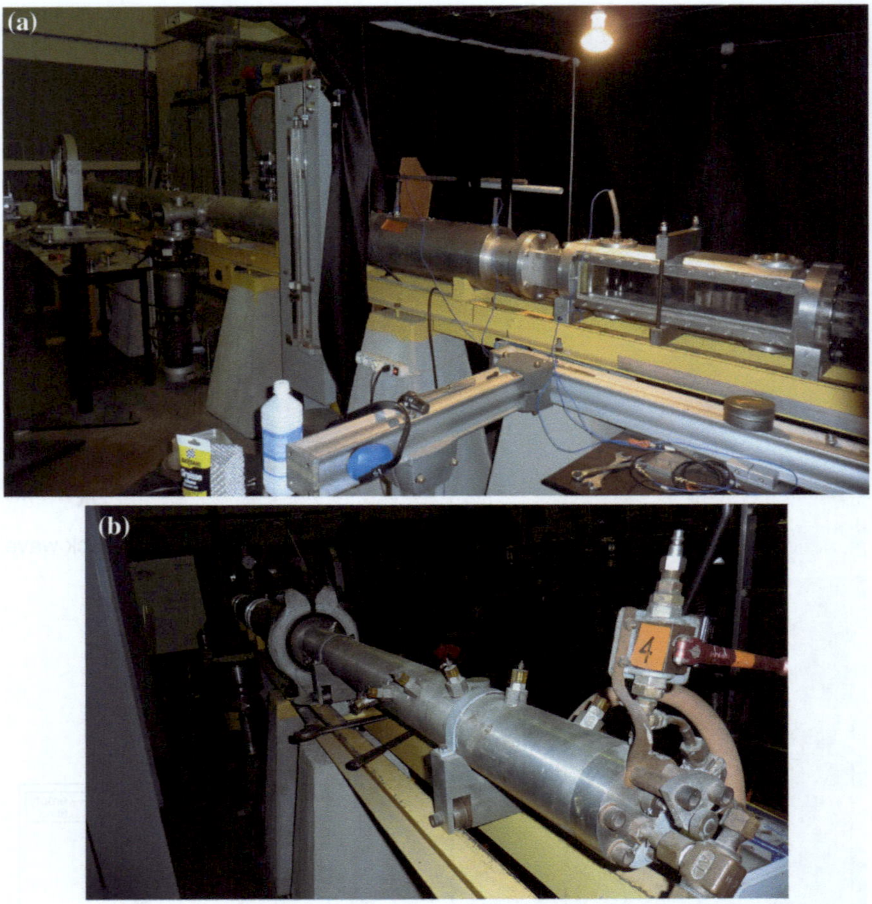

Fig. 28 **a** View of the T85 high Mach number IUSTI shock tube, its test-section appears on the *right* and its diffusion pump and special device to pump out and introduce gases into both sides of the low pressure area are visible in the *centre*. **b** View of the IUSTI T85 high pressure chamber equipped with spark plugs

Other driver gases used are hydrogen or nitrogen according to the desired strength of the generated shock wave.

An example of study undertaken in the T85 shock tube is shown in Figs. 29 and 30. A laser absorption diagnostic method is undertaken to directly obtain the density profile within a Co_2/A_r two-gas Richtmyer-Meshkov mixing. The theory behind the method allowing measuring both temperature and density within a two-fluid mixing created by the shock wave acceleration of their initial common interface is available in Fortes et al. [7]. In short, when a laser beam crosses a gaseous medium (containing CO_2 as the absorbing gas in the present example) part of its energy is absorbed by the molecules present along the path of the laser beam according to the following relation (assuming thermo-dynamical equilibrium):

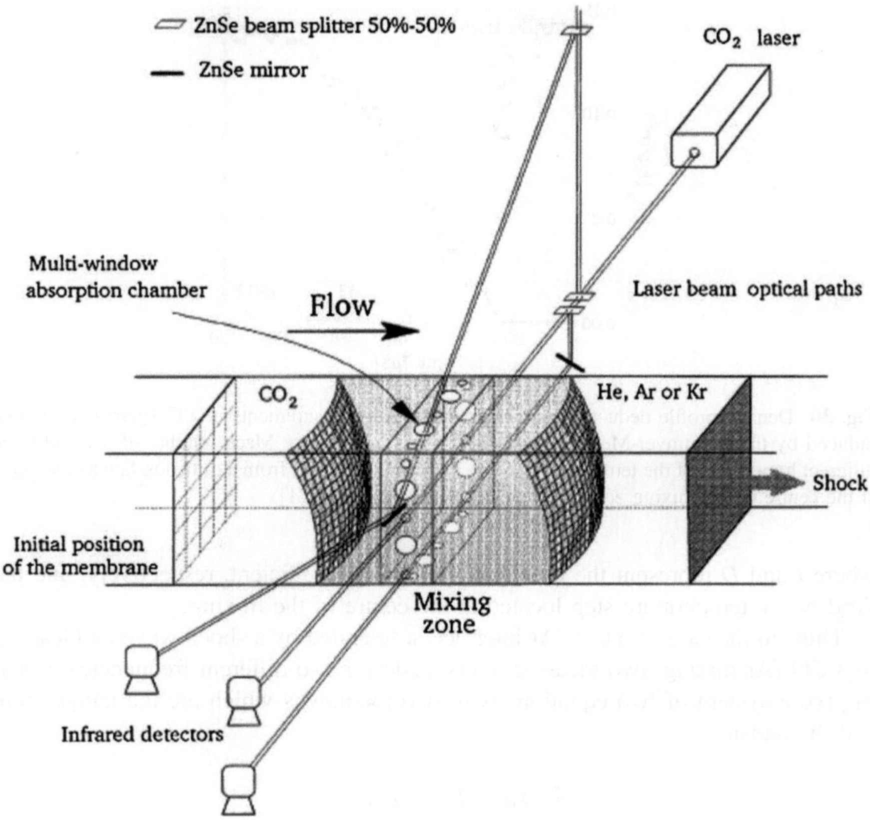

Fig. 29 Experimental set-up of a three-directional laser absorption method for density measurements in a CO_2 monatomic gas mixing zone generated behind a shock wave. (Taken from Jourdan et al. [20])

$\alpha_{\nu_i} = f_{\nu_i}(T, \rho)$ where α is the absorption coefficient of the CO_2 molecule (for a frequency ν_i), f is a function depending on the mean temperature and density of the CO_2 within the mixing, and T and ρ are the mean temperature and density within the mixing, respectively.

Improvement of the theory and the diagnostic set-up in the case when a weak temperature gradient exists in the mixing, allows the hypothesis that a well defined temperature profile has developed; for details see in Jourdan et al. [20] and in Jourdan et al. [21]. For example, the temperature variation between the two gases, CO_2 (1080 K) and argon (630 K) was 450 K, while assuming three temperature profiles. The first one is linear; the second is based on a diffusion law of the type,

$$T = T_{CO_2} + [T_{Ar} - T_{CO_2}]e^{-Dt^2}$$

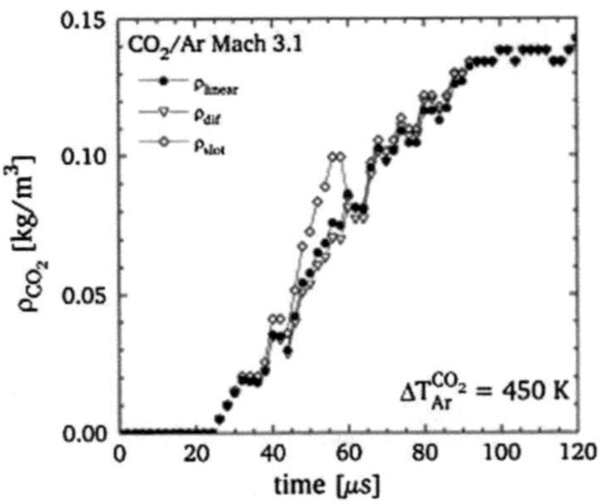

Fig. 30 Density profile deducted from laser absorption measurements in a CO_2/Ar mixing zone induced by the Richtmyer-Meshkov instability for a shock wave Mach number of 3.1 and three different hypothesis of the temperature profile i.e. linear, deduced from a diffusion law and stepped at the centre of the mixing zone. (Taken from Jourdan et al. [21])

where *t* and *D* represent the time and a diffusion coefficient, respectively, and the third was a temperature step located at the centre of the mixing.

Thus, in the case of a CO_2/Ar interface accelerated by a shock wave and leading to a CO_2/Ar mixing, two measurements made for two different frequencies ν_1 and ν_2 give a system of two equations with two unknowns which are the temperature and the density:

$$\alpha_{\nu_1} = f_{\nu_1}(T, \rho_{CO_2})$$
$$\alpha_{\nu_2} = f_{\nu_2}(T, \rho_{CO_2})$$

This system is easily solved and an example of results obtained is shown in Fig. 30. It was shown that with a suitable limitation of the temperature variation domain ($\Delta T \sim 450$ K), the influence of the temperature is weak and the absorption method looks like a technique for measuring the density.

3.6.4 Blast Wave Generation in a Shock Tube

The T32 is a 32 mm inner diameter PVC shock tube especially built to create blast waves (see Fig. 31). Its driver and driven sections are 210 mm and 945 mm long respectively, but for such a type of shock tube these values can be changed keeping a certain ratio without modifying the nature of the expected blast wave. A blast wave is easily achieved when working with an open-end shock tube. In such a case, the shock wave emitted from the tube and the pressure jump it creates is followed by a strong rarefaction wave, resulting in pressure below the initial pressure level, i.e., a typical blast wave is generated at the tube exit.

Fig. 31 Schematic description of the T32 IUSTI PVC shock tube

An example from an investigation conducted with the T32 producing blast waves at the shock tube end-wall is presented in Figs. 32 and 33 (see Rodriguez et al. [28]. In Fig. 32a the pressure developed behind the incident shock wave, before leaving the tube is shown. As expected, a fairly uniform constant pressure zone prevails behind the shock front. This is not the case for the pressure prevailing behind the exiting wave as is evident from recorded pressure shown in Fig. 32b; this is a clear pressure signature of a blast wave. A photograph of this emerging blast wave is shown in Fig. 33.

The aim of this experimental study was to understand the origin of the instability that creates fingers when a blast wave accelerates a gas and granular medium interface. In that experiment, the gas was air and the granular medium was flour particles, 10 μm in diameter.

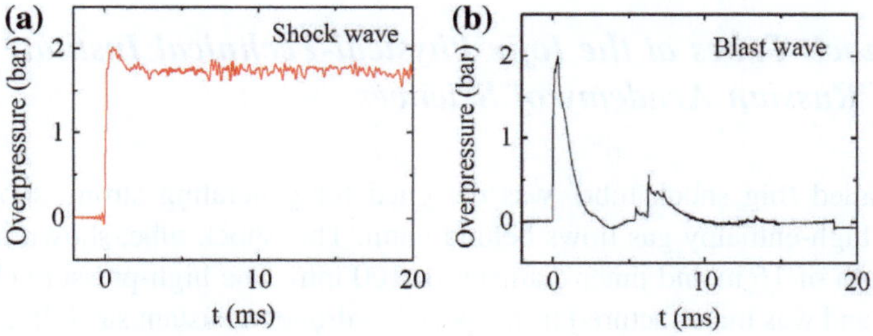

Fig. 32 Examples of recorded pressure histories during a run in the T32 shock tube. **a** Pressure recorded inside the tube and **b** pressure recorded at the tube exit. In both cases the incident shock wave mach number is 1.25

Fig. 33 Schlieren photograph of the blast wave generated at the T32 shock tube exit

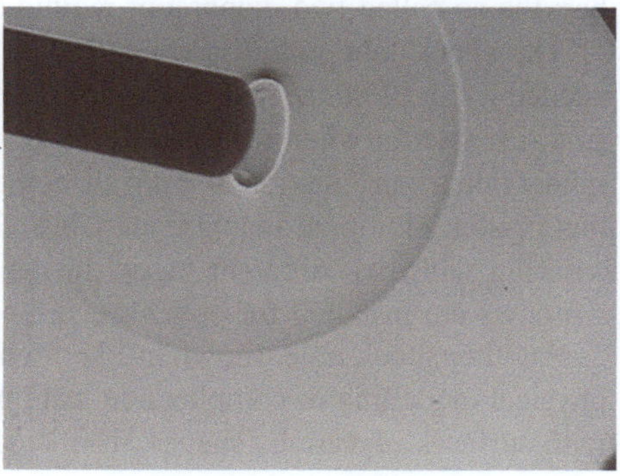

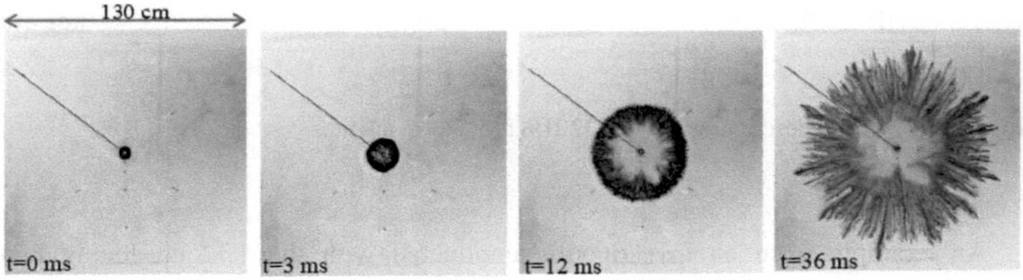

Fig. 34 High speed direct photography of the evolution of a flour particle ring exposed to a blast wave. The particle ring had external and internal initial diameters of 60 and 20 mm, respectively. It was initially positioned between two Hele-Shaw plates 4 mm apart

As T32 is usually employed having its exit open; its range of operation is limited due to safety reasons. However, this shock tube produces genuine blast waves and allows investigating phenomenon like that shown in Fig. 34, blast wave generation at the center of a particle ring. These experiments have shown that particle "fingers" first appear propagating inward with respect to the particle ring and later reverse their motion and propagate outward.

3.7 *Shock Tubes at the Ioffe Physical-Technical Institute of Russian Academy of Sciences*

The so-called 'big shock tube' was designed for generating strong shock waves inducing high-enthalpy gas flows behind them. This shock tube, shown in Fig. 35, has a length of 16 m and inner diameter of 100 mm. The high-pressure chamber is 3 m long and was manufactured from special hydrogen-resistant steel. It is equipped with a 30 kW Ohmic electric heater. This enables employing hydrogen as the driving gas at temperatures up to 750 K and pressures up to 500 atm.

The low-pressure channel of this shock tube can be divided into several sections by diaphragms that enable one to vary the driven section length. Thereby, implementing different regimes of operation of the shock tube with intermediate chamber; the so-called two-diaphragm regime of operation.

The shock tube outlet is connected to a tank having a volume of 6 m^3 and if needed it can be connected to additional vessel of 25 m^3. This tank can also be used as a test section when the shock tube is operated as a shock tunnel. In the shock tunnel option, installation of a nozzle at the driven section end is made. When using this option additional diaphragm is installed at the nozzle inlet, thereby allowing working with two different gases. In the tank and in the test-section walls glass windows are installed for optical recording of the generated flow field.

Another shock tube, shown in Fig. 36 has a rectangular cross-section, is used for physical modelling of complex non-stationary processes of shock waves interaction with surfaces of various shapes. The shock tube consists of high-pressure section

Fig. 35 The 'big shock tube'

(driver), low-pressure section (driven), test section and a dump tank. The driver and the driven sections of the shock tube are made from identical steel welded sections having rectangular cross sections of 50 mm by 150 mm segments, each is 1 m long.

The inner walls are polished and covered with chromium in order to ensure a perturbation free post-shock flow. The total length of the shock tube is 10 m. As the shock tube consists of identical sections, the relative lengths of the driver and the driven sections of the shock tube can be easily varied depending on the required regime of operation. For generation of strong shock waves; Mach number M > 7, the gas inside the driver can be heated up to temperatures between 200 and 400 °C.

Fig. 36 Shock tube with a rectangular cross section

The original structure of the shock tube enables one to generate single shock waves with the desired post-shock properties and also to generate two shock waves that move one after the other.

The shock tube shown in Fig. 37 was build for investigating magnetohydrodynamic (MHD) flows. It has an inner diameter of 50 mm. At its end wall of the driven section an MHD disk is installed. The driver section consists of two identical sections, each is 0.64 m long. This enables one to vary the chamber length. In addition, the high-pressure chamber is equipped with an external electric heater for heating the driver gas up to 700 K at a pressure of 100 atm.

The driven section is 3 m long and it is connected to the disk channel which is shaped by two Plexiglas disks of 0.3 m in diameter. The disks are positioned normally to the tube axis at a distance of 0.01 m from each other. After passing through the disk channel, the gas flows out into the receiver tank. The disk channel is located between poles of an electromagnet. Energizing the magnet is implemented from a source of constant current. The induced magnetic field in the pole gap attains 1.6 T. Typical type of investigations conducted in this facility are:

1. Investigation of flow conditions inside a disk channel leading to optimal regime of operation of the Faraday MHD-generator and the Hall MHD-generator.
2. Investigations of various type of instabilities generated in non-equilibrium plasma flows.

Fig. 37 Shock tube equipped with an MHD disk

Inside the disk channel the following flow parameters are measured: velocity, pressure, concentration, the electron's temperature, plasma conductivity and current density.

The next shock tube to be presented was designed for conducting experimental investigation into interaction of magnetic field with weakly ionized non-equilibrium plasma moving through channels as well as in external flows around various models. A photo of this shock tube is shown in Fig. 38. The driver chamber of this shock tube is 1 m long and 0.05 m in diameter. As driver gases either helium or hydrogen could be use having pressure up to 20 atm. The low-pressure channel is 4.3 m long; it is connected to a vacuum chamber 0.44 m long and 0.18 m in height. Inside the vacuum chamber there is a supersonic diffuser; the walls of the nozzle could be either plain or equipped with a system of electrodes, placed along the nozzle surfaces for implementation of an MHD impact on the flow. Pulsed magnetic field is established by two electromagnetic coils positioned between the vacuum chamber walls. The attained magnetic induction field is 1.5 T. At the lateral walls of the vacuum chamber optical glasses of 200 mm diameter are installed for flow visualization using a schlieren technique.

The Ludwieg Tube of the Ioffe Physical-Technical Institute is shown in Fig. 39. Its driver section has inner diameter of 156 mm and a length varying from 0.5 to 1.2 m. The critical cross section of the supersonic axisymmetric nozzle has a diameter of 60 mm, and its outlet diameter is ~ 140 mm. When the length of the high-pressure chamber is maximum, the time of efflux of nitrogen flow reaches

Fig. 38 Shock tube equipped with a pulsed magnet

Fig. 39 Ludwieg shock tube

2 ms. Supersonic nitrogen flow at the nozzle outlet corresponds to Mach number M ~ 3.

The setup is destined for testing new problems (pilot projects) with the aim of estimating their efficiency and choosing technical equipment for solving them. In particular, it was studied supersonic nitrogen flow around models capable of ejecting plasma jets or micrometeorites upwind.

In the table shown below basic dimensions and maximum permissible pressure and temperature of the driver gas used in the driver section of the Ioffe Physical-Technical Institute are given.

Setup	Dimensions	Limiting parameters of the driver gas		
		Gas	P, atm	T, K
Big shock tube	Dia 100 mm, length 16 m	H_2	500	750
Shock tube having rectangular cross section	50 × 150 mm^2, length 10 m	H_2	100	700
Shock tube equipped with the disk MHD-channel	Dia 50 mm, length 4 m	H_2	100	700
Shock tube equipped with a pulsed magnet	Dia 50 mm, length 5 m	H_2	20	300
Ludwieg tube	Dia 156 mm, length 1.2 m	N_2	20	300

4 Diagnostics

While describing different shock tubes in the previous text, examples of obtained results from experiments conducted in these shock tubes were presented. Subsequently a brief description of methods/techniques used for obtaining these and similar results are presented.

4.1 Pressure Measurements

Shock tubes are usually equipped with high response time (about 1 μs) dynamic pressure gauges. They are generally of two types: piezo-electric, like for example PCB transducers, and piezo-resistive, like for example ENDEVCO transducers.

Some natural crystals produce an electric potential difference between their faces when they are submitted to pressure variations. The recorded potential difference is proportional to the pressure difference and the crystal response is quite linear. It can be transformed into voltage variation for piezo-electric gauges, or a variation of electric resistance for piezo-resistive gauges. Knowing the distance between two used pressure gauges and the time passed between their triggering allow measuring the velocity and the Mach number, of both, the incident and reflected shock waves in the tube, as well as the different pressure values reached. The Mach number of the shock wave is defined as

$$M_i = \frac{W_{sw}}{a_i}$$

where W_{sw} is the velocity of the shock wave and a_i is the speed of sound in the gas at state i.

If P_1 is the gas pressure into which the incident shock wave propagates and γ_1 is its specific heat capacity ratio, P_2 the pressure behind the shock wave and M_1 is the shock Mach number, then as shown in Sect. 2.1,

$$\frac{P_2}{P_1} = \frac{2\gamma_1 M_1^2 - (\gamma_1 - 1)}{\gamma_1 + 1}.$$

Accordingly, when W_{sw} or P_2 are known, the Mach number of the shock wave can easily be deduced.

Dynamic pressure transducers are positioned at different locations along the shock tube walls and thereby enable construction of the (x, t) diagram of the considered flow field. Figure 40 shows an example of PCB piezo-electric and ENDEVCO piezo-resistive transducers and two signal pressures recorded during the same run by the two transducers, placed face to face at the same location.

4.2 Shadowgraph Visualization

The shadowgraphy technique is based on the dependence of the gaseous medium refractive index on its density and thereby it detects density variations. When a parallel light beam crosses the test-section, it is deflected from its original path, and the deflection angle is proportional to the density gradient of the medium through which the light beam passes. According to this principle, the shock tube gas flow can easily be visualized. The advantages are the absence of disturbing probes and the recording of conditions throughout an extended flow region. However, the method records an integrated or averaged data along the light path. Its usefulness is ultimately limited by an insufficient density change or gradient. A basic typical shadowgraph arrangement is shown in Fig. 41 where lenses can be replaced by mirrors in the part where the light path has to be parallel. Note that, in theory, only one lens, or mirror, is needed but in practice we avoid the possible imperfection of an optic by using two identical ones.

Black-and-white shadowgraph photographs greatly assisted in understanding shock-tube flows, but in the case of complex flow structures, where monochrome pictures might be unclear, one is able to interpret the phenomenon with the aid of

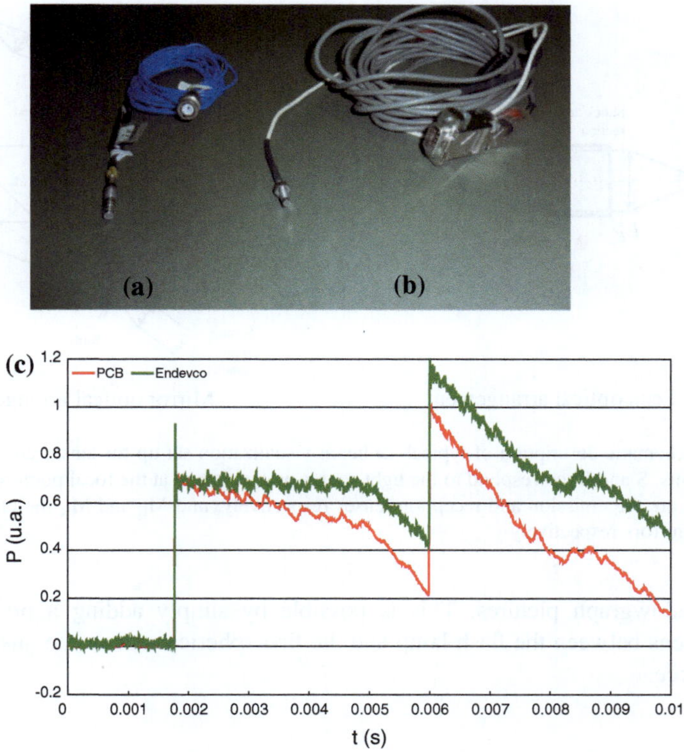

Fig. 40 **a** Photograph of PCB piezo-electric; and **b** ENDEVCO piezo-resistive transducers; and **c** their recorded signals for face to face location during the same run. Note that PCB transducers are useful for detecting shock or blast wave passages, but if the accurate pressure level is needed, then a 0.5 mm layer of silicone coverage is necessary. (Taken from Rodriguez et al. [28]

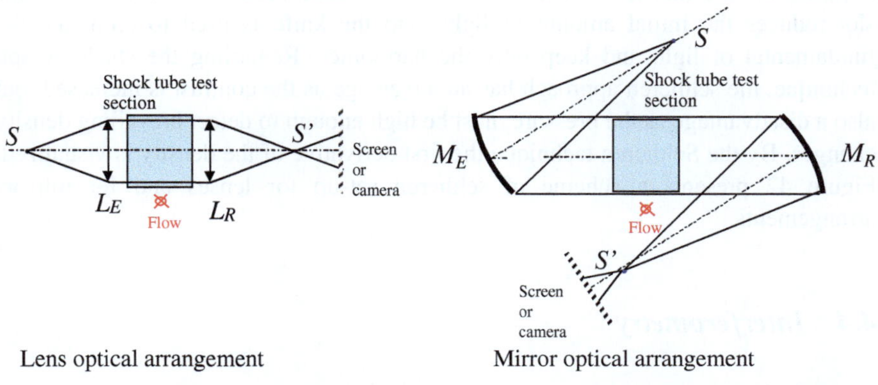

Fig. 41 Schematic drawing showing shadowgraph set-ups for lenses and mirrors arrangements. S and S' correspond to the light source and its image at the focal point respectively, L_E and L_R are the emission and reception lenses respectively, and M_E and M_R the emission and reception mirrors, respectively

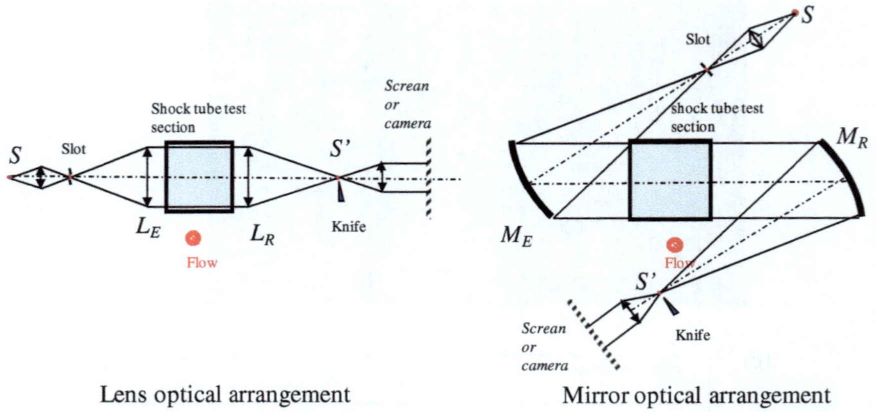

Fig. 42 Schematic description of typical schlieren visualization set-up for lenses and for mirrors arrangements. S and S' correspond to the light source and its image at the focal point respectively, L_E and L_R are the emission and reception lenses respectively, and M_E and M_R the emission and reception mirrors respectively

color shadowgraph pictures. This is possible by simply adding a prism and a suitable lens between the flash lamp and the first spherical mirror, i.e. just after the light source.

4.3 Schlieren Visualization

The schlieren visualization method is finer than shadowgraphy because it allows enhancement of the contrast of the image. Two identical lenses, or two mirrors, are required and the addition of a slot and an appropriate knife is needed. In theory, the slot reduces the initial amount of light, and the knife is used to eliminate the fundamental of light and keep only the harmonics. Regarding the shadowgraph technique, the schlieren approach has an advantage as the contrast is increased, but also a disadvantage as the pressure must be high enough to detect prevailing density changes. By the Schlieren technique the first derivative of the density is visualized. Figure 42 presents a scheme of schlieren set-up for lenses and for mirrors arrangements.

4.4 Interferometry

While both the shadowgraph and the schlieren methods provide only qualitative information of the investigated flow field, using a Mach-Zehnder interferometer provides quantitative density measurements in the investigated flow field. A Mach-Zehnder interferometer is composed of four optically flat glass plates set parallel to each other as shown in Fig. 43. P_1 and P_2 are reflecting mirrors set at a fixed position. P_3 and P_4 are 50 % beam splitters and 50 % reflecting mirrors that can be rotated around horizontal and/or vertical axes. A monochromatic light source (laser light) is collimated and it partly passing and partly reflecting through/from P_4 reaches P_3. As a result part of the light source passes through the test section while the other part passes through the compensation chamber having identical windows thickness to those in the test section. The two beams are combined after passing/reflecting through/from P_3 and focused on a screen; see Fig. 43.

Before conducting an experiment the distance passed by the two beams must be equal, identical to within a fraction of a wavelength. This requirement requires some skill in adjusting the facility. While conducting experiments in a shock tube the refractive index in the test section is changed according to changes in the flow density. The relation between changes in the gas refractive index and changes in the optical path is given by: $\Delta S = \int_0^l (n - n_0)dz$, where n is the local refractive index and n_O is the undisturbed refractive index; l is the test section width. Once the distribution of n is known, the gas density, ρ is determined using the Gladstone-Dale equation: $\frac{n-1}{\rho} = K$ where K is the Gladstone-Dale constant; for air K = 0.2274 cm^3/gm; depending on the light frequency and the temperature. Alternatively, one can write $(n - n_0) = K(\rho - \rho_0)$.

Recorded interferograms shows line of fringes like those shown in Figs. 5, 6 and 7. Prior to flow generation the lines are parallel to each other, upon flow initiation a fringe shift occurs as seen in Figs. 5, 6 and 7. The magnitude of the measured fringe shift, S_{ij} is proportional to the changes in the flow refractive index and thereby

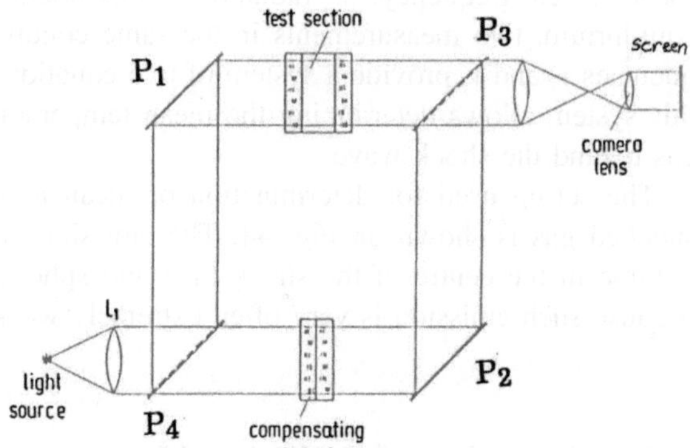

Fig. 43 Schematic description of a Mach-Zehander interferometer

proportional to the gas density, i.e., $S_{ij} = \frac{l}{\lambda}\{(n_j - 1) - (n_i - 1)\}$ where S_{ij} is the nondimensional fringe shift measured from region I to region j, l is the test section width and λ is the light source wavelength. Using the Gladstone-Dale relation the last equation can be rewritten as: $S_{ij} = \frac{l}{\lambda}\{K(\rho_j - \rho_i)\}$.

4.5 *Emission of Radiation Measurements*

Since the last century, diagnostic techniques in high speed flows tend more and more towards spectroscopic methods. In shock-tube investigations, one can use the properties of an emissive gas as one constituents of the gaseous flow. The others must be gases which are not emissive in the domain of the shock wave intensity. Accordingly, for instance, carbon dioxide (CO_2) can be used as the emissive gas, because of its spectroscopic properties in the infrared domain (4.3 μm wavelength). The measurement of the infrared emission coming from the shocked CO_2, using adequate photovoltaic detectors, allows the determination of the mean temperature and density of the gas behind the shock wave. This is possible because, assuming thermo-dynamical equilibrium and for a frequency v_i, the intensity of the shocked CO_2 infrared emission recorded by a photovoltaic detector is given by (see Houas et al. [12])

$$X_{CO_2} = \rho_{CO_2} f_{v_i}(T_{CO_2})$$

where X_{CO_2} is the CO_2 infrared emission intensity, ρ_{CO_2} and T_{CO_2} are the mean density and temperature of the shocked CO_2 respectively, and f_{v_i} is a function of T_{CO_2}. We easily see that the measured emission intensity is proportional to the mean density of the emissive gas multiplied by a function related to its mean temperature, for a given frequency of radiation. Consequently, assuming thermo-dynamic equilibrium, two measurements in the same conditions and for two different frequencies v_i and v_j provide a system of two equations with two unknowns. Solving this system allows determining the mean temperature and density of the shocked gas behind the shock wave.

The set-up used for determination of mean temperature and density within a shocked gas is shown on Fig. 44. The emission is recorded from an elementary volume in the centre of the shock tube and spherical mirrors are usually needed because such emission is very often extremely weak.

4.6 *Absorption of Radiation Measurements*

It is well known [9] that the absorption of gaseous molecules for an isolated frequency v_i is related to the absorption coefficient α_{v_i} by the relation $\alpha_{vi} = \ln\left[\frac{I_{0v_i}}{I_{v_i}}\right]^l$

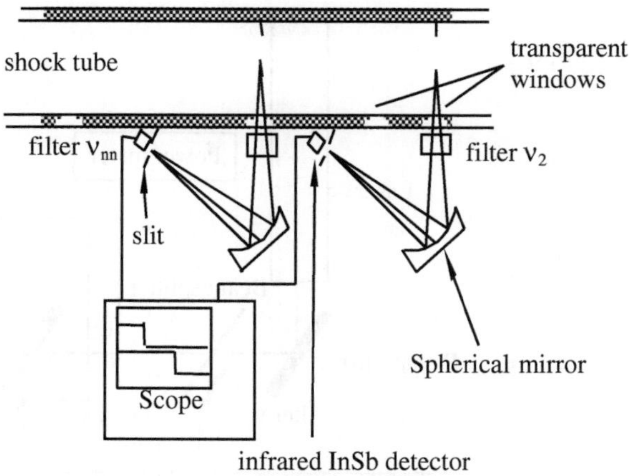

Fig. 44 Experimental set-up for determination of mean temperature and density of a shocked gas moving behind a shock wave in a shock tube by two simultaneous emission of radiation measurements for two different frequencies ν_1 and ν_2

where I_0 and I_{ν_i} are the laser intensities with and without absorption and l is the path length of the laser beam, that is, in the present case, the width of the shock tube test-section. Assuming that the laser beam has sufficiently low power (less than 10 mW), the emitted laser line coincides with the centre of the absorption line of the absorbing gas, there is no overlap between the absorption lines, and there is a local temperature equilibrium, then the absorption coefficient is a function of the mean density and temperature of the absorbing medium

$$\alpha_{\nu_i} = f_{\nu_i}(T, \rho).$$

Therefore, two measurements of two isolated lines of frequencies ν_1 and ν_2, from two different runs (one photovoltaic detector technique) or during the same run (two photovoltaic detectors technique) are necessary to solve the following density and temperature two-equation system

$$\alpha_{\nu_1} = f_{\nu_1}(T, \rho)$$
$$\alpha_{\nu_2} = f_{\nu_2}(T, \rho)$$

from which the mean temperature and density are deduced.

A scheme of the experimental set-up of the absorption of radiation measurement for two photovoltaic detectors is given in Fig. 45.

Additional information regarding the theory and the experimental method can be found in the work of Fortes et al. [7] in the case of a CO_2 absorbing medium. Note that the present method is strongly dependant on the shock wave intensity. When the shock wave is not strong enough, the absorption is too weak and the signal versus noise ratio of the recorded absorption is insufficient to be processed. On the other hand, if the shock wave Mach number is too high, dissociation of the absorbing gas behind the shock appears and the expected absorption lines change.

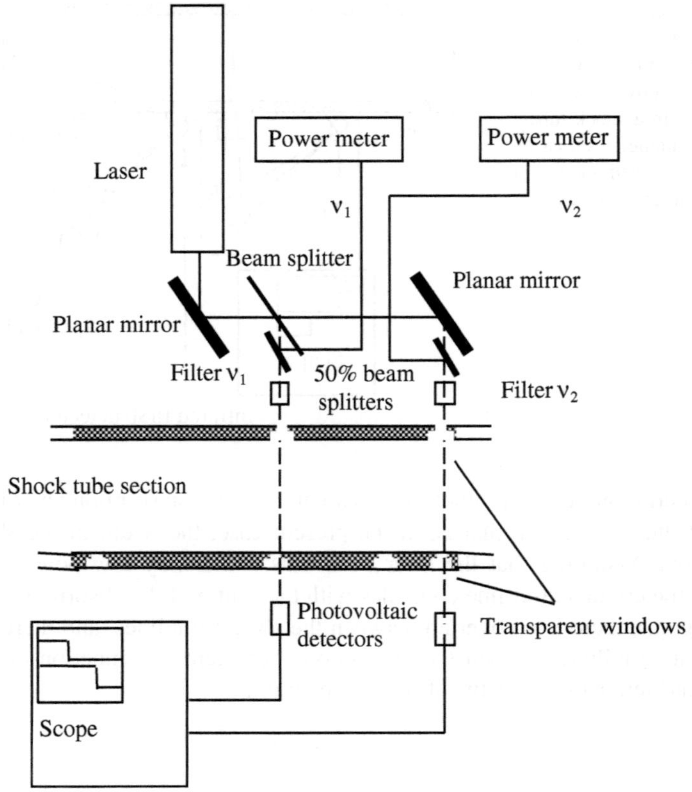

Fig. 45 Lay-out showing a two detector technique, used in radiation absorption measurement method for detecting the mean temperature and density behind a shock wave

5 Shock Tunnels, a Brief Outlook

An alternative option for operating a shock tube is the shock tunnel option. In this option the shock tube end-wall is replaced by a wall having a small opening at its center; a converging-diverging nozzle is attached to this opening. In this tunnel two diaphragms are used; one separating between the driver and the driven sections, the second blocks the nozzle intake. When the incident shock wave reaches the driven section end-wall it reflects backward and the high pressure prevailing behind the reflected shock breaks the second diaphragm and the prevailing high pressure (P_5) acts as a high pressure reservoir feeding the nozzle. Thereby, a supersonic/hypersonic flow is achieved at the nozzle exit having flow duration longer than the one obtained in the regular shock tube. But sometimes the measuring test time is smaller depending on the test conditions. This operational procedure is shown schematically in Fig. 46. A detailed description of shock tunnel operation is given in following chapters.

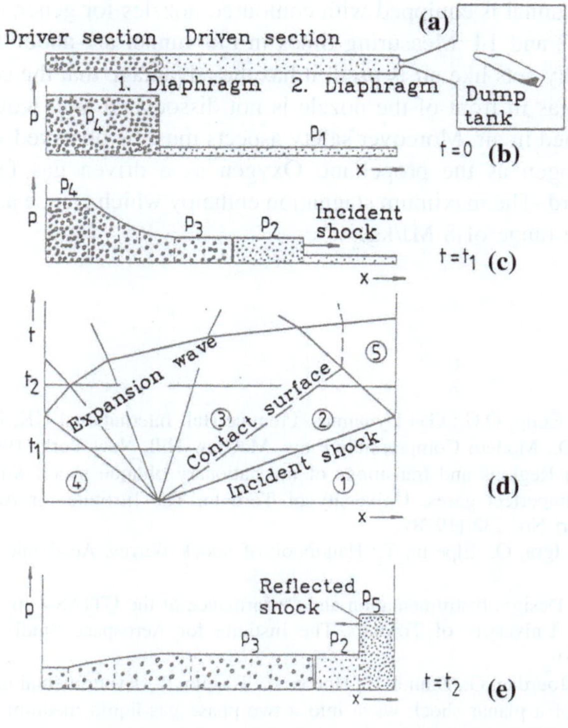

Fig. 46 Schematic description of the generated flow field inside a shock tunnel

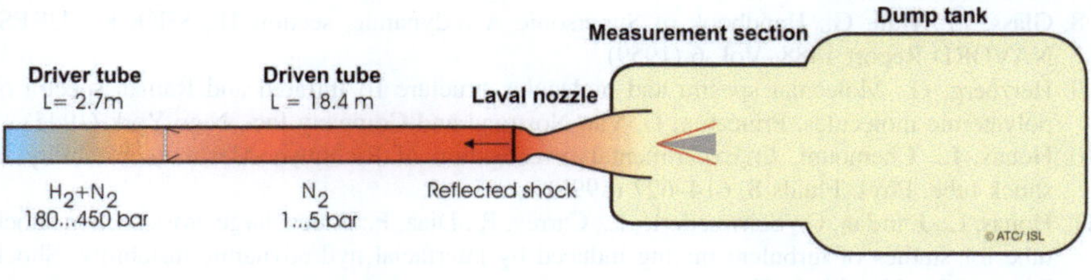

Fig. 47 Set-up of shock tunnel STB of ISL

One example of such a shock tunnel is the one available in the French-German Research Institute of Saint-Louis (ISL), France. A brief description of this facility follows; a detailed description and discussion of the shock tunnels at ISL is given in the Chapter Shock Tunnels at ISL.

The ISL has a shock tube laboratory in which two shock tubes, called STA and STB, are operated in the supersonic and in hypersonic flow domains for simulating real atmospheric flight conditions. Especially the STB is structured as a shock tunnel for flow modeling up to Mach number 14 and at altitude of H = 70 km. STA and STB consist essentially of three parts, a driver section, a driven section and a nozzle mounted at the end of the driven section. A schematic description of the STB shock tunnel is shown in Fig. 47.

This shock tunnel is equipped with contoured nozzles for generating flow Mach numbers 10, 12 and 14. Measuring times in this tunnel are about 1 μs. Nitrogen, which practically acts like air is used; it has the advantage that the compressed and heated driven gas in front of the nozzle is not dissociated as it would be with the oxygen contained in air. Moreover safety aspects must be observed when operating with the hydrogen as the propellant. Oxygen as a driven gas fails because of explosion hazard. The maximum stagnation enthalpy which can be achieved in STB tunnel is in the range of 8 MJ/kg.

References

1. Aksel, M.H., Eralp, O.C.: Gas Dynamics. Prentice Hall International UK, London (1994)
2. Anderson, J.D.: Modern Compressible Flow. McGraw-Hill, New York (1990)
3. Ben Dor, G.: Regions and transitions of nonstationary oblique shock wave diffractions in perfect and imperfect gases. University of Toronto, The Institute for Aerospace Studies, UTIAS Report No. 232 (1978)
4. Ben-Dor, G., Igra, O., Elperin, T.: Handbook of Shock Waves. Academic Press, New York (2001)
5. Boyer, A.G.: Design, Instrumentation and Performance of the UTIAS 4 in x 7 in Hypersonic Shock Tube. University of Toronto, The Institute for Aerospace Studies, UTIAS Report No. 99 (1965)
6. Chauvin, A., Jourdan, G., Daniel, E., Houas, L., Tosello, R.: Experimental investigation of the propagation of a planar shock wave into a two-phase gas-liquid medium. Phys. Fluids **23**, 113301 (2011)
7. Fortes, J., Ramdani, A., Houas, L.: CO_2 absorption measurements of temperature and density in the shock-induced Richtmyer-Meshkov mixing zone. Phys. Rev. E **50**, 3041–3049 (1994)
8. Glass, I.I., Hall, G.: Handbook of Supersonic Aerodynamic, section 18, SHOCK TUBES. NAVORD Report 1488, Vol. 6 (1959)
9. Herzberg, G.: Molecular spectra and molecular structure II, Infrared and Raman spectra of polyatomic molecules. Princeton, D. Van Nostrand and Company Inc., New York (1945)
10. Houas, L., Chemouni, I.: Experimental investigation of Richtmyer-Meshkov instability in shock tube. Phys. Fluids **8**, 614–627 (1996)
11. Houas, L., Jourdan, G., Schwaederlé, L., Carrey, R., Diaz, F.: A new large cross-section shock tube for studies of turbulent mixing induced by interfacial hydrodynamic instability. Shock Waves **12**, 431–434 (2003)
12. Houas, L., Farhat, A., Ramdani, A., Fortes, J., Brun, R.: Concentration and temperature profiles in a shocked gaseous interface. In: Grönig H. (ed.) Proceedings of the 16th International Symposium on Shock Tubes and Waves, Aachen (1987)
13. Igra, O., Falcovitz, J., Reichenbach, H., Heilig, W.: Experimental and numerical study of the interaction process between a planar shock wave and a square cavity. J. Fluid Mech. **313**, 105–130 (1996)
14. Igra, O., Hu, G.Q., Falcovitz, J., Heilig, W.: Blast wave reflection from wedges. ASME J. Fluids Eng. **124**, 510–519 (2003)
15. Igra, O., Wang, L., Falcovitz, J., Amann, O.H.: Simulation of the starting flow in a wedge-like nozzle. Shock Waves Int J Shock Waves, Detonations Explosions **8**, 235–242 (1998)

16. Igra, O., Wu, X., Falcovitz, J., Meguro, T., Takayama, K., Heilig, W.: Experimental and theoretical studies of shock wave propagation through double-bend ducts. J. Fluid Mech. **437**, 255–282 (2001)
17. Igra, D., Igra, O.: Attenuating shock waves by barrier having different orientations; a numerical investigation. In: The 29th International Symposium on Shock Waves, Madison, Wis. USA (2013)
18. Igra, O. An experimental investigation of a nonequilibrium corner-expansion flow of ionized argon. University of Toronto, The Institute for Aerospace Studies, UTIAS Report No. 161 (1970)
19. Ikui, T., Matsuo, M., Yamamoto, Y.: A study of a quick opening valve for shock tube operations. Bull. JSME Ser. B **42**, 2127–2132 (1976)
20. Jourdan, G., Fortes, J., Billiotte, M., Houas, L.: Laser absorption density direct measurements with limitation of temperature variation field in shock-induced gaseous mixing. Shock Waves **7**, 43–47 (1997)
21. Jourdan, G., Houas, L., Billiotte, M.: Density evolution within a shock accelerated gaseous interface. Phys. Rev. Lett. **78**, 452–455 (1997)
22. Jourdan, G., Houas, L., Schwaederlé, L., Layes, G., Carrey, R., Diaz, F.: A new variable inclination shock tube for multiple flow investigations. Shock Waves **13**, 501–504 (2004)
23. Krehl, P.O.K.: History of Shock Waves Explosions and Impacts, A Chronological and Biographical Reference. Springer, Berlin (2009)
24. Mariani, C., Vandenboomgaerde, M., Jourdan, G., Souffland, D., Houas, L.: Investigation of the Richtmyer-Meshkov instability with stereo-lithographed interfaces. Phys. Rev. Lett. **100**, 254503 (2008)
25. Mazor, G., Igra, O., Ben Dor, G., Mond, M., Reichenbach, H., Head-On Collision of Normal Shock Waves with a Rubber Supported Wall. Philos. Trans. R. Soc. Lond. Ser. A **338**, 237–269 (1992)
26. Oguchi, H., Funabiki, K., Sato, S.: A new type of shock valve and its characteristic performance. University of Tokyo, ISAS RN 20 (1976)
27. Ohtomo, F., Ohtani, K., Takayama, K.: Attenuation of shock waves propagating over arrayed baffle plates. Shock Waves **14**, 379–390 (2005)
28. Rodriguez, V., Saurel, R., Jourdan, G., Houas, L.: On solid particle jet formation under shock wave acceleration. Phys. Rev. E (2013) (accepted)
29. Suzuki, T., Sakamura, Y., Igra, O., Adachi, T., Kobayashi, S., Kotani, A., Funawatshi, Y.: Shock tube study of particles motion behind a planar shock wave. Measur. Sci. Technol. **16**, 2431–2436 (2005)
30. Vielle, P., Sur la vitesse de propagation d'un movement dans un milieu en repos. C. R. Acad. Sci. Paris **126**, 31–33, **127**, 41-43 (1898)

Author's Biography

Ozer Igra has devoted most of his research activities to studying various aspects of shock and blast waves phenomena. His studies include both experimental and numerical investigations of strong (ionizing shocks in argon) and moderate to weak shock/blast wave in gases and in suspensions. Results of his investigations can be found in his many publications available in leading professional journals dealing with gas-dynamic flows and shock waves. He received his B.Sc. and M.Sc. degrees from the Department of Aeronautical Engineering of the Technion, Israel Institute of Technology and his Ph.D. from the Institute for Aerospace Studies, University of Toronto, Canada. He Joined the Ben Gurion University of the Negev in 1971. There he established the Shock Waves Laboratory, supervised many masters, doctorate and post-doc students and served as the chairman of the Department of Mechanical Engineering and thereafter as the Dean of the Faculty of Engineering. He is on the International Advisory Committee of the International Symposium on Shock Waves (ISSW) and the International Symposium on Shock Interactions (ISIS), and on the editorial board of the Shock Waves Journal.

Lazhar Houas Started his shock waves studies in 1982 at the Aix-Marseille University. He entered the French National Center for Scientific Research (CNRS) in 1989. In 1981 he received his PhD, in the field of magneto-hydro-dynamics, and his Doctorate of Science in 1988 from the same University. He is active in developing and using diagnostics suitable for investigation of complex shock wave generated flows, e. g. Richtmyer-Meshkov instabilities. His experimental studies covered variety of shock/blast wave induced flows. He is a member of the International Advisory Committee of the International Symposium of Shock Waves. He, together with Prof. Georges Jourdan headed and controlled the creation of Marseille's Shock Tubes Laboratory. This laboratory has six different shock tubes equipped with suitable diagnostics enabling a wide range of experiments. Results from their wide range of conducted experiments appeared in the professional literature.

Measurement of the Physical Properties of Blast Waves

John M. Dewey

1 Introduction to Blast Waves

A blast wave is formed in an ambient atmosphere when there is a rapid release of energy from a concentrated source. Examples of such sources are the detonation of an exothermic material such as trinitrotoluene (TNT); nuclear fission or fusion; the rupture of a pressurized container; a spark, or the rapid heating caused by a focused pulsed laser. The sudden release of energy causes the material of a symmetrical centered source to expand rapidly as a spherical piston. This piston produces a compression wave in the ambient gas. If the speed of the piston is fast enough, or of long enough duration, the compression wave develops into a shock wave. A shock wave is characterized by the very rapid increase, within a distance of about ten mean-free-path lengths, of all the physical properties of the ambient gas, namely, the hydrostatic pressure, density, particle velocity, temperature and entropy. Immediately behind the shock front the properties decay in an exponential fashion, and in the cases of hydrostatic pressure and density will fall below the values of the ambient atmosphere. The particle velocity also decreases until it comes to rest and begins to move in the opposite direction. A typical time-history of a physical property such as hydrostatic pressure, density, particle velocity and dynamic pressure at a fixed point in a blast wave is shown in Fig. 1.

The period when the physical properties are above the ambient value is known as the positive phase, and the period when the properties are below the ambient value is the negative phase. The duration of the positive phase is slightly different for each of the physical properties. Close to the minimum of the negative phase a second

J.M. Dewey (✉)
Department of Physics and Astronomy, University of Victoria, Victoria,
BC V8P 5C2, Canada
e-mail: jdewey@uvic.ca

J.M. Dewey
Dewey McMillin and Associates Ltd, Victoria, BC, Canada

O. Igra and F. Seiler (eds.), *Experimental Methods of Shock Wave Research*,
Shock Wave Science and Technology Reference Library 9,
DOI 10.1007/978-3-319-23745-9_2

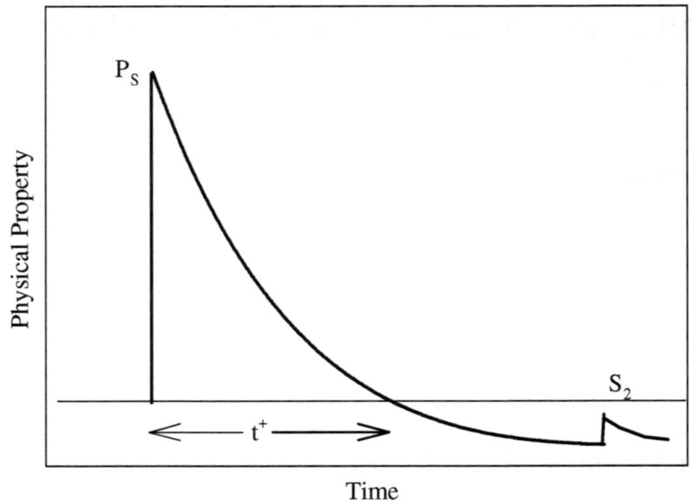

Fig. 1 The form of a time-history of hydrostatic pressure, density, particle velocity or dynamic pressure at a fixed point in a free-field blast wave. P_S is the peak value immediately behind the primary shock, S_2 the second shock, and t^+ the duration of the positive phase

shock arrives, produced by the over expansion and subsequent implosion of the detonation products or source materials.

2 The Physical Properties of Blast Waves

A uniform chemical detonation rapidly produces a high-pressure, high-temperature sphere of gas, which expands in the ambient medium to produce a spherical shock wave. The contact surface between the detonation products and the ambient gas soon becomes irregular and there is considerable mixing of the two gases (Brouillette) [1], but only in extreme cases, such as when there has been a non-symmetrical detonation, does this affect the uniformity of the expanding spherical shock. The release of energy from a detonation is rapid, but not instantaneous, and the rate of energy release has a small but measurable effect on the physical properties of the resulting blast wave. Some of the energy from an explosion may be released as radiation: about 5 % from a TNT detonation, and approximately 50 % from a nuclear explosion. If the explosion is on or close to the ground some of the energy will be disbursed as seismic waves and to excavate a crater. These are some of the reasons why blast waves from different sources and at different distances from the ground, are not identical even though the nominal energy release may be the same.

A characteristic of a shock is that it causes a change of entropy in the gas through which it is passing. As the spherical shock produced by a centred explosion expands in three dimensions, it monotonically decays in strength and leaves the air in a state of radially decreasing entropy and temperature. This means that there are no simple thermodynamic relationships between the physical properties of the gas passing a fixed point in a blast wave. In other words, if, for example, the time-history of hydrostatic pressure is measured at a fixed point, it is not possible to calculate the time-histories of the density or temperature from that measurement. To fully describe the physical properties of a gas element in a blast wave it is necessary to independently measure at least three of the physical properties, such as, hydrostatic and total pressures, and density or particle velocity.

More detailed information about the physical properties of blast waves is provided by Needham [2] and Dewey [3].

2.1 Definitions of the Physical Properties

In order to interpret and relate the measured physical properties of a blast wave it is important to have a precise understanding of the definitions of those properties. A powerful tool in the study of blast waves is the fact that for most explosives the physical properties accurately scale for charge mass, over several orders of magnitude, and for a large range of ambient atmospheric conditions. The appropriate Hopkinson's and Sachs' scaling laws are described in Sect. 4. In order for the physical properties to be used in these scaling laws they are usually reported in terms of the values of those properties in the ambient atmosphere, e.g. P/P_0, ρ/ρ_0 and T/T_0, where P is the hydrostatic pressure, ρ the density, and T the absolute temperature, and the suffix 0 indicates the value in the ambient atmosphere. The strength of the shock front at the leading edge of the blast wave is usually stated in terms of its Mach number relative to the speed of sound in the ambient atmosphere, *i.e.* $M_S = V_S/a_0$, where M_S is the shock Mach number, V_S shock velocity, and a_0 the ambient sound speed.

2.1.1 Hydrostatic Pressure

The hydrostatic pressure of a gas is defined as the force per unit area on a surface caused by the random motion of the molecules in the gas. It does not include any component due to the translational movement of the gas, and therefore can only be measured by a transducer that is flush mounted in a surface that is parallel to the flow in a blast wave. It is a scalar quantity. In the case of a blast wave the most important feature of the hydrostatic pressure maybe that part that is greater than the ambient pressure. This is called the overpressure, viz. $OP = P - P_0$. After the passage of a blast wave the hydrostatic pressure quickly returns to the ambient value because any pressure gradients equilibrate at the rate of the local sound speed.

Hydrostatic pressure is the easiest of the physical properties of a blast wave to measure, and therefore is the property most usually quoted when defining the magnitude or other features of the wave. Unfortunately, it is also the property that gives the least information about the blast wave because any changes of pressure dissipate at the local sound speed. An example of this is in the boundary layer which forms as the blast wave moves over the ground surface. The boundary layer in the blast wave from a large explosion is known to reach a height of about 30 cm [4]. Over that vertical distance there is no discernable change of the hydrostatic pressure, but the particle velocity, and therefore the dynamic pressure, changes from zero at the surface to what may be a very high value at the top of the boundary layer. Also, a hydrostatic pressure measurement does not detect the passage of a contact surface, such as that between the detonation products and the air, although there may be large changes in other physical properties such as density and temperature across that surface. If a hydrostatic pressure transducer does detect the passage of a contact surface, this indicates that it is also sensitive to temperature changes so that the pressure measurements may not be valid.

2.1.2 Density

The density of a gas is defined as its mass per unit volume. The time-history of density in a blast wave is similar to that for hydrostatic pressure but the initial rate of decay after the primary shock and the positive duration are not the same. The passage of a blast wave leaves the ambient atmosphere at an elevated temperature, and therefore at a density lower than that of the original ambient atmosphere. There is a residual density gradient with the lowest density at the explosion centre. The buoyancy effect of this density gradient creates the up-draught that produces the phenomenon referred to as the mushroom cloud.

2.1.3 Temperature

The temperature of a gas is defined as a measure of the kinetic energy of the random molecular motion, and its gradient indicates the direction that heat will flow. After the passage of a blast wave the gas is left in a state of radially decreasing temperature, higher than that in the original ambient atmosphere. As a result, the time-history of temperature in a blast wave is dissimilar from that shown in Fig. 1. The high temperature immediately behind the primary shock begins to decay but then starts to increase again as the warm air shocked closer to the centre of the explosion flows past the measurement location. The temperature is normally expressed in Kelvin, and as a ratio of the ambient temperature, viz. T/T_0.

2.1.4 Particle Velocity

A feature of a blast wave, unlike a sound wave, is that there is a net translation of the gas within the wave. For scaling purposes, the particle velocity within the blast wave is normally recorded relative to the sound speed in the ambient atmosphere, viz. u/a_0, where u is the particle velocity and a_0 is the ambient sound speed. However, strictly speaking, this is not a Mach number because the Mach number of the flow is u/a, where a is the local speed of sound within the blast wave. The time-history of particle velocity in a blast wave is similar to that shown in Fig. 1, but the initial rate of decay and the positive duration are not the same as those for hydrostatic pressure and density. For explosives, such as TNT, that have significant after-burning of the detonation products as they mix with atmospheric oxygen, the particle velocity may have an extended positive, or outward flow, as shown by Dewey [5].

2.1.5 Dynamic Pressure

For the study of blast waves, dynamic pressure is defined as $P_D = ½ \rho u |u|$, where ρ is the density and u the particle velocity. Defining dynamic pressure in this way makes it into a vector rather than a scalar property. The dynamic pressure in a compressible flow is a mathematical rather than a physical property of the flow, because it is a property that cannot be directly measured. However, it can be calculated from independent simultaneous measurements of the hydrostatic and total pressures, as described in Sect. 3.8, or it can be calculated from independent measurements of the density and particle velocity. If the dynamic pressure is reported in non-dimensional units it has the form

$$\frac{P_D}{P_0} = \frac{\gamma}{2}\frac{\rho}{\rho_0}\frac{u}{a_0}\left|\frac{u}{a_0}\right| \tag{1}$$

where γ is the ratio of specific heats, which for air is 1.4, and the other parameters are as previously defined. The dynamic pressure time-history in a blast wave is similar to that shown in Fig. 1, and the positive duration will be identical to that for particle velocity.

The loading on a structure caused by the translational motion of the gas within the blast wave, sometimes called the drag loading, is usually stated as the product of the dynamic pressure and a drag coefficient. In the rapidly varying dynamic pressure of a blast wave the drag coefficient may not have a constant value, and may be a function of the local Mach number and/or the Reynolds number. This problem is discussed, relative to the loading on a cylindrical structure, by van Netten and Dewey [6]. Most of the damage caused by a blast wave to above ground structures is due to the drag loading, although the damage criteria are usually related to the peak value and the positive-duration impulse of hydrostatic pressure.

2.1.6 Reflected Pressure

When the primary shock of a blast wave strikes a plane surface that is face-on to the blast it is reflected normal to the surface, and the gas behind the shock is brought to rest non-isentropically so that the translational kinetic energy is added to the hydrostatic pressure. The resulting pressure on the surface is known as the reflected pressure. The reflected pressure is the largest loading force that can be produced by a blast wave, but its duration at a point on the reflecting surface is determined by the distance of that point from the closest edge of the structure. If the plane reflecting surface is finite in size, the reflected pressure, P_R, will be relieved by a rarefaction wave generated as the reflected shock diffracts around the boundary of the reflecting surface, as illustrated in Fig. 2. The rarefaction wave moves at the speed of sound behind the reflected shock, $a_{R,}$ (16). After the arrival of the head of the rarefaction wave, the pressure on the reflecting surface decreases to what is known as the total pressure.

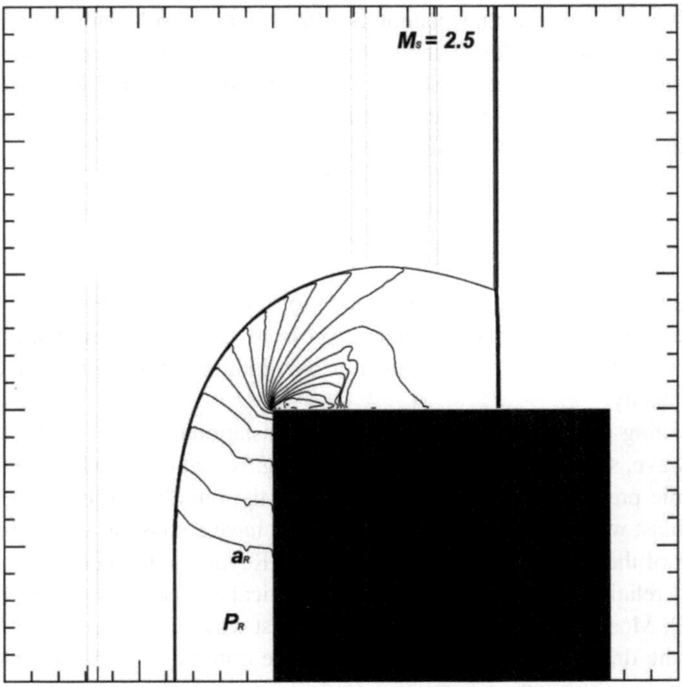

Fig. 2 Numerical simulation of a Mach 2.5 shock incident on a rigid structure. The lines shown between the reflected shock and the structure are isopyknics. (*Courtesy* A.A. van Netten)

2.1.7 Total Pressure

Total or stagnation pressure, P_T, is defined as the pressure exerted on a surface which is face-on to the flow, as a result of the flow being brought to rest isentropically. Work is done both to bring the gas to rest and to compress it adiabatically. After the reflected pressure has been relieved by the arrival of the rarefaction wave from the closest edge of the reflecting surface, the reflected shock moves back into the incoming flow generated by the blast wave. If the Mach number of the primary shock of the blast wave is less than 2.0681, the reflected shock will continue to move backwards against the flow, becoming increasingly weaker. If the Mach number of the primary shock of the blast wave is greater than 2.0681, the reflected shock will move back until its speed is matched by that of the incoming flow, and it will form a bow shock around the structure. In this case, the gas flow impinging on the surface and causing the total pressure will have passed through the bow shock, which will change the physical properties of the gas. As result, the relationship of the total pressure to the physical properties of the blast wave depends on whether the Mach number of the primary shock is greater or less than 2.0681 [see (13) & (14)].

2.1.8 Loading Pressure

The loading pressure is the pressure on any surface of a structure exposed to a blast wave. It can be measured by a pressure transducer flush mounted in the surface and is strongly influenced by the orientation of the surface, and the shape and size of the structure.

3 Measurement Techniques

The sections below describe many of the methods that have been used to measure the physical properties of blast waves. Reisler et al. [7] provides, in three volumes, a complete compendium of all the blast wave measurement techniques and instrumentation used by the United States and several of its NATO allies between 1943 and 1993. The previous reference does not contain information about photogrammetric methods applied to the measurement of blast waves, and these techniques are described in Dewey [8–10].

3.1 Primary-Shock Velocity Analysis

One of the easiest measurement techniques to implement, and which gives accurate values of the physical properties immediately behind the primary shock, is an analysis of the primary shock velocity using the Rankine-Hugoniot equations [11].

There are several ways by which the times-of-arrival of the primary shock can be measured at a series of positions at known distances from the charge centre. The pressure transducers described in Sect. 3.2 can provide this information. Alternatively, a series of simple time-of-arrival detectors connected by a single cable to a recording device can be used as described by Dewey [12]. The highest density of time-of-arrival data can be obtained from high-speed photography of the refractive image of the expanding blast as described in Sect. 3.9.1. In order to use these radius-time data to calculate the physical properties immediately behind the shock it is necessary to calculate the shock velocity. The average velocity between two relatively close detectors can be calculated by dividing the distance between the detectors by the difference between the two times of arrival. When a continuous series of time-of-arrival data is available, the shock velocity at any position can be determined by differentiation of an equation fitted by least squares to the radius-time data.

The primary-shock velocity decays monotonically with radius, from a Mach number of about three as it emerges from the detonation products, until it asymptotically approaches the ambient speed of sound at larger radii. It is important that any equation used to describe the radius-time measurements reflect this monotonic decay. For this reason, a polynomial equation is not suitable because it will have a non-monotonically decaying slope. Dewey [13] proposed the following equation for this purpose, and it has been used successfully since that time to describe the radius-time trajectories of the primary shocks of the blast waves from a variety of centred explosive sources:

$$R_s = A + a_0 t_s + C \ln(1 + a_0 t_s) + D\sqrt{\ln(1 + a_0 t_s)}, \tag{2}$$

where R_S is the shock radius, t_S the time-of-arrival of the shock at that radius, a_0 the speed of sound in the ambient air and A, C and D are the least-squares fitted coefficients.

If it is applied over a wide range of distances, the relative error of measuring R_S and t_S at larger radii may be greater than the error for measurements closer to the charge. In this case, the goodness of the least squares fit may be better at large rather than small radii. This problem can be overcome by weighting the data by $1/R_S^2$ to emphasise the fit at small radii, and by $1/R_S$ at intermediate distances. The implementation of this technique is illustrated in Kleine et al. [14].

The time derivative of Eq. (2) is

$$\frac{dR_s}{dt_s} = a_0\left(1 + \frac{C}{1 + a_0 t_s} + \frac{D}{2(1 + a_0 t_s)\sqrt{\ln(1 + a_0 t_s)}}\right) \tag{3}$$

so that as $t_S \rightarrow \infty$, the speed of the shock approaches the ambient sound speed, and the Mach number of the shock, M_S, at any other time is given by

$$M_s = \frac{1}{a_0}\frac{dR_s}{dt_s} \tag{4}$$

This value of the shock Mach number can be used in the Rankine-Hugoniot equations to determine the values of all the physical properties immediately behind the shock, as follows, where the ambient atmosphere is assumed to be air with a ratio of specific heats $\gamma = 1.4$, and the properties are given as ratios of the ambient values. The suffices $_s$ and $_0$ respectively indicate the value immediately behind the shock and the ambient value.

Hydrostatic pressure:

$$\frac{P_s}{P_0} = \frac{7M_s^2 - 1}{6}; \tag{5}$$

hydrostatic overpressure:

$$\frac{OP_s}{P_0} = \frac{7\left(M_s^2 - 1\right)}{6}; \tag{6}$$

density:

$$\frac{\rho_s}{\rho_0} = \frac{6M_s^2}{M_s^2 + 5}; \tag{7}$$

particle velocity:

$$\frac{u_s}{a_0} = \frac{5}{6}\left(\frac{M_s^2 - 1}{M_s}\right); \tag{8}$$

absolute temperature:

$$\frac{T_s}{T_0} = \frac{\left(7M_s^2 - 1\right)\left(M_s^2 + 5\right)}{36M_s^2}; \tag{9}$$

sound speed:

$$\frac{a_s}{a_0} = \frac{1}{6M_s}\sqrt{\left(7M_s^2 - 1\right)\left(M_s^2 + 5\right)}; \tag{10}$$

local sound speed:

$$\frac{u_s}{a_s} = \frac{5\left(M_s^2 - 1\right)}{\sqrt{\left(7M_s^2 - 1\right)\left(M_s^2 + 5\right)}}; \tag{11}$$

dynamic pressure:

$$\frac{P_{Ds}}{P_0} = \frac{35}{12}\frac{\left(M_s^2 - 1\right)^2}{\left(M_s^2 + 5\right)}; \tag{12}$$

total overpressure (subsonic case, $M_S < 2.0681$):

$$\frac{OP_{Ts}}{P_0} = \frac{7M_s^2 - 1}{6}\left[\frac{12M_s^2\left(M_s^2 + 2\right)}{\left(M_s^2 + 5\right)\left(7M_s^2 - 1\right)}\right]^{3.5} - 1; \tag{13}$$

total overpressure (supersonic case, $M_S > 2.0681$):

$$\frac{OP_{Ts}}{P_0} = \frac{67920.1\left(M_s^2 - 1\right)^7}{\left(M_s^2 + 5\right)\left(42M_s^4 - 96M_s^2 + 45\right)^{2.5}} - 1; \tag{14}$$

reflected overpressure:

$$\frac{OP_{Rs}}{P_0} = \frac{\left(7M_s^2 - 1\right)\left(4M_s^2 - 1\right)}{3\left(M_s^2 + 5\right)} - 1; \tag{15}$$

reflected sound speed, a_r: (a_r is the sound speed behind the reflected shock and therefore the speed at which the rarefaction wave from the edge of a structure will move across the surface, reducing the reflected pressure[1])

$$\frac{a_R}{a_0} = \frac{\sqrt{\left(M_s^2 + 2\right)\left(4M_s^2 - 1\right)}}{3M_s}; \tag{16}$$

increase of entropy, ΔS, of the gas passing through the shock:

$$\Delta S = 8.3143 \ln\left[\frac{\left(7M_s^2 - 1\right)^{2.5}}{46656}\left(\frac{M_s^2 + 5}{M_s^2}\right)^{3.5}\right] \text{ J mole}^{-1}\text{K}^{-1}. \tag{17}$$

The above Rankin-Hugoniot relationships are derived from the two possible states of a compressible fluid for which the mass, momentum and energy are conserved, and provide values of the physical properties of the gas immediately behind the shock that are completely reliable, such that they are the preferred method of calibrating electronic transducers, such as those discussed below. For strong shocks with a Mach number greater than about 3.5, corresponding to a peak

[1]Knowing a_R and the distance from a point on a normally reflecting surface to the nearest edge of the surface, permits a calculation of the duration of the reflected pressure at that point before it is relieved by the arrival of the rarefaction wave from the edge of the surface.

hydrostatic overpressure greater than 13 atm, additional degrees of freedom of the gas are excited and for exact results it is necessary to use a real-gas equation of state to correct the above relationships.

3.2 Hydrostatic Pressure Gauges

Hydrostatic pressure is the easiest physical property to measure directly in a blast wave, and is therefore the one most widely used. A pressure transducer is normally a metal diaphragm that is displaced by the loading of the blast wave, and the amount of displacement is sensed by a piezo-electric crystal, a capacitor or a strain transducer. The signals from the transducer may be pre-amplified by a device close to the gauge before being transmitted by cable to an amplifier and storage device. A piezoelectric transducer produces a change of voltage but is unable to generate a significant current, and so the pre-amplifier for such a device needs to be a charge amplifier. A pressure gauge used to monitor a blast wave may be exposed to thermal radiation from the explosion, and heated by the blast wave itself. It is important therefore that it be insensitive to these thermal effects, and this can be tested by exposing the gauge to a thermal source, such as an infrared lamp, and ensuring that no signal is generated by this exposure. Also, the gauge must be isolated from the structure in which it is mounted so that stresses in the mount do not strain the case of the gauge and generate spurious signals.

To measure the hydrostatic pressure the transducer must be mounted so that the surface of the diaphragm is parallel to the flow within the blast wave. For a hemi-spherical surface-burst explosion, or in the Mach reflection region of an airburst explosion, this can be done by flush mounting the gauge in the ground surface. The gauge will be in the boundary layer, but there is no change in hydrostatic pressure through a boundary layer. To measure the hydrostatic pressure in the blast wave from a spherical airburst explosion the gauge must be mounted in a disc-shaped baffle with a sharp-edged rim, sometimes called a lollipop or pancake mount, or in the side of a sharp pointed cylinder. To minimise edge effects, experience has shown that the radius of the disc, or the distance of the cylinder mounted gauge from the pointed tip, should be at least 10 times the thickness of the disc or the diameter of the cylinder. In both cases the bevel length should be at least seven times the thickness or diameter. The disc or cylinder must be mounted on a long arm to minimise the effects of shocks, compression and rarefaction waves reflected from the stand holding the gauge. These waves will travel back towards the transducer against the blast wave flow at a speed of $a - u$, where a is the local sound speed and u the particle velocity within the blast wave. This type of baffle must be carefully aligned with the centre of the charge because any misalignment will cause the primary shock to strike the transducer at an angle, thus not measuring the hydrostatic pressure, and will generate additional noise due to the non-symmetry of the blast as it passes the trailing edge of the baffle. An array of gauge-mounts such as those described above, deployed on a large scale test, is shown in Fig. 3.

Fig. 3 An array of pressure gauges deployed on a large scale test. From lower left to right: a pair of face-on time-of-arrival detectors to measure the shock velocity; a lollipop or pancake mount; a cylindrical mount; another pair of time-of-arrival detectors, and a rectangular baffle [7]

Pressure transducers supplied by a manufacturer will normally be accompanied by a calibration curve relating the imposed pressure to the output voltage. Ideally, this relationship should be as linear as possible. The calibration may change with time, and particularly with use, and so the gauge should be regularly recalibrated, ideally before and after each exposure to a blast wave, and with the pre-amplifier, amplifier and recorder used for the exposure. With some calibration systems a measured pressure is applied to the transducer and then rapidly released. However, this monitors an outward deflection of the diaphragm rather than an inward deflection as would be caused by the passage of a blast wave. A preferred method of calibration is to use a simple shock tube, with the transducer flush mounted in the wall of the tube, and with a sensor on each side of the transducer to measure the shock velocity, from which the shock Mach number can be calculated. This Mach number is then used in (5) or (6) to determine the change of pressure as the shock moved across the transducer. Pressure gauges can be calibrated in situ on a test site using a portable, flexible, valve driven shock tube with the open end of the tube being held firmly over the transducer. The Mach number of the calibrating shock can be measured using a pair of sensors installed close to the end of the tube, and the reflected pressure to which the transducer is subjected can be calculated using (15).

Some companies that supply pressure transducers and auxiliary equipment for the measurement of blast and shock waves are: PCB (www.pcb.com), Endevco (www.endevco.com) and Kistler (www.kistler.com).

The recorded signal from a pressure transducer is converted to a pressure-time history by application of the pressure-voltage calibration relationship. The changing pressure at the shock front is virtually instantaneous because the thickness of the shock is estimated to be about 10 mean-free-path lengths. In order to measure this very rapid pressure change the transducer needs to have a high frequency response, but this may result in signal over-shoot and ringing. To overcome this affect the manufacturer may introduce mechanical and/or electronic damping of the signal, but if too much damping is applied the peak signal may be rounded and not record the maximum value. The manufacturer must make a compromise between these two effects, but in any event the estimation of peak pressure from a pressure-time history needs careful interpretation.

The decay of pressure behind the primary shock usually has an exponential form, in which case the peak can be estimated by least squares fitting a section of the pressure-time data to an exponential decay, viz.

$$P = P_s e^{-\alpha t}, \tag{18}$$

where P and t are respectively the measured values of pressure and time measured from the time of arrival of the primary shock, in the region close to the shock, and P_S and α are the least-squares fitted coefficients. P_S is a good estimate of the peak pressure immediately behind the shock. If the pressure-time history from the primary shock to the second shock has the classic form similar to that shown in Fig. 1, the measured overpressure data may be well described by a least squares fit to the modified Friedlander equation, viz.

$$OP = OP_s e^{-\alpha t}\left(1 - \frac{t}{t^+}\right), \tag{19}$$

where P_S, α and t^+ are the fitted coefficients to the measured P, t data. Usually, P_S and t^+ are good estimates of the peak pressure and the positive duration, respectively. Equation (19) can be integrated between $0 < t < t^+$ to give the overpressure impulse in the positive phase, viz.

$$I = \int_0^{t^+} OP_s e^{-\alpha t}\left(1 - \frac{t}{t^+}\right) dt = OP_s \frac{1}{\alpha^2 t^+}\left(e^{-\alpha t^+} - 1 + \alpha t^+\right) \tag{20}$$

3.3 Total (Stagnation) Pressure

Total pressure, as defined in Sect. 2.1.7, can be measured by the same type of electronic transducer as used to measure hydrostatic pressure, but with the surface face-on to the blast. There should be no flow across the surface of the transducer, which is held in a pencil or cylindrical mount, and is slightly recessed by a few millimetres to eliminate any trans-surface flow. The gauge must be supported on a narrow arm in order to minimise the effect of waves reflecting from the main

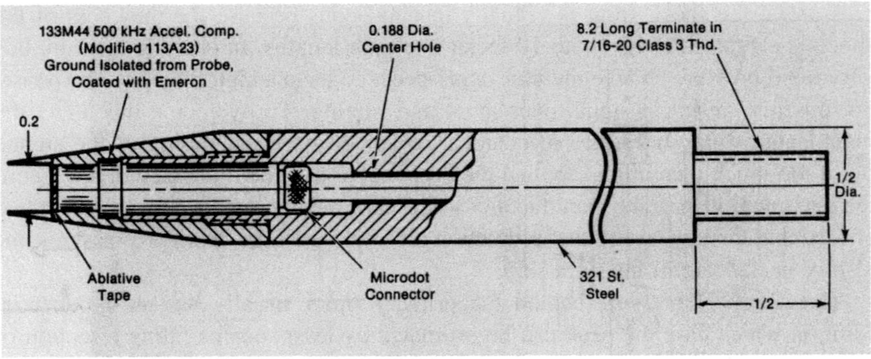

Fig. 4 A drawing of the design of a total-pressure gauge [7]. The units of the dimensions shown are inches. The front surface of the pressure transducer is set back 0.2 in. from the tip of the probe

support. It is important that the gauge be pointed directly at the charge centre to eliminate any transverse flow across the surface of the transducer. A pressure transducer can also be flush mounted in the side of the cylinder holding the total-pressure transducer to obtain a simultaneous measurement of the hydrostatic pressure. A diagram of the design of a total-pressure gauge is shown in Fig. 4.

The total-pressure transducer will initially record the reflected pressure (Sect. 2.1.6), but this will be relieved in the few micro-seconds that it takes for the rarefaction wave, created by the diffraction of the reflected shock around the edge of the mount holding the transducer, to reach the surface of the transducer. This short duration reflected-pressure measurement will appear as a spike at the beginning of the total-pressure measurement, but because of its very short duration may not be a reliable measure of the reflected pressure.

3.4 Loading Pressure

The loading pressure time-history at any point on the surface of a structure exposed to a blast wave can be measured by a pressure transducer flush mounted in the surface at that point.

3.5 Density

The density of the gas within a blast wave has been measured by devices that monitor the soft β radiation from radioactive sources such as erbium, ^{169}Er, and promethium, ^{147}Pm, (Dewey & Anson [15]; Ritzel [16]). The radioactive source in such a device is mounted in a sharp edged baffle a few centimetres from the radiation detector which is mounted in a similar baffle. The density gauge can be calibrated by passing aluminium and plastic foils of known aerial density between the source and the detector. The baffles holding the source and the detector in the gauge described by Dewey and Anson [15] had sharp leading edges but rectangular trailing edges. As a result, the shocks and rarefaction waves generated by the primary shock as it diffracted around the blunt ends of the baffles travelled back between the baffles against the blast wave flow and disturbed the measurement of the density-time history. This error was corrected in future versions of the density gauge by using baffles with sharp leading and trailing edges [16].

The density gauge described by Slater et al. [17] was incorporated in a blast gauge station that, in addition to measuring the density-time history, also measured the hydrostatic and total pressures, and the speed of the shock as it traversed the gauge (Fig. 5).

3.6 Temperature

There appear to have been no successful techniques to accurately and directly measure the temperature of the air in a blast wave. A temperature transducer, such as a thermistor or a resistor, must be supported on a substrate that will have a high thermal capacity and the transducer measures the temperature of that substrate rather than the air in the blast wave. In addition, it is difficult to shield a temperature transducer from the thermal radiation of the detonating charge without affecting the properties of the blast wave passing over the transducer. The temperature is calculated, therefore, from the other physical properties that can be more easily measured, viz.

$$\frac{T}{T_0} = \frac{P}{P_0}\frac{\rho_0}{\rho}, \tag{21}$$

where T is the absolute temperature, P the hydrostatic pressure, ρ the density and the suffix 0 indicates the values in the ambient atmosphere.

3.7 Particle Velocity

The particle velocity throughout a blast wave flow field has been measured using high-speed photography of an array of flow tracers established immediately before the detonation of the charge [5]. The technique for doing this is described in more detail in Sect. 3.9.2. The time-history of particle velocity in free-field blast waves from centred charges of explosives such as nitro-glycerine, Pentolite, C4 and ammonium-nitrate fuel-oil (ANFO) is well described by the modified Friedlander

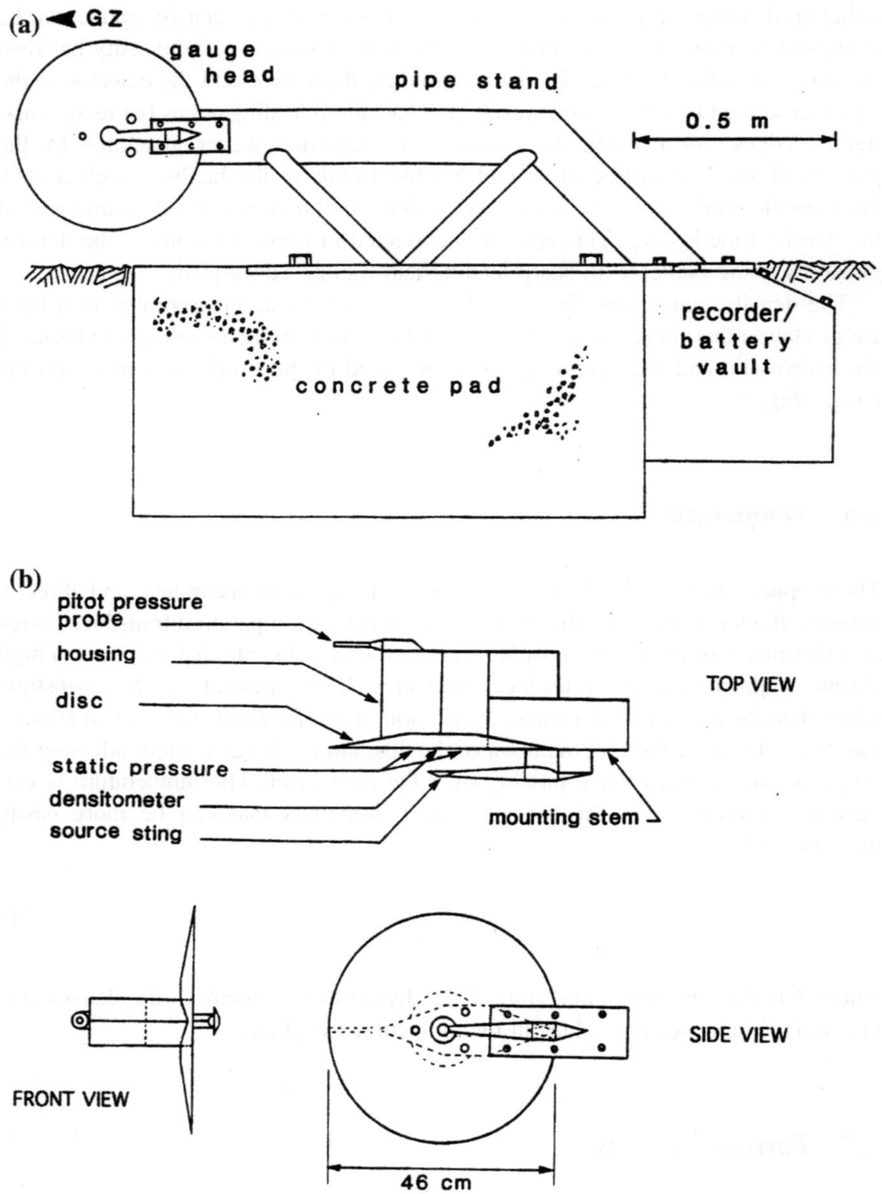

Fig. 5 Diagram of Suffield Blast Station [17] to make simultaneous, independent measurements of hydrostatic and total pressures, density and shock velocity

equation (19), but due to the after-burning of TNT, Dewey [5] suggested that the time history of particle velocity for that explosive was better described by

$$u = u_s e^{-\alpha t}(1 - \beta t) + A\ln(1 + \beta t) \tag{22}$$

where u is the particle velocity at time t after the arrival of the primary shock, u_S is the peak particle velocity immediately behind the shock, and α, β and A are least-squares fitted coefficients.

3.8 Dynamic Pressure

The dynamic pressure in a compressible fluid, as defined in 2.1.5, is a mathematical rather than a physical property of the flow in that it cannot be directly measured by a transducer inserted into the flow. The dynamic pressure in a blast wave can be calculated by inserting the independently measured particle velocity, u, and density, ρ, into (1), or from independent and simultaneous measurements of the total and hydrostatic pressures, P_T and P, respectively.

The functional relationship to calculate the dynamic pressure from the total and hydrostatic pressures depends on whether the flow is locally sub-or super-sonic, that is, whether $M_u \leq 1$ or > 1, where $M_u = u/a$, u is the particle flow speed, and a the local sound speed. Since $a^2 = \gamma P/\rho$,

$$M_u^2 = \frac{\rho u^2}{\gamma P} = \frac{2P_D}{\gamma P} \tag{23}$$

and therefore

$$P_D = \frac{\gamma}{2} P M_u^2 \tag{24}$$

where P_D is the dynamic pressure and γ the ratio of specific heats. For $\gamma = 1.4$ and $M_u \leq 1$

$$P_T = P\left(0.2M_u^2 + 1\right) \tag{25}$$

which gives

$$M_u^2 = 5\left[\left(\frac{P_T}{P}\right)^{0.286} - 1\right] \tag{26}$$

and this value of M_u^2 can be inserted in (24) to give the dynamic pressure.

For $\gamma = 1.4$ and $M_u > 1$,

$$P_T = \frac{PM_u^2(3.58 \times 10^{-2})}{\left[1.167 - \left(\frac{0.167}{M_u^2}\right)\right]^{2.5}} \tag{27}$$

There is no simple algebraic solution of (27) to give M_u^2 in terms of P_T and P, and the equation must be solved iteratively or graphically to give values of M_u^2 that can be inserted in (24). Equation (27) is called the Rayleigh supersonic pitot formula. Equations (26) and (27) are derived in Liepmann and Roshko [18].

If the blast wave under study has the classical form shown in Fig. 1, the time-histories of density and particle velocity may be well described by the modified Friedlander equation (19). However, experience has shown, that the time-history of dynamic pressure is not precisely described by the modified Friedlander equation. This is to be expected because the product of two or more Friedlander equations does not have the algebraic form of the Friedlander equation, and this may lead to small errors if it is used to determine the peak dynamic pressure, the positive duration and impulse, as described in Sect. 3.2 for hydrostatic pressure. An equation of the form

$$P_D = A + Be^{-Ct} + Dt \tag{28}$$

where *A, B, C* and *D* are the least-squares fitted coefficients to the measured values of dynamic pressure, P_D, and time, *t*, may better describe the dynamic pressure time history than the modified Friedlander equation. (28) can be integrated to obtain the dynamic pressure impulse in the positive phase, viz.

$$I_{PD} = \int_0^{t^+} \left(A + Be^{-Ct} + Dt\right)dt = At^+ - \frac{B}{C}e^{-Ct^+} + \frac{D}{2}t^{+2} + \frac{B}{C} \tag{29}$$

where t^+ is the dynamic pressure positive phase duration. Equations (28) and (29) have been successfully used to determine the peak values, positive durations and impulses of dynamic pressure for a blast wave created by a propane/oxygen explosion [19], but to date, have not been evaluated for other explosive sources.

Setting $t = 0$ in (28) gives $P_{DS} = A + B$, where P_{DS} is the peak dynamic pressure immediately behind the primary shock. Setting $P_D = 0$ in (28) gives

$$Be^{-ct^+} + Dt^+ = -A \tag{30}$$

where t^+ is the positive phase duration of dynamic pressure. There is no simple algebraic solution of (30) to give t^+ as a function of the other parameters, and (30) must be solved iteratively or graphically. The positive phase duration of dynamic pressure must equal that for particle velocity because of the way in which dynamic pressure is defined.

3.9 *Flow Visualization Techniques*

High-speed photographic techniques have been used extensively to measure the physical properties of blast waves from explosive sources ranging in size from megaton nuclear weapons to as small as 0.5 mg. These techniques have been described in general terms by Dewey [3, 8–10]. The two methods used most extensively have been a modification of shadow photography, known as refractive image analysis (RIA), and flow visualisation of particle tracers, known as particle trajectory analysis (PTA). The RIA and PTA techniques are illustrated in Fig. 6.

3.9.1 Refractive Image Analysis

The sudden increase of density at the primary shock of a blast wave causes an intense gradient of the refractive index of the ambient air, so that light passing through the shock is sharply refracted producing a distortion of the observed background, as shown in Fig. 6. This distortion can be photographed with a high-speed camera to produce a series of images from which the radius of a spherical shock can be measured as a function of time. On the lower left in Fig. 6 is an array of striped backdrops to enhance the refractive image of the shock close to the ground. The larger backdrops are 15 m high. Fiducial markers are also shown in the lower left.

The required framing rate of the camera depends on the size of the explosion. In the case of a nuclear explosion 100 frames per second is adequate [20]. For explosions in the range of 1 kt, a framing rate of about 1000 is required [21, 22], and for micro-scale charges a framing rate of the order of 1 million is needed [14].

Fig. 6 A 200 t hemispherical surface-burst explosion showing the refractive image of the shock front and an array of particle-flow smoke tracers

High-speed film cameras do not run at a constant speed because of the inertia of the film spools as they are brought up to speed and then decelerated. Such cameras will normally incorporate a constant-rate timing light that will leave an image on the edge of the film. Some may leave a coded image that is related to Universal Time. The timing light is normally not immediately adjacent to the film frame that is being exposed so that there will be a displacement on the film between the exposed frame and the appropriate timing mark. This displacement must be known in order to relate the exposed frame to the correct timing mark. Digital cameras can be expected to run at the set framing rate, but this should be calibrated from time to time. Digital cameras do not provide an image with the spatial resolution of film cameras and must be used with a smaller field of view to match the resolution of film cameras.

The natural background may contain enough contrast to visualise the refractive image of the primary shock, but when very precise measurements are required, such as the accurate visualisation of the triple point region of a Mach reflection, an artificial background can be created. Such a background usually consists of a series of black and white stripes painted on boards or large canvas sails which are mounted behind the charge at a sufficient distance that they will not be struck by the blast wave until the complete blast wave in the object plane has been recorded. The precision added with such a background usually enables the second shock to also be visualised, and in rare cases the third shock. On many US and British nuclear tests such backgrounds were created to high altitudes using rockets to produce arrays of white smoke trails [20].

The refractive image of a spherical shock is recorded in the film plane of the camera that is parallel to an object plane containing the charge centre. As a result, the radius of the image measured in the film plane, r', is not directly related to the radius of the shock, R, but to a distance, R', in the object plane, as shown in Fig. 7 and related in (31).

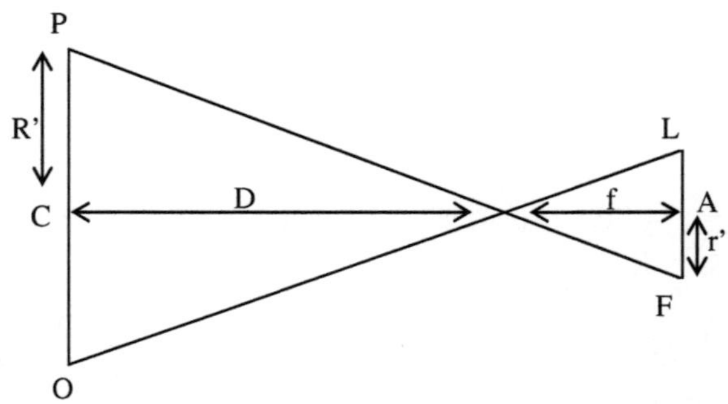

Fig. 7 Relationship between the object and film planes of a camera. *CA* is the central axis of the camera; *OP* is the object plane containing the charge centre, which is parallel to the film plane, *FL*; *D* the distance of the centre of the object plane to the camera; *f* the focal length of the camera; r' a distance measured in the film plane, and R' the corresponding distance in the object plane

$$R' = \frac{r'D}{f} \tag{31}$$

where R' and r' are defined above, D is the distance of the camera lens from the charge centre and f is the focal length of the camera. The radius of the shock is then given by

$$R = D \sin\left[\tan^{-1} \frac{\cos\theta}{\left(\frac{D}{R'} - \sin\theta\right)}\right] \tag{32}$$

where θ is the offset angle of the optical axis of the camera from the charge centre, as shown in Fig. 8. This simple relationship is for the special case when the shock centre, the centre of the image plane, the shock image, and the camera are coplanar. In other situations, such as measuring the displacement of the Mach stem shock produced by an air-burst explosion, more complex three-dimensional geometric corrections must be used.

In Fig. 8, O is the centre of the spherical or hemispherical shock, S; C the camera lens; OC = D, the distance between the shock centre and the camera; CA the central axis of the camera, which is perpendicular to OP the object plane through the centre of the explosion parallel to the image plane of the camera; ∠ ACO = θ; ∠ QCA = ∠ QOT = α; OT = R the shock radius, and OQ = R′ the projection of R on the object plane. As the spherical shock expands the locus of T is a hemi-circle with diameter OC.

In order to ensure the accuracy of the measurements of the shock radius, a series of fiducial markers should be deployed in the field-of-view of the camera. Ideally, the markers should be placed in the object plane of the camera containing the charge centre, but this would significantly interfere with the blast wave flow. The fiducial markers must therefore be placed in planes beyond the charge and between the charge and the camera. The markers must be accurately survey relative to the

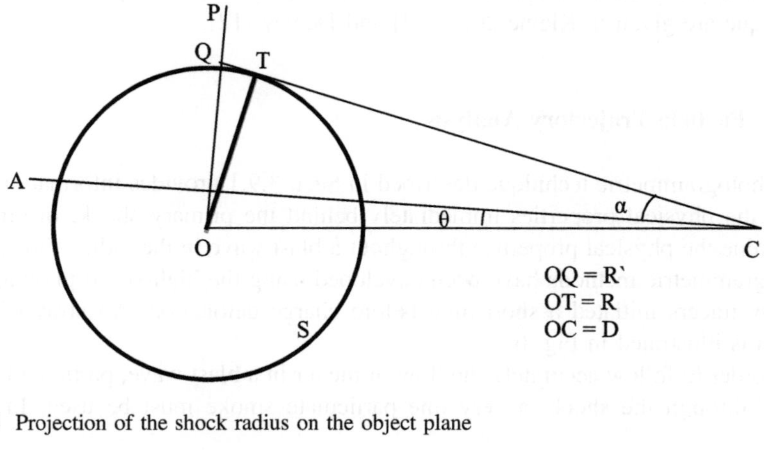

Fig. 8 Projection of the shock radius on the object plane

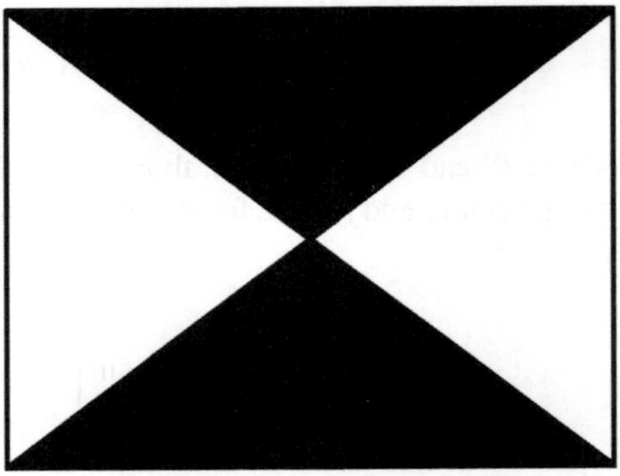

Fig. 9 Suggested design of a fiducial marker used to provide a length scale in the photogrammetry of the refractive image and flow tracers of blast waves

charge centre and the camera so that their positions can be geometrically projected on to the object plane. This permits confirmation of the scale used to convert the displacements measured in the film plane, r′, to the displacements in the object plane, R′, without relying entirely on (31). The design of a fiducial marker that has been found to be most effective is that shown in Fig. 9.

The survey of the fiducial markers may contain errors, and this will become apparent when the positions of the markers are projected onto the object plane and used to calculate the image scaling factor. If a single marker appears to be at fault it can be ignored. If the survey errors appear to be random, the surveyed distance of the camera from the centre of the explosion and the orientation angle of the camera can be adjusted to minimise the apparent errors in the fiducial marker positions. It is useful in the image analysis process to have fiducial markers close to the camera, which will appear in the lower left and lower right corners of the image, because these will not be concealed by dust or detonation products, and thus provide correct orientation of the image at later times in the recording.

The primary shock radius-time data obtained from high-speed photography can be used with the techniques described in Sect. 3.1 to determine the shock Mach number as a function of radius, and thus the values of all the physical properties immediately behind the shock using (5)–(17). Examples of the application of this technique are given in Kleine et al. [14] and Dewey [12].

3.9.2 Particle Trajectory Analysis

The photogrammetric technique described in Sect. 3.9.1 provides information only about the physical properties immediately behind the primary shock. In order to determine the physical properties throughout a blast wave in the radius-time plane, photogrammetric methods have been developed using the high-speed photography of flow tracers initiated a short time before charge detonation. An array of such tracers is illustrated in Fig. 6.

In order to follow accurately the flow of the air in a blast wave, particularly as it passes through the shock, a very fine particulate smoke must be used. In early studies, such as US and British nuclear tests, a chloro-sulphuric acid mixture was used, released in a trail from a rocket or mortar shell, as described by Dewey [5]. This mixture is toxic and highly corrosive, and therefore difficult to use. In air, the mixture forms small droplets of sulphuric acid which vaporise at high temperatures so that the trail disappears for a short time behind very strong shocks. In more recent studies, smoke trails and puffs have been generated using carbon-black to produce black trails and fumed silica to produce white trails. These powders have sub-micron size particles, which remain suspended in air by the Brownian movement and have proved to be excellent tracers of the flows within blast waves, in that their velocities immediately behind a shock agree exactly with the Rankine-Hugoniot values. The smoke tracers close to the charge need to be generated a short time, e.g. 0.5 s., before the charge is detonated, and shortly before the arrival of the primary shock for tracers at greater distances. The exact time depends on the size of the explosion. The colour of the smoke that is used depends on the background against which it will be photographed. Frequently, alternate black and white tracers have been used to allow for variability of the background.

The smoke tracers should be generated in the object plane of the camera that contains the charge centre. This provides a simple relationship between measurements in the film plane of the camera and displacements of the tracers in the object plane. If the tracers are not in the object plane, a more complex geometrical manipulation is required to determine their displacements. In any event, the plane containing the smoke tracers and the charge centre must be known accurately. Accurately surveyed fiducial markers must be deployed throughout the field of view, as described in Sect. 3.9.1.

The flow tracers are observed within the varying density field of a spherical blast wave and this will deflect the optical path from a flow tracer to the camera. However, this deflection appears to be very small and not to have a significant effect on the displacement of the tracer measured from the photographic record.

Measurement of the flow-tracer trajectories in blast waves was first used only to measure the particle velocities in the radius-time plane [5]. Subsequent analyses showed that by measuring the relative distance between two adjacent flow tracers the density field could also be measured. Measuring the density at a fixed point in a blast wave does not permit calculation of the hydrostatic pressure because of the varying entropy of the gas passing that point. However, when flow tracers are used, the density is measured along the particle trajectories which are loci of constant entropy between the primary and secondary shocks. As a result, the simple adiabatic relationship between the density and pressure can be used in the Lagrangian coordinate system. Assuming that both the density and pressure immediately behind the shock, ρ_S and P_S, are known from a measurement of the shock velocity, the pressure, P, can be determined from the measured density, ρ, from

$$\frac{P}{P_s} = \left(\frac{\rho}{\rho_s}\right)^{\gamma} \tag{33}$$

where γ is the ratio of specific heats, and the subscript $_S$ indicates the values immediately behind the primary shock. These procedures map the hydrostatic pressure, density and particle velocity throughout the flow field in the radius-time plane, and these three properties can be used in the thermodynamic relationships to calculate all the other physical properties such as temperature, local sound speed and dynamic pressure (Dewey)[13].

Subsequently, it has been found that the measurement of the particle trajectories in a blast wave can be used more easily to map all the physical properties by using the measurements in conjunction with a hydro-code, as described in Sect. 3.11.

3.9.3 Background Oriented Schlieren

A new technique has been developed for measurement of the density field in a compressible flow, called background oriented schlieren (BOS) [23, 24]). This technique uses a digital still camera, a structured background, and inverse tomographic algorithms which can extract two-dimensional slices from a three-dimensional flow. This has been applied to obtain the density field for an axisymmetric supersonic flow over a cone-cylinder model. At this time it has not been conclusively demonstrated that the technique can be used to measure the density profile along a radial of a spherically-symmetric centred blast wave, or along any radial of an axially symmetric blast wave such as a muzzle blast, but this may be achieved in the future.

3.10 Passive Methods

Passive techniques for the measurement of the physical properties of blast waves involve the use of simple non-electronic devices that are permanently deformed or displaced by the passage of the blast wave such that the deformation or displacement can be used to estimate some property of the wave. The concept of such devices arose from the observations of the effects of explosions for which there was no, or very limited access to direct measurements. For example, when Lord Penny toured the scenes of the Hiroshima and Nagasaki nuclear events he noted that metal posts were bent by the blast waves in a uniform fashion, and that the displacement of standard Japanese gravestones was related to their distance from ground zero. On early nuclear tests, limited electronic measurements were available and some blast properties could be inferred from the degree of crushing of beer and jerry cans that had been left on the test sites. Such methods are still be used in the evaluation of accidental explosions when no direct measurements are available. As a minimum, observation of the degree of damage to simple structures can be used to estimate the relative energy release of an explosion. Details of the various passive gauges used during the US nuclear testing program are given in Reisler et al. [7].

3.10.1 Cantilever Gauges

The passive device that has been used and studied most extensively is the cantilever gauge. This was based originally on the observations of the bending of metal posts by Penny [25] which he used to estimate the explosive yields of the Hiroshima and Nagasaki nuclear explosions. Early devices by Ewing and Hanna [26] and Baker et al. [27] used cantilevers with rectangular cross sections which therefore needed careful alignment with the centre of the charge. Dewey [28] used the deformation of solder cantilevers to compare the energy yields of TNT and ANFO explosions. Some researchers have added spheres and cylinders to the tops of cantilevers in order to make them more sensitive to the drag forces, but such additions add to the complexity of the analysis of the cantilever's motion, and when rectangular or cylindrical appendages are used, careful alignment of the cantilever with the centre of the charge is required. Simple cylindrical cantilevers are the easiest to deploy and analyse. Because of their cylindrical symmetry they do not need to be accurately aligned with the centre of the charge, and in fact, the direction of bending provides an accurate measure of the direction of the charge centre, so that the centre of an explosion can be identified from the directions of bending of two cantilevers at different azimuthal positions. van Netten and Dewey [6] provides the most detailed study of cantilevers exposed to blast waves; how they may be used to determine some properties of blast waves, and measure the energy yield of an explosion as it relates to TNT and ANFO.

Because of its small cross-section, a cylindrical cantilever is essentially insensitive to the hydrostatic pressure and responds only to the dynamic or drag forces. Two types of material have been used to make cantilever gauges: brittle, and ductile. Brittle cantilevers fail almost immediately after the passage of the primary shock, and can therefore be used to assess the peak drag force immediately behind the shock. Ductile cantilevers respond more slowly and are subjected to the total duration of the drag force so that their final bend angle can be related to the drag impulse.

Brittle cantilever gauges have been made from glass rods for measurements relatively close to the explosion, and from pencil leads for use at greater distances. Both of these materials are readily available and are consistent in their properties, especially the pencil leads. Brittle cantilevers are deployed at a location in an array of different lengths. The longer cantilevers are subjected to greater loading, thus placing greater stress at the base of the cantilever where it is held in its mount. As a result, for an ideal measurement, the longer cantilevers will fail while the short ones will remain intact. The length of the shortest cantilever to fail is used to assess the peak drag force at that location.

Ductile cantilevers have been made from steel, aluminium and solder. All of these materials are easily available in cylindrical form with a range of diameters, and are consistent in their properties if the appropriate metal type is chosen. Short, e.g. 20 cm long, 2.5 cm diameter steel rods have been used at very high pressures close to the rim of the crater of a large surface burst explosion where they were buried by the ejecta from the crater, and were strong enough not to be affected by the falling

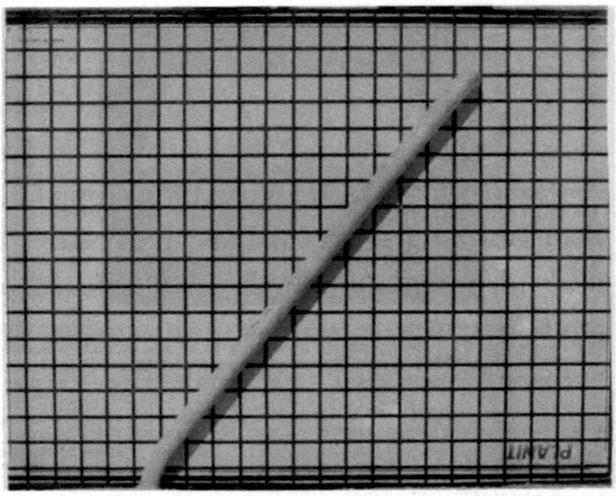

Fig. 10 A 2.5 cm diameter aluminium rod cantilever after exposure to a blast wave, showing that the rod bends only at the end where it was held in concrete

debris. Aluminium rods of various lengths and diameters have been used to study blast waves at intermediate distances, see Fig. 10, and solder has been used in regions of low loading pressure. The rod shown in Fig. 10 was photographed against a rectangular grid orientated horizontally with a spirit level so that the degree of bending could be measured from the photograph. The length, diameter and material of the cantilever were chosen so that the predicted blast wave properties at the location of the gauge would produce a bend angle of approximately 45°.

Examples of the use of such cantilever gauges are given by van Netten and Dewey [29, 30]. Methods of choosing the appropriate size and material for a cantilever gauge, based on the estimated properties of the blast wave to which it is to be exposed are given by van Netten and Dewey [6].

Cantilever gauges are sensitive to the dynamic or drag forces within a blast wave which change significantly in the boundary layer over the ground, from effectively zero at the ground surface to the free-field levels above the boundary layer. At high and intermediate loading levels, where the boundary layer is relatively thin, cantilever gauges can be mounted in the ground, but at intermediate levels it is preferable to use relatively long cantilevers so as to minimise the boundary layer effect. At lower loading levels cantilever gauges should be mounted on stands at least 30 cm above the ground, which seems to be a typical boundary level height in blast waves.

At high and intermediate pressure levels cantilevers can be installed in the ground embedded in concrete. At lower pressure levels where cantilevers need to be supported on stands to avoid the boundary layer, they should be mounted in sharp edged baffles with a smooth transition from the bevelled edge to the level surface where the cantilevers are mounted. The cylindrical cantilevers are inserted in holes drilled in the baffle and may need a small amount of glue to ensure that they are not loose.

Horizontal cantilevers are an excellent way of determining the height of the boundary layer in a blast wave. For this purpose they are mounted in a vertical baffle with sharp leading and trailing edges, which needs to be carefully aligned with the centre of the explosion. The cantilevers are mounted horizontally in the baffle in a vertical array with a spacing of a few centimeters. Both sides of the baffle can be used. After the passage of the blast wave the decrease in bending of the cantilevers close to the ground relative to those at higher levels clearly indicates the extent of the boundary layer. This is illustrated in van Netten and Dewey [30].

3.10.2 Estimating Blast Properties from Damage and Injury

In the case of accidental explosions normally there will be no direct measurements of blast wave properties, but gross measures of properties, such as the TNT equivalence of an explosion, can often be estimated from the distances at which certain damage or injury has occurred. Such assessment becomes more difficult if there have been multiple explosions. Most blast damage is caused by the drag or dynamic pressure loading rather than the hydrostatic pressure, but because of the difficulty of measuring dynamic pressure most blast damage and injury criteria are expressed in terms of the peak hydrostatic pressure and its impulse over the positive phase. Except for cases of rapidly responding structures such as eardrums or thin glass windows, blast damage depends on both the peak loading and on the loading duration. This is related to the amount of energy released in an explosion, and the rate at which the energy was released. A massive structure may sustain little damage from a small explosive device, even though it is placed very close to the structure where the peak loading pressure may be many atmospheres, but the loading duration is only a few milliseconds. That same structure may be totally destroyed by the blast from a nuclear weapon when the peak loading pressure is less than 1 atm, but the duration of the loading is several seconds.

Because any specific level of damage or injury occurs at a high pressure from a small explosion and at a lower pressure but longer duration from a larger explosion, the relationship between the peak pressure and the impulse for that damage level has the approximate form of a rectangular hyperbola. An idealised damage criterion curve is shown in Fig. 11. The curve approximates a rectangular hyperbola in the peak-hydrostatic-pressure—hydrostatic-pressure-impulse plane. If at a location in a blast wave the point represented by the hydrostatic peak pressure and impulse lies above the criterion curve, damage/injury is likely to occur. If it lies below the curve, damage/injury is unlikely to occur. The pressure asymptote indicates the minimum pressure, which if continuously applied, would produce the damage or injury. The impulse asymptote indicates the minimum duration below which damage or injury would not occur, no matter how large the pressure.

Damage criterion curves can only be used to give a rough estimate of the likelihood of damage because of the limited amount of information available for a specific type of damage or injury over a wide range of charge sizes. Many of the existing damage and injury criteria were developed for nuclear explosions, or, in the case of intermediate size explosions, from studies of damage produced during the Blitz in the United Kingdom during WWII. Much less information is available for smaller explosions, which are the main concern at the present time. There seems to be no published compendium of blast wave damage and injury criteria, but information can be found by searching the Internet. The application of damage and

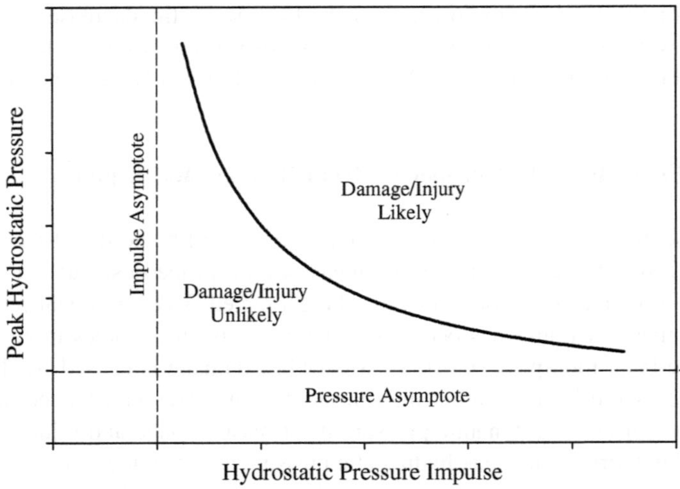

Fig. 11 An idealized damage/injury criterion curve

injury criteria curves is discussed in some detail in Dewey and van Netten [31], and specifically for cantilever bending in van Netten and Dewey [6]. In particular, the latter reference describes how a certain level of damage can be used to estimate the TNT mass equivalence of an explosion.

3.11 Measurement Methods Using Numerical Simulation

For several decades numerical simulation has become a major tool in the estimation of the physical properties of blast waves. The calculation of the physical properties within an established blast wave is relatively straightforward, but a calculation of the detonation process requires an accurate knowledge of the complex chemical and physical processes in such an event. Few hydro-codes have this ability. The code that seems to have been most successful in solving this problem is SHAMRC (see www.ara.com). Numerical simulation techniques need to be carefully validated by comparison with physical measurements of the blast wave properties. It must be stressed that comparison with measurements of the hydrostatic pressure only has proved to be unreliable, and validation comparisons must be made with measurements of such properties as density, particle velocity and particle displacement.

A powerful measurement technique is a combination of direct physical measurement with a numerical simulation. In this technique the starting processes of the simulation are iteratively adjusted until the calculated physical properties match a series of measured properties as accurately as possible. An example of such a process is the piston path method (PPM). Taylor [32] showed that a blast wave from a centred source could be generated by a spherical piston with an appropriate radius-time trajectory: the piston path. This trajectory can be determined experimentally by high-speed photography of smoke tracers initiated close to an explosive charge immediately before its detonation. Examples of the use of this technique are given by Dewey and McMillin [21, 22]. In these examples, the results from the piston path calculations were compared with many hundreds of measurements of hydrostatic pressure, dynamic pressure and density made using electronic transducers, and the excellent agreement validated the technique.

More recently, it has been shown that the piston-path technique can be used to determine all the physical properties of a centred blast wave in the radius-time plane using only the measured radius-time trajectory of the primary shock (Dewey and Dewey [19]. In this method an arbitrary piston-path is used in a hydro-code to generate a blast wave, the primary-shock trajectory of which can be compared with the measured trajectory. The piston path was defined by a series of nodes. The nodes were iteratively adjusted until a piston path was obtained that generated a blast wave whose primary-shock trajectory exactly matched the measured values.

A similar method has been used in which the numerical simulation is driven by a sphere of high-temperature high-pressure gas. The initial pressure and temperature can be set to approximate the expected energy yield of the source. The energy yield and the rate of energy release can be adjusted until a blast wave is obtained that matches as closely as possible a set of measured values.

4 Blast Wave Scaling Laws

The physical properties of a spherically expanding blast wave, expressed either as wave profiles at specific times or as time histories at specific distances, are not self-similar. This means that a profile or time history measured at one time or distance cannot be scaled to a profile or time history at another time or distance. However, the physical properties are subject to precise scaling laws relating to the amount of energy released at the source and the properties of the ambient gas into which the shockwave is expanding.

Hopkinson [33] and Cranz [34] originally defined the scaling laws for blast waves. They showed that the distance from the centre of an explosive charge of a physical property's peak value immediately behind the primary shock, and the positive duration of that property are proportional to the linear dimension of the spherical charge. The linear dimension of the charge is also proportional to the cube root of the charge mass. If a property value or positive duration occurs at a distance R_1 from a charge of mass W_1 and at a distance R_2 from a charge of the same material with a mass W_2 then

$$\frac{R_2}{R_1} = \left(\frac{W_2}{W_1}\right)^{1/3} \tag{34}$$

Normally, W_1 is set at a unit charge mass of 1 kg or 1 kt. The charge mass W can be replaced by the calculated energy content of the source, E, but this may be less reliable than using the charge mass because it is not known what percentage of the energy is released as thermal or other forms of radiation, or, if the explosion is close to the ground, how much of the energy is used in making a crater, or is dissipated as seismic waves.

Sachs [35] augmented the cube root scaling law for mass to account for differences in the ambient atmospheric pressure, P_0, and temperature, T_0, of the gas into which the shockwave is moving. Sachs stated that: (a) the magnitudes of hydrostatic pressure, density and absolute temperature scale in proportion to their values in the ambient gas; (b) distances and durations will scale in inverse proportion to the cube root of the ambient pressure, and (c) durations and times of arrival will be proportional to the sound speed in the ambient gas, and thus to the square root of the ambient absolute temperature. Equation (34) therefore becomes

$$\frac{R_2}{R_1} = \left(\frac{W_2}{W_1}\right)^{1/3} \left(\frac{P_{01}}{P_{02}}\right)^{1/3} \tag{35}$$

where P_{01} and P_{02} are the ambient pressures for the two explosions. Times of arrival of the primary shock and times of duration within the blast wave scale as

$$\frac{t_2}{t_1} = \left(\frac{W_2}{W_1}\right)^{1/3} \left(\frac{P_{01}}{P_{02}}\right)^{1/3} \left(\frac{T_{01}}{T_{02}}\right)^{1/2} \tag{36}$$

where T_{01} and T_{02} are the ambient absolute temperatures in degrees Kelvin, viz. T° K = T° C + 273.16. To scale measured values of distance and time to those expected for a unit charge in an ambient atmosphere at Normal Temperature and Pressure (NTP), W_1 is set at 1 kg, P_{01} = 101.325 kPa and T_{01} = 288.16 K. Measurements scaled in this way provide a comparison with the blast waves from other explosions of different masses of the same explosive in different ambient atmospheric conditions. The validity of these scaling procedures has been demonstrated by Dewey and Sperrazza [36] and Dewey [5].

5 Summary and Discussion

The electronic pressure transducer remains the principal device for monitoring the physical properties of blast waves, but its limitations, described in Sects. 2.1.1 and 3.2, must be remembered. In particular, the hydrostatic pressure measured by a flush mounted gauge does not detect the passage of a contact surface between two gases which may have very different densities and temperatures. Neither does it detect variations of flow velocity, nor therefore the dynamic pressure, that occur in the boundary layer as a blast wave passes over the ground or a structure. Pressure transducers flush mounted in the surface of a structure are required to measure the loading pressures on the structure exposed to a blast wave.

Passive gauges such as cantilevers can be valuable tools to monitor blast waves, even though they do not provide precise measures of the physical properties. Some of their principal advantages are that they are cheap; can be left in place for long periods without attention; provide an accurate reading of the direction towards the centre of the blast, and can be quickly and easily interpreted. Therefore, they are particularly useful for deployment in an area in which there is the possibility of an accidental or unscheduled explosion. They can also be used to rapidly compare the blast waves from a sequence of explosions carried out in quick succession.

High-speed photography remains a powerful tool to monitor blast waves both qualitatively and quantitatively. Its unique advantage is that it provides information about large regions of the blast wave and not at just a few specific locations. Refractive image analysis is undoubtedly the most accurate method for measurement of the position and velocity of the primary shock front. In this regard, it is more accurate for measurements of the peak values immediately behind the primary shock than electronic transducers, the signals from which do not provide an accurate measure because of overshoot, ringing or damping.

All of the physical properties of a centred blast wave can be determined from high-speed photography of the time-resolved displacement of flow tracers established shortly before the arrival of the primary shock, and is one of the few techniques that provides a complete mapping of all the physical properties in the radius-time plane. Background oriented schlieren (BOS) shows promise in being able to map the density field in an axisymmetric or centred blast wave but the technique has not been fully validated, and the resolution of the density field close to the primary shock has yet to be demonstrated. Also, this technique maps only the density field and does not provide information about other physical properties such as hydrostatic pressure and particle velocity.

Numerical simulation of blast waves and their interaction with structures, using suitable hydro-codes, is a powerful technique that maps all the physical properties but must be validated using direct measurements, ideally of other physical properties in addition to pressure. This is particularly important when using numerical simulation to study phenomena such as blast mitigation. Most numerical simulations assume that there is no energy loss in a blast-structure interaction. The energy of a blast wave has two components, one associated with the hydrostatic pressure and the other with the dynamic pressure. If, therefore, the hydrostatic pressure is reduced there must be a corresponding increase of the dynamic pressure, which is usually the more damaging of the two components. It is important that the dynamic pressure be calculated using the form of Eq. (1) so that the dynamic pressure is a vector rather than a scalar quantity. In this way it can be determined if a reduction in hydrostatic pressure has resulted in an increase of any component of the dynamic pressure. Only if the hydrostatic pressure and the components of the dynamic pressure have been reduced can mitigation of the blast be said to have occurred.

The most powerful technique of measuring all the physical properties of a centred or axisymmetric blast wave is undoubtedly one that uses a combination of direct measurement and numerical simulation, such as particle trajectory analysis (PTA) described in Sect. 3.9.2. The simplest way of doing this may be that described by Dewey and Dewey [19] in which the numerical simulation is iteratively adjusted until one set of calculated properties exactly matches the corresponding set of measurements. The easiest and cheapest set of such measurements is the times of arrival of the primary shock, which can be obtained from an array of simple switches, the signals from which can be transmitted through a single cable to a timed recording channel.

It is important that all measurement devices, particularly pressure transducers, be calibrated regularly, ideally before and after each use. Most pressure transducers will be accompanied by a manufacturer's calibration curve relating the applied over- or under-pressure to the output signal over the range of pressures to which the transducer is to be submitted. Ideally this curve will be linear, but that should not be assumed. Wherever possible, the calibration should be carried out using the transmission cabling, the amplifiers and the recording channel that will be used on the test. The calibrations over the range of expected over-pressures are best done in a shock-tube type device in which the shock velocity can be measured by a pair of time-of-arrival detectors as it passes the transducer. The ambient pressure and temperature must also be measured so that the shock Mach number can be calculated and used in the Rankin-Hugoniot relationship (5) to determine the pressure immediately behind the shock. Calibration in the under-pressure range can be done using a device similar to the driver section of a shock tube in which an applied pressure is rapidly released by the rupture of a diaphragm or fast acting valve.

In order that the measured physical properties of a blast wave can be compared with those from other explosions, the properties should be expressed in terms of the ambient pressure, temperature and density at the time of the explosion and scaled to those from a unit charge in an atmosphere at NTP, as described in Sect. 4. To do this, the ambient values of pressure, temperature and wind velocity should be measured as close to the explosive source and time of detonation as possible. The ambient temperature is usually measured at a height of about 2 m to avoid the large temperature gradients that may occur close to the ground, and at higher altitudes for large charges. Scaling of blast waves also requires knowledge of the charge mass, and this should be measured as accurately as possible. It may be unreliable to use the manufacturer's value for charge mass which is often related to the volume of the mould in which the charge was made, rather than a direct measurement of mass. The charge mass provided by the manufacturer should be treated with particular suspicion if it exactly matches the specified or nominal mass. The charge mass used in scaling must include the mass of any booster charge.

The locations of all measurement devices, cameras, fiducial markers used to scale photographic measurements, and the positions of structures being exposed to the blast must be surveyed accurately relative to the centre of the explosion, and to each other. Experience has shown that survey errors are not uncommon, and for tests on which there is a large amount of instrumentation it may be appropriate to have dual, independent surveys so that such errors can be identified.

References

1. Brouillette, M.: The Richtmyer-Meshkov instability. Annu. Rev. Fluid Mech. **34**, 445–468 (2002)
2. Needham, C.E.: Blast Waves. Springer, Heidelberg (2010)
3. Dewey, J.M.: Expanding spherical shocks (blast waves). In: Ben-Dor G, Igra O, Elperin T (eds.) Handbook of Shock Waves, Vol. 2, pp. 441–481. Academic Press, San Diego (2001)
4. Dewey, J.M., McMillin, D.J., van Netten, A.A., Ethridge, N., Keefer, J., Needham, C.E.: A study of the boundary layer in a large scale blast wave over a natural surface. In: Proceedings of 14th International Symposium on Military Aspects Blast and Shock (MABS14), Las Cruces, NM, USA, Defense Threat Reduction Agency, pp. 1–28 (1995)
5. Dewey, J.M.: The air velocity in blast waves from t.n.t. explosions. Proc. Roy. Soc. A **279**, 366–385 (1964)
6. van Netten, A.A., Dewey, J.M.: A study of the blast wave loading on cantilevers. Shock Waves **7**, 175–190 (1997)
7. Reisler, R.E., Keefer, J.H., Ethridge, N.H.: Air blast instrumentation, 1943-1993, measurement techniques and instrumentation: vol. 1, the nuclear era, 1945-1963; Vol. 2 the high explosive era, 1959-1963, and Vol. 3 air blast structural target and gage calibration. In: MABS Monograph, Defense Nuclear Agency (now Defence Threat Reduction Agency), Alexandria, VA, USA
8. Dewey, J.M.: Blast wave measurement techniques: photo-optical methods. In: Proceedings of 13th International Symposium on Military Applications of Blast Simulation (MABS 13), pp. 1–15, Division Defence Technology, Royal Institution of Engineers in the Netherlands (Klvl), The Hague, The Netherlands (1993)
9. Dewey, J.M.: Shock waves from explosions. In: Ray SF (eds.) High-Speed Photography and Photonics. Focal Press, Oxford (1997)
10. Dewey, J.M.: Explosive flows: shock tubes and blast waves. In: Yang W-J (eds.) Handbook of Flow Visualization, 2nd Edn. Hemisphere, New York (1997)
11. Dewey, J.M.: The Rankine-Hugoniot equations: their extensions and inversions related to blast waves. In: Proceedings of 19th International Symposium on Military Aspects Blast and Shock (MABS19), Can. Def. Res. Suffield, Alberta, Canada, P008 (2006)
12. Dewey, J.M.: The TNT equivalence of an optimum propane-oxygen mixture. J. Phys. D Appl. Phys. **38**, 4245–4251 (2005)
13. Dewey, J.M.: The properties of a blast wave obtained from an analysis of the particle trajectories, Proc. Roy. Soc. Lond. A. **324**, 275–299 (1971)
14. Kleine, H., Dewey, J.M., Ohashi, K., Mizukaki, T., Takayama, K.: Studies of the TNT equivalence of silver azide charges. Shock Waves **13**, 123–138 (2003)
15. Dewey, J.M., Anson, W.A.: A blast wave density gauge using beta-radiation. J. Sci. Instr. **40**, 568–572 (1963)
16. Ritzel, D.V.: Blast-wave density measurements. In: Bershader, D., Hanson, R. (eds.) Proceedings of 15th International Symposium on Shock Waves and Shock Tubes, Stanford Univ., Berkley, California, 578–585 (1985)
17. Slater, J.E., Boechler, D.E., Edgar, R.C.: DRES measurement of free-field airblast. In: Minor Uncle Symposium Report, Defense Nuclear Agency, POR 7453-4, Vol. 4, 2, pp. 1–98 (1995)
18. Liepmann, H.W., Roshko, A.: Elements of Gasdynamics. Wiley, New York (1957)
19. Dewey, J.M., Dewey, M.C.: The physical properties of the blast wave produced by a stoichiometric propane/oxygen explosion. Shock Waves, **24**, 593–601 (2014)
20. Harvey, R.B., Lutz, H.B.: Photo-optical studies of nuclear airblast, operations BUFFALO (1956) and ANTLER (1957): measurements of elapsed time, Vol. 4, Pt 2, TTCP Panel TLG3/0004, Atomic Weapons Establishment, Ministry of Defence, Aldermaston, UK (1988)
21. Dewey, J.M., McMillin, D.J.: Smoke puff photo-diagnostics. In: MISERS GOLD Symposium Report, Defence Nuclear Agency, Alexandria, VA, POR 7352-2, Vol. II (1991)

22. Dewey, J.M., McMillin, D.J.: Smoke trail photo-diagnostics. In: MINOR UNCLE Symposium Report, Event Diagnostics and Calculations, Vol. 2, pp. 2-1–2-48, Defence Nuclear Agency, Alexandria, VA, POR 7453-2 (1995)
23. Venkatakrishnan, L., Meier, G.E.A.: Density measurements using the background oriented Schlieren technique. Expt. Fluids **37**, 237–247 (2004)
24. Ramanah, D., Raghunath, S., Mee, D.J., Rösgen, T., Jacobs, P.A.: Background oriented schlieren for flow visualisation in hypersonic impulse facilities. Shock Waves **17**, 65–70 (2007)
25. Penny, Lord: The nuclear explosive yields at Hiroshima and Nagasaki. Roy. Soc. Phil. Trans. **266**, 357–424 (1969)
26. Ewing, W.O., Hannah, J.W.: A Cantilever for Measuring Air Blast, Ballistic Research Laboratories. Tech, Note 1139 (1957)
27. Baker, W.E., Ewing, W.O., Hannah, J.W.: Laws for large elastic response and permanent deformation of model structures subjected to blast loading. Ballistic Research Laboratories, Rept. 1060 (1958)
28. Dewey, J.M.: Surface burst of a 100 ton TNT hemispherical charge: Wire drag gauge measurements, Suffield Experimental Station, Technical Note 80 (1962)
29. Van Netten, A.A., Dewey, J.M.: Cantilever gauges. In: Proceedings of DISTANT IMAGE Symposium, Defense Nuclear Agency, POR7379-5, p. 5 (1992)
30. Van Netten, A.A., Dewey, J.M.: Horizontal cantilever gauges and cantilever gauges. In: Proceedings of MINOR UNCLE Symposium, Defense Nuclear Agency, POR7453-5, p. 5 (1995)
31. Dewey, J.M., van Netten, A.A.: Calculating blast effects distances in urban environments. In: Proceedings of 17th International Symposium on Military Aspects Blast and Shock (MABS17), Blast and Shock Effects, US Department of Defense, p. 8 (2002)
32. Taylor, Sir G.I.: The air waves surrounding an expanding sphere. Proc. Roy. Soc. Lond. A **186**, 273–292 (1946)
33. Hopkinson, B.: British Ordnance Board Minutes, 13565 (1915)
34. Cranz, C.: Lehrbuch der Ballistik. Springer, Berlin (1926)
35. Sachs, R.G.: The dependence of blast on ambient pressure and temperature. BRL Report 466, Aberdeen Proving Ground, Maryland, USA (1944)
36. Dewey, J.M., Sperrazza, J.: The effect of atmospheric pressure and temperature on air shock, BRL Report 271, Aberdeen Proving Ground, Maryland, USA (1950)

Author Biography

John Dewey received his B.Sc. and Ph.D. in Physics from the University of London. He has studied the physical properties of blast waves since 1956, first at the Defence Research Establishment Suffield, where he was Head of the Aerophysics Section, and then at the Department of Physics and Astronomy at the University of Victoria, Canada, where he established and ran the Shock Studies Laboratory for over 30 years. He specialized in the use of high-speed photographic techniques for the study of shock and blast waves in shock tubes and on many hundreds of field experiments involving explosive charges ranging from 1 mg to 5 kt. He has published more than 135 journal papers, articles and chapters.

Shock Wave Experiments in a High-Temperature Gas Flow

Michio Nishida

1 Introduction

To simulate the aerodynamic heating environment for a hypervelocity vehicle, various ground test facilities have been developed and constructed. An arc-heated wind tunnel is one of these facilities and features long operational time; thus the arc-heated wind tunnel has been used mainly for the development and evaluation of a thermal protection system for space transportation systems. Several types of arc heaters (such as Huels, Segment, Constrictor and MPD) have been adopted as a plasma generator, and their electric power scales are in the range of 10 kW–100 MW. Thus generated plasma flows can be employed for the experiments of aerodynamic heating and heat protection systems but also to reveal the physical phenomenon across a shock wave in a high-temperature plasma flow.

In this chapter, a 20 kW-class arc-heated wind tunnel operated at Department of Aeronautics and Astronautics, Kyushu University is depicted and shock wave experiments in an arc-heated high-temperature flow are discussed.

2 Arc-Heated Wind Tunnel and Arc Heater

A schematic of an arc-heated wind tunnel is illustrated in Fig. 1, which consists of a constrictor-type arc heater, a cylindrical test chamber whose diameter and length are 100 mm, respectively, vacuum pumps, and a water-cooling system. The test chamber is connected to the vacuum pumping system, which is composed of an oil-sealed rotary vacuum pump and a mechanical booster pump.

M. Nishida (✉)
Kyushu University, Fukuoka 819-0395, Japan
e-mail: michio.nishida@sct.bbiq.jp

O. Igra and F. Seiler (eds.), *Experimental Methods of Shock Wave Research*,
Shock Wave Science and Technology Reference Library 9,
DOI 10.1007/978-3-319-23745-9_3

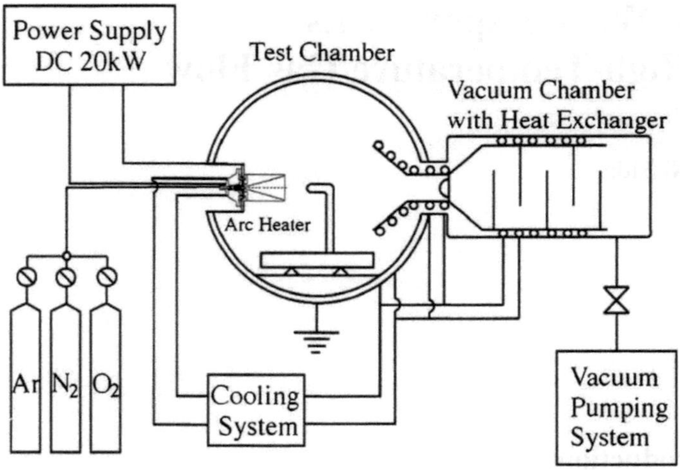

Fig. 1 Schematic of an arc-heated wind tunnel

Figure 2 shows the arc heater, which is composed of a 2 % thoriated tungsten cathode and a copper anode forming a conical nozzle with a semi-angle of 15°. The nozzle throat and exit are 5 and 100 mm in diameter, respectively, which lead to a nominal Mach number of 9.4 for diatomic gas. The schematic of the arc heater is shown in Fig. 3. Argon or nitrogen can be used as a test gas, which is supplied into the discharge section with swirling and then heated by a DC arc discharge. An arc-heated gas expands into the test chamber through the nozzle. Besides the argon and nitrogen arcjets, an air arcjet also can be generated. In the experiment of the air arcjet, nitrogen gas is injected first into the discharge section and heated by the discharge. Then, oxygen gas is seeded into the heated nitrogen plasma downstream of the cathode to prevent oxidation of the cathode. Thus air plasma can be generated. Arc-heated gas is not only partially dissociated and partially ionized, but also in thermochemical nonequilibrium in which translational, rotational, vibrational and electron temperatures are different from each other.

Fig. 2 Arc heater

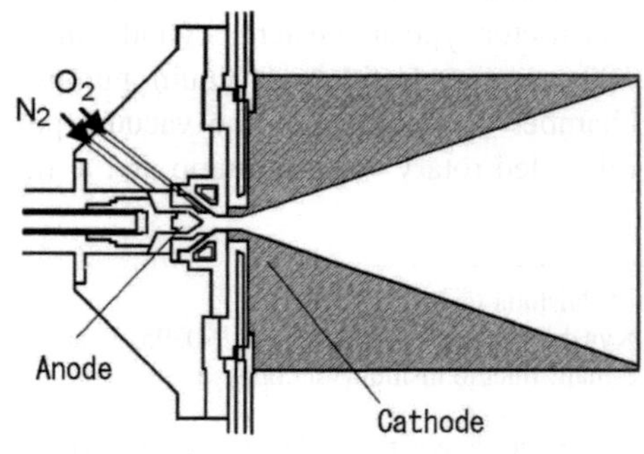

Fig. 3 Schematic of an arc heater

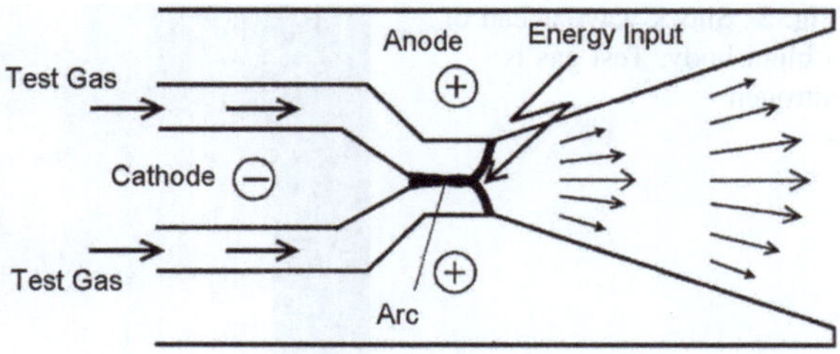

The arc-heated wind tunnel can serve the following shock wave experiments in a high-temperature gas flow:

1. dark space,
2. emission spectra from a shock layer,
3. spallation.

In this chapter, the results of the above experiments carried out in an arc heated facility are shown and discussed.

3 Dark Space

When a blunt body is placed in an ionized gas flow, a region of reduced self-luminosity can be observed ahead of the detached shock wave (see Fig. 4). This region is called "dark space". The dark space phenomenon is typical only in an argon plasma flow and not observed in such a diatomic gas plasma flow as nitrogen (see Fig. 5). The mechanism of the dark space was theoretically and experimentally studied in detail in [1, 2].

According to Nishida and Nakajima ([1, 2]), the elevation of electron temperature ahead of a shock wave causes the dark space. Namely, there exists a broad zone of elevated electron temperature ahead of the ionized shock wave, caused by

Fig. 4 Shock wave ahead of a blunt body. Test gas is argon

Fig. 5 Shock wave ahead of a blunt body. Test gas is nitrogen

the high-thermal conductivity of the electron gas. The electron density being unchanged in this region, the elevated electron temperature reduces the equilibrium population densities of highly lying electronic levels. This reduction propagates down to lower levels through electron-collisional de-excitation. Therefore, radiative de-excitation between lower levels becomes weak, which leads to dark space. Thus, the dark space can be observed only in the plasma where highly lying level populations are in equilibrium with free electrons and where the transitions between highly lying levels are controlled by electron collisions.

The self-luminosity is caused by radiative transitions between levels of bound electrons. For argon, the strongest visible spectral lines result from the transitions 5p → 4s and 4p → 4s, so that the reduction in the number of atoms excited in the electronic levels 5p and 4p is expected to make the dark space.

The experiments were carried out using an arcjet facility with argon as test gas. Argon, after being heated and partially ionized by an arc heater, expands through a convergent-divergent nozzle into a test section. A blunt body placed in the arcjet flow is electrically insulated from the tunnel wall, so that it is at a floating potential. The surface temperature of the body is controlled by cooling the body with nitrogen gas, which is kept within 402 and 412 K.

Emission spectra measurements were carried out on the lines at 430, 415.9 and 763.5 nm at 20 stations between the nozzle exit and blunt body. The population densities of the states 5p[5/2], 5p[3/2] and 4p[3/2] are deduced from the data of 430, 415.9 and 763.5 nm, respectively. The optical system is calibrated for absolute intensity measurements by using a standard tungsten ribbon filament lamp. After the Abel inversion for radial data is made, the spatially resolved population densities are determined.

In Fig. 6, the population density distributions of the sates 5p[3/2] and 5p[5/2] along the centerline with and without the presence of the blunt body are shown. In a case of the presence of the blunt body the population densities evidently fall below those in its absence. This is the dark space. The population densities of both states rapidly increase behind the dark space. This is due to the compression by shock wave.

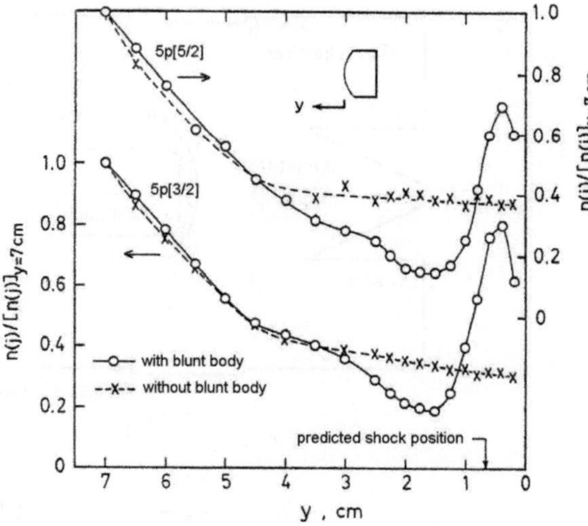

Fig. 6 On-axis distribution of population density of 5p

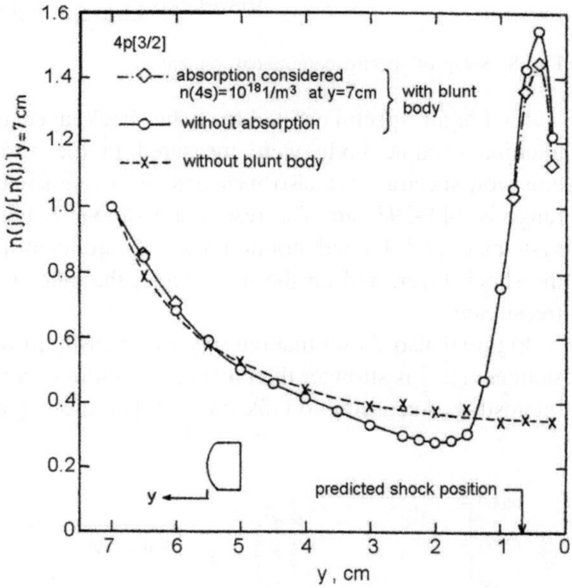

Fig. 7 On-axis distribution of population density of 4p

The population distribution of the state of 4p[3/2] is shown in Fig. 7, in which the dark space also can be recognized.

4 Spectra from a Shock Layer

Emission spectra from the shock layer of a flat-face body were measured. As shown in Fig. 8, a blunt body is placed in the arc-heated nonequilibrium flow and emission spectra were caught by a spectrometer through a fiber glass tube.

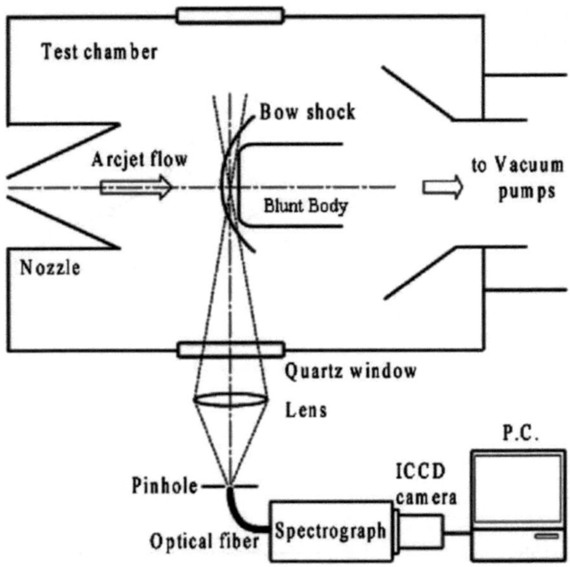

Fig. 8 Setup of spectroscopic measurement

First of all, spectra emitted from the shock layers of a stainless steel body and an alumina ceramic body were measured in the nitrogen arcjet. For comparison, emission spectra were also measured in a free stream. The measured wavelength range is 343–393 nm. The results are shown in Fig. 9, in which molecular band systems of $N_2^+(1-)$ and atomic lines of N are identified. The former is dominant in the shock layer, and on the other hand, the latter is clearly identified only in the freestream.

Figure 9 also shows that the spectral intensity measured in the shock layer of the stainless steel is stronger than that of the alumina ceramic. This fact is significantly interesting. According to Gökçen et al. [3], these quantitative characteristics of the

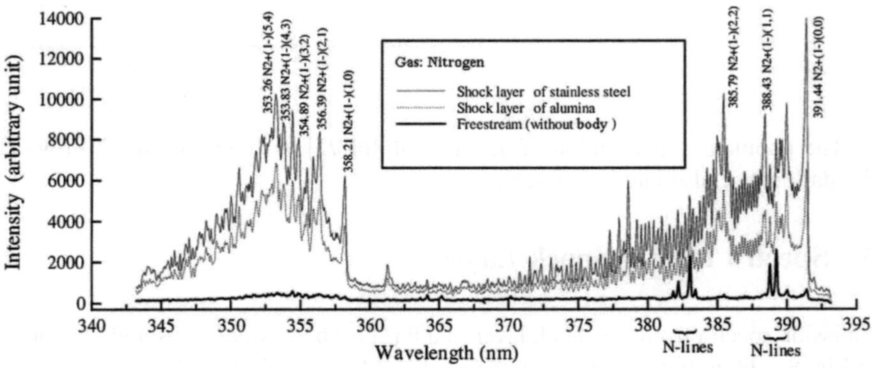

Fig. 9 Comparison of measured spectra from the shock layers of stainless steel and alumina ceramic, and from the free stream (without body). Gas is nitrogen

spectral intensity is derived from an increase in the number density of N_2^+ and/or to a lesser degree from an increase in temperature. Moreover, Doihara [4] calculated species mole fractions on the stagnation streamline of a shock layer for FCW (fully catalytic wall) and NCW (noncatalytic wall) considering the atmospheric entry conditions of a super orbital capsule at an altitude of 64 km. This work showed that the number density of N_2^+ for FCW is much larger than that for NCW very near the surface of the body, that is derived from a series of chemical reaction calculations considering catalytic recombination reactions on the wall. In consideration of this, it is concluded that an increase in the spectral intensity of N_2^+ for the stainless steel is caused by an increase in the number density of N_2^+ because of the catalytic effect of the stainless steel. That is, stainless steel would promote the atomic recombination reaction.

Similar measurements were done in the shock layer of an air arcjet. As shown in Fig. 10, the results are very similar to those of the nitrogen shock layer but atomic lines of N are not identified. It may be mentioned that the reaction $N + O \rightarrow NO$ decreases atomic nitrogen in the air arcjet.

Finally, measurement was conducted in the argon shock layer to confirm whether the stainless steel surface actually affects atomic recombination reaction. Argon is a monatomic gas; therefore atomic recombination does not occur on the body surface. It is thus expected that the spectral intensities would be identical for both the stainless steel and alumina ceramic. The results are indicated in Fig. 11. Although the intensity for the stainless steel is appreciablly weaker than that for the alumina ceramic, no remarkable differences are observed. If an obvious difference is observed even in the result for the argon arcjet, the phenomenon shown in the results for the nitrogen and air arcjets would be caused by an effect other than the catalytic effect of the stainless steel surface. No remarkable difference is observed, however, and this confirms that the stainless steel acts catalysis on high-temperature nitrogen.

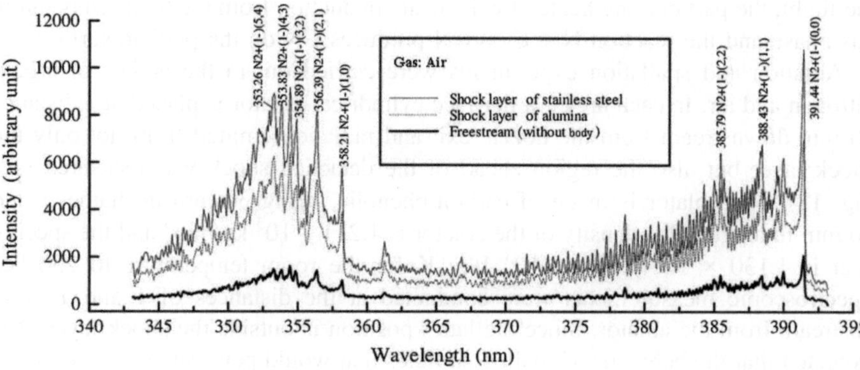

Fig. 10 Comparison of measured spectra from the shock layers of stainless steel and alumina ceramic, and from the free stream (without body). Gas is air

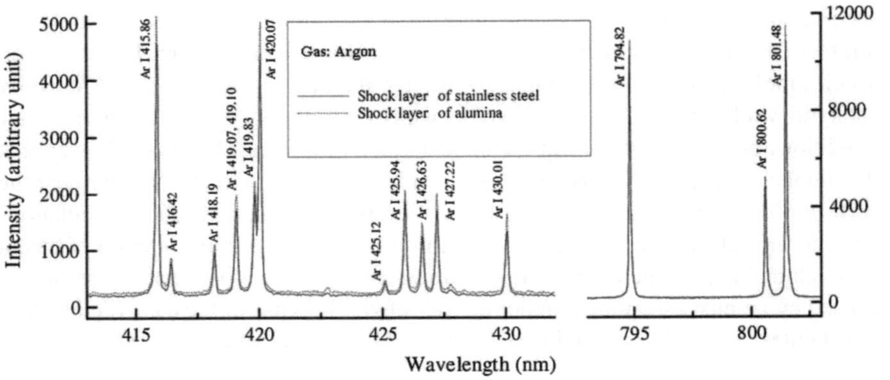

Fig. 11 Comparison of measured spectra from the shock layers of stainless steel and alumina ceramic. Gas is argon

5 Spallation

During ablation experiments in an arcjet facility, the presence of spalled particles was observed upstream of a detached shock. Yoshinaka [5] made spectroscopic measurements outside the stagnation boundary layer of the carbon-phenolic ablator placed in an arc-heated subsonic air flow and identified CN violet band spectra and CI line spectrum. He concluded that these spectra are attributed to solid particles ejected from the ablator, because gaseous carbon cannot reach outside of the boundary layer of the ablator. Similar experiments were conducted by Yoshinaka et al. [6] and observed CN violet band spectra and C_2 swan band spectra upstream of the shock layer of the ablator.

The spallation phenomenon can be likely explained as follows: When a carbonaceous ablator is exposed to a high temperature gas flow, a char layer is produced and at the same time pyrosis gas is generated. Carbonaceous solid particles are ejected by the pyrosis gas pressure and flies away upstream of a shock layer. During the flight, the particles are heated by the heat conduction from the high-temperature gas phase and the reaction $N + C \rightarrow CN$ produces CN on the particle surface.

Ablation and spallation experiments were carried out in the arcjets of argon, nitrogen and air. In each arcjet, a flat-face cylindrical ablator is placed at a distance 80 mm downstream from the nozzle exit and radiation emitted from not only the shock layer but also the region ahead of the detached shock was measured (see Fig. 12). The ablator is made of carbon-phenolic, being 30 mm in diameter and 50 mm in length. The density of the ablator is 1.211×10^3 kg/mm^3 and the specific heat is 1.130×10^3–1.677×10^3 J/kg K for the room temperature to 250 °C. Spectroscopic measurements were conducted at the distances of 1 and 19 mm upstream from the ablator. Since the latter position is outside the shock layer, it is expected that the behavior of spalled particles that would penetrate the shock layer be revealed. For reference, spectroscopic measurements were also conducted by replacing the ablator by a copper body of the same size as the ablator.

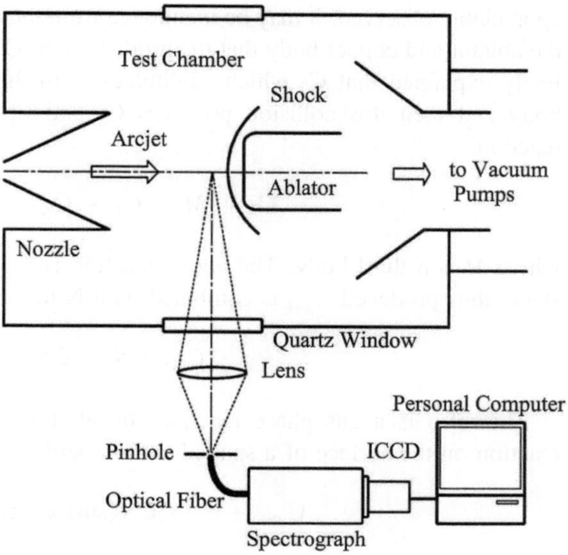

Fig. 12 Setup of ablation and spallation measurements

Figure 13 shows the spectra over the wave lengths from 360 to 400 nm that were observed at a distance 1 mm from the ablator, along with the result from the copper body. The measurement position is inside the shock layer. Similar measurements were conducted in a free stream by withdrawing the ablator. The result from the free stream is also illustrated in the figure. Very intense spectra of CN violet (0,0), (1,1), (2,2) and (3,3) band can be identified in the shock layer of the ablator. However, these spectra disappear when the ablator was replaced by a copper body. Instead the spectra of N_2^+(1-)(0,0), (1,1), (2,2) band can be identified in the result of the copper, which are typical in high-temperature air. The spectra from the free stream are similar to those from the copper body. Namely, the band spectra of N_2^+ are

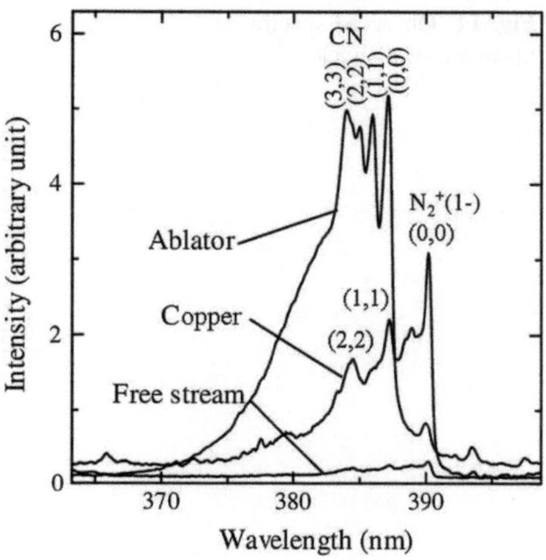

Fig. 13 Observed spectra in shock layers of ablator and copper body, and in free stream

appreciably observed. It may be mentioned from the comparison of the results from the ablator and copper body that the production of CN results from the ablator. It is likely explained that C_3 which sublimates from the ablator collides with a third body and then this collision produces C_2 and C_{gas} as shown in the following reaction:

$$C_3 + M \rightarrow C_2 + C_{gas} + M$$

where M is a third body. The above reaction is a gas phase reaction. In the next stage, thus produced C_{gas} is combined with N to generate CN.

$$C_{gas} + N \rightarrow CN$$

This also is a gas phase reaction. In addition, CN also is generated by the reaction on the surface of a spalled carbon particle as follows:

$$C_{solid} + N \rightarrow CN\,(\text{surface reaction})$$

Thus, CN is produced by both gas phase reaction and surface reaction.

Intense spectra of CN violet are also identified at 360 nm in the air arcjet, though that is not shown here. Although the spectra of CN violet were also observed in the nitrogen arcjet, they were not identified in the argon arcjet. This evidence emphasizes that CN is attributed to the ablation. In addition, the line spectrum of C was identified at 247.85 nm both in the air and nitrogen arcjets.

Emission spectra over the wavelengths from 350 to 370 nm observed at the position of 19 mm upstream from the ablator are illustrated in Fig. 14. By numerical calculation and picture, it was found that the shock stand-off distance was 13 mm, so that the measurement position was upstream of the shock. In addition to the band spectra of N_2^+(1-) which is typical in the air arcjet, intense CN spectra were

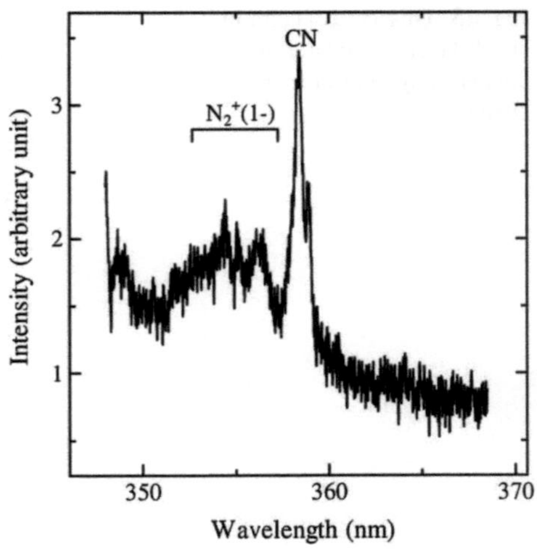

Fig. 14 Observed spectra upstream of a shock

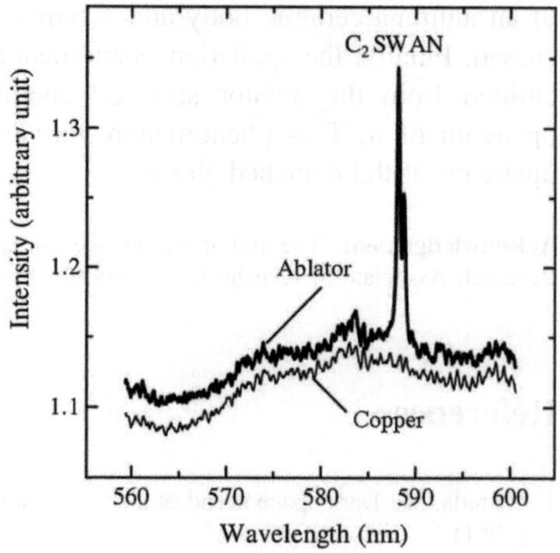

Fig. 15 Observed spectra upstream of a shock

identified. Since CN is not contained in the original test gas, it may be mentioned that this species is caused by carbon flying from the ablator. Thereby, we reach the conclusion that carbon emitted from the ablator surface penetrates the shock layer and reaches the upstream of the shock layer. The existence of N in the free streams of arc-heated nitrogen and air was identified by the observation of line spectra of NI at the wavelength of 742.36, 744.23 and 746.83 nm. Hence, possible explanation is that the reaction $C + N \rightarrow CN$ occurs on the surface of spalled carbon particles.

Figure 15 shows emission spectra over the wavelengths from 560 to 600 nm observed at a distance of 19 mm upstream from the ablator. In this region, C_2 swan (6,8) band can be identified with ablator. However, no such spectrum can be identified with copper body. We have no satisfactory explanation of the C_2 production. However, it is supposed that the following reactions produce C_2.

$$C + C + M \rightarrow C_2 + M$$
$$C_3 + C \rightarrow 2C_2$$
$$C_3 + M \rightarrow C_2 + C + M$$

where M is a third body. In the above reactions, the source of C and C_3 are to be solved. It is most probable that C is supplied from spalled carbon particles and that C_3 sublimates from the spalled particles.

6 Concluding Remarks

An arcjet facility can generate a thermochemical nonequilibrium high-temperature flow with a long operational time, so that it has been used to reveal the interesting phenomenon across a shock wave in the high-temperature plasma flow such as dark space. Also, by analyzing emission spectra from the high-temperature shock layer of an alumina ceramic body and a stainless steel body, surface catalysis was discussed. Finally, the spallation phenomenon was shown, in which carbon particles emitted from the ablator surface penetrate the detached shock and reach the upstream of it. This phenomenon was discussed based on the spectra from the upstream of the detached shock.

Acknowledgement The author would like to express his deep gratitude to Mr. Hisashi Kihara, Research Associate of Kyushu University, for his help in preparation of figures.

References

1. Nishida, M.: Dark space ahead of a shock in ionized gas flows. Z. Naturforsch. A **36**, 980–991 (1981)
2. Nishida, M., Nakajima, A.: Spectroscopic studies of the dark space ahead of a shock wave in an argon plasma flow. Z. Naturforsch. A **38**, 802–807 (1983)
3. Gökçen, T., Park, C.S., Newfield, M.E., Fletcher, D.G.: Computational simulation of emission spectra from shock-layer flows in an arcjet facility. J. Thermophys. Heat Transfer **12**, 180–189 (1998)
4. Doihara, R.: Thermochemical nonequilibrium flow around a super orbital reentry capsule. Doctor of Engineering Thesis, Kyushu University (2001)
5. Yoshinaka, T.: Spallation measurements at the ablator plasma wind tunnel tests. NASDA-TMR-970006E, National Space Development Agency of Japan (1998)
6. Yoshinaka, T., Mizuno, M., Morino, Y.: Detection of ablator spallation phenomena at arc wind tunnel test. In 26th International Electric Propulsion Conference, Kitakyushu, Japan, IEPC 99-025, 17–21 Oct 1999

Author Biography

Michio Nishida has been engaged in the experimental and computational research of high-velocity flows with strong shock wave. With a three-temperature model in which translational-rotational, vibrational and electron-translational temperatures are different from each other, he revealed the high-temperature shock layer flow on the catalytic wall of an atmospheric entry vehicle. He also computed a high-temperature nozzle flow of air operated by an arc discharge, using not only the three temperature model but also dissociation and recombination of high temperature air. His research field is so wide as arc jet thruster, shock focusing, spallation phenomena and gas-solid two phase flow. He received Doctor of Engineering from Kyoto University in Japan. After he had been working as associate professor at Kyoto University, he was invited to the chair of fluid mechanics at the Department of Aerospace Engineering of Kyushu University in Japan. After the retirement from Kyushu University, he was awarded Emeritus Professor of Kyushu University.

Laser Focusing

Akihiro Sasoh

1 Background

When some amount of energy is released into matter during a short period, its temperature and pressure increase; even if the matter is initially in its condensed phase, the phase can change to gas. Due to the mechanical imbalance between the high-temperature gas and the surrounding media, the former expands, driving the latter outward. This situation is equivalent to what happens in the so-called 'piston problem,' in which a piston pushes the gas in front of it, sending a shock wave ahead. In the present case, the contact surface between the high-pressure gas and the surrounding media acts as a piston. When the matter has its own intrinsic chemical energy, and is ignited during a short period, such an event is termed an 'explosion.' Even if the matter does not have such chemical energy, energy can be deposited from an external source. The typical example of such energy source is a pulse laser. The shape and behavior of the shock wave depends on the initial shape of the matter and the processes of the energy release or deposition.

Since the amount of energy is finite, the pressure imbalance is alleviated during the expansion of the matter. The contact surface is decelerated, and finally its velocity vanishes when the mechanical imbalance terminates. Due to this motion of the contact surface, the shock wave is accompanied by expansion waves behind. When a static pressure is measured at a certain location, it jumps at the moment of the shock wave passage and decreases due to the subsequent expansion waves. Such a shock wave which is accompanied by expansion waves behind it, which appears after explosion, is termed a 'blast wave.'

Blast waves appear on various occasions, for example, by vapor explosion of a volcano, or in a hazardous accident in atomic plant, or following detonation of explosive, and even by supernova explosion. Of course such events cause serious

A. Sasoh (✉)
Department of Aerospace Engineering, Nagoya University, Nagoya 464-8601, Japan
e-mail: sasoh@nuae.nagoya-u.ac.jp

O. Igra and F. Seiler (eds.), *Experimental Methods of Shock Wave Research*,
Shock Wave Science and Technology Reference Library 9,
DOI 10.1007/978-3-319-23745-9_4

disaster. On the other hand, if such event could be controlled, it can act favorable functions as will be seen later.

A representative method of generating such a 'controlled' blast wave is to use a pulse laser. When a laser pulse is focused in gas or in liquid medium, a small number of free electrons experience resonant oscillation to the applied electromagnetic waves, thereby absorbing the energy of the laser beam. If the electrons obtain sufficient energy to cause inelastic collisions against heavy particles, the electrons excite the internal energy state of the heavy particles, then dissociate and ionize them. After ionization, another electron is released from the heavy particle. Once such events occur, the rate of the ionization is abruptly increased due to the increase in the electron number density. These processes are termed as 'electron avalanche.' If the electron avalanche occurs during a period that is much shorter than the laser pulse duration period, the electrons efficiently absorb the proceeding laser energy, generating a laser-pulse-induced blast wave. Moreover, the laser focusing generates not only a blast wave but also a laser-heated, high-temperature, low-density gaseous region. Hereafter, the latter will be referred to as a 'laser-pulse-induced bubble.' As will be shown later, it involves an important and interesting flow phenomena in compressive fluid dynamics.

Laser-pulse-induced blast waves and bubbles often provide us with better access to the investigation of blast wave and associated flow phenomena because of their better accessibility; even a table-top experiment is possible. The good controllability of a small explosion gives us useful and important advantages.

2 Laser-Pulse-Induced Blast Wave

The laser-pulse-induced flow is utilized for various applications. Also, basic research of the behavior of a laser-induced blast wave has been conducted, for example by Jiang et al. [1]. Sasoh et al. [2] investigate the interactions between a laser-induced blast wave and a low-density bubble which is generated by the same laser pulse as well. Figure 1 illustrates the arrangement of the experiment to observe

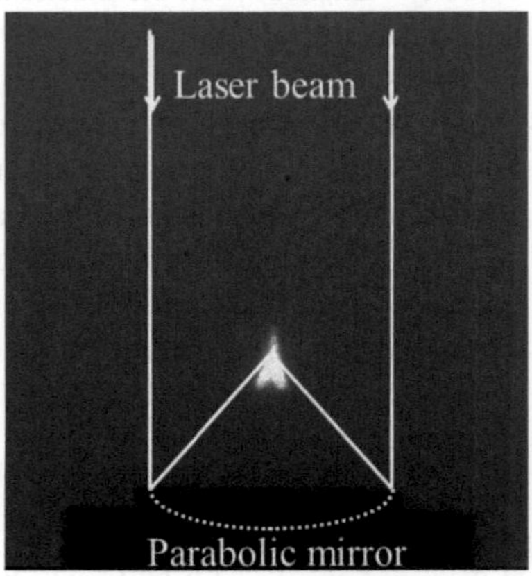

Fig. 1 Optical arrangement for the experiment of the interaction between a laser-induced shock wave and a bubble induced by the same laser pulse

the behavior of a laser-pulse induced blast wave and bubble. In the vacuum chamber, a parabolic aluminum mirror is set so that its axis is aligned to the axis of a TEA (transversely excited atmospheric) CO_2 laser beam of wavelength 10.6 μm. The diameter of the parabolic mirror is 42 mm, the focal length 21 mm. The laser beam is sent through a ZnSe window. The output laser beam had a diameter of 50 mm. The full width at half maximum of the primary laser power peak was 0.2 μs; 90 % of the total energy is irradiated in 3 μs. The time, t, starts at the initiation of the laser pulse.

Figure 2 shows sequential schlieren images of the event. At $t = 16$ μs, an almost conical region with intense radiation emission is observed. In this region, the laser

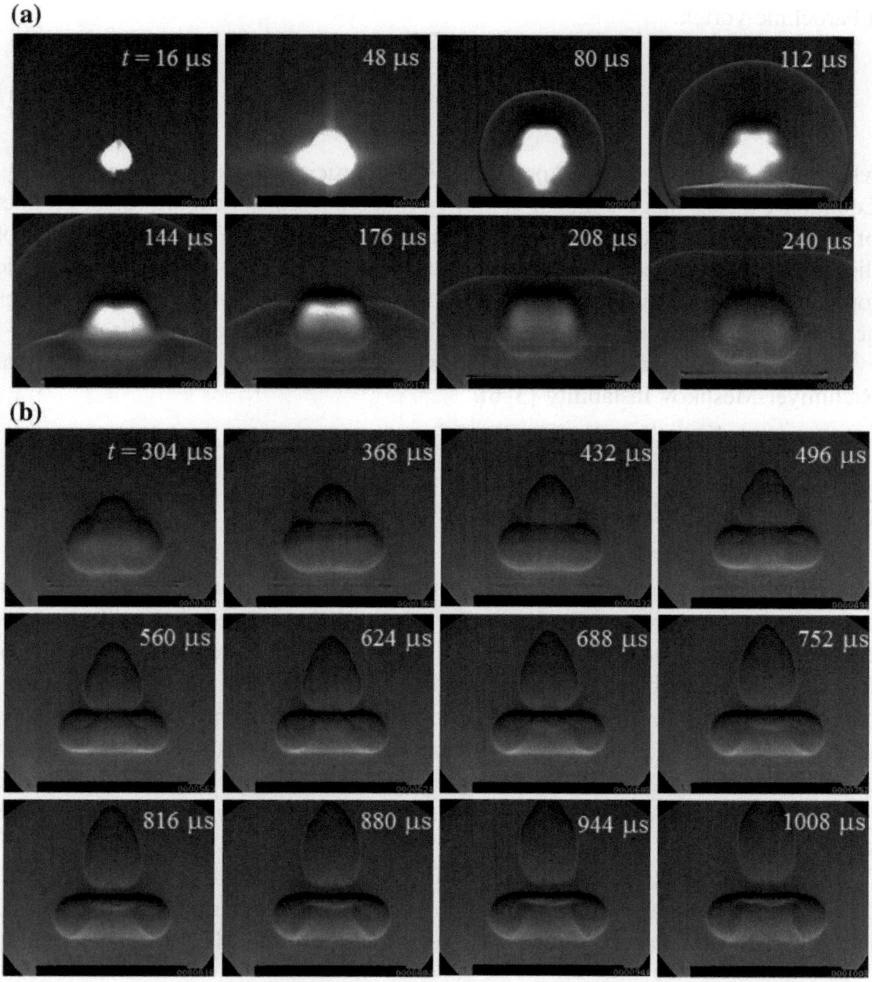

Fig. 2 Framing of schlieren images showing laser-pulse induced shock waves and flow. The test gas is Kr; laser pulse energy, 0.78 J; ambient pressure, 10 kPa; laser beam diameter, 18 mm. **a** 16 μs ≤ t ≤ 240 μs, framing rate, 32 μs. **b** 304 μs ≤ t ≤ 1008 μs, framing rate, 32 μs

pulse energy is directly absorbed. The difference between the shock wave and the contact surface cannot be resolved. Then, the laser-pulse-heated bubble expands, with its front approaching a spherical shape (16 μs ≤ t ≤ 112 μs). At t = 80 μs, the spherical contact surface at the outermost boundary of the bubble is clearly seen, and separated from the spherical shock wave, which barely touches the parabolic mirror. The image of the central part of the bubble still is over-exposed by its radiation emission. At t = 112 μs, the shock wave reflected from the parabolic mirror is seen. Because the almost spherical shock wave is reflected from the parabolic surface, the shape of the reflected wave is almost plane. During 112 μs ≤ t ≤ 176 μs the reflected shock wave interacts with the bubble. Since the bubble forms a density discontinuity and through the shock wave a pressure discontinuity exists, this interaction results in a baroclinic vortex.

$$\left.\frac{d\omega}{dt}\right|_{\text{baroclinic}} = \frac{1}{\rho^2}\nabla\rho \times \nabla P \tag{1}$$

where P, t, ω and ρ denote pressure, time, vorticity and density, respectively. Equation 1 means that the rate of vorticity production scales with the vector product of the density gradient and the pressure gradient. Due to this effect, the lower half of the bubble is deformed so that its curvature changes its sign, changing the bubble topology. At t = 304 μs, the interface of the bottom surface of the bubble extrudes across the upper surface. This 'inverted' bubble grows with elapse of time.

The observed phenomena are closely related to the initial processes of Richtmyer-Meshkov instability [3–6].

3 Application to Space Propulsion

The interaction between the laser-pulse induced blast wave and the solid wall can be applied to aerospace propulsion. Kantrowitz [7] proposed that utilizing laser energy to launch a space vehicle can yield a significant increase in the payload ratio. By breathing the ambient air, propellant does not need to be loaded onboard. Proof of the principle of the air-breathing laser propulsion experiments were done by Pirri et al. [8]. Myrabo [9] launched a 'Lightcraft' by using a repetitive-pulse CO_2 laser to an altitude of 211 m. Sasoh [10] proposed a 'laser-driven in-tube accelerator (LITA),' in which a projectile is accelerated in a tube filled with inert working gas see Fig. 3.[1]

Figure 3 shows the schematic illustration of LITA50 launch tube and projectile. The LITA50 has a launch tube with an inner diameter of 50 mm, length of 4 m, and is made of acrylic plastic. Figure 4 shows snapshots of the LITA launch operation

[1]Later, they developed a different type of LITA in which the tube was in vacuum, a solid-state ablator is set on the wall of the launch tube (wall-propelled LITA, [11]).

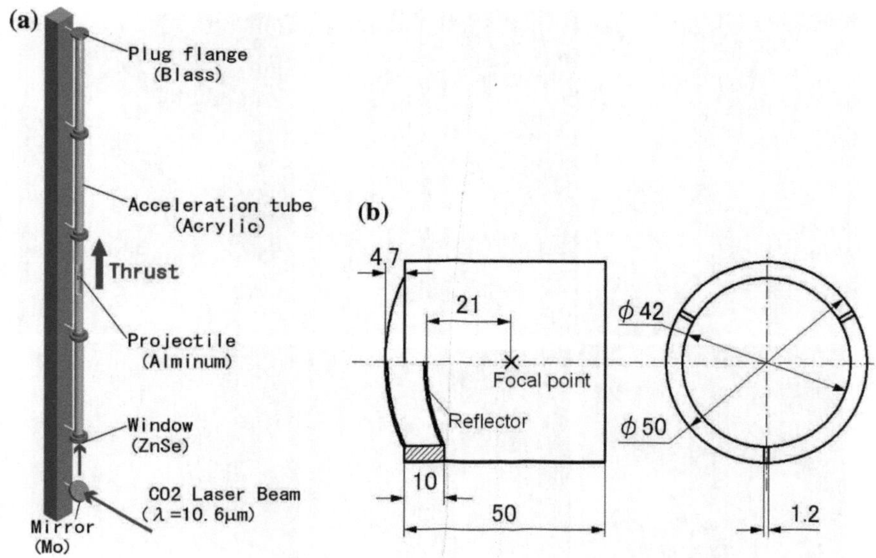

Fig. 3 Schematic illustration of LITA50. **a** Launch tube. **b** Projectile

which was recorded using a commercially-available video camera. The projectile was successfully launched to the top of the launch tube. Its position is indicated by an arrow.

4 Laser-Pulse-Induced Low-Density Bubble

As mentioned, laser focusing causes not only the generation of a blast wave but also the generation of a low-density bubble. This bubble is found to be useful to modulate the flow field over a supersonic body [2]. This method is referred to as 'energy deposition.' Energy deposition over a supersonic body using laser pulses has been studied with the hope of vastly improving the aerodynamic performance [12–15]. Although a single laser pulse causes a transient moderation of the flow field, the time-averaged aerodynamic performance is modified only by repetitively depositing laser pulse energies. Tret'yakov et al. [16] and Kim et al. [17] demonstrated that with a high laser pulse repetition frequency, the shock layer over a model experiences significant pressure and flow pattern modulations. Along the central axis a virtual spike comprised of vortex rings is built up. The vortex rings are generated after the interaction between the laser-heated bubble and the bow shock wave [2]. Tret'yakov et al. [16] obtained 45 % drag reduction over a cone. However, the corresponding efficiency of energy deposition was estimated to be lower than unity. Kim et al. [17] obtained 21 % drag decrement over a flat head cylinder with the efficiency of energy deposition higher than 5 (500 %).

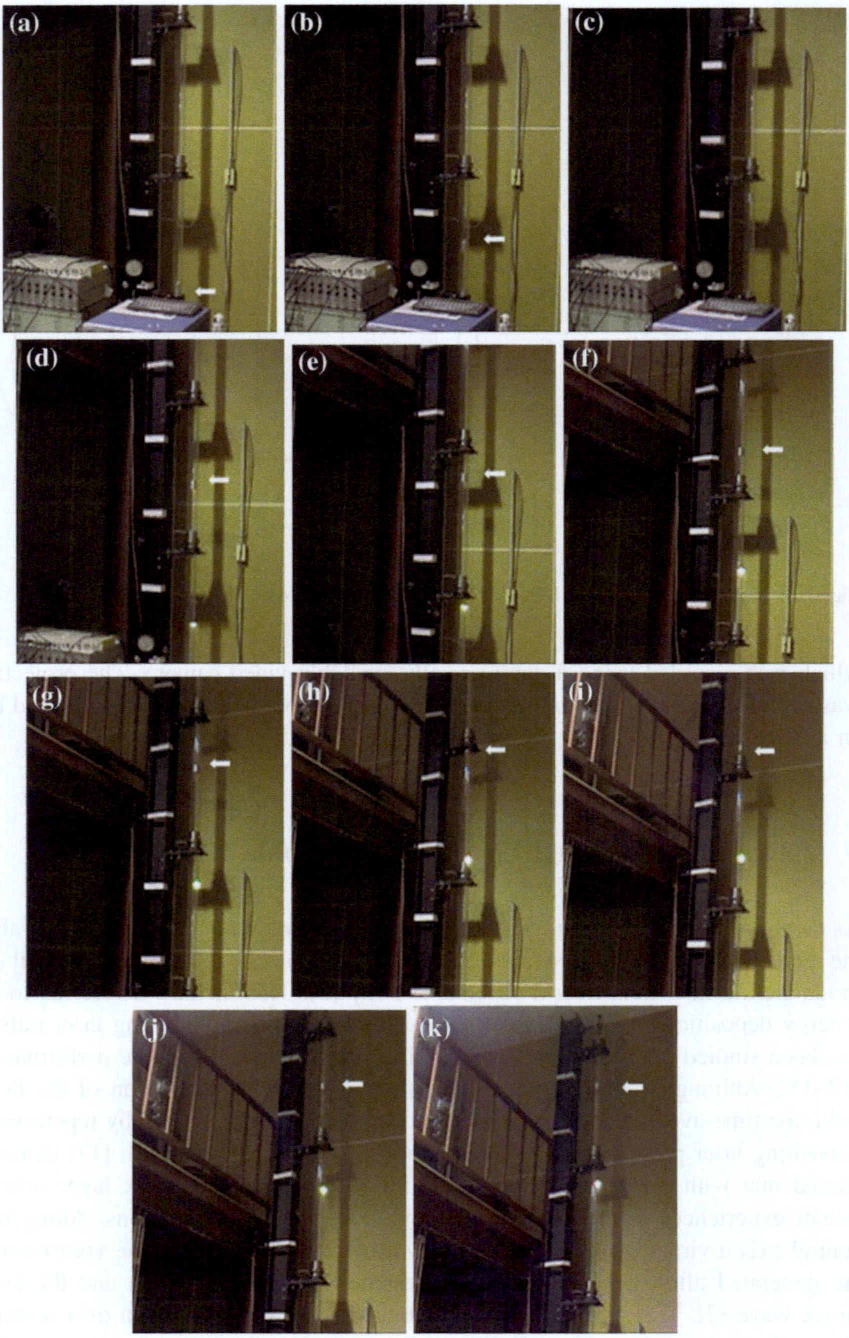

Fig. 4 A series of video frames showing LITA operation. The framing rate is 0.3 s, argon is used as propellant; the laser pulse energy is 8.3 J/pulse and the laser pulse repetition frequency is 50 Hz. **a** 0 s. **b** 0.3 s. **c** 0.6 s. **d** 0.9 s. **e** 1.2 s. **f** 1.5 s. **g** 1.8 s. **h** 2.1 s. **i** 2.4 s. **j** 2.7 s. **k** 3.0 s

Experiments of drag performance modulation over a body in supersonic flow are conducted using the experimental system documented in Kim et al. [17, 18]. The experimental facility is comprised of an in-draft supersonic wind tunnel with cross-section 80 mm × 80 mm; test time 10 s, a laser, and optics for the energy deposition and diagnostic systems. The Mach number, static pressure and temperature of the test flow are 1.94, 13.8 kPa and 163 K, respectively. A repetitive-pulse Nd: YVO4 laser (wavelength of 1.047 nm; pulse duration of 10 ns; maximum repetition rate of 100 kHz with deposited energy 4 mJ) is used to deposit laser pulse energies. The laser beam is expanded to a 12 mm × 12 mm square and sent through a focusing lens (focal length 60 mm) and then a BK-7 side window of the wind tunnel towards a spot in front of a model at its central axis. The drag force is measured by using the single-axis force balance system developed by Sasoh et al. [14]. The accuracy of drag measurements is better than 4 %.

The shock layers over the models without laser pulses are shown in Fig. 5. Except for the case of the sharp cone model with d_f/d = 0.0 (d_f, diameter of truncation; *d*, diameter of cylinder body), a bow shock wave is formed ahead of the models. Figure 6 presents Schlieren images with laser pulse energy depositions (f = 80 kHz, E = 5.0 mJ). Comparing Figs. 5 and 6, other than the sharp cone, the effect of energy depositions on the shock wave shape is significant; the normalized shock stand-off distance, $^{\delta}/_{d_f}$, is increased by 50 % for d_f/d = 1.00 and 35 % for d_f/d = 0.50. In front of the flat head of the models an acting spike which is

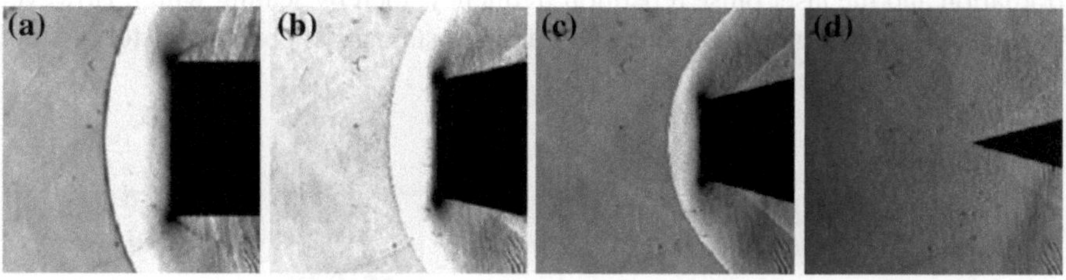

Fig. 5 Schlieren images of shock layers without laser pulse, all with an equal magnification [19]. **a** d_f/d = 1.0. **b** d_f/d = 0.75. **c** d_f/d = 0.5. **d** d_f/d = 0

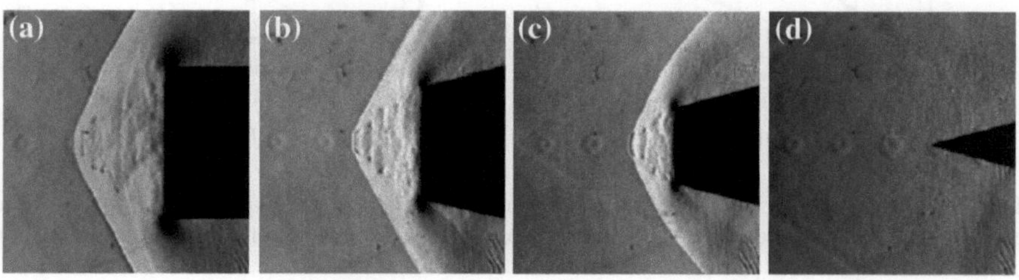

Fig. 6 Instantaneous schlieren images with repetitive laser pulses, f = 80 kHz, E = 5.0 mJ, all with the equal magnification as in Fig. 2. **a** d_f/d = 1.0. **b** d_f/d = 0.75. **c** d_f/d = 0.5. **d** d_f/d = 0

comprised of baroclinically-generated vortex rings is formed [19]; the bow shock wave is moderated to a conical shock wave. The time-averaged effective half apex angle of the acting spike equals 40° while that of the conical shock wave is 60°. The Taylor–MacColl solution for the conical shock with a half apex angle of 60° with $M_s = 1.94$ yields a half apex cone angle of 37°. Considering that the visualized outline of the acting spike is not equivalent to a solid wall due to differences in the induced velocity field, these values roughly agree with each other. Based on the Taylor–MacColl solution, the static pressure on the cone is estimated to be 49 kPa. Without laser energy depositions, the stagnation pressure at the centerline on the model head corresponds to a Pitot pressure of 74 kPa. Therefore, the acting spike decreases a representative pressure by 33 %. Over the conical model ($d_f/d = 0.0$; Fig. 6d), a vortex ring is not observed.

Hereafter, let ΔD be defined as the drag decrement. This is a function of the laser pulse energy, E, the repetition frequency of laser pulse depositions, f, and the normalized truncation diameter, d_f/d. This decrement is normalized by the baseline drag $D_0 = 21.5 \pm 0.45$ N that is obtained with $d_f/d = 1.00$ and $f = 0$ kHz: Drag reduction characteristics with a constant value of E as a function of f are shown in Fig. 7. With $E = 5.0$ mJ, air breakdown was stably obtained in the range 10 kHz $\leq f \leq$ 80 kHz; with $E = 7.2$ mJ, the range is 10 kHz $\leq f \leq$ 50 kHz. In the case of the sharp cone ($d_f = 0$), the drag decrement is very small, 1.7 % of D_0 at most. For other models, the drag decrement is much larger and approximately scales with f. If the interactions within laser-heated gases are weak, the drag decrement would be equal to the product of the impulse which is induced by a single laser pulse irradiation and the laser pulse repetition frequency, thereby scaling with f. However, for $f \geq 5$ kHz the visualized shock layer and the stagnation pressure history exhibit strong pulse-to-pulse interactions [19].

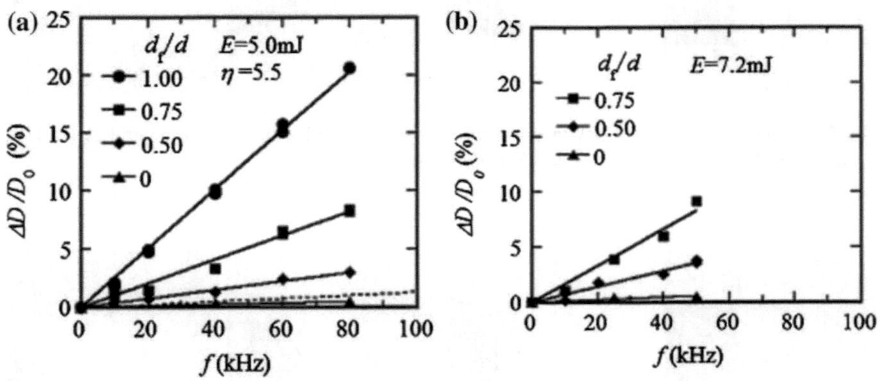

Fig. 7 The dependence ΔD versus f. **a** $E = 5.0$ mJ, **b** $E = 7.2$ mJ; for $d_f/d = 1$ *solid lines* are fit to experimental data, η denotes the efficiency of energy deposition, the ration of the saved propulsion power to the input laser power

This linearity leads to a constant efficiency of energy deposition as defined by Knight [13]. The linearity of $\Delta D/D_0$ with respect to f is equivalent to η having a constant value. The slope of fitted lines in Fig. 7 scales with η.

As seen in this section, laser focusing widens the variety and possibility in the development of experimental methods to investigate controlled blast waves and associated flows.

References

1. Jiang, Z., Takayama, K., Moosad, K.P.B., Onodera, O., Sun, M.: Numerical and experimental study of a micro-blast wave generated by pulsed-laser beam focusing. Shock Waves **8**, 337–349 (1998)
2. Sasoh, A., Ohtani, T., Mori, K.: Pressure effect in a shock-waveplasma interaction induced by a focused laser pulse. Phys. Rev. Lett. **97**, 205004 (2006)
3. Jacobs, J.W.: Fluids A **5**, 2239 (1993)
4. Meshkov, E.E.: Izv. Akad. Nauk. SSSR. Mekh. Zhidk. Gaza **4**, 151 (1969)
5. Meshkov, E.E.: Sov. Fluid Dyn. **4**, 101 (1969)
6. Richtmyer, R.D.: Commun. Pure Appl. Math. **13**, 297 (1960)
7. Kantrowitz, A.: Propulsion to orbit by ground-based lasers. Astronaut. Aeronaut. **10**, 74–76 (1972)
8. Pirri, A.N., Monsler, M.J., Nebolsine, P.E.: Propulsion by absorption of laser radiation. AIAA J. **12**, 1254–1261 (1974)
9. Myrabo, L.N.: World record flights of beam-riding rocket lightcraft: demonstration of "disruptive" propulsion technology. AIAA Paper 2001-3798 (2001)
10. Sasoh, A.: Laser-driven in-tube accelerator. Rev. Sci. Instrum. **72**, 1893–1898 (2001)
11. Sasoh, A., Suzuki, S., Matsuda, A.: Wall-propelled, in-tube propulsion with repetitive-pulse laser ablation. J. Prop. Power **25**, 540–542 (2009)
12. Adelgren, R.A., Yan, H., Elliott, G.S., Knight, D.D., Beutner, T.J., Zheltovodov, A.A.: Control of Edney IV interaction by pulsed laser energy deposition. AIAA J. **43**, 256–269 (2005)
13. Knight, D.: Survey of aerodynamic drag reduction at high speed by energy deposition. J. Prop. Power **24**, 1153–1167 (2008)
14. Sasoh, A., Sekiya, Y., Sakai, T., Kim, J.-H., Matsuda, A.: Wave drag reduction over a blunt nose with repetitive laser energy depositions. AIAA J. **48**, 2811–2817 (2010)
15. Zudov, V.N., Tret'yakov, P.K., Tupikin, A.V., Yakovlev, V.I.: Supersonic flow past a thermal source. Fluid Dyn. **38**, 782–793 (2003)
16. Tret'yakov, P.K., Garanin, A.F., Grachev, G.N., Krainev, V.L., Ponomarenko, A.G., Tishchenko, V.N., Yakovlev, V.I.: Control of supersonic flow around bodies by means of high-power recurrent optical breakdown. Physics-Doklady **41**, 566–567 (1996)
17. Kim, J.-H., Matsuda, A., Sakai, T., Sasoh, A.: Wave drag reduction with acting spike induced by laser-pulse energy depositions. AIAA J. **49**, 2076–2078 (2011)
18. Kim, J.-H., Matsuda, A., Sasoh, A.: Formation of a virtual spike built-up with vortex rings generated by repetitive energy depositions over a bow shock layer. Phys. Fluids **23**, 021703 (2010)
19. Sasoh, A., Kim, J.-H., Yamashita, K., Sakai, T.: Supersonic aerodynamic performance of truncated cones with repetitive laser pulse energy depositions. Shock Waves **24**, 59–67 (2014)

Author Biography

Akihiro Sasoh
Department of Aerospace Engineering, Graduate School of Engineering, Nagoya University
Specialization: Compressible fluid dynamics, Non-chemical space propulsion
Education: 1984 Bachelor, Department of Aeronautics, University of Tokyo
1986 Master, ditto
1989 Doctor, ditto
Academic carrier:
1989 Research Associate, Department of Aeronautics, Nagoya University
1991 Associate Professor, Institute of Fluid Science, Tohoku University
2003 Professor, ditto
2006 Professor, Department of Aerospace Engineering, Nagoya University

Part II
Shock Tunnels

The Aachen Shock Tunnel TH2 with Dual Driver Mode Operation

Herbert Olivier

1 Introduction

After more than 50 years, high-enthalpy facilities suitable for aerodynamic testing in the hypersonic regime are still mostly based on the shock tunnel principle. Recent use of these facilities involves the development of space planes, reentry vehicles and scramjets for studying the complex aerothermochemistry associated with these vehicles.

The effects of thermal and chemical relaxation in air become important for flight velocities higher than approximately 3 km/s. In addition, high-enthalpy, aerodynamic-impulse facilities are not only used to study high-temperature effects but are also suitable for generating high Mach number flows at moderate Reynolds numbers to investigate viscosity-dominated effects also at low-enthalpy conditions.

At present, three categories of shock tunnels are in use. The so-called conventional shock tunnel is operated with a driver that is filled up to the desired driver pressure by a preferably light driver gas like helium or hydrogen [1]. In detonation driven shock tunnels [2] the driver section is filled with a detonable gas mixture which is ignited either at the main diaphragm station (backward driver mode) or at the end wall of the driver section (forward driver mode). A detonation wave running through the driver section leaves a highly compressed and heated driver gas behind it. Piston driven shock tunnels employ a nearly adiabatic compression and heating of the driver gas accelerating a heavy piston in the driver towards the main diaphragm station. Especially the last two techniques enable achieving very high shock Mach numbers at reasonably high filling pressures in the driven section, which makes them suitable for aerodynamic high-enthalpy testing. The Aachen shock tunnel TH2 is operated both as conventional and detonation driven shock tunnel.

H. Olivier (✉)
Shock Wave Laboratory, RWTH Aachen University, 52056 Aachen, Germany
e-mail: olivier@swl.rwth-aachen.de

O. Igra and F. Seiler (eds.), *Experimental Methods of Shock Wave Research*,
Shock Wave Science and Technology Reference Library 9,
DOI 10.1007/978-3-319-23745-9_5

2 Operation Principle of the Conventional Shock Tunnel

The elementary shock tube performance is well documented in a number of text books [1, 3–5]. It will thus be sufficient to bring into attention the most important relations and the usual notations of the different flow fields and kind of shock reflections.

A shock tunnel consists of a driver and driven section, a nozzle and a dump tank with a test section. Driver and driven section are separated by the main diaphragm. The driver section is filled with the so-called driver gas, usually helium, at an initial pressure p_4 and a temperature T_4, while the driven section is at an initial pressure p_1 and a temperature T_1 (see Fig. 1). Typical driver pressures are between 10 and 150 MPa and driven gas pressures between 0.01 and 1 MPa. Before the test starts the dump tank is evacuated. Driven section and dump tank are separated by a second thin diaphragm to avoid the inflow of the test gas into the dump tank before the test starts. A double diaphragm technique allows to precisely fill driven and driver sections and to start the shock tunnel run in a controlled manner. Figure 1 shows schematically the setup of the tunnel and the pressure distribution within the shock tube for various times during the run.

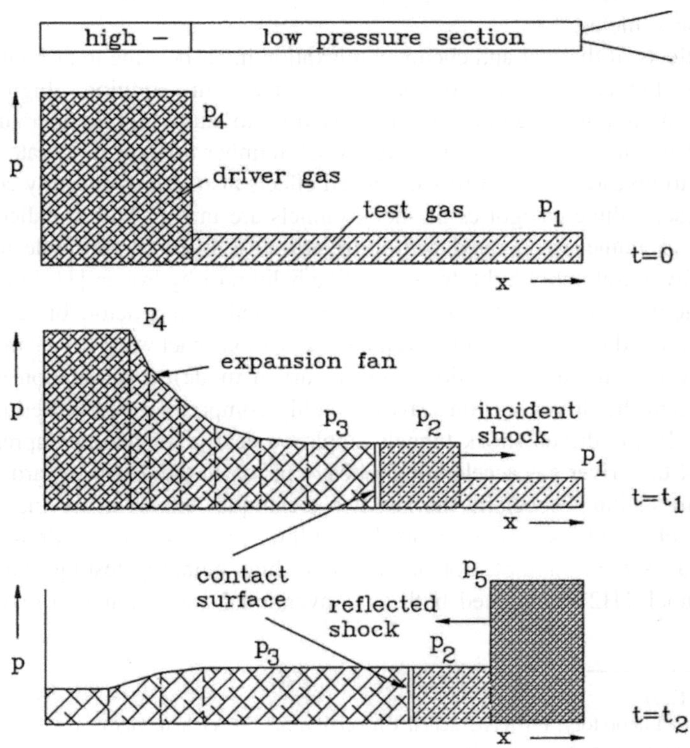

Fig. 1 Schematic setup and operation principle of a conventional shock tunnel

In Fig. 1 ideal diaphragm opening and ideal wave behaviour are assumed (no damping, no boundary layer influence). After rupture of the main diaphragms a primary shock wave propagates into the test gas (1) and compresses it to a pressure p_2 at a temperature T_2 in the region (2). Behind this shock wave a contact surface follows which separates the driver and driven gases. The driver gas acts like a piston which compresses the test gas rapidly. Density and sound velocity on both sides of the interface are different, but pressure and flow velocity are equal in (2) and (3). This pressure balance is produced by an expansion wave the head of which moves upstream into the driver section reducing the pressure continuously from its initial value p_4 to p_2.

A few milliseconds after the bursting of the diaphragm the primary shock wave arrives at the end wall of the driven section. The Aachen shock tunnel operates in the reflected mode, i.e. the primary shock wave is reflected at the end wall due to its relatively small central nozzle inlet and after that propagates upstream (see Fig. 2). During this reflection process the second diaphragm between driven section and nozzle bursts, and the nozzle flow starts. To a first approximation the flow velocity behind the reflected shock is zero if the outflow from the nozzle is neglected. The complete kinetic energy of the shock heated gas in region (2) is thus converted to high values of temperature and pressure in region (5) behind the reflected shock. The compressed test gas having a temperature of a few thousand degrees centigrade subsequently expands through the nozzle. Thus gas of high stagnation enthalpy is converted into a free stream of high velocity in the test section.

After a short time the reflected shock interacts with the contact surface. During this interaction new waves are usually generated which propagate again downstream towards the end wall while the reflected shock moves with changed velocity further upstream (see Fig. 3). In the case of a special combination of the initial gas variables the reflected shock penetrates the contact surface without generating waves of finite amplitudes, the so-called tailored interface case (see Figs. 2 and 3).

Fig. 2 Wave diagram for tailored interface operation

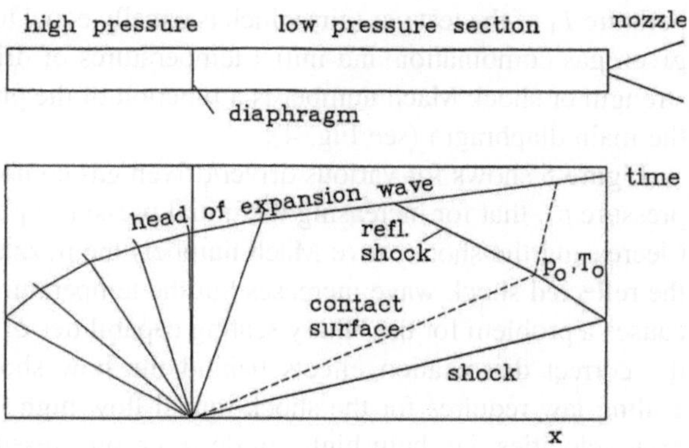

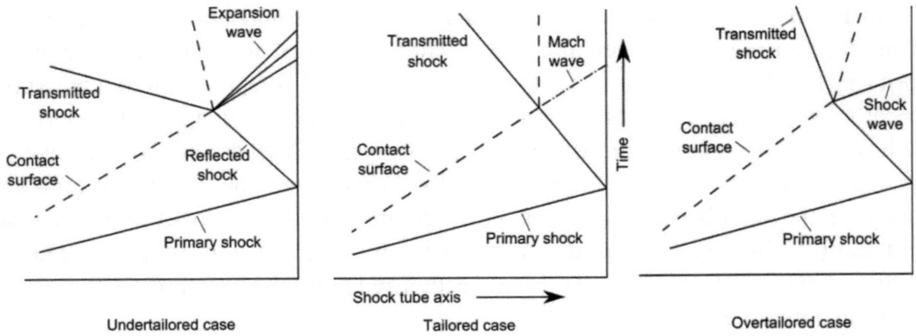

Fig. 3 Interaction of the reflected shock with the contact surface

In the tailored interface case the nozzle reservoir pressure and temperature produced by the initial reflection of incident shock at the end wall persists for a relatively long period.

Figure 3 shows the wave diagram for the tailored and both the under-tailored and over-tailored cases; in the under-tailored case the incident shock wave Mach number is lower than in the tailored case. In this case an expansion wave is generated when the reflected shock and the contact surface interact. In the over-tailored case the incident shock wave Mach number is higher than in the tailored case. This leads to a secondary reflected shock wave and a transmitted shock wave which are generated by the interaction of the initial reflected shock wave with the contact surface.

The secondary reflected shock again reflects at the end wall, and in this way a region of multiple shock reflections develops which leads to an equilibrium nozzle reservoir pressure [6]. The shock tunnel TH2 is generally operated in the tailored and over-tailored interface mode but under-tailored operation is also possible.

The temperature of the test gas behind the reflected shock in the driven section of the shock tube (corresponding to the nozzle "reservoir" or "supply" temperature) is a function of the strength or velocity of the incident shock wave and the initial temperature T_1 of the test gas (air) which is usually equal to the ambient temperature. For given gas combination and initial temperatures of driver and driven gas the shock strength or shock Mach number is a function of the pressure ratio $p_4/p_1 = p_{41}$ across the main diaphragm (see Fig. 4).

Figure 5 shows for various driver/driven gas combinations and for a fixed driver pressure p_4, that for increasing the initial pressure p_1 up to a specific initial pressure (decreasing the shock wave Mach number) the nozzle reservoir pressure p_5 behind the reflected shock wave increases but the temperature T_5 decreases. This generally causes a problem for the binary scaling capabilities of shock tunnels. For simulating the correct dissociation effects behind the bow shock of a re-entry vehicle this scaling law requires for the shock tunnel flow high free stream densities and high flow velocities, i.e. both high nozzle reservoir pressures and temperatures.

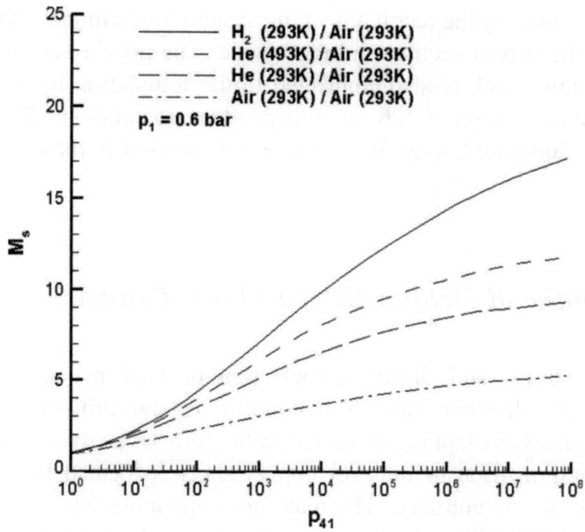

Fig. 4 Shock Mach number M_s as function of pressure ratio p_{41} for different gas combinations; driver gases are assumed as ideal gases, driven gas air as 9 component equilibrium

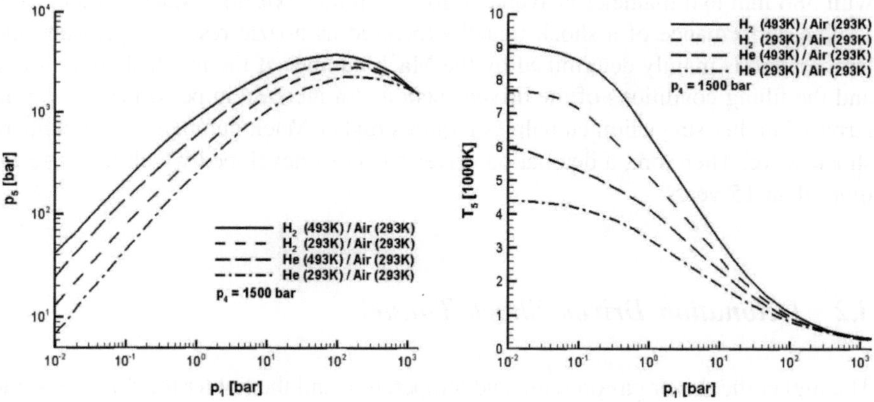

Fig. 5 Influence of the driven gas pressure p_1 on the pressure and temperature behind the reflected shock (p_4 = 1500 bar). Driver and driven gas combinations with their initial temperatures are indicated

3 Setup of the Aachen Shock Tunnel TH2

The shock tube of the Aachen shock tunnel has an inner diameter of 140 mm and a wall thickness of 80 mm. The building which houses the shock tunnel was built especially for the use of such tunnels. A 800 mm steel-enforced concrete wall which separates the driver room from the driven section serves as a protecting wall but is also used for supporting the recoil absorbing system of the tunnel. There is a sliding joint between the driven section and the nozzle. The nozzle and dump tank compose a single unit which is also connected to the foundation by a recoil damping system. The model support has an independent foundation. Thus even if the receiver or the dump tank may move, the model support is fixed to the laboratory foundation.

3.1 Conventional Helium Driven Shock Tunnel

Driver (length 6 m) and driven section (length 15.4 m) are separated by a double-diaphragm chamber which at maximum pressure utilizes two 10 mm thick stainless steel plates as diaphragms scored in the form of a cross by a milling cutter. Another diaphragm made of brass or copper sheet is located between the driven section and the nozzle entrance. The maximum operating (steady) pressure of the complete tube is 150 MPa. The driver can be electrically heated to a maximum T_4 of 600 K. The exit diameter of the conical nozzle with 5.8° semi-angle amounts to 586 mm. The nozzle throat diameter and therefore the test section flow Mach number can also be changed by inserting different throat pieces. A contoured nozzle with 586 mm exit diameter is available for a nominal exit flow Mach number of 7.

The performance of a shock tunnel expressed as nozzle reservoir pressure and temperature is mainly determined by the Mach number of the incident shock wave and the filling conditions of the driven section. An increase in performance, e.g. in terms of higher stagnation enthalpies requires higher Mach numbers of the incident shock wave. Therefore, a detonation driver has been developed which is in use for more than 15 years.

3.2 Detonation Driven Shock Tunnel

The higher the driver gas pressure and temperature and the lighter the driver gas, the stronger is the incident shock wave for a given filling pressure p_1 in the driven section. The conventional helium driver reaches its operational limits for a driver pressure of about 150 MPa and tube temperatures of about 600 K. To further increase the shock tunnel performance a detonation driver has been developed [7, 8] which alternatively to the helium driver can easily be connected to the driven section.

The detonation driver consists of a 9.4 m long detonation tube and a 6.4 m long damping section which is placed upstream of the detonation section (see Fig. 6). The whole tube has an inner diameter of 140 mm and its maximum working pressure is fixed at 150 MPa. The detonable gas mixture in the detonation section is ignited at the main diaphragm station by an exploding wire. Details about the initiation system and more are given in [8]. After a fast deflagration detonation

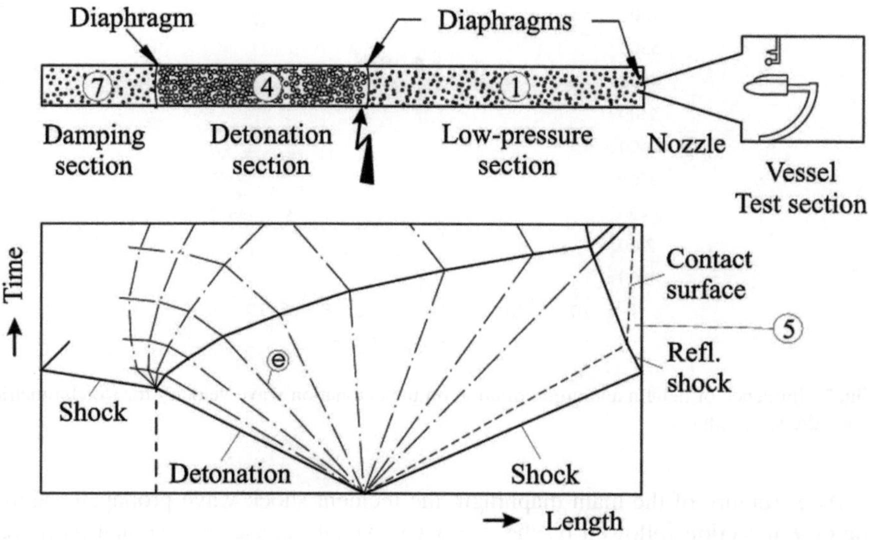

Fig. 6 Principle setup and wave diagram for detonation driven shock tunnel in the upstream mode

transition the detonation wave travels towards the driver section end wall. The damping section is needed for extinguishing the detonation wave before it reflects at the tube end wall. For this, it is filled with nitrogen at relative low pressures. The detonation and damping section are separated by a stack of thin plastic diaphragms which are totally burnt by the detonation wave. The damping section not only reduces the mechanical loading of the tube during the reflection process, it also leads to a longer period of constant driver conditions. In case of no damping section, the detonation wave would reflect as shock wave which propagates towards the main diaphragm station. In case of a damping section, at the location of the diaphragm between detonation and damping section a rarefaction wave develops, the tail of which also travels towards the main diaphragm section. Because the tail of the rarefaction wave travels slower than a reflected shock, in case of a damping section the time of constant flow conditions at the main diaphragm station is longer.

The detonation wave forms as so-called Chapman Jouguet (C-J) detonation which is highly reproducible and stable. This detonation wave is characterized by a high pressure increase across the leading shock front, followed by a Taylor expansion which in case of a closed tube decelerates the flow again to zero velocity. The strength of the detonation wave not only depends on the initial filling pressure but also on the chemical composition of the detonable gas mixture. For the Aachen shock tunnel stoichiometric oxyhydrogen is used with varying admixtures of helium and argon to obtain a wide range of test conditions. Up to a certain limit helium increases the speed of sound of the detonation gas products leading to higher velocities of the detonation wave. Increasing the amount of argon significantly reduces the speed of the detonation wave (see Fig. 7).

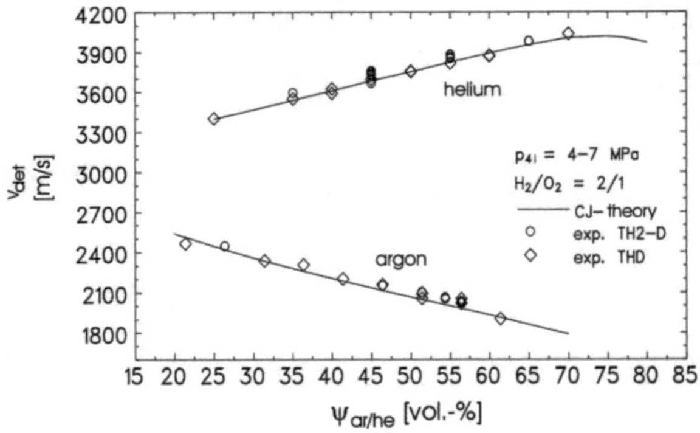

Fig. 7 Influence of helium and argon dilution on the detonation wave velocity for stoichiometric oxyhydrogen mixtures

After rupture of the main diaphragm the incident shock wave propagates along the driven section followed by the driver gas which in case of a detonation driver mainly consists of gaseous water. After the useful running time of the facility this gaseous water would pass through the nozzle leading to strong erosion and water condensation in the test section. To avoid associated problems caused by the gaseous water flow and condensation in the test section, a fast acting valve is in use which originally was developed at the Cornell Aeronautical Laboratory [9]. It works without any external triggering signal or power supply. A sketch of this valve is shown in Fig. 8. When the incident shock wave passes the conical forebody, its overpressure accelerates a piston which via a spring is connected to a plug which after an adjustable time delay closes the nozzle entrance. The central body of the valve is fixed in the shock tube by four thin struts. The operation of this valve is highly reliable and repeatable. The influence of the blockage effect associated with

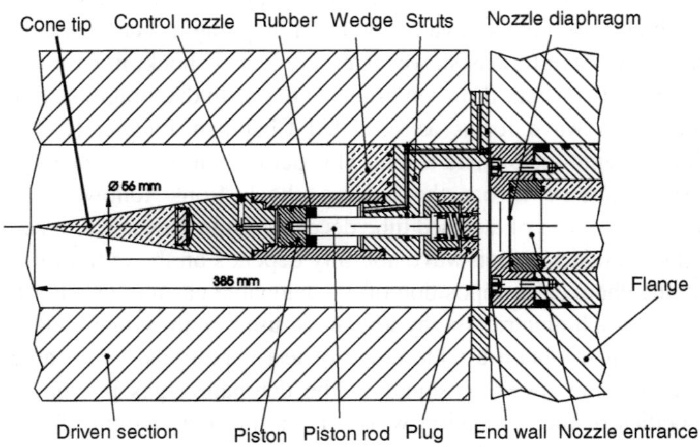

Fig. 8 Fast acting nozzle valve

the nozzle valve has been tested by comparing the reservoir and Pitot pressure histories in experiments conducted with and without a valve. It was found that there is only a very weak influence at the beginning of the run which is not surprising because of the very low flow velocities in front of the shock tunnel nozzle.

3.3 Shock Reflection and Driver Gas Contamination

For reflected shock tunnels a great concern exists concerning premature driver gas contamination of the test gas. There are numerous papers in the literature about this and how to measure or estimate the amount of the driver gas contaminating the test gas.

The mechanism which leads to driver gas contamination is due to bifurcation of the reflected shock at the shock tube wall. Because of the interaction of the reflected shock with the shock tube boundary layer close to the wall, the reflected shock bifurcates into two oblique shocks (see Fig. 9). Using oblique shock wave theory it is straightforward to show that these two shocks do not decelerate the gas to velocity zero as does the straight part of the reflected shock. Therefore, especially after the reflected shock wave has passed the contact surface, a wall jet of cold driver gas forms which penetrates into the region of the shock heated test gas close to the shock tube end wall. This finally leads to a change of the chemical composition of the test gas, but more important to a reduction of the total enthalpy because of the mixing with the cold driver gas. This phenomenon of bifurcation of the reflected shock wave in a shock tube was thoroughly studied by Mark [10] already in the late fifties.

It is obvious that an important criterion determining the total amount of the driver gas penetrating into the test gas is given by the height of the bifurcated shock, i.e. of the triple point. Therefore, the height of the triple point has been determined by flow visualization in a rectangular shock tube which allows the visualization of its entire cross section. In the context of detonation drivers it was of special interest to study this phenomenon for the interaction of the reflected shock with driver gas consisting of gaseous water. As an example, Fig. 10 shows some schlieren pictures

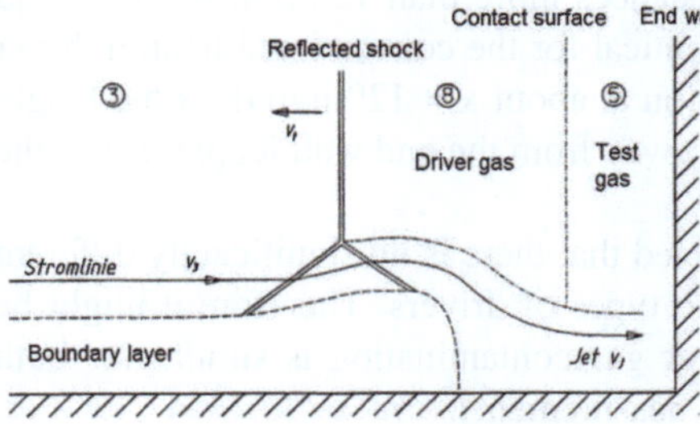

Fig. 9 Sketch of bifurcation of reflected shock

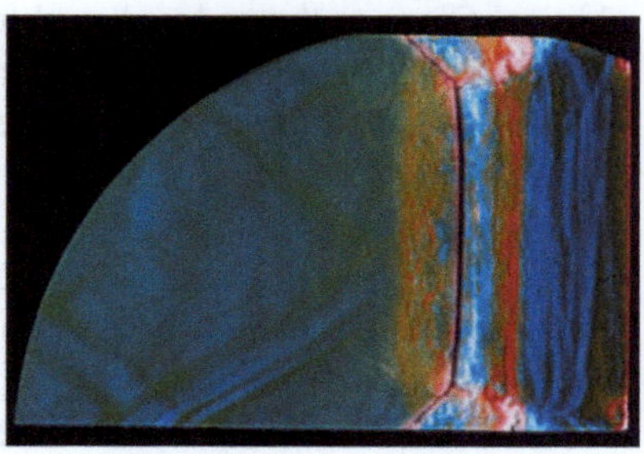

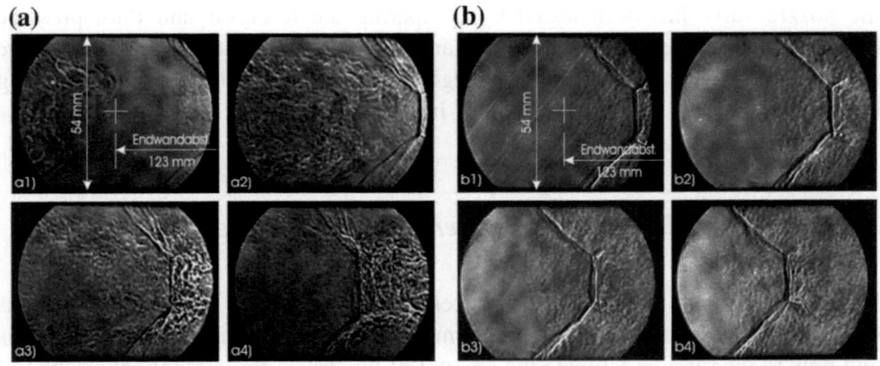

Fig. 10 Interaction of a reflected shock with a contact surface, $Ms = 7.6$, $\Delta t = 12$ µs; **a** helium driver, **b** detonation driver

of the interaction of the reflected shock with a contact surface for the case of a conventional helium driver and a detonation driver. The shock tube end wall is at the right side of the picture and the contact surface becomes visible, especially in the first picture in case of the helium driver. As expected, it is not a planar surface but wrinkled and showing a strong shock tube boundary layer influence. In the case of the detonation driver no contact surface could be visualized, the reason for which is the small density difference between the shock heated test gas and the still relatively hot driver gas.

Nevertheless, calculations of the shock tube flow prove that for the visualized region in front of the shock tube end wall as in the case of the helium driver the interaction with the contact surface takes place. Due to strength limitations of the windows and the shock tube, the initial pressure p_1 for these experiments was limited to 0.4 kPa which resulted in a thick shock tube boundary layer and therewith in a large shock bifurcation. However, the comparison between the results for the helium and detonation driver allows a first statement whether the detonation driver leads to a stronger shock bifurcation or not. From these figures the height of the bifurcated shock has been measured and plotted versus the distance from the end wall (see Fig. 11). For the detonation driver two results are given, one for a stoichiometric hydrogen/oxygen mixture diluted with helium and one for argon as additive. Close to the end wall and for distances more than 120 mm from the end wall the heights of the triple points are identical for the conventional helium driver and the detonation driver. Only in the region at about x = 120 mm does the height decrease for the helium driver, but further away from the end wall it approaches the value for the detonation driver.

From these experiments it can be expected that there is no significantly different behaviour in shock bifurcation for the two types of drivers. Therefore it might be expected that also the magnitude of driver gas contamination is similar for both cases which is confirmed by free stream measurements.

Fig. 11 Height of triple point of reflected shock for helium and detonation driver

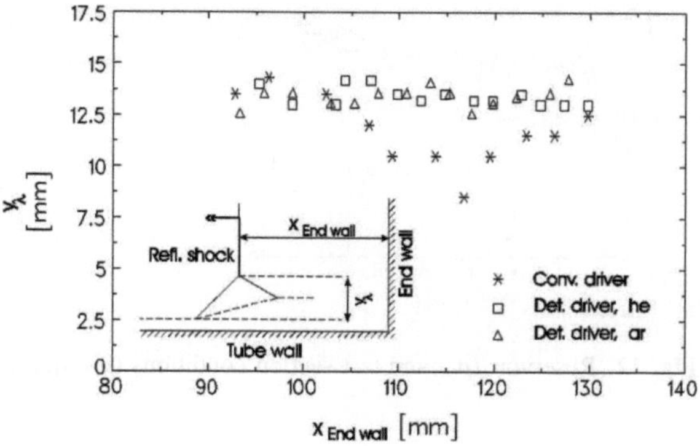

The developing shock tube boundary layer behind the incident shock leads to a continuous reduction of its speed and therewith of its strength. This shock attenuation should be proportional to the boundary layer thickness and the flow Mach number behind the incident shock. Of course, the tube diameter has an inverse influence. The attenuation of the shock speed has been measured for a variety of test conditions close to the end wall of the shock tube. It turns out that for the shock tunnel TH2 with 140 mm inner diameter the shock attenuation is nearly constant and in the average amounts to 1.2 % per meter [8]. This shock attenuation as well as losses caused by the diaphragm opening and other dissipative mechanisms lead to a loss of performance of the shock tube compared to an ideal shock tube process. The losses can easily be determined by comparing the computed pressure and enthalpy behind the reflected shock based on the measured shock speed with the values for an ideal inviscid, adiabatic shock tube process. As for the shock attenuation, the total pressure and enthalpy losses are almost independent of the considered shock tube conditions with respect to the ideal values. For both, the helium and detonation driven mode the relative losses of the nozzle reservoir pressure and total enthalpy amount to 30 %. This is a quite realistic value for shock tunnels of this size and has to be taken into account when choosing driver gas conditions.

3.4 *Nozzle Reservoir and Free Stream Flow Conditions*

A conical nozzle with 5.8° half opening angle and 586 mm nozzle exit diameter accelerates the shock-compressed, heated gas to hypersonic velocities in the test section. Behind the nozzle exit a permanent Pitot and static pressure probe, as well as a small sphere for measuring the stagnation point heat flux are installed. From these signals the time-resolved free stream conditions, especially u_∞ and ρ_∞ are resolved employing a suitable data reduction technique which is based on applicable gas dynamic relations [11].

As examples, Figs. 12 and 13 show the time histories of the nozzle reservoir pressure p_0, the Pitot pressure p_{t2}, the free stream static pressure p_∞ and of the

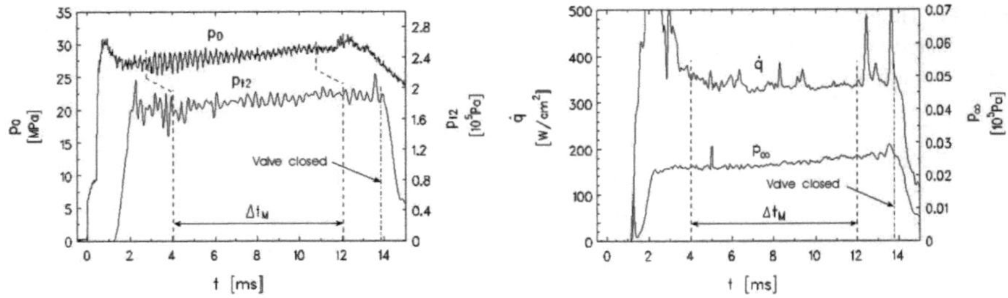

Fig. 12 Reservoir (p_0) and test section conditions (p_{t2}, $\dot{q}$, p_∞) for h_0 = 2.4 MJ/kg total enthalpy

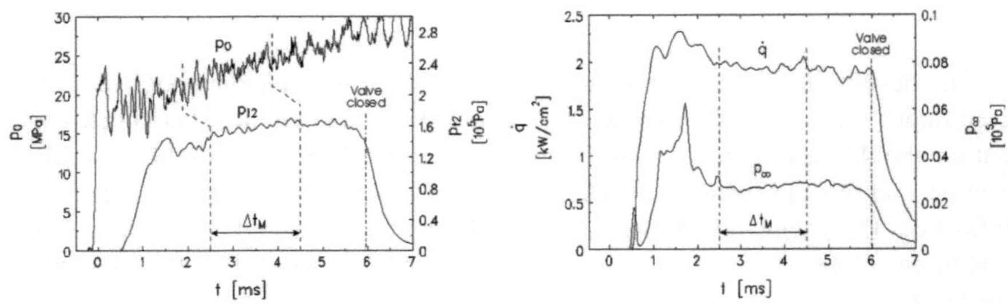

Fig. 13 Reservoir (p_0) and test section conditions (p_{t2}, $\dot{q}$, p_∞) for h_0 = 14.4 MJ/kg total enthalpy

stagnation point heat flux $\dot{q}$ measured at a small sphere of 20 mm diameter in the test section. These measurements have been performed in the detonation driven mode for a relatively low enthalpy and for the maximum one of 14.4 MJ/kg at which the detonation driver is used so far.

For the low enthalpy case a useful testing time of about 8–10 ms can be identified, whereas for the high-enthalpy case it reduces to 2 ms. After 14 and 6 ms, respectively, the fast acting valve closes the nozzle entrance which causes the rapid drop of the flow parameter in the test section. The high-enthalpy condition is slightly over-tailored which causes the weak pressure rise in the nozzle reservoir.

The free stream has been calibrated by Pitot rake measurements at various positions downstream of the nozzle exit. The Pitot rake was equipped with a total of 38 small spheres, each instrumented with a combined coaxial thermocouple allowing a heat flux and Pitot pressure measurement at the stagnation point of the sphere. Additionally, a free stream static pressure probe was installed on the rake. From this data, the Pitot pressures as well as the total temperature profiles have been deduced, employing for the latter one the formula of Fay and Riddell (Figs. 14 and 15). The Pitot pressure profiles show a typical core flow region of constant pressure and a drop across the nozzle boundary layer, i.e. in the free shear layer downstream of the nozzle exit. For the most upstream position, i.e. 20 mm downstream of the nozzle exit, the dashed line in the Pitot pressure profile indicates the boundary layer thickness according to the Edenfield correlation [12] for nozzle flows. The agreement with the measured thickness is very good which holds for both the low- and

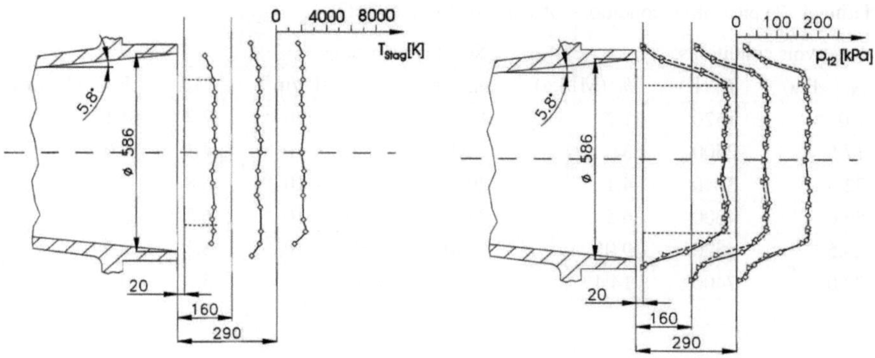

Fig. 14 Pitot pressure and total temperature profiles for three axial positions downstream of the nozzle exit, $h_0 = 2.4$ MJ/kg

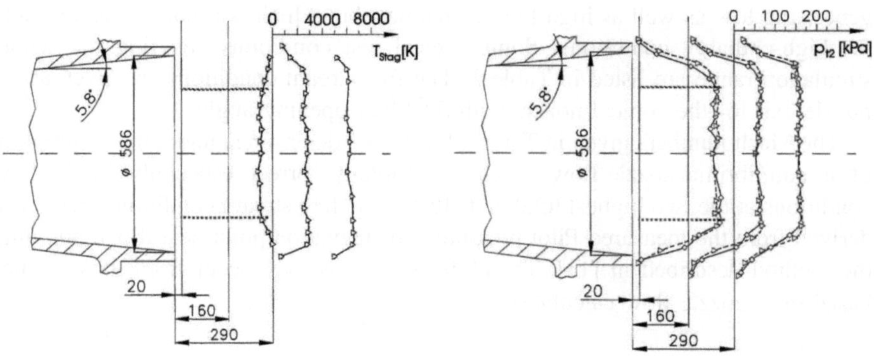

Fig. 15 Pitot pressure and total temperature profiles for three axial positions downstream of the nozzle exit, $h_0 = 14.4$ MJ/kg

high-enthalpy flow. Within the free shear layer the deduced total temperatures are questionable because the Mach number drops significantly and the validity of Fay and Riddell's formula is violated.

As known, the conical nozzle leads to axial flow gradients which become apparent by a decrease of the Pitot and free stream static pressure in downstream direction. For the considered nozzle this results in a Mach number increase of 4.2 % along a distance of 300 mm. This is of some disadvantage compared to contoured nozzles and has to be taken into account for long model geometries or the investigation of Mach number sensitive phenomena. However, a conical nozzle easily allows varying the free stream Mach number and thereby other flow parameters by changing the throat diameter. Furthermore, the shape of conical nozzles does not lead to a focusing of pressure disturbances towards the nozzle axis, as it happens for contoured nozzles.

Table 1 Typical flow conditions of the shock tunnel TH2

Reservoir conditions			Nozzle exit values			
p_0 (MPa)	T_0 (K)	h_0 (MJ/kg)	u_∞ (m/s)	ρ_∞ (kg/m^3)	M_∞	Re_∞ (10^6/m)
7.0	1520	1.7	1745	2.1×10^{-2}	7.7	4.1
17.5	2500	3.0	2350	2.7×10^{-2}	7.4	4.0
32.5	3640	4.1	2910	3.2×10^{-2}	6.8	3.7
56.0	4600	6.5	3400	3.6×10^{-2}	6.5	3.7
26.5	5520	9.0	3950	1.3×10^{-2}	5.8	1.3
22.0	7400	14.4	4850	0.76	5.3	0.8

The shock tunnel TH2 is operated for low as well as medium to high total enthalpies ranging from 1.7 to 14.4 MJ/kg. Up to 6.5 MJ/kg the conventional helium driver is used which in this case is filled with helium at 120 MPa and the whole driver is electrically heated to 570 K. The detonation driver is capable of generating low as well as high total enthalpies, but which is usually operated only for high-enthalpy conditions. Some typical test conditions covering the whole simulation range are listed in Table 1. The free stream conditions are given at the nozzle exit for the conical nozzle with 5.8° half opening angle.

The Mach numbers given in Table 1 have been determined under the assumption of an equilibrium nozzle flow which is not totally correct, especially for the flow conditions at the two highest total enthalpies. The free stream conditions have been derived from the measured Pitot pressure and stagnation point heat flux employing the method described in [11]. Therefore deviations may occur from those values based on a nozzle flow calculation.

3.5 Driver Gas Detection by the Free Stream Static Pressure

Reliable measurements in reflected shock tunnels require a detailed knowledge of the amount of driver gas contamination of the test gas. The mechanism leading to driver gas contamination is due to bifurcation of the reflected shock as described above. The useful testing time of shock tunnels should not be defined by the time period of constant reservoir or Pitot pressure. Even the measured time history, for example, of the stagnation point heat flux in the free stream might not be adequate to determine a useful testing time window because parameters like the heat flux, Pitot pressure etc. mainly depend on the free stream density and velocity. A contamination of the test gas by cold driver gas leads to a decrease of the free stream velocity caused by a reduced total temperature. On the other hand the reduced temperature leads to higher densities so that flow parameters like Pitot pressure, stagnation point and heat flux might not be strongly influenced by driver gas contamination. This effect is visible in the experimental data of reflected shock tunnels. The mixing of the hot test gas with the cold driver gas leads not only to lower total temperatures but also to a change of the isentropic exponent. For example, in case of helium as driver gas the isentropic exponent increases. This partially offsets the effect of the reduced total temperature on the free stream density and velocity.

However, the equations of quasi-onedimensional, isentropic flow show that for a fixed nozzle area ratio and nozzle reservoir pressure an increase of the isentropic exponent leads to higher flow Mach numbers and therewith to lower free stream static pressures. The proposed method of driver gas detection by the free stream static pressure measurement is based on this effect. In literature other methods are described as well. Paull developed a sensor for driver gas detection in which with increasing driver gas amount an oblique shock steepens up and finally blocks a channel flow [13]. The time of the blockage indicates a certain level of driver gas contamination. This sensor has been modified and used in the T5 shock tunnel by Sudani and Hornung [14]. Mass spectrometry to determine the arrival time of helium in the test gas has also been used [15] but this is much more complicated compared to sensor based methods.

Experiments have been performed with the conventional helium driver measuring the free stream static pressure with a small static pressure probe [16]. Figure 16 shows the histories of the nozzle reservoir pressure p_0, of the Pitot pressure p_{t2} and of the free stream static pressure p_∞ for an over-tailored test condition. Whereas the Pitot pressure nicely follows the temporal behaviour of the reservoir pressure, the free stream static pressure shows a substantial decrease after its initial rise. For tailored-interface conditions the free stream static pressure signal follows the Pitot pressure as expected. This static pressure drop is caused by driver gas contamination leading to a temporal rise of the isentropic exponent. The mixture of air ($\gamma = 1.4$) and helium ($\gamma = 1.66$) results in an increase in γ of the test gas. For a nozzle with constant area ratio $A/A*$ ($A*$ is the throat area), an increase of

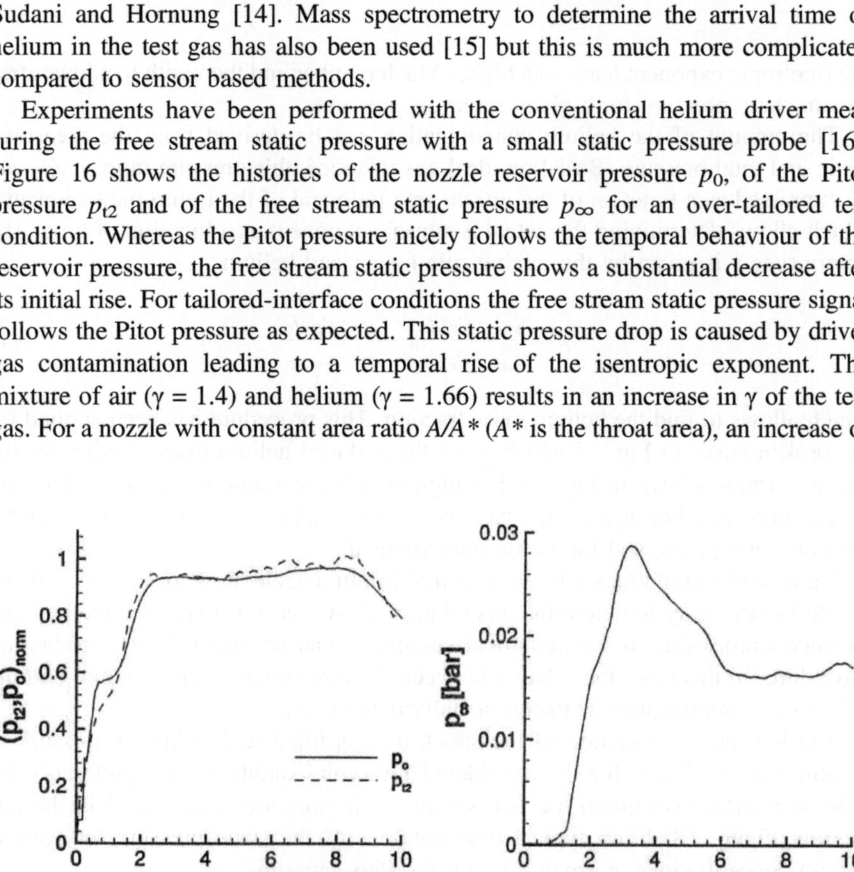

Fig. 16 Nozzle reservoir (p_0), Pitot (p_{t2}) and free stream static pressure (p_∞) histories for strongly over-tailored test conditions

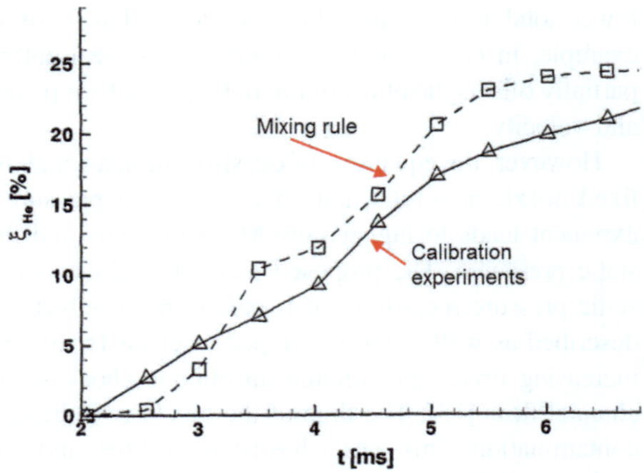

Fig. 17 Temporal increase of helium concentration in the test gas for over-tailored test condition

the isentropic exponent leads to a higher Mach number and therewith to a lower free stream static pressure for a given reservoir pressure.

The amount of the helium contamination can be derived from the measured static and total pressure. Based on ideal gas relations, this pressure ratio p_∞/p_0 can be expressed as a function of the nozzle area ratio and of the isentropic exponent γ which allows determining the actual isentropic exponent as a function of time. The appropriate γ is given by the mixing rule for air and helium:

$$\gamma = \frac{\xi_{air} \cdot c_{p,air} + \xi_{He} \cdot c_{p,He}}{\xi_{air} \cdot c_{v,air} + \xi_{He} \cdot c_{v,He}} \tag{1}$$

which allows to find the helium mass fraction. This procedure has been applied for the broken curve in Fig. 17 which gives the deduced helium mass fraction for the pressure traces shown in Fig. 16. For high-enthalpy conditions this method would be cumbersome, because in this case no simple relation is available between the pressure ratio p_∞/p_0 and the isentropic exponent.

Numerical simulations of the nonequilibrium nozzle flow or other methods would be necessary to determine this relation. However, this extensive method and its uncertainties due to the numerical simulation can be avoided by a calibration procedure. In this case, the relation between the free stream static pressure and the driver gas contamination is experimentally determined.

The low pressure section of the shock tube is filled with a known mixture of helium and air. Then, for the considered reservoir conditions and preferably for tailored interface operation the free stream static pressure is measured in the test section. Figure 18 shows typical time histories of the static pressure for various helium concentrations, normalized with the Pitot pressure.

From Fig. 18 it is obvious that with increasing helium mass fraction the ratio p_∞/p_{t2} decreases. This data can be used as calibration data to obtain the level of helium concentration from the measured static and Pitot pressure. This procedure has been applied for the dashed line in Fig. 17. The agreement between the two

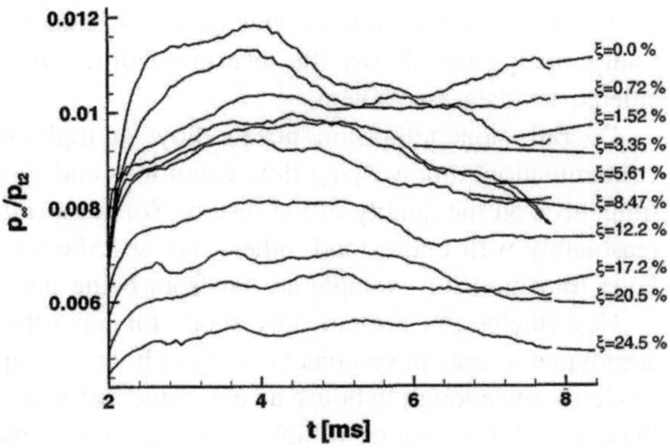

Fig. 18 Time histories of the pressure ratio p_∞/p_{t2} for various helium concentrations

methods is reasonably good and the accuracy achieved is comparable to that of other methods. After the useful testing time of about 4 ms the helium concentration amounts to about 20 % which is also comparable to values given in the literature (see e.g. [14]).

For pure air as the test gas or for a very low helium concentration the pressure ratio history in Fig. 18 should show a more or less flat or constant behaviour as for the higher helium concentrations. But after 4 ms for the very low helium concentrations a decrease of the pressure ratio occurs which indicates that even for this nominal tailored-interface condition the interaction of the reflected shock with the contact surface causes an additional driver gas contamination. This results in a higher helium concentration than the chosen nominal premixed one. Therefore in this case the pressure ratio has been determined from the measured signals just after flow establishment where the additional driver gas contamination due to the reflected shock is still negligible.

The advantage of the described method of driver gas detection based on free stream static pressure measurements is that it does not need any additional measuring device except a permanently installed free stream static pressure probe in the test section. Therefore it can be run in parallel to any other measurements or wind tunnel model testing.

4 Summary

Shock tunnels working in the reflected mode are in worldwide use for studying hypersonic flows from the low- to the high-enthalpy regime. They mainly differ in their driver principle which comprises the classical driver utilizing pre-compressed gas, combustion and detonation driver and the free piston driver. With the largest existing shock tunnels, the detonation driven facility JF12 in China and the free piston driven shock tunnel HIEST in Japan, today's available technology seems to be at an upper limit because of material problems and substantial operational requirements.

The working principle of shock tunnels leads to the reality that the higher the total enthalpy the shorter the measuring time. Other side effects have to be considered for data assessment.

Overall, nonequilibrium nozzle flow at high-enthalpy conditions, driver gas contamination, time varying flow conditions and short running times have different influences on the quality of the results. Some aspects of shock tunnel operation are reasonably well understood, others like specific shock tunnel effects on boundary layer transition for example are far from being understood.

Nevertheless, at present only shock tunnels offer the possibility of performing aerodynamic tests of sufficient quality at high-enthalpy conditions with wind tunnel models large enough to house an extensive instrumentation. In recent years not only the external flow around vehicles has been considered, but even the internal flow in large scramjet models including injection and combustion of different types of fuel has been successfully examined. For studying wall temperature sensitive flow phenomena the drawback of shock tunnels of low model wall temperatures can be avoided, or at least reduced, by heating the model wall up to the desired temperature prior to the experiment.

In summary, shock tunnels are a powerful tool for studying high-enthalpy, hypersonic flows in ground based facilities. As has been proven, careful operation and data reduction methods result in reliable data with acceptable uncertainty margins.

References

1. Hertzberg, A., Witliff, C.E., Hall, J.G.: Summary of Shock Tunnel Development and Application in Hypersonic Research. AFOSR TR 60–139. Cornell Aeronautical Laboratories, New York (1961)
2. Yu, H.R., Esser, B., Lenartz, M., Groenig, H.: Gaseous detonation driver for a shock tunnel. Shock Waves **2**(4), 245–254 (1992)
3. Thompson, P.: Compressible Fluid Dynamics. McGraw-Hill Book Company, New York (1972)
4. Liepmann, H.W., Roshko, A.: Elements of Gasdynamics. Wiley, New York (1960)
5. Glass, I.I., Sislian, J.P.: Nonstationary Flows and Shock Waves. Oxford University Press, Oxford (1994)
6. Copper, J.A.: Experimental investigation of the equilibrium interface technique. Phys. Fluids **5**, 844–849 (1962)
7. Yu, H.R.: Oxyhydrogen combustion and detonation driven shock tube. Acta. Mech. Sin. **5**(2), 97–107 (1999)
8. Olivier, H., Zonglin, J., Yu, H.R., Lu, F.K.: Detonation-driven shock tubes and tunnels. In: Lu, F., Marren, D. (eds.) Advanced Hypersonic Test Facilities. Progress in Astronautics and Aeronautics, vol. 198, pp. 135–203. AIAA Inc, USA (2002)
9. Bird, K.D., Martin, J.F., Bell, T.J.: Recent developments in the use of the hypersonic shock tunnel as a research and development facility. In: Proceedings 3rd Hypervelocity Techniques Symposium, Denver, USA (1964)
10. Mark, H.: The Interaction of a Reflected Shock Wave with the Boundary Layer in a Shock Tube. NACA TM 1418, (1958)

11. Olivier, H.: An improved method to determine free stream conditions in hypersonic facilities. Shock Waves **3**, 129–139 (1993)
12. Edenfield, E.E.: Contoured Nozzle Design and Evaluation for Hotshot Tunnels, pp. 68–369. AIAA, USA (1968)
13. Paull, A.: A simple shock-tunnel driver gas detector. In: Sturtevant, B., Shepherd, J.E., Hornung, H.G. (eds.) Proceedings 20th International Symposium on Shock Waves, pp. 1557–1562, Pasadena, USA (1995)
14. Sudani, N., Hornung, H.G.: Detection and reduction of driver gas contamination in a high-enthalpy shock-tunnel. In: Boyce, R.R. et al. (eds.) Proceedings 21st International Symposium on Shock Waves, pp. 543–548, Great Keppel Island, Australia (1997)
15. Skinner, K.A.: Mass spectrometry in shock tunnel experiments of hypersonic combustion. Ph.D thesis, University of Queensland, Australia (1994)
16. Kindl, H., Olivier, H., Zhao, H., Muylaert, J., Wong, H., Walpot, L.: Conventional flow diagnostics in shock tunnels. In: Boyce, R.R. et al. (eds.) Proceedings 21st International Symposium on Shock Waves, pp. 493–498, Great Keppel Island, Australia (1997)

Author Biography

Herbert Olivier is working on high speed flows and shock tube technology at the Shock Wave Laboratory of RWTH Aachen University. His research activities cover trans-, super- and hypersonic flows, ignition phenomena of fuels, and various topics of interdisciplinary applications of shock waves for industrial use. As editor of the Shock Waves Journal and reviewer of most important international journals he is involved in the reviewing process of scientific papers related to various fields of fluid and gas dynamics. He is member of the International Advisory Committee of the Shock Wave Symposium and of other international organizations. He received his Diploma degree in aeronautics from RWTH Aachen University in Germany and his 'Dr.-Ing'. degree from the same university. Since 1996 he is head of the Shock Wave Laboratory at RWTH Aachen University. Current interests among others are related to film cooling and Görtler vortices in super- and hypersonic flows, unsteady transonic airfoil flow, ignition phenomena in hydro-carbon fuels, development of fast thermocouples.

Shock Tunnels at ISL

Patrick Gnemmi, Julio Srulijes, Friedrich Seiler, Berthold Sauerwein, Myriam Bastide, Christian Rey, Pierre Wey, Bastien Martinez, Hermann Albers, Gunther Schlöffel, Robert Hruschka and Thibaut Gauthier

1 Introduction

Studies of flying vehicles under real flight conditions are essential when both atmospheric pressure and temperature have significant effects on their behaviors, which is the case for high-speed vehicles. Therefore, the use of shock tunnels for aerodynamic and aerothermodynamic investigations of high-speed vehicles is particularly interesting as the real flight conditions can be correctly reproduced [1]. Additionally, this can be done at very low cost compared to experiments conducted in wind tunnels which should have an air heater in order to reach the right temperature value.

P. Gnemmi (✉) · J. Srulijes · F. Seiler · B. Sauerwein · M. Bastide · C. Rey · P. Wey · B. Martinez · H. Albers · G. Schlöffel · R. Hruschka · T. Gauthier
French-German Research Institute of Saint-Louis (ISL), 68301 Saint-Louis Cedex, France
e-mail: patrick.gnemmi@isl.eu

J. Srulijes
e-mail: juliosrulijes@yahoo.com

F. Seiler
e-mail: friedrich.seiler@kit.edu

B. Sauerwein
e-mail: berthold.sauerwein@isl.eu

M. Bastide
e-mail: myriam.bastide@isl.eu

C. Rey
e-mail: christian.rey@isl.eu

P. Wey
e-mail: pierre.wey@isl.eu

O. Igra and F. Seiler (eds.), *Experimental Methods of Shock Wave Research*,
Shock Wave Science and Technology Reference Library 9,
DOI 10.1007/978-3-319-23745-9_6

The hypersonic shock-tunnel facilities of ISL and their performances are presented in the second section of this chapter. The measurement techniques used for high-speed flow investigations are shortly described in the third section. The methodologies for the design of Laval-contoured nozzles and for the nozzle-flow qualification are presented in the fourth section. The fifth section deals with the main most recent research subjects studied in the former "Aerothermodynamics & Shock-Tube' laboratory, now part of the "Aerodynamics, Measurements and Simulations" department. Studies of side-jet missile control, of free-flight trajectory measurements for aerodynamic coefficient estimation, of projectile control by plasma discharge, of heat-flux measurements at missile noses to predict heat loads on missile surfaces and structures and of the atmospheric dispersion of droplets are summarized.

2 Hypersonic Shock-Tunnel Facilities

The laboratory has two high-energy shock tubes (STA and STB) able to supply up to 8 MJ/kg to carry out high-speed flow experiments [2–4]. The shock-tube inner diameter is 100 mm and each facility is about 30 m long. During the past fifteen years, the ISL's shock tubes are mainly used as hypersonic shock tunnels located in parallel in the same laboratory (Figs. 1 and 2).

The shock tunnel is a very-short-test-time wind tunnel consisting of a shock tube connected to a supersonic nozzle, a measurement chamber and a dump tank. The shock tube itself is divided into a high-pressure driver tube and a low-pressure driven tube, as shown in Fig. 3. The STA driver tube is 3.6-m long, the STB one is 4.0-m long and the driven tube is 18.4-m long for both facilities. The driven tube is followed by the nozzle that is connected to the test section and followed by the dump tank.

Preferably a light driver gas is compressed into the driver tube, up to 450 bar. The steel membrane separating the high-pressure from the low-pressure tubes is designed to burst at a determined pressure dependent on the required experimental

B. Martinez
e-mail: bastien.martinez@isl.eu

H. Albers
e-mail: hermann.albers@isl.eu

G. Schlöffel
e-mail: gunther.schloeffel@isl.eu

R. Hruschka
e-mail: robert.hruschka@isl.eu

T. Gauthier
e-mail: thibaut.gauthier@isl.eu

Fig. 1 Picture of the ISL's shock tunnel STA

Fig. 2 Picture of the ISL's shock tunnel STB

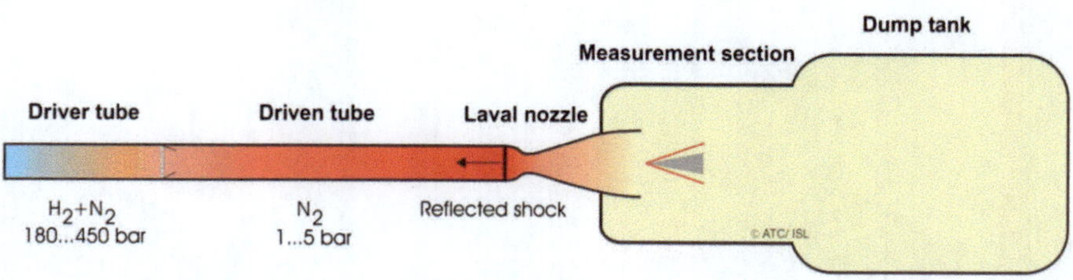

Fig. 3 Schematic of the shock tunnels of ISL

conditions. Upon the diaphragm rupture, a shock wave propagates through the driven tube where the test gas (usually nitrogen) is contained at a pressure up to a maximum of 5 bar. As a consequence, the driven test gas is compressed and heated by the incident shock wave as well as accelerated towards the end of the shock tube. Simultaneously, an expansion wave propagates in the opposite direction and is reflected on the driver-tube end. When the shock wave reaches the entrance of the nozzle, it is reflected and as a consequence the test gas is compressed and heated once again. The upstream moving reflected shock wave brings the gas at rest and it remains almost stationary for a very short time forming the stagnation conditions in front of the nozzle. Then, the driven gas expands through the nozzle, resulting in a quasi-stationary supersonic or hypersonic flow inside the test section. The test section contains a model to be studied and the shock-tube gases after the termination of the experiment. The gases are also stored inside the dump tank attached to the test section and they are evacuated into the atmosphere after the experiment ended. The dump tanks have a volume of about 10 m^3 and 20 m^3 for STA and STB, respectively.

Figure 4 shows different parts of the shock tunnels. The 1000-bar compressor compresses the hydrogen-driver-gas in a bottle of 20 liters located in a special room

1000-bar compressor room

driver-tube room of STB

device for the analysis of the gas composition

STA and STB command room

Fig. 4 Different pictures of ISL's shock tunnels

next to the driver-tube room. Before an experiment, the hydrogen is injected into the driver tube up to a maximum pressure of 450 bar. The 1000-bar compressor room and the driver-tube room are completely and extremely secured in order to avoid accident from possible hydrogen leaks. During an experiment, the staff stays in the reinforced concrete command room having about 80 cm thick walls and reinforced doors. After the driver tube is filled with a mixture of hydrogen and nitrogen, the gas mixture is analyzed in order to have its real composition needed for the experiment analysis. In 2012, all electronic components have been replaced by new ones and therefore, the software for the management of the shock tunnels has been completely re-actualized; all these modifications improve the safety of the facilities.

After each shot, the free-stream flow conditions are recalculated using a one-dimensional shock-tube/shock-tunnel code. The code requires the measured shock-wave speed in the driven tube as input. This is determined by using the temporal signals of pressure gauges (or thin film gauges) placed along the driven tube [5]. By varying the driver and driven tube pressures, the nozzle flow can be adjusted to reproduce the flow conditions present in the atmosphere. Real atmospheric flight conditions can be produced in these facilities from about ground level up to a flight altitude of 70 km depending on the Mach number *M*, as shown in Fig. 5. The experimental flow conditions, i.e. ambient pressure and temperature, are based on the US Standard Atmosphere (1976) model. Experiments reported in this paper were conducted either in the STA shock tunnel or in the STB one at various Mach numbers and diverse simulated altitudes. Nozzles having a Laval contour are available for experiments at Mach numbers of 3, 4.5, 6, 8 and 10. Divergent nozzles are used for Mach numbers of 3.5, 10, 12 and 14.

Figure 6 depicts diagrams of the Reynolds number and the flow enthalpy with respect to the flight altitude for different Mach-number flows, for experiments carried out in the ISL's shock tunnels.

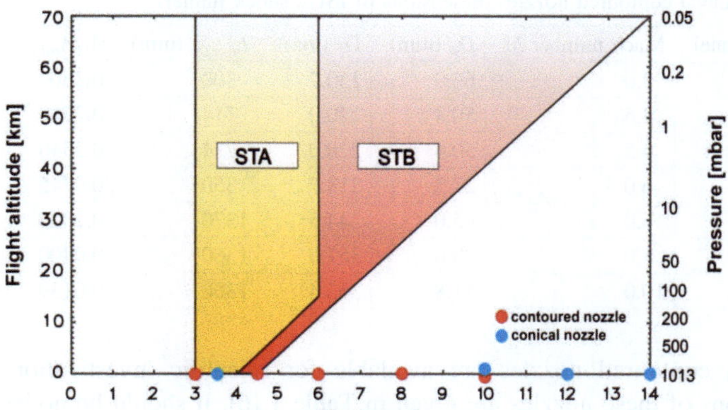

Fig. 5 Red and overlapped yellow areas representing the working range of respectively the STB and STA shock tunnels of ISL

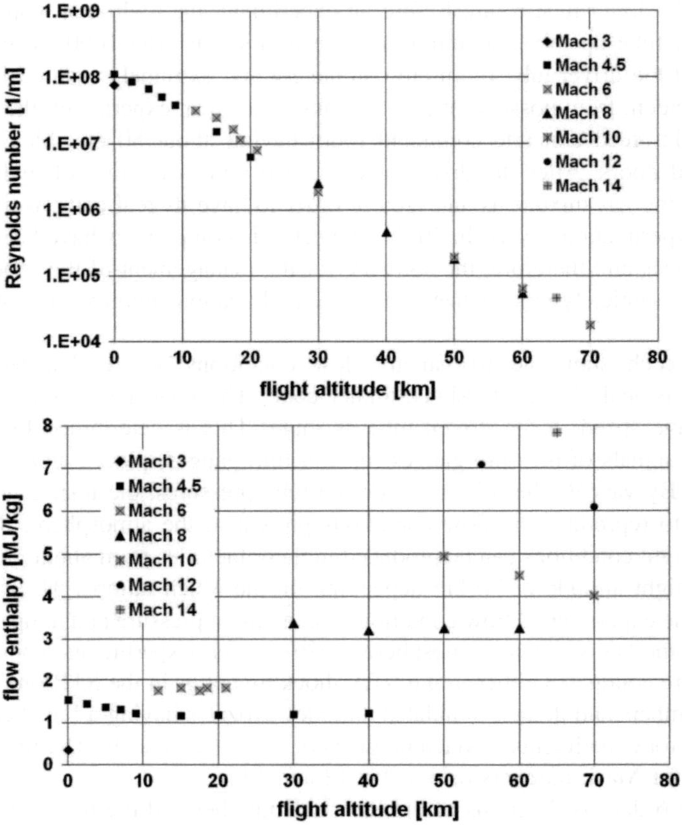

Fig. 6 Reynolds number and flow enthalpy with respect to the flight altitude for different Mach numbers

Table 1 Laval contoured nozzles dimensions of ISL's shock tunnels

Shock tunnel	Mach number M	D_0 (mm)	D_e (mm)	L_{nozzle} (mm)	A_0/A_{tube}	A_{exit}/A_0
STA	3.0	60.0	130.7	300	0.3600	4.7480
STA	4.5	50.4	220.0	714	0.2540	19.0539
STB	4.5	50.4	220.0	714	0.2540	19.0539
STA	6.0	27.3	218.7	850	0.0745	64.1759
STB	6.0	45.0	344.5	1370	0.2025	58.6075
STB	8.0	20.0	357.0	1360	0.0400	318.6225
STB	10.0	11.8	363.4	1360	0.0139	948.5356

Seven contoured nozzles are available for scientific investigations and the dimensions of these nozzles are given in Table 1 [6]. It should be noticed that for the high altitude conditions the exit Mach number is smaller than the design Mach number, since the effective area ratio decreases, due to the increase of the nozzle boundary-layer thickness. In the graphs, the data given for Mach numbers ranging from 3 to 10 are those obtained with the Laval contoured nozzles during investigations carried out in the past.

3 Measurement Techniques for High-Speed Flows

In the experiments, the model is fixed in the test section of the shock tunnel and when the flow is established around the model, it is quasi-stationary for about 1 to 4 ms and the measurements have to be carried out during this time interval.

3.1 Basic Measurements for Flow Characterization

After an experiment has been carried out, the real flow conditions are determined using the one-dimensional shock-tube/shock-tunnel code fed by systematic measurements on shock tunnels.

Figure 7 shows an example of raw data recorded during an experiment conducted in the shock tunnel STA. The study concerns the interaction of lateral jets which expand into the cross-flow of a generic missile model. The figure has two graphical zones. Starting from the top in the upper part of Fig. 7, the first purple curve depicts the synchronization signal for the activation of one lateral jet; the jet is activated before the cross-flow is produced. The second magenta curve indicates the acquisition rate of visualizations taken by a high-speed camera. Then, the third green curve shows that the horizontal lateral jet is not activated because the total pressure measured in the distribution chamber remains equal to zero. The next blue curve proves that the vertical lateral jet expands into the cross-flow at a constant total pressure of 20 bar. The yellow curve depicts the evolution of the static pressure measured at the exit of the nozzle of the shock tunnel; the pressure indicates that the experiment is carried out for conditions prevailing at about 20 km of altitude. The light blue curve proves that the cross-flow is quasi constant during at least 4 ms, which is more than sufficient for the data acquisition in that series of experiments. In the lower part of Fig. 7, curves show the pressure evolution at different locations in the shock tunnel: this leads to the following of the shock-wave propagation in the driven tube and to the determination of the shock-wave velocity. These measured data serve after the experiment ended, for the verification that the pre-set experimental conditions were really obtained.

In other experimental configurations such as free-flight tests for force and pitching moment measurements, the measurement duration of 4 ms is not long enough. Therefore, because the Mach number only depends on the nozzle geometry, it remains constant over an additional time period of about 15 ms, until the driver gas arrives into the test section. During this extended measurement time, it is necessary to know how the history of the flow conditions (e.g. velocity and density) changes at the nozzle exit. Therefore, the transient velocity change is measured with a Laser-Doppler-Velocimeter (LDV) using titanium-dioxide particles for the flow

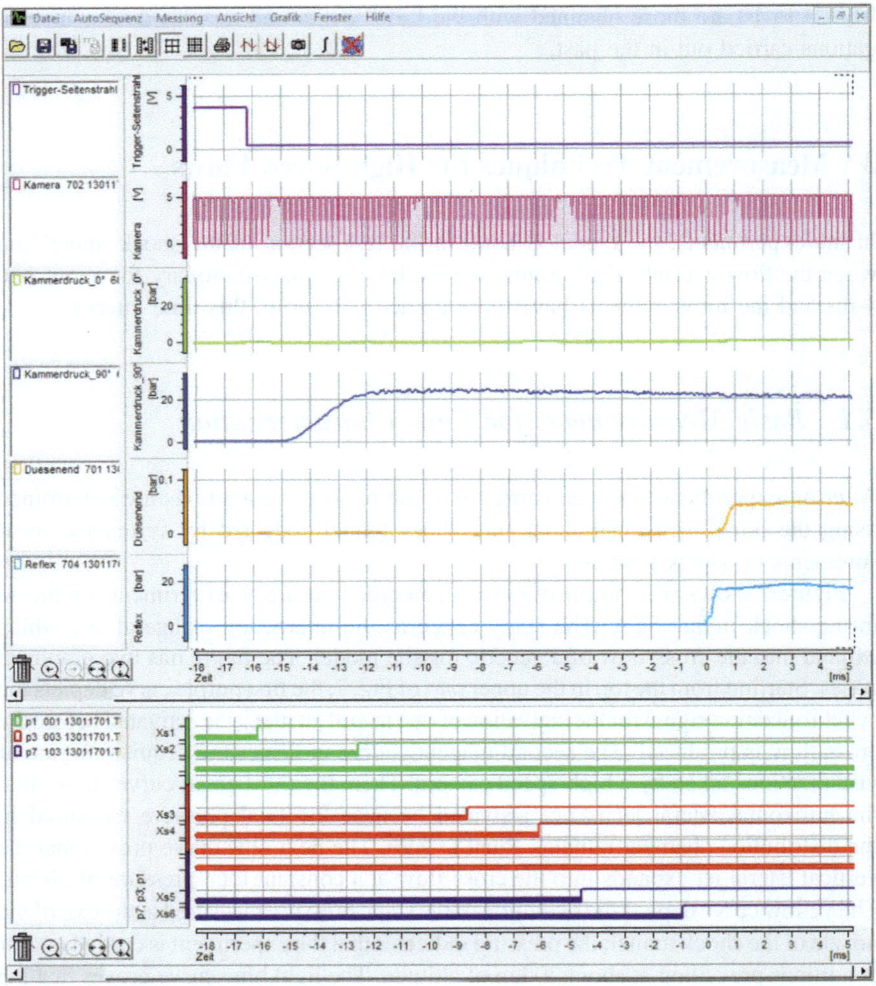

Fig. 7 Raw data recording during an experiment in the shock tunnel STA

seeding [7] (see Sect. 3.5). The density is obtained from both, the static pressure measured at the nozzle wall close to the nozzle exit and the LDV-velocity measurement at constant Mach number.

3.2 *High-Speed Visualizations*

The flow field is illuminated by a halogen or a LED lamp and observed through a differential interferometer (DI) system; the flow-field structure is then visualized and recorded by high-speed CCD cameras. The best cameras available at ISL in

2013 are Photron Fastcam SA-X which are able to record images at a rate of 12,500 frames per second having a spatial resolution of 1024 pixels by 1024 pixels (full sensor resolution) or 50,000 frames per second for a spatial resolution of 512 pixels by 384 pixels.

The DI used as a flow-visualization technique [8] is based on detecting the density-gradient prevailing in the investigated flow field in order to gather information about the flow pattern around the model. Two perpendicularly polarized widened light bundles which are separated by a Wollaston prism pass through the flow field in parallel. They are separated from each other by a certain distance (some tenths of millimeter). The light beams brought together by a second Wollaston prism create interferences and they are analyzed by a polarizer and are focused on the CCD camera by a lens system (Fig. 8).

The density changes occurring in the flow field generate different optical path lengths between the two split light bundles, producing a change in the interference pattern on the CCD camera. The DI can be adjusted to produce fringe patterns or an infinite fringe pattern. In the latter, the density-gradient field in the gas flow is visualized in terms of light-intensity distribution shown on interferogram pictures. They look like schlieren pictures, but they are produced on the basis of light interference instead of light refraction.

Figure 9 shows the arm of the DI setup located on the side of observation. From the front to the back of the picture, the parabolic mirror (1), the laptop (2) with the picture as the result of the DI technique named interferogram, the Wollaston prism (3) located in front of a mirror (4), the CCD camera (5) located behind the mirror (4) and the test section (6) inside which the model is placed, are highlighted.

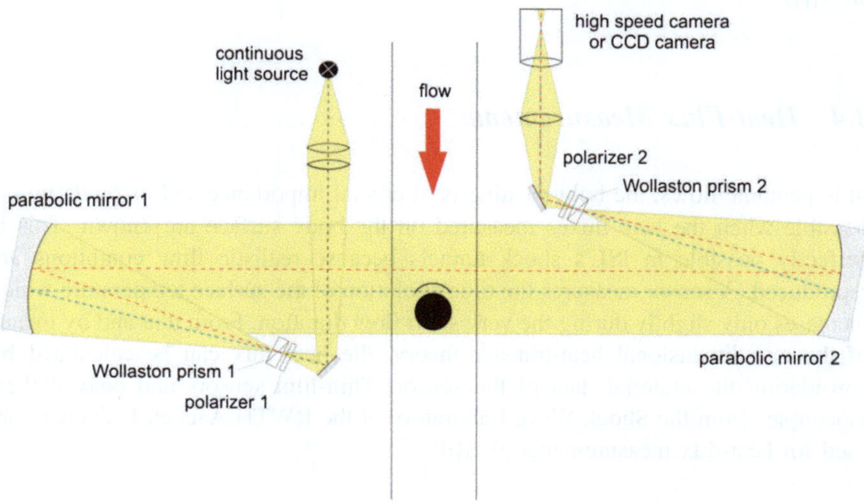

Fig. 8 Differential interferometry principle

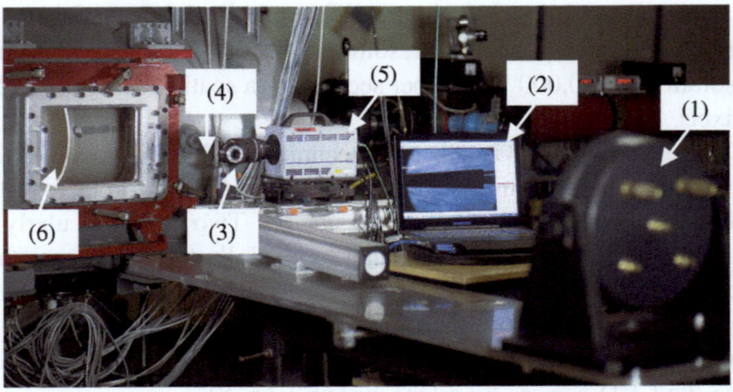

Fig. 9 Picture of a differential interferometry setup installed near the STB test section

3.3 *Surface-Pressure Measurements*

Kulite transducers manufactured by Kulite Semiconductors Products, Inc. (Leonia, USA) are used for measuring the pressure at the nozzle exit. Both total pressure (Pitot pressure) in the flow and static pressure acting on the model surface are measured. The transducer is a semi-conductor sensor based on the piezoresistive effect with passive temperature compensation. The transducer used for surface-pressure measurement has an eigen frequency of 150 kHz, which is appropriate for transient experiments. The chosen measurement range of the transducer depends on the experiment. The electrical signal from the measurement is amplified through a transient recorder TransAS distributed by MF Instruments GmbH (Alpstadt, Germany). It is a 12/14-bit A/D-converter for 100 channels at a maximum rate of 40 MHz.

3.4 *Heat-Flux Measurements*

In hypersonic flows, the body heating is of crucial importance and its prediction is possible when the heat fluxes measured on the body surface are known. This is perfectly possible in ISL's shock tunnels because realistic flow conditions are reproduced. A sensor measures the time evolution of the surface temperature which increases only slightly during the very short flow duration. From that and by means of the one-dimensional heat-transfer theory, the heat flux can be calculated by considering the material data of the sensor. Thin-film sensors and coaxial thermocouples from the Shock Wave Laboratory of the RWTH Aachen University are used for heat-flux measurements [9, 10].

3.5 Laser Doppler Velocimetry (LDV)

The LDV-technique developed at ISL [7] allows a direct velocity measurement of particles for estimating the most probable flow velocity at the nozzle exit. The shock-tunnel flow is seeded shortly before the experiment with titanium-dioxide (TiO_2) particles having a mean diameter of about 0.3 μm. An Ar^+ laser sends a monochromatic light through one window into the test section in front of the nozzle. The light is scattered and Doppler shifted by the moving tracer particles passing the measurement point. A bundle of that light is collected by an objective located behind the opposite window and transmitted through a light fiber to the LDV-velocimeter which is located further on. The velocimeter essentially consists of a phase-stabilized Michelson interferometer used as a sensitive spectrometer. An electro-optical circuit ensures the phase stabilization that results in a voltage signal independent of the scattered light intensity and proportional to the mean velocity of the particles at the measurement point. Because of the very short response time (1 μs) of the used LDV system, the latter gives a continuous real-time signal of the particle acceleration. A typical flow velocity signal—in relation with the light intensity distribution—is shown in Fig. 10. The data was measured at the Mach-3 nozzle-exit flow using an acquisition time step of 16 μs. The flow velocity is about 1100 m/s at the beginning of the flow and remains nearly constant during 3.5 ms. This technique is very important and applied when flow conditions have to be known for a length of time higher than 3.5 ms, which is the case in that application. This experiment duration can be extended up to 15 ms because the driver gas reaches the test section later on at this Mach number compared to higher Mach numbers.

3.6 Particle Image Velocimetry (PIV)

A double-frame/single-exposure digital PIV system is used for gathering the most probable velocity field around models. The light source is provided by a

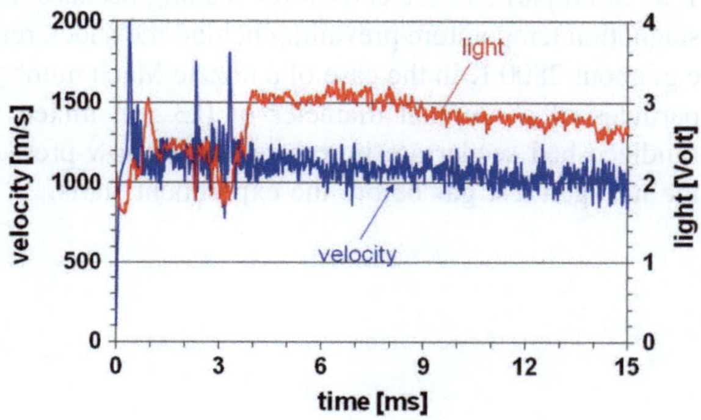

Fig. 10 Flow velocity at the nozzle exit at $M = 3$

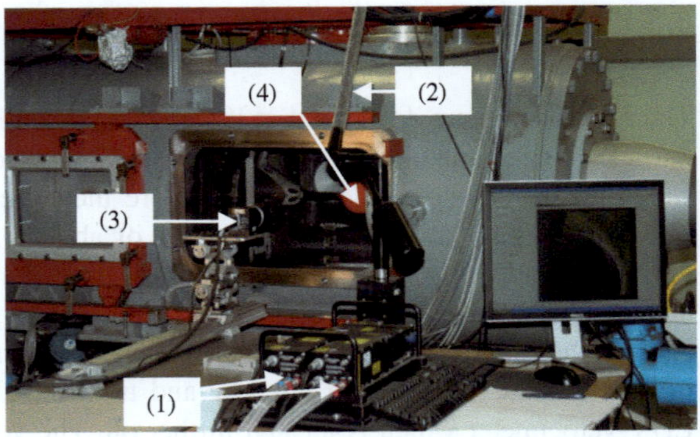

Fig. 11 PIV system installed at STB for flow-velocity measurements

frequency-doubled Nd:YAG double-pulse laser CFR Ultra 200 from Quantel (les Ullis, France). It delivers a nominal pulse energy of 200 mJ and a duration of about 5 ns per pulse. The vertical laser-light sheet illuminates the investigated flow perpendicularly to the nozzle axis; it is formed by means of a TSI light arm distributed by TSI France Inc. (Marseille, France). The use of the light arm considerably simplifies the optical setup of the PIV system.

The CCD camera is mounted on the horizontal axis to view the illuminated flow field behind the nozzle exit and can acquire two images within a pulse delay of 0.3 μs. The experiments are carried out with a PowerView Plus 4MP PIV camera distributed by TSI. Figure 11 depicts the pair of lasers (1) in the foreground, the light arm (2), the CCD camera (3) and a spherical model (4) in the shock-tunnel test section in the background.

Currently, the camera spatial resolution is of 2048 pixels by 2048 pixels. The PIV images are analyzed after the experiment by an intercorrelation algorithm included in the ISL software and described in [11, 12]. The efficiency of the process proposed in the software was demonstrated in previous studies at a Mach number of 6 [11, 13].

Nikon zoom lenses are mounted on the camera for changing the observed size of the field of view. The laser and the camera are synchronized and triggered by a heat-flux sensor placed on the shock-tube wall. The synchronizer separation time is checked by a fast-response photodiode and timing errors were found to be less than 1 %. Solid particles are chosen for seeding because they have to withstand the high stagnation temperature prevailing behind the shock reflection at the nozzle entrance, e.g. about 2800 K in the case of a nozzle Mach number of 8. For this reason, Al_2O_3 particles of a nominal diameter of 0.3 μm mixed with Aerosil dispersed by a fluidized-bed seeder are introduced in the low-pressure driven tube together with the nitrogen test gas before the experiment starts.

3.7 *Pressure Sensitive Paints (PSP)*

ISL has started the study of a new generation of Pressure-Sensitive Paints (PSP) by using nanotechnology for surface-pressure measurements on models. The nanostructure could fulfill the requirements imposed by unsteady supersonic flows: reducing the time response by increasing the paint porosity; increasing the pressure range by using a certain type of nanostructure which allows the sensitivity of the paint to oxygen to be controlled; improving the accuracy by increasing the luminescence density; solving the erosion problem by making a hard structure surface and also solving the paint thickness problem by depositing a nanoPSP having a thickness smaller than 150 nm. The shock tunnels are currently used for conducting a feasibility study and the first results presented are very encouraging [14, 15] and an application in wind tunnel is also interesting [16].

A picture of the PSP experimental setup (a typical experiment in STA) is shown in Fig. 12. The driven tube is visible in yellow and the test-section door in red. The model (1) is located at the exit of the nozzle. Two high-speed cameras (2) are placed horizontally and vertically in front of the test-section windows and the corresponding LED lamps (3) are next to these cameras. Two laptops used for recording the PSP images during the experiment are also visible.

Unfortunately, no results obtained in shock tunnels can be presented in the studies discussed in Sect. 5 due to confidentiality reasons.

3.8 *Free-Flight Force Measuring Technique (FFM)*

Extraction of aerodynamic coefficients and derivatives is the primary goal of aeroballistic range facilities which are used to observe short trajectories typically

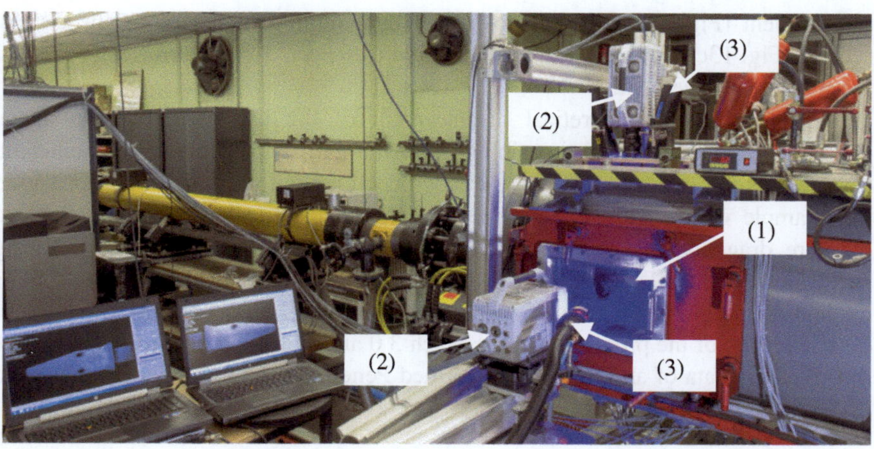

Fig. 12 Experimental setup for PSP measurements at STA

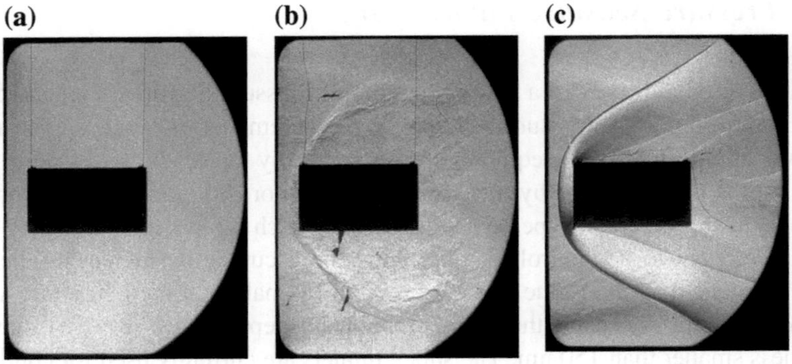

Fig. 13 Free-flight motion of a cylinder at $M = 4.5$, **a** $t = 0$, **b** $t = 1.0$ ms, **c** $t = 1.6$ ms

ranging from tens to hundreds of meters. The basic challenge addressed here can be summarized as follows: is it possible and relevant to carry out aeroballistic data reduction based on ultra-short trajectories with a flight distance of about 10 cm and a time of flight of about 10 ms? The promising solution called "Free-Flight Force Measuring Technique (FFM)" is based on shock-tunnel experiments in supersonic regime. This technique was originally described in [17] and has been further developed at ISL to observe six degrees of freedom (6 DOF) motions of rigid bodies. Three tasks are required in performing the data reduction: (1) measurement of the time-dependent flow parameters as a basis to compute a theoretical motion, (2) accurate observation of the translational and angular motions of the body and (3) estimation of the aerodynamic coefficients using a least-square fit of the measurements to predict the theoretical motion. The method also allows checking the dynamic stability of a projectile model, a missile model or other vehicles.

Before the experiment, the model is suspended in front of the nozzle, e.g. by paper disks and by threads (Fig. 13a) at a given angle of attack. After bursting of the shock-tube diaphragm, the test gas reaches the model and breaks its suspension arrangement (Fig. 13b), leading to a free-flight motion of the model in the test section (Fig. 13c). A high-speed visualization (see Sect. 3.2) is performed to record the model displacement.

By adjusting the theoretical displacement of the test model to the actual experimental trajectory, it is possible to determine the appropriate drag coefficient. As cited before, this technique has already been known for a long time and is used for example in wind tunnels and in shock tubes at quite low Mach numbers to measure drag and lift forces. Based on these previous studies, tests are being performed at the ISL shock tunnel to demonstrate the applicability of this FFM method for drag and lift measurements at higher Mach numbers [18].

The results of the process tested at Mach 3.0 and Mach 4.5 against the reference aerodynamic data of an Explosively Formed Penetrator (EFP) model are presented in Sect. 5.

4 Nozzle Design for Laval-Contoured Nozzles and Nozzle-Flow Qualification

Before carrying out experiments in which the flow field around models can be studied, the flow field generated inside the nozzles of the shock tunnels must be validated. This is of great importance in order to correctly design models and to locate them in the test-section core flow.

4.1 Laval Contoured Nozzle Design

The Mach-4.5 nozzle is the first contoured one built at ISL. The contour has been determined by the ISL's numerical code "Tuyere" [19] which is based on the method of characteristics [20] with a correction for the existing boundary layer thickness [21]. The diameter of the nozzle throat has been set to equal half of the driven-tube diameter (that is 50 mm) in order that a pronounced reflection of the incident shock wave reaches the driven tube end wall. The nozzle inlet conditions have been calculated using the ISL one-dimensional shock-tube/shock-tunnel code "Stklavga" [22] which is also able to account for real gas effects. The nozzle-exit diameter is determined using Eq. (1) [23] by assuming a constant ratio of the specific-heat capacities $\gamma = Cp/Cv$. In Eq. (1), A is the nozzle-exit area and $A*$ the nozzle-throat area.

$$\frac{A}{A^*} = \frac{1}{M^*}\left(1 + \frac{\gamma - 1}{\gamma + 1}\left(M^2 - 1\right)\right)^{\frac{\gamma+1}{2(\gamma-1)}} \tag{1}$$

The code has been set so that a Mach number M of 4.5 is established at the nozzle exit with the given area ratio $A/A*$ at the desired nozzle length of about 700 mm. It should be noticed that the contour at the outlet of the nozzle is still diverging. This gives the boundary layer the possibility to grow without compromising the core flow. The contoured nozzles for Mach numbers of 3 and 6 have also been design following the same process.

A main geometrical physical constraint exists for the design of a contoured nozzle for the shock tunnel STB. As an example, the nozzle length should not exceed 1.36 m leading to a diameter of 0.36 m in order to achieve a Mach number of 8 at the nozzle exit. Therefore the "Tuyere" code has been adjusted in order to design this nozzle. The nozzle flow was also calculated using CFD (Computational Fluid Dynamics). The axisymmetric Reynolds-Averaged Navier-Stokes simulation using a Goldberg *Rt* turbulence model resulted in a nozzle-exit Mach number of 7.5. The difference between the analytical solution and the numerical one is attributed to the assumption of constant specific heat in "Tuyere", whereas it is variable in the CFD. At the end, the nozzle contour has been adjusted to a larger exit diameter to achieve the desired Mach number of 8.

The last contoured nozzle for achieving a flow Mach number of 10 was designed on the basis of the contour given by DLR (German Aerospace Center) in Göttingen and scaled down to the physical limits of STB. After a scaling process, the nozzle flow has been verified by means of CFD investigations which are briefly presented in the next section.

4.2 *Nozzle-Flow Qualification*

After the installation of the nozzle in one of the shock-tunnels, a series of experiments have been conducted to determine the real nozzle flow. A radial Pitot-pressure distribution is measured at different positions along the flow axis by means of a rake for different flow conditions. The velocity field is also determined by means of PIV; currently, this measurement is possible up to a Mach-number flow of 8. All measurements are then compared with the one-dimensional shock-tube/shock-tunnel code "Stklavga" for the nozzle-flow qualification.

The first example is the qualification of the Mach number 8 Laval-contoured nozzle. Among four altitudes ranging from 30 to 60 km, the chosen shock-tunnel conditions are those of the atmospheric altitude near 30 km (Fig. 6). The Pitot-pressure measurements are carried out by using a rake equipped with 11 Kulite piezoresistive pressure transducers (Fig. 14, left). During the experiments the flow field around the Pitot-pressure rake is visualized by grey patterns obtained while using differential interferometry ensuring the visualization of the density-gradient field (Fig. 14, right). The detailed description of the experiments and the complete set of results are reported in [24].

Fig. 14 Rake equipped with Pitot-pressure transducers (*left*) in front of the Mach number 8 Laval-contoured nozzle and interferogram picture of the flow around the transducers (*right*)

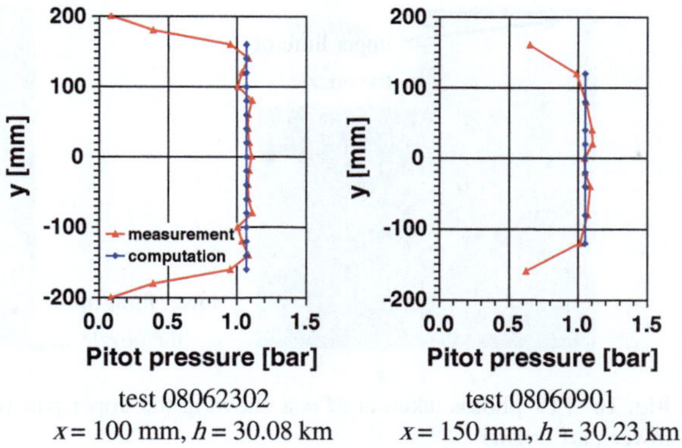

Fig. 15 Pitot-pressure measurement profiles at two different distances from the nozzle exit, $M = 8$

Figure 15 depicts the Pitot-pressure profiles measured and computed at two distances x from the nozzle exit. The calculated values are determined by the shock-tunnel measurements and the "Stklavga" code. The Pitot pressure is measured with an accuracy of about ±5 % compared to the calculated value. The measurements point out that the flow core has a diameter of about 260 mm.

The shock-tunnel conditions are equal to those of the atmosphere at an altitude near 30 km for PIV investigations. The Mach number 8 nozzle flow field is recorded with a field of view of 380 mm by 140 mm by using a camera having a focal length of 105 mm and an aperture of 5.6. Taking into account the pulse delay of 0.3 μs and the optical calibration factor of 115 μm per pixel, a maximum particle displacement of about 6 pixels is obtained for a velocity of 2500 m/s. The pair of images is analyzed using ISL's PIV software with a correlation window of 64 pixels by 64 pixels and a grid step of 16 pixels, which led to a spatial resolution of 1.84 mm by 1.84 mm. This large correlation window is selected because the particle displacement is quite long.

The qualification of the flow field in front of the nozzle is performed by carrying out two different tests: the first one consisted of the PIV measurement of the upper half part of the nozzle flow field and the second one dealt with the PIV measurement of the lower one. The measurements are carried out in such a way that an overlapping of the maps occurred. On the left hand side of Fig. 16, the PIV photo of the upper half part of the nozzle flow field shows a fairly homogeneous seeding density except for the upper region of the nozzle. On the right hand side of the figure, the PIV picture of the lower half part of the nozzle flow field shows a much more homogeneous seeding density, particularly in the flow-field core.

Figure 17 shows the obtained results for the longitudinal and transverse components of the velocity in front of the Mach number 8 nozzle. The detailed measurement results show a homogeneous flow in the nozzle core, which has a diameter of nearly 260 mm, corresponding to the size found when using Pitot-pressure measurements. The flow is parallel in that core and the average values of the velocity components as well as their mean standard deviations are calculated. The average horizontal and vertical components in the flow core are 2481 ± 35 m/s and

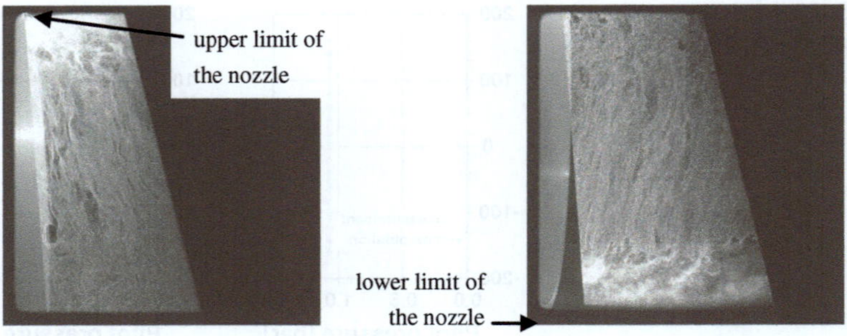

Fig. 16 PIV photos taken at $M = 8$ showing the upper part (*left*) of the nozzle flow field and its lower part (*right*)

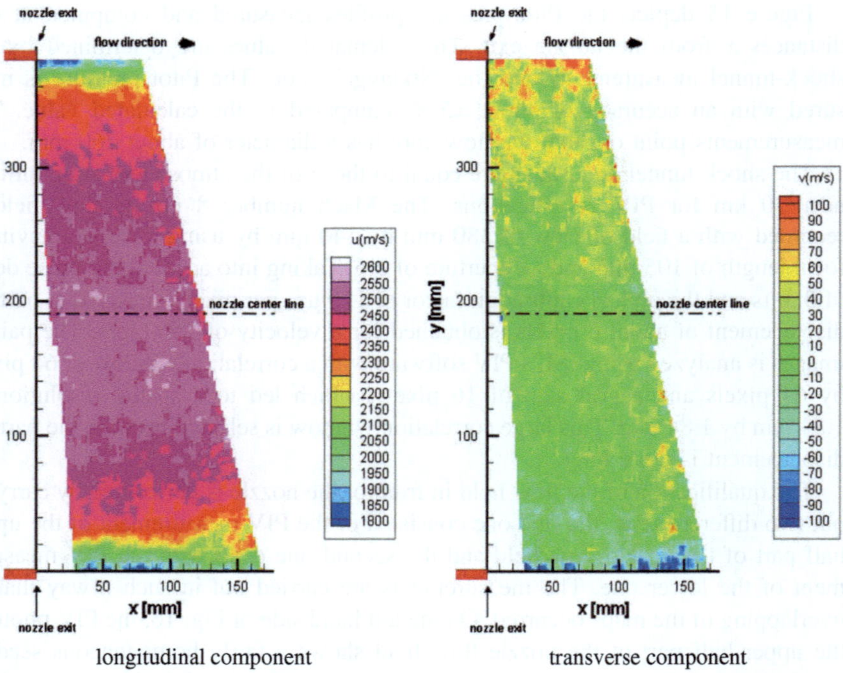

Fig. 17 Tests 08100101 and 08093002: velocity field in front of the Mach number 8 Laval-contoured nozzle

28 ± 15 m/s respectively, in the upper part. They are equal to 2510 ± 31 m/s and 13 ± 13 m/s respectively, in the lower part. In the flow core, the vertical component should theoretically be zero; the very small amplitude of that component compared to the horizontal one proves that the flow is practically parallel with an accuracy better than 1°.

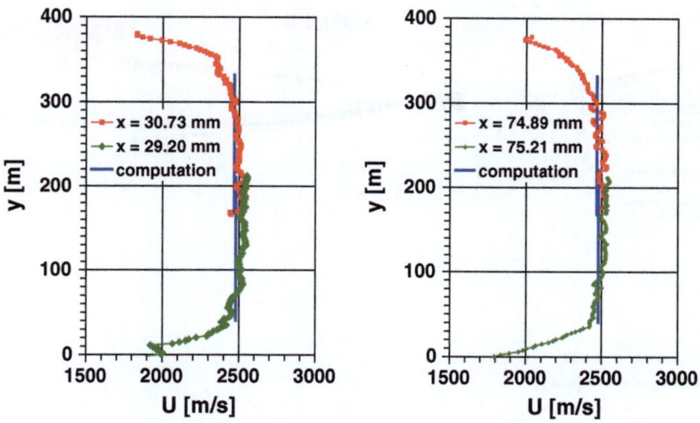

Fig. 18 Tests 08100101 and 08093002: velocity profiles in front of the Mach number 8 Laval-contoured nozzle

Figure 18 presents the horizontal velocity profiles measured and computed at different distances from the nozzle exit. The calculated value is also determined by using the shock-tunnel measurements and the "Stklavga" code. The velocity is measured with an accuracy better than ±2 % compared to the calculated value.

The second example is the qualification of the Mach number 10 Laval-contoured nozzle. In order to complete the knowledge of the nozzle flow field, axisymmetric numerical simulations of conditions prevailing at 70-km altitude are conducted. The inlet conditions are calculated on the basis of the initial fill conditions of the shock tube, the shock speed and the measured pressure at the end of the shock tube. All simulations are run with second order spatial discretization and a nozzle wall at 300 K. The simulation has to cope with two special challenges of this flow field: firstly, the flow is turbulent in the upstream parts of the nozzle and becomes again laminar during the expansion. The calculation of this configuration would be difficult, as turbulence intensities at the nozzle inlet are not known. Instead, the flow field is calculated as being fully laminar as well as fully turbulent, using the *k-ω-SST* turbulence model in a Reynolds-Averaged Navier Stokes simulation. Secondly, the rapid expansion causes the vibrational energy of the nitrogen molecules to "freeze" at a high value not far downstream of the nozzle throat. The vibrational energy is hence no longer available to drive the expansion, which leads to an approximately 8 % lower exit velocity and about 15 % higher density compared with the thermal equilibrium case. The thermal non-equilibrium flow is simulated by the Landau-Teller two-temperature model [25] implemented by means of user-defined functions into the FLUENT code which is distributed by ANSYS, Inc. (Canonsburg, USA).

As a result, the vibrational temperature at the nozzle exit is only 150 K lower than the nozzle supply stagnation temperature of 3300 K determined with the "Stklavga" code. The radial distribution of the Pitot pressure 50 mm downstream of the nozzle exit is shown in Fig. 19 left. While the turbulent simulation fails to

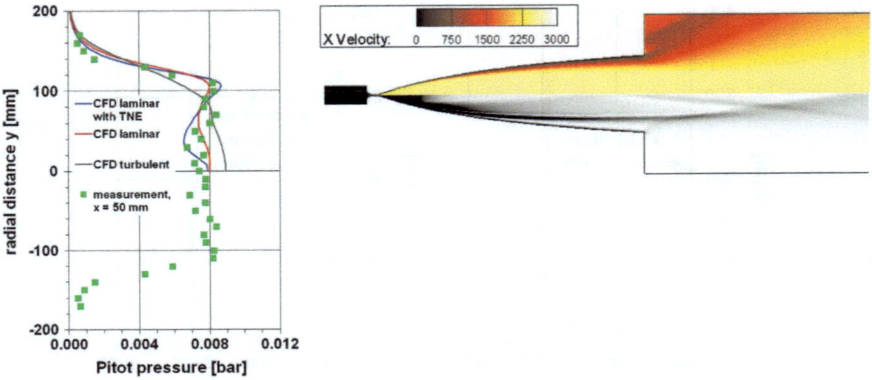

Fig. 19 Distribution of Pitot pressure 50 mm downstream of the Mach number 10 contoured nozzle (on the *left*); CFD velocity field (*top*) and Schlieren picture (*bottom*) of the Mach number 10 contoured nozzle for 70-km altitude (on the *right*)

predict the boundary-layer profile and the pressure at the centerline of the flow core, both laminar equilibrium as well as non-equilibrium (TNE) simulations agree reasonably with the measurements. However, as mentioned before, larger discrepancies occur for other flow parameters. On the right of Fig. 19 the simulated velocity field is shown in the upper half of the nozzle-exit flow and a numerical Schlieren picture is presented in its lower half.

A weak oblique shock is visible close to the centerline, which explains the small pressure maximum at the centerline. The TNE simulation provides a centerline-exit velocity of 2560 m/s, a density of 1.16×10^{-4} g/m^3 and a (trans-rotational) temperature of 160 K, which results in a Mach number of 9.9.

5 Most Recent Research Subjects

5.1 Missile Control by Side Jets

The side or lateral jet control is an important function for the guidance of existing and future missiles, especially if rapid trajectory changes are required at low dynamic pressures. At high altitudes, the side jet technology has considerable advantages compared to conventional guidance techniques, e.g. flaps or flexible fins and rudders. The interaction between a side jet and an external flow is extremely complex and highly nonlinear; this largely influences the control efficiency of the missile. Therefore, these aerodynamic effects must be known before predicting the behavior of a missile [26, 27].

In case the jet thruster is switched on, the gases blow out of the thruster and cause an interference with the cross-flow around the missile. The complex flow field developed by the interaction of the laterally blowing gas jet with the supersonic

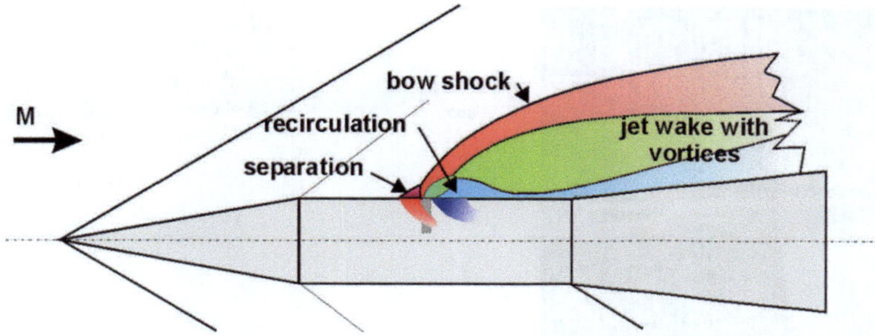

Fig. 20 General sketch of the interaction of a lateral jet with a supersonic cross-flow

cross-flow is sketched in Fig. 20. The resulting jet plume is deflected and acts as a massive obstacle located on the surface of the missile. Thus, a strong bow shock occurs in the supersonic flow in front of the jet obstacle. A separation zone develops upstream of the jet, as the boundary-layer separation causes a forward-facing separation shock [28]. In this separation zone, the pressure acting on the body surface becomes higher than the static cross-flow pressure, whereas it decreases in the recirculation zone formed downstream of the lateral jet. As a consequence, the resulting forces and moments exerted upon the missile are influenced by the interaction of the jet flow with the cross-flow. Indeed, while in some specific orientations and flow conditions, the interaction effects have been found to amplify the lateral force and the moment acting on the missile. Under a number of other flow and orientation conditions, the jet may cause adverse effects like a deamplification of the lateral force or of the moment. Finally, for a given missile design, the general question to be addressed is the amplitude of these resulting forces and moments in order to quantify the efficiency of the side-jet control system.

Since about a decade, many experiments involving different missile geometries, cross-flow conditions, thruster shapes and thruster-working conditions were done mostly in the STB shock tunnel. These investigations were carried out mainly in cooperation with MBDA-Germany and MBDA-France. Some of the experimental results allowed the validation of MBDA-Germany numerical computations to be done [29, 30].

As an example, detailed experimental investigations were conducted on a generic kill vehicle (KV) equipped with a lateral thrust system for evaluating the effects of the interaction between multiple side jets and the cross-flow of a missile [31–33]. For this purpose, experiments at Mach numbers of 6, 8 and 10 were carried out with a model adapted to the STB shock tunnel. On the left hand side of Fig. 21, the model having five thruster nozzles is located in front of the Mach number 10 nozzle. On the right hand side of Fig. 21, the geometry with the numbering of thruster nozzles is shown. 30 pressure transducers were embedded longitudinally in the missile model and marked by blue points on the sketch of Fig. 21. Thirteen transducers were aligned along the meridian passing through thrusters designated as

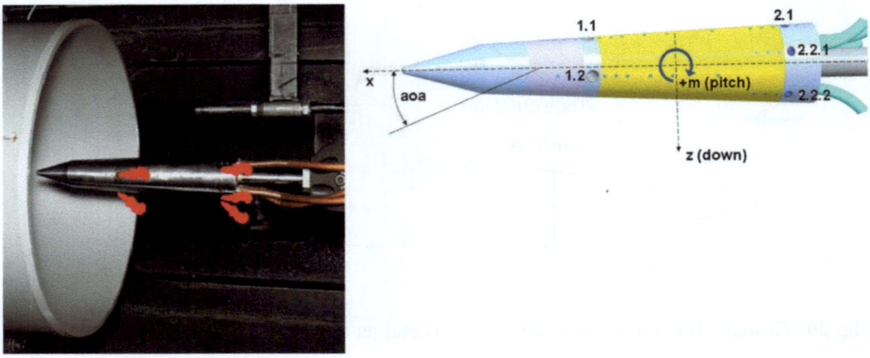

Fig. 21 Generic missile model located in STB (*left*); numbering of five thruster nozzles sketch of the interaction of a lateral jet with a supersonic cross-flow

Nos. 1.1 and 2.1. Four, nine and four transducers were distributed along meridians passing through thruster Nos. 2.2.1, 1.2 and 2.2.2, respectively.

Among more than 130 experiments carried out in the year 2012, four were conducted at a Mach number of 10 without any angle of attack and for 50-km altitude conditions. The experimental mean values of the cross-flow are as follows: static pressure of 8240 Pa and static temperature of 259.4 K leading to a density of 0.00107 kg/m^3, a velocity of 3057.6 m/s and a Mach number of 9.32.

The uncertainties in these values are less than 5 %. The lateral jet is produced by nitrogen at stagnation temperature corresponding to the laboratory temperature which is 298 K.

Figure 22 depicts the pressure distribution along the four meridians for three identical experiments without jet. The figure shows both the transducer locations on the model and the prevailing pressure at these locations. The abscissa indicates the transducer location and the ordinate represents the surface pressure. The pressure variations are due to measurement errors and to the presence of the thruster holes; the accuracy in the pressure measurements is less than 1 %.

Figure 23 presents the same kind of pressure distribution along the four meridians for three identical experiments in a configuration in which only the 1.1 jet is activated near the middle of the model. A mean value of 100.9 is obtained for the ratio R_{0J} which corresponds to the jet stagnation pressure over the cross-flow static pressure. The interaction of the lateral jet with the cross-flow produces an over-pressure on the transducer located in front of the jet exit indicating the boundary-layer separation due to the adverse jet pressure. It generates also an under pressure behind the jet corresponding to the recirculation region and the pressure increase corresponds to the shock reattachment. That overpressure region extends up to 90° because it is visible on transducers located along the 2.1-thruster meridian.

Figure 24 shows the configuration in which both jets 1.1 and 1.2 located forward in the same longitudinal section of the model are activated. Both jets are supplied

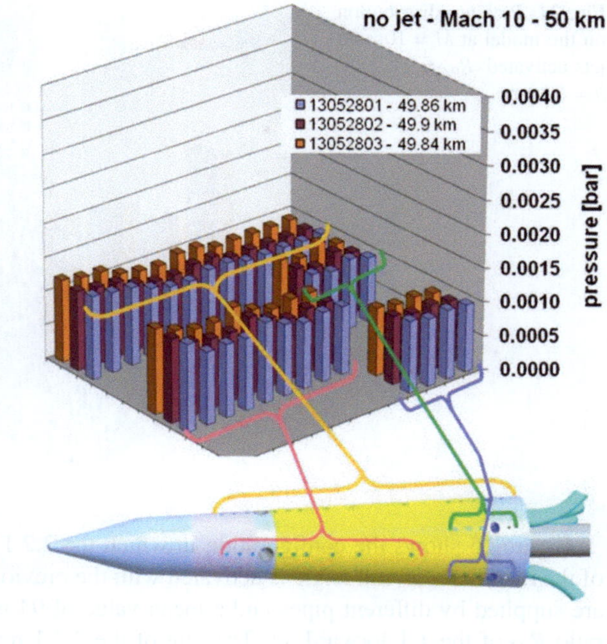

Fig. 22 Pressure distribution on the model at $M = 10$; without jet; $h = 50$ km

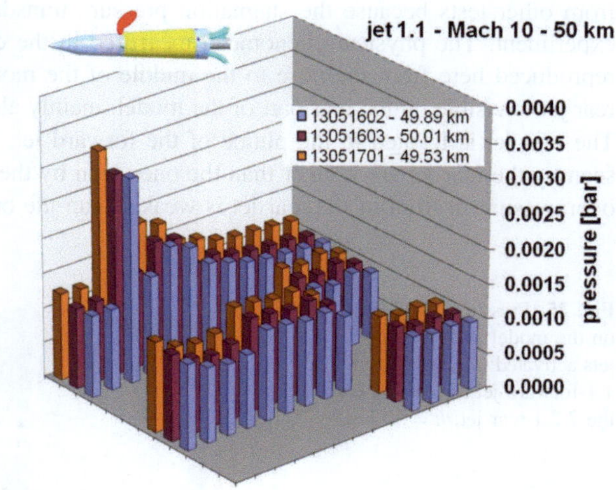

Fig. 23 Pressure distribution on the model at $M = 10$; one jet activated: $R_{0J} = 100.9$; $h = 50$ km

by the same pipe and a mean value of 93 is obtained for the pressure ratio R_{0J}. The physical phenomena measured in the case of the isolated jet are reproduced here for both jets. The overpressure in front of the isolated jet is about the same in the case of the jet pair; there is no significant interaction between the two jets. The under pressure behind the jets increases a little bit. Both configurations of the isolated jet and the jet pairs located in the same longitudinal model section produce a pitching moment whose amplitude depends on the pressure balance.

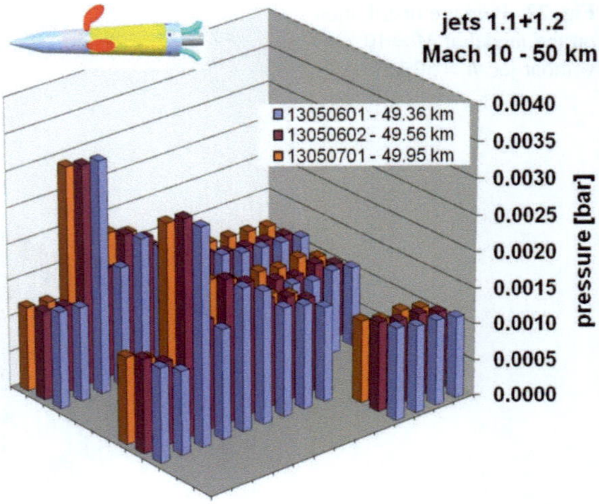

Fig. 24 Pressure distribution on the model at $M = 10$; two jets activated: $R_{0J} = 93$; $h = 50$ km

Figure 25 shows the configuration in which the 2.2.1 jet located at the rear part of the model at 45°-roll angle is activated with the previous isolated jet 1.1. The jets are supplied by different pipes and a mean value of 94 is obtained for the pressure ratio R_{0J} of the 1.1 forward jet. The one of the 2.2.1 rear part is evaluated to 134 from other tests because the stagnation pressure transducer was defective in this experiment. The physical phenomena measured in the case of the isolated jet are reproduced here from the nose to the middle of the model. The interaction of the rear jet is visible on the rear part of the model, mainly along the 2.2.1-jet meridian. The rear jet is located in the plume of the forward jet. The cross-flow conditions seen by the rear jet are weaker than the ones seen by the forward jet, therefore, the overpressure in front of the rear jet is weaker than the one of the forward jet. This

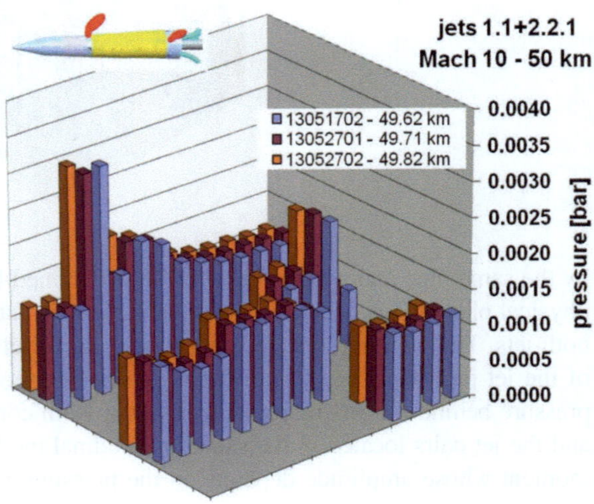

Fig. 25 Pressure distribution on the model at $M = 10$; two jets activated: $R_{0J} = 94$ for the 1.1-forward jet, $R_{0J} = 134$ for the 2.2.1-rear jet; $h = 50$ km

last configuration of the jets located in different longitudinal model sections and different azimuth sections produces a pitching moment and a roll moment which amplitudes depend on the pressure balance.

The experimental results with lateral jet interactions are measured with a mean uncertainty smaller than 3.5 %. The maximal uncertainty which is smaller than 9 % are located in the overpressure region and they are due to the unsteadiness of the boundary-layer separation.

5.2 Aerodynamic Coefficients from Free-Flight Trajectory Measurements

As presented in Sect. 3.8, the FFM technique allows the determination of aerodynamic coefficients of a rigid body. This technique applied to the Explosively Formed Penetrator (EFP) detailed in [34] is resumed in the present section as an example.

A series of experiments was conducted in the shock tunnel STA by using a very light model of the EFP. This model was chosen for the following reasons: the supersonic regime of the shock-tunnel flow, the small size of the full scale model, the aerodynamic stability of the model [35] and its manufacturing simplicity. The reference model is a 12.8-mm caliber EFP. It is composed of an ogive, a cylindrical part, a flare having a conical angle of 17° and a second one with a conical angle of 40.8°. Two models were manufactured in order to have different locations of their center of mass: the first one was made of steel (model 1) while the second one consisted of a steel body with a tungsten nose screwed on the body (model 2). The full aerodynamic data of this supersonic projectile was provided for Mach numbers 3.2 to 5.5 using free-flight analyses, wind tunnel measurements and CFD codes [36, 37].

As mentioned in Sect. 3.8, before the experiment the model is suspended in front of the nozzle of the shock tunnel, e.g. by paper disks and by threads at a given angle of attack (Fig. 34 in the next section). After the membranes of the shock tube burst, the airflow is accelerated up to the desired pressure, temperature and flight velocity, leading to the mean values given in Table 2. The threads and paper disks break and the projectile model can fly freely in the test section. The time history of the dynamic pressure is recorded in order to later compute the theoretical motion of the model.

During the experiment the nozzle flow velocity is measured during 15 ms as explained in Sect. 3.5: it is recalled that the extended measurement duration is

Table 2 Nozzle-flow conditions

M	Ps (kPa)	ρs (kg/m^3)	Ts (K)
3.06	108	1.26	290
4.49	86	1.01	286

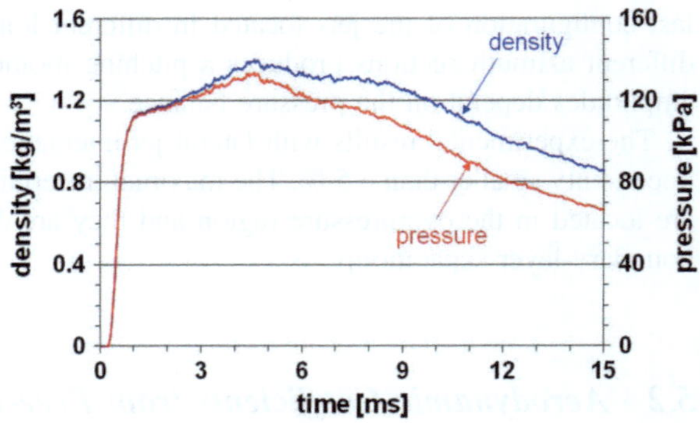

Fig. 26 Nozzle-wall pressure and nozzle-flow density for experiments at $M = 3$

possible at that Mach number. A typical flow velocity signal obtained in one of the experiments is shown in Fig. 10. Considering a fixed Mach number, the nozzle-flow density can be directly computed using the velocity and the wall pressure measured at the nozzle exit (Fig. 26). The density ranges from 1.17 to 0.85 kg/m^3, passing through a maximum value of 1.35 kg/m^3.

As the main interest focuses on the projectile-model behavior, movies are recorded in order to measure the projectile displacement at a rate of 12,500 frames per second with an exposure time of 2 μs. Two sets of pictures are taken during each experiment: one series of pictures of the displacement in the horizontal plane and one series in the vertical plane. The image of the model is sharply focused onto the camera using a uniform illumination of the whole field of view in order to increase the contrast level between the model and the background.

Figure 27 presents a sequence of four pictures extracted from the high-speed movie that describes the downstream motion of the steel-EFP model observed in the horizontal plane at Mach number 3 without initial angle of attack α_0. Figure 28 shows the same sequence for an initial angle of attack of 3°. Both sequences correspond to the same testing time interval (8 ms) and prove the perfect flight stability of the models. The observed displacement is 73.1 mm for $\alpha_0 = 0°$ and 75.1 mm for $\alpha_0 = 3°$. This slight difference results from changes in the initial conditions (including the support removal) and in the total drag due to the angular motion.

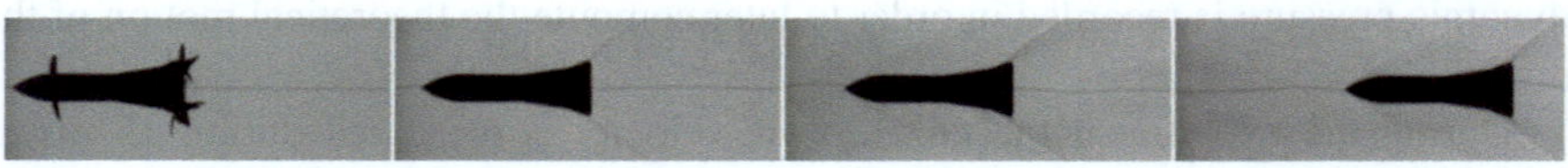

Fig. 27 Downstream motion of the EFP model made of steel, $M = 3$, $\alpha_0 = 0$

Fig. 28 Downstream motion of the EFP model made of steel, $M = 3$, $\alpha_0 = 3°$

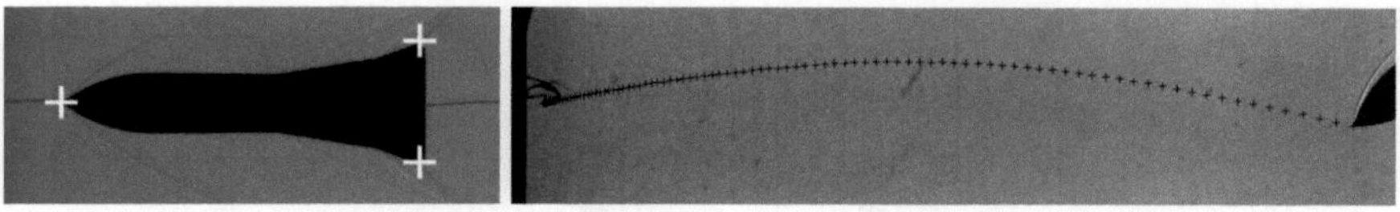

Fig. 29 Reference points (left); trajectory of the nose tip, $M = 3$, $\alpha_0 = 6°$ (right)

The image processing and the trajectory tracking follow the experiments. The goal of the image-processing step is to track the trajectory of at least three reference points attached to the model. Considering the model shape, the reference points are the nose tip and the edges of the flare as shown in Fig. 29 left. Data is extracted according to the Harris method [38] which is based on the local contrast detection. This algorithm tests the local presence of an edge or a corner. At the beginning of the process, a first guess is given by the user with a mouse click and at the end the process gives the trajectory of each reference points. Figure 29 displays the trajectory of the nose tip for an initial angle of attack of 6°, as an example.

The trajectories of three reference points allow the determination of the trajectory of the center of mass and of the angular motion of the model. Besides, since the motion is observed with two orthogonal video systems, this technique is currently being improved in order to get the full 6 DOF motion of the model. The aerodynamic coefficients related to the drag force and the overturning moment (pitching and pitch damping motions) are extracted from these data. By knowing these coefficients for both types of models, the lift-force coefficient and the static margin are easily determined using the value of the mass-center shift between the models. The coefficients are estimated by comparing the theoretical motion of the model with the observed data. In each case, the theoretical motion is computed with a second order Runge-Kutta integration of the equations of motion using the measured time history of the dynamic pressure.

Data extraction of the total drag-force coefficient is carried out for both models at Mach numbers 3 and 4.5, using observed motions with an initial angle of attack ranging from 0° to 6°. The testing time ranged from 8 to 16 ms. Table 3 and Fig. 30 summarize the results in which M states for the Mach number. The extracted results (Ext. C_D) are compared to the reference zero yaw-drag coefficient (Ref. C_{D0}), neglecting the influence of the quadratic yaw-drag coefficient. This is justified because the resulting increase in the total drag force coefficient varies from 0.25 % ($M = 3$, $\alpha_0 = 3°$) to 1.26 % ($M = 4.5$, $\alpha_0 = 6°$).

Table 3 Drag-force coefficient

M	Ref. C_{D0}	EFP	α_0 (°)	Ext. C_D
3.0	1.35	Steel	6	1.34
		St/T	3	1.34
4.5	1.08	Steel	0	1.18
		Steel	6	1.20
		St/T	6	1.17

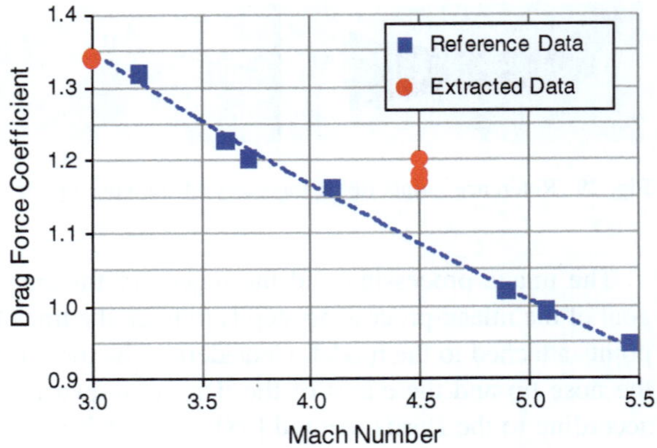

Fig. 30 Drag-force coefficient

The data compares extremely well at Mach number 3 with a discrepancy of less than 1 %. The comparison is less satisfying at Mach number 4.5 with discrepancies ranging from 8 to 11 %. The influence of the Reynolds number is firstly investigated and CFD results clearly show no significant influence of the Reynolds number on the drag coefficient. The overall dispersion in the measurements is also investigated and it comes out that the level of precision in the drag-coefficient estimation at Mach number 4.5 is poor. Therefore, further investigations are to be carried out in order to check the dependency between the Mach number and the magnitude of the measurement errors.

Pitching coefficients are determined starting with equations describing the angular motion of the model. These equations are integrated with the following initial conditions: $\omega_0 = 0$ and α_0 chosen in the neighborhood of the observed value, ω being the angular velocity of the model. In this case, the extracted parameters are α_0, $C_{m\alpha 0}$, $C_{m\alpha 2}$ and C_{mq}. $C_{m\alpha 0}$ is the linear pitching-moment coefficient, $C_{m\alpha 2}$ is the cubic pitching-moment coefficient and C_{mq} is the pitch-damping moment coefficient. The curve fitting method consists in cycling through a predefined set of parameter values and to select the solution that minimizes the sum of the squared errors.

Two data extractions are carried out at Mach number 3 (model 1, $\alpha_0 = 6°$ and model 2, $\alpha_0 = 3°$) and three at Mach number 4.5 (steel model, $\alpha_0 = 4°$ and 6° and steel-tungsten model, $\alpha_0 = 6°$). Table 4 summarizes the results and Fig. 31 displays the reference and extracted values of $C_{m\alpha 0}$. The dashed blue lines represent regressions of reference data. Note that for model 2, the reference value of $C_{m\alpha 0}$ is only an estimate that was obtained by shifting the center of mass of the steel model and using the reference values of the static margin and lift force coefficient. For model 2, no reference values are estimated for $C_{m\alpha 2}$ and C_{mq}.

The overall discrepancy ranges from 6.7 to 25 % for $C_{m\alpha 0}$ and $C_{m\alpha 2}$. Regarding the value of the pitch-damping coefficient C_{mq}, the extracted data is consistent with the reference data (discrepancy between 20 and 40 %) which is a fairly good result considering the very brief testing time limited to one half of the angular motion cycle. The best result of model 1 is obtained at Mach number 4.5 when $\alpha_0 = 6°$: the

Table 4 Pitching moment and pitch damping moment coefficients

M	EFP	Reference data				Extracted data		
		$C_{m\alpha 0}$	$C_{m\alpha 2}$	C_{mq}	α_0 (°)	$C_{m\alpha 0}$	$C_{m\alpha 2}$	C_{mq}
3.0	Steel	−1.00	−40	−102	6.00	−1.25	–	–
	St/T	−6.60	–	–	2.70	−5.30	−100	−200
4.5	Steel	−2.25	−150	−107	4.20	−2.80	−110	−150
	Steel				6.40	−2.10	−140	−130
	St/T	−6.60	–	–	5.90	−7.30	−100	−400

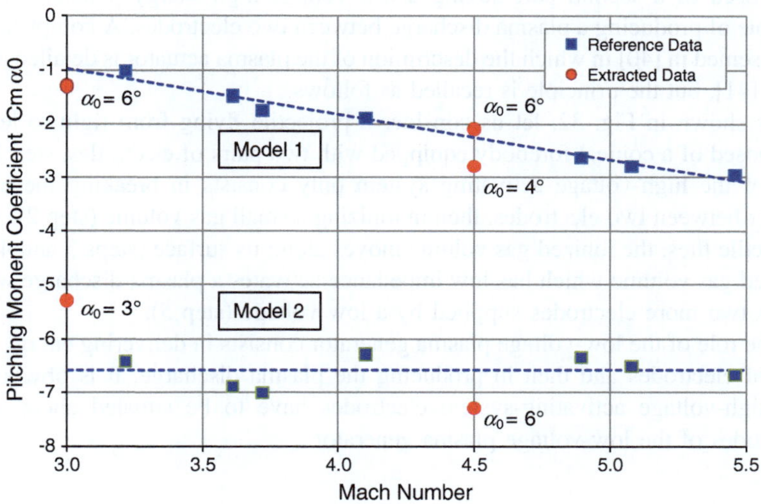

Fig. 31 Linear pitching moment coefficient

extracted values of $C_{m\alpha 0}$ and $C_{m\alpha 2}$ compare very well with the reference data since the discrepancy is only 6.7 %. The discrepancy increases up to 25 % when $\alpha_0 = 4°$ mainly because the noise of the observed angular data increases rapidly when the angular magnitude decreases. At Mach number 3, the discrepancy is also of 25 % for $C_{m\alpha 0}$: this may be due to the low stability conditions that are very sensitive to any kind of perturbations. Regarding model 2, the discrepancy ranges from 10.6 % at Mach number 4.5 to 19.7 % at number Mach 3. In this case, the interpretation of the results is more complex. As a matter of fact, the total error can be explained by the transformation of the reference data and by the manufacturing process of the model (aerodynamic or inertial asymmetries can be created by the insertion of the tungsten nose). This point militates in favor of using real firing data as reference data.

5.3 Projectile Control by Plasma Discharge

The generation of a plasma discharge on a projectile surface is a means of producing a pressure imbalance in order to divert a projectile from its initial trajectory [39]. Thus, some experiments dealing with the steering of a supersonic projectile by using a plasma actuator are under study in the STA shock tunnel. The plasma discharge is produced by a low-voltage generator capable of delivering a power-controlled electric discharge between the electrodes flush with the projectile surface. The plasma-discharge actuator is composed of a first part having a high-voltage low-energy activating system which is able to break the dielectric barrier. It is composed of a second part having a low-voltage high-energy plasma generator capable of producing a plasma discharge between two electrodes. A complete study is presented in [40] in which the description of the plasma actuator is detailed as well as in [41], but the principle is recalled as follows.

As shown in Fig. 32, let us consider a projectile flying from right to left and composed of a conical forebody equipped with two pairs of electrodes, step 1. The role of the high-voltage activating system only consists in breaking the electric barrier between two electrodes, then in ionizing a small gas volume (step 2). As the projectile flies, the ionized gas volume moves along its surface (steps 3 and 4). The ionized gas volume which has low impedance activates a plasma discharge when it meets two more electrodes supplied by a low voltage (step 5).

The role of the low-voltage plasma generator consists in delivering energy to the pair of electrodes and then in producing the plasma discharge. It is obvious that the high-voltage activating-system electrodes have to be situated ahead of the electrodes of the low-voltage plasma generator.

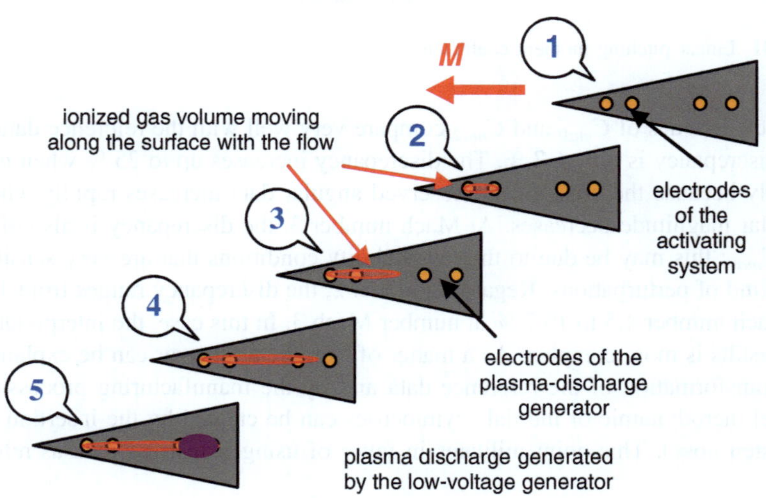

Fig. 32 Principle of the activation of a low-voltage plasma-discharge actuator

The high-voltage activating system is composed of a low-voltage supply transmitting little energy to the ionizing supply and to the impulse generator. The ionizing supply and the impulse generator are connected to a step-up transformer generating the high-voltage. The transformer is connected to the pair of electrodes. An external signal allows triggering the activating system. The transformer is the main part of the activating system. In the experiments presented hereafter a 320 V/5000 V transformer is used; however, the plasma-actuator design could be adapted to any projectile flight conditions.

The low-voltage plasma-discharge generator is composed of a capacitor connected to the pair of electrodes. The capacitor is charged by a low-voltage supply. Aluminium electrolytic capacitors meet the requirements for the present application; indeed, they have a large capacity/volume ratio and a low equivalent series resistance (ESR), allowing the use of a large discharge current. As an example, a capacitor with a 35-mm diameter and a 50-mm length supplied with 550 V has a stored energy of 50 J. Figure 33 shows the plasma-discharge actuator embedded inside the 50-mm-diameter model for testing purposes.

A series of experiments was conducted in the STA shock tunnel using the same model of an Explosive Formed Penetrator (EFP) mentioned in the previous section. The model is made of aluminium alloy except of the electrode support which is made of PVC. The electrodes are embedded near the junction of the ogive to the cylindrical part.

Figure 34 shows the EFP model (No. 1) suspended in the test section of the shock tunnel by means of two very thin and small disks made of paper (No. 2) linked to threads which are fixed on the test-section wall. The plasma discharge is produced by using the low-voltage actuator located outside the shock tunnel due to the dimensions of the actuator and of the projectile model. The electric wires (No. 3) connected to the plasma-discharge actuator are very flexible and contained inside a small tube (No. 4) fixed inside the test section. The displacement amplitude of the model is of the order of the model length. The Pitot-pressure probe (No. 5) contributes to the determination of the flow conditions.

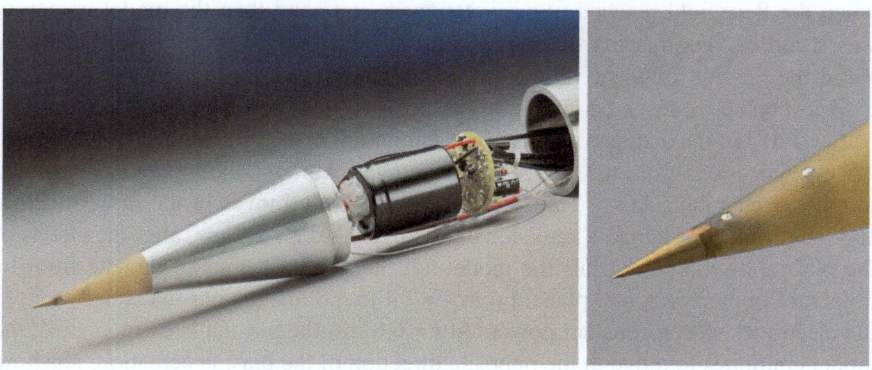

Fig. 33 Plasma-discharge actuator inside a 50-mm-diameter test model and a zoom on electrodes

Fig. 34 EFP model suspended in the test chamber

The aim of the experimental study consists in recording the free-flight motion of the projectile model subjected to a plasma discharge using the CCD high-speed camera.

A first test aimed at verifying the free-flight stability of the projectile model without any plasma discharge. When the membranes of the shock tube burst, the airflow is accelerated up to the desired pressure, temperature and flight velocity, leading to the Mach number of 4.5 at altitude of 2.5 km (Fig. 6). The threads and paper disks break and the projectile model can fly freely in the measurement section. As the main interest focuses on the projectile-model behavior, only photographs are taken by the CCD camera from which the projectile displacement can be recorded and analyzed at a rate of 10,000 frames per second with an exposure time of 2 μs. Two sets of photos are taken during each experiment: one photo for recording the displacement in the horizontal plane and one in the vertical plane. Figure 35 presents the complete flight of the projectile in six pictures recorded during that first test; the model moves along its longitudinal axis, thus proving its flight stability. It is important to point out that the electric wires used for the plasma discharge do not disturb the behavior of the model during its flight.

In the next experiment a plasma discharge was generated on the projectile nose when the flow field reached a quasi-stationary state and then the model trajectory was recorded. The plasma discharge is produced in the horizontal plane so as to avoid the gravity influence on the model trajectory.

A test result is presented in Fig. 36 for a plasma discharge having a stored energy of $E = 120$ J in the actuator and a current limitation of 100 A. A plasma duration of 4.5 ms is deduced from the current measurement. At $t = 0$, the model is suspended in the test chamber. At $t = 1.2$ ms, the model slightly moves in the flow direction and the plasma discharge is activated on the ogive. At $t = 2.4$ ms, after the effects of the plasma discharge reaches the model tail, the model begins to deviate from its trajectory in the horizontal plane. From $t = 2.4$ to 4.8 ms, the plasma is delivered to the electrodes with a constant power. At $t = 6.0$ ms, the plasma is extinguished, the model continues to move in the opposite direction of the plasma actuation, reaching an angle of attack of about −11°. The direction of model displacement in the

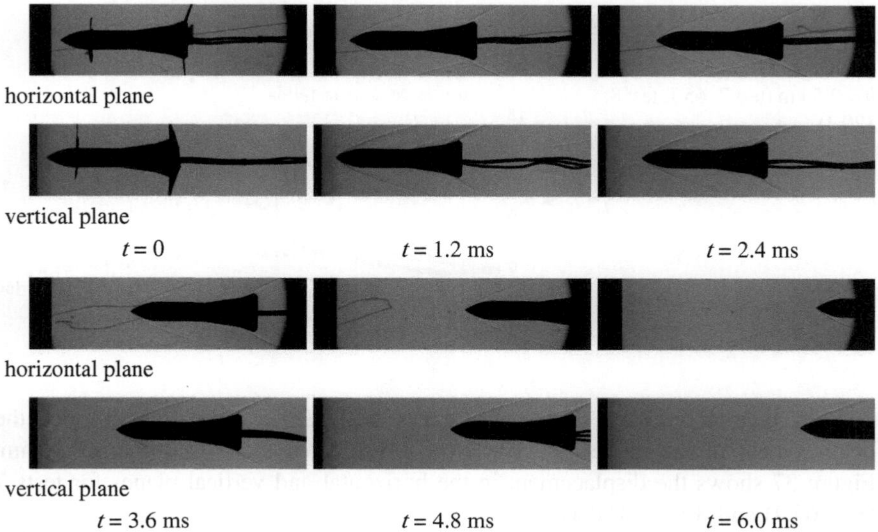

Fig. 35 Visualization of the displacement of the free-flight EFP model, $M = 4.5$, $h = 2.5$ km

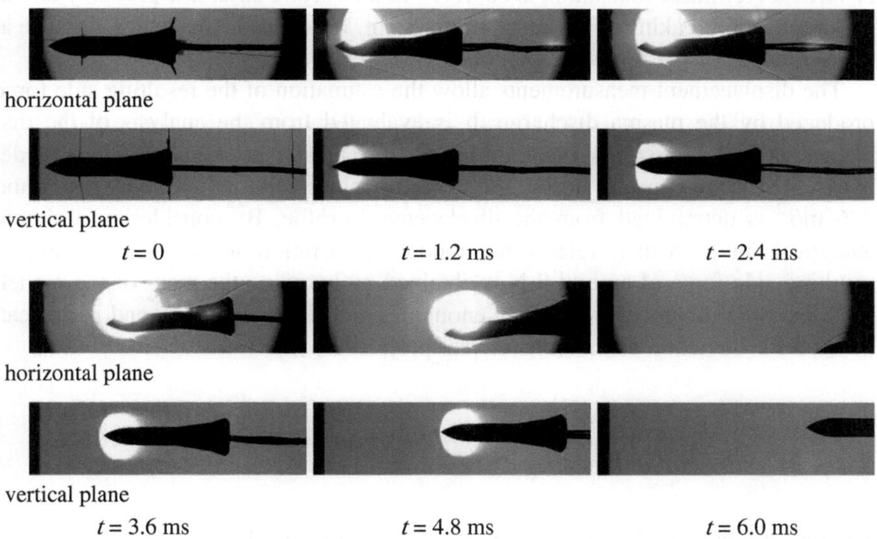

Fig. 36 Visualization of the displacement of the free-flight EFP model subjected to a plasma discharge, $M = 4.5$, $h = 2.5$ km, $E = 120$ J

horizontal plane tends to show that the plasma discharge creates an overpressure, which is transported along the model. Surface-pressure measurements on a projectile model have to be done in the future in order to confirm this assumption.

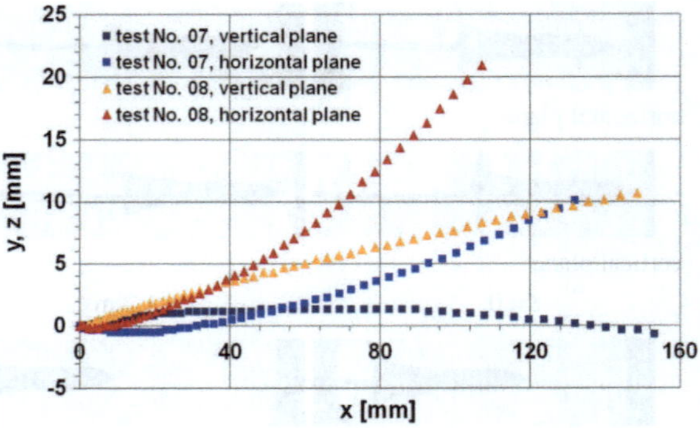

Fig. 37 Displacement measurement during a plasma discharge, $M = 4.5$, $h = 2.5$ km (test 7: 65 J, test 8, 120 J)

Each picture recorded during the test is analyzed in order to reproduce the displacement of the model tip. The error is estimated to be less than ±0.3 mm. Figure 37 shows the displacement in the horizontal and vertical planes for tests 7 ($E = 65$ J) and 8 ($E = 120$ J).

A lateral displacement of 10 mm is obtained after a projectile-model translation of about 135 mm in the case of a 65-J energy. A value of 20 mm is reached after a projectile-model translation of about 110 mm in the case of a 120-J energy. A vertical rectilinear translation is observed in the second case; it is probably due to a non-uniform breaking of the sustaining system, which leads to a lateral impulse at the beginning of the projectile-model displacement.

The displacement measurements allow the estimation of the resulting side force produced by the plasma discharge. It is evaluated from the analysis of the displacement in the horizontal plane during test 8. The side acceleration of the model is assumed to be constant during the model displacement and its value of around 446 m/s^2 is determined from the displacement profile. By considering a simple analytical model which relates the model acceleration to its displacement, a resulting side force of around 9 N is obtained by knowing the mass of the model. This first order approach is accurate enough to model the side force and it also can be used for deviation estimations of real projectiles.

5.4 Missile-Nose Heating

The flow around a high-speed missile is strongly influenced by the bow shock wave at the tip of the missile. Especially the front part of such a missile has to withstand extremely high heat fluxes. As a consequence, the temperature of the missile nose can increase during the flight and reach the melting temperature, leading to the missile destruction. The flight altitude, i.e. the atmospheric density, the flight speed and the flight time are the main parameters influencing the heating. Additionally, the geometry, material and dimensions of the missile and especially the nose-cone angle and the nose-tip contour are also significant parameters influencing the heating rate on the frontal part of the missile. Therefore, the design procedure of all high-speed and hypersonic missiles has to consider the heat loads acting on missiles, by selecting appropriate geometries, structures and materials. As the heat-transfer rate depends on flow conditions and missile-design characteristics, it is essential to measure reliable heat fluxes to correctly predict the prevailing heat loads on the missile surfaces and structures.

The analytical method reported in [42] for the calculation of the surface heat flux proves to be accurate enough to be used as a fast prediction tool in early design phases. In later development phases, CFD codes have to be used for modelling the geometrical details of the missile and reaching a higher precision in local temperature predictions [43, 44].

Comprehensive information on the heating scales encountered in flight must be available to predict the structural load capacity of a high-speed missile and to protect the surface, especially the nose, with suitable materials. A generic design of a missile nose was selected to conduct experiments and theoretical studies under realistic flow conditions. In this way, a better insight into the heating processes could be gained and reliable data for less critical missile parts could be also obtained. The ISL's shock tunnels are used because surface-temperature measurements are very difficult to be carried out in flight [44].

As a first example, the conical missile forebody is fixed in front of the shock-tunnel nozzle inside the test section. Heat fluxes are measured with special fast-reacting thin-film temperature gauges positioned on the nose surface of a blunt cone and a sharp cone (see Sect. 3.4). Figure 38 shows the blunt shock-tunnel model and a Pitot probe placed in front of the Mach number 6 Laval-contoured nozzle.

Experimental results are compared with theoretical calculations based on the boundary-layer theory for laminar and turbulent boundary-layer formation on a sharp conical nose [42]. Measurements and corresponding calculations are

Fig. 38 Blunt nose-cone model and Pitot probe in front of the Mach number 6 Laval-contoured nozzle

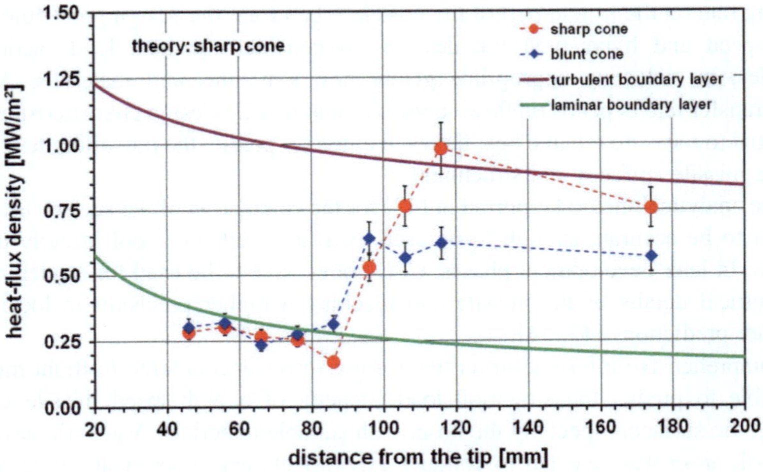

Fig. 39 Heat-flux density for the 7.125° cone at $M = 6$, $h = 15$ km $\alpha = 0°$

performed for Mach numbers 3.5, 4.5, 6, 8 and 10 for flight-altitude conditions from sea level up to 60 km. The detailed study is reported in [45].

For moderate flight altitudes the Reynolds number ranges from zero at the nose tip to more than $Re \approx 6 \times 10^6$ at the end of the cone. Therefore, transition from laminar to turbulent boundary layer may occur in these flight scenarios somewhere downstream the missile tip, which is predicted at $Re_{TR} \approx 3 \times 10^6$. At very high-altitude flights, i.e. at low atmospheric densities, only laminar boundary layer formation may occur along the surface of the cones.

Figure 39 shows the heat flux measured along the surface of the cones at an altitude of 15 km. The experimental values of the cross-flow are as follows: static pressure of 12,200 Pa and static temperature of 222 K leading to a density of 0.186 kg/m^3, velocity of 1800 m/s yielding a Mach number of 5.93. The laminar and turbulent boundary-layer computations are performed for the sharp cone only. The measured heat flux matches quite well the laminar boundary-layer solution for $x < 90$ mm. For $x > 90$ mm, the transition to a turbulent boundary layer starts both for the sharp and the blunt cones. Finally, the heat-flux density of the sharp cone is about three times the amount of the laminar one; the heat-flux density of the blunt cone is lower than that of the sharp cone.

The transition Reynolds number Re_{TR} ranges from 2.05×10^6 to 2.80×10^6 at the cone surface in both cases. There is quite a good agreement between the measured values and the turbulent boundary-layer solution for the sharp cone within the given error bars of about ±10 %. A temperature overshoot at the laminar to turbulent boundary-layer transition, as often assumed, is not really achieved, probably due to an insufficient measurement resolution in the length direction [45].

For flight conditions at an altitude of 21 km, the experimental values of the cross-flow are as follows: static pressure of 4700 Pa and static temperature of 215 K leading to a density of 0.074 kg/m^3, velocity of 1778 m/s yielding a Mach number

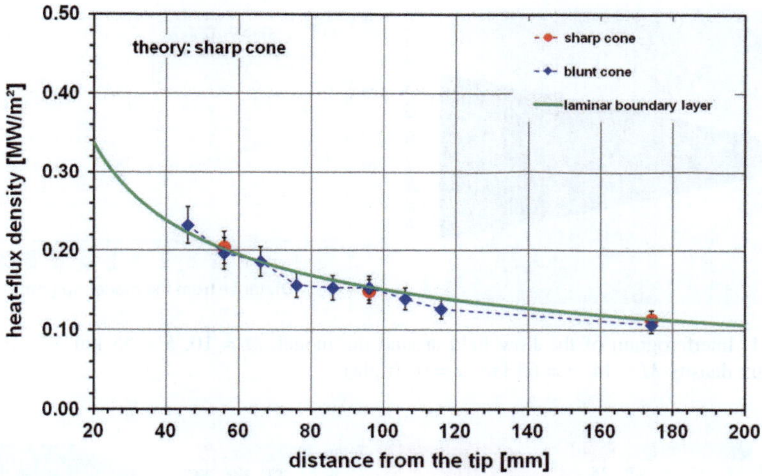

Fig. 40 Heat-flux density for the 7.125° cone at $M = 6$, $h = 21$ km, $\alpha = 0°$

of 5.95. As depicted in Fig. 40, the experimental and theoretical heat-flux density results show that the boundary layer is fully laminar along the model surface [45].

As a second example, another conical missile forebody with a blunt nose is studied at higher Mach numbers. The model is equipped with 11 thin-film gauges distributed on its surface and 1 coaxial thermocouple integrated in the nose tip. Figure 41 left shows the interferogram taken at Mach number 10 for 55-km altitude conditions. For this Mach number, the detached bow shock is still clearly visible at the considered altitude conditions, whereas it is undistinguishable at higher altitudes. Figure 41 right depicts the heat flux along the surface of the model measured during two tests at Mach number 14 for an altitude of 65 km. The detailed sets of data mainly measured for the validation of computational codes are reported in [46, 47].

In the same contractual study as last example, another generic missile model composed of a blunt nose, a cylindrical part and a flare is studied. The shock-tunnel experiments were investigated by DSMC (Direct Simulation Monte Carlo) in order to provide a comparison between numerical results and experimental findings reported in [48]. While the one-dimensional shock-tube code coupled to a nozzle code was used to determine the shock-tunnel free-stream parameters, the axisymmetric flow around the model was simulated using DSMC. The DSMC method is an algorithm for determining the velocity-distribution function as a solution to the Boltzmann equation which should be used for studying rarefied flows. The flow conditions were used as inlet boundary conditions for the axisymmetric code DS2V (V.4.5), developed by Bird [49].

Figure 42 shows four variable distributions in the symmetry plane of the flow for a DSMC computation carried out at Mach number 8 for 50-km altitude: the Mach

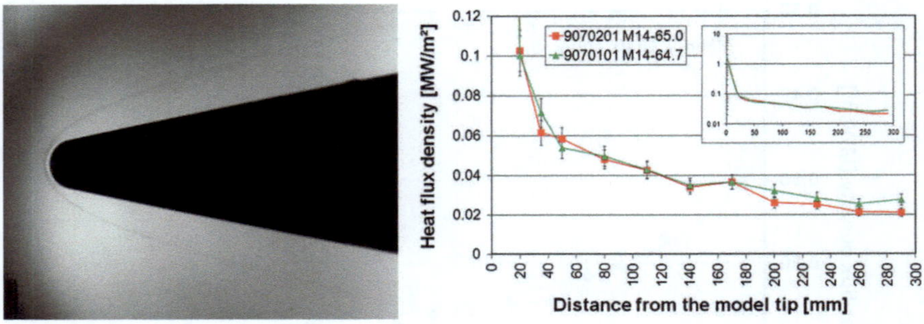

Fig. 41 Interferogram of the flow field around the model, $M = 10$, $h = 55$ km, $\alpha = 0°$ (left); heat-flux density $M = 14$, $h = 65$ km, $\alpha = 0°$ (right)

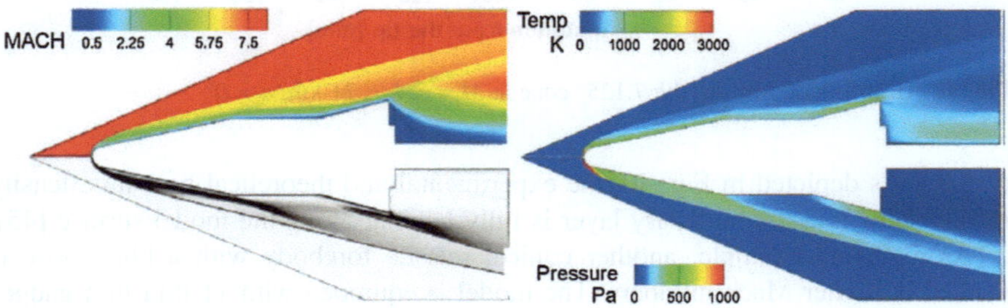

Fig. 42 Results of a DSMC computation, $M = 8$, $h = 50$ km, $\alpha = 0°$

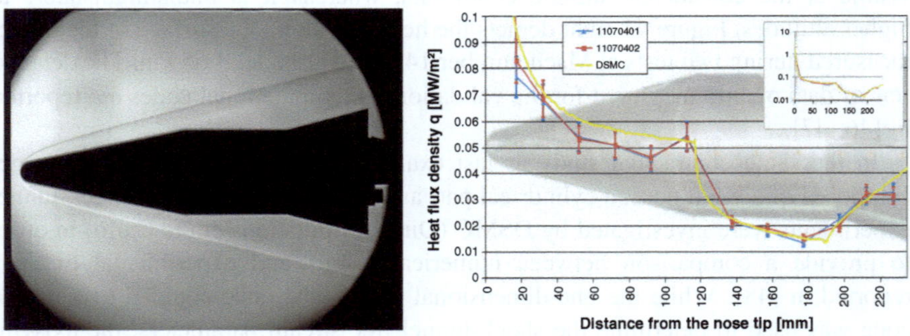

Fig. 43 Interferogram of the flow field around the model (*left*), heat-flux density (*right*); $M = 8$, $h = 50$ km, $\alpha = 0°$

number (left top), the numerical Schlieren (left bottom), the temperature (right top) and the pressure (right bottom) are presented.

Figure 43 left shows the interferogram taken for the same Mach number and for the same altitude conditions. Figure 43 right depicts the corresponding heat flux along the surface of the model measured during two tests. It also presents the corresponding DSMC calculated heat flux which is compared to the measurements; the computational result agrees with the measured one within the range of the measurement accuracy.

5.5 *Atmospheric Dispersion of Droplets*

The development of hypersonic flight applications is making substantial progress. Therefore, more and more flight vehicles will be able to fly through the atmosphere at Mach number 5 and higher. In the case of an accident, a liquid substance carried by the vehicle, e.g. propellants or other toxic liquids, could be released into the atmosphere during the hypersonic flight. Following such a scenario, a bulk of a liquid substance would suddenly be exposed to the hypersonic atmospheric flow surrounding the vehicle. As a result of the instantly developed interaction between the liquid and the atmospheric flow, the liquid would be decelerated, fragmented, atomized into drops and evaporated to some extent during a very quick dynamic process. Since liquid fragmentation and atomization in gaseous flows has been of interest for a long time, an abundant literature devoted at least to some aspects of this matter exists, including some comprehensive reviews listed in [50].

First comprehensive experiments and calculations on liquid fragmentation and atomization in subsonic and supersonic regimes were carried out at ISL in 1998 [51]. An overview of the former work carried out at ISL on liquid fragmentation as well as a thorough introduction to the subject was published in 2009 [52], when the subject became of high priority.

Observing the fragmentation and atomization of a liquid bulk in atmospheric hypersonic flows requires an experimental set-up capable of reproducing the relevant hypersonic atmospheric flow conditions, a method to bring the liquid into the flow and appropriate measurement techniques to observe the liquid behavior. The study employs the STA shock tunnel to duplicate a real atmospheric hypersonic flow. A magnetic piercing mechanism shown in Fig. 44 was constructed and

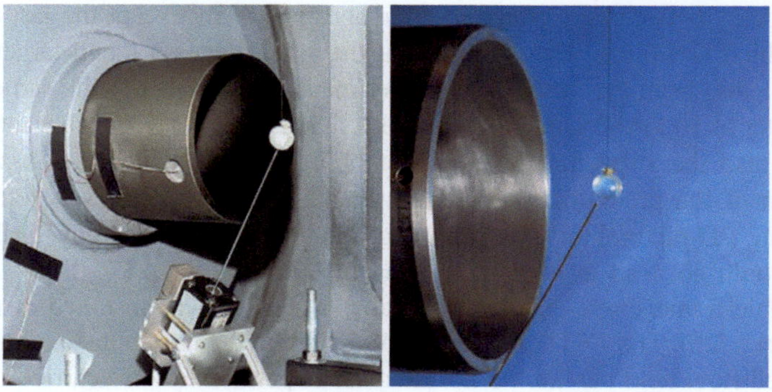

Fig. 44 Magnetic piercing mechanism with balloon filled with liquid in front of the nozzle

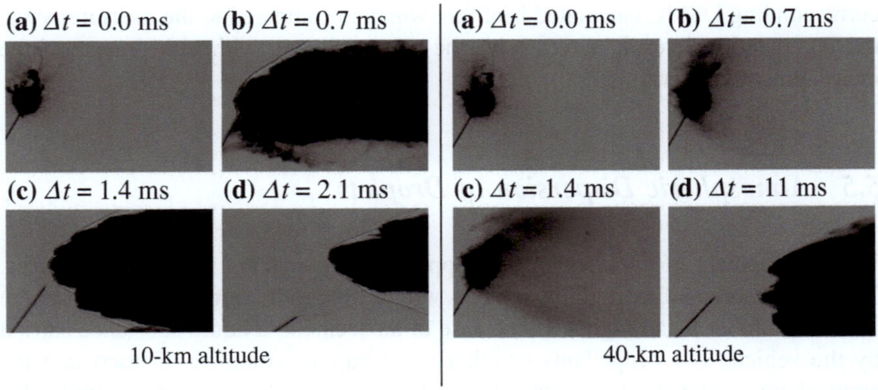

Fig. 45 Fragmentation of water at $M = 4.5$; series of shadowgrams by high-speed photography. **a** $\Delta t = 0.0$ ms; **b** $\Delta t = 0.7$ ms; **c** $\Delta t = 1.4$ ms; **d** $\Delta t = 2.1$ ms; **a** $\Delta t = 0.0$ ms; **b** $\Delta t = 0.7$ ms; **c** $\Delta t = 1.4$ ms; **d** $\Delta t = 11$ ms

installed at the nozzle exit; it pierces a latex balloon filled with the particular liquid substance to be studied when triggered just before the flow sets in the test section.

The piercing and the destruction of the balloon could obviously influence the contained bulk of liquid. Preliminary tests without flow and with the help of backlight illuminated high-speed camera images have proven that this influence remains very small, i.e. the balloon vanishes in less than 1 ms and leaves the bulk of liquid in front of the nozzle. After the demonstration at Mach number 3 of the applicability of the magnetic piercing mechanism, two experiments with water were analyzed using backlight illuminated photography (Fig. 45): the first at 10-km and the second at 40-km altitude conditions.

A macroscopic bag building is observed in each experiment. The comparison of the temporal development of the macroscopic fragmentation effects illustrates the strong influence of the flow density on the process. The flow at 10-km altitude exhibits a density three orders of magnitude greater than that at 40 km. Thus, the bulk of liquid at 40 km moves much slower, i.e. it would need much more time to decelerate in a real flight scenario, than at lower altitudes. In the two experiments, the time delay between the consecutive images is 0.7 ms except the last image for 40 km which is 9.6 ms. This reveals that the water needs a much larger delay at 40 km (11 ms) from the onset of the experiment to cover nearly the same distance as the water at 10 km (2.1 ms). It is obvious that due to this very different temporal behavior also the structural development of ligaments is different. A systematic discussion of the preliminary results on the basis of the pictures is left, because interpreting 2D-pictures of a complicated 3D-structure without any further information would not lead to fundamental insights.

A double-frame/single-exposure digital PIV system was installed at the shock tunnel (Sect. 3.6). The CCD camera was mounted on the horizontal axis to view the illuminated flow field behind the nozzle axis. Experiments were carried out with a TSI PowerView Plus 4MP PIV camera. A 100-mm Zeiss objective was mounted in

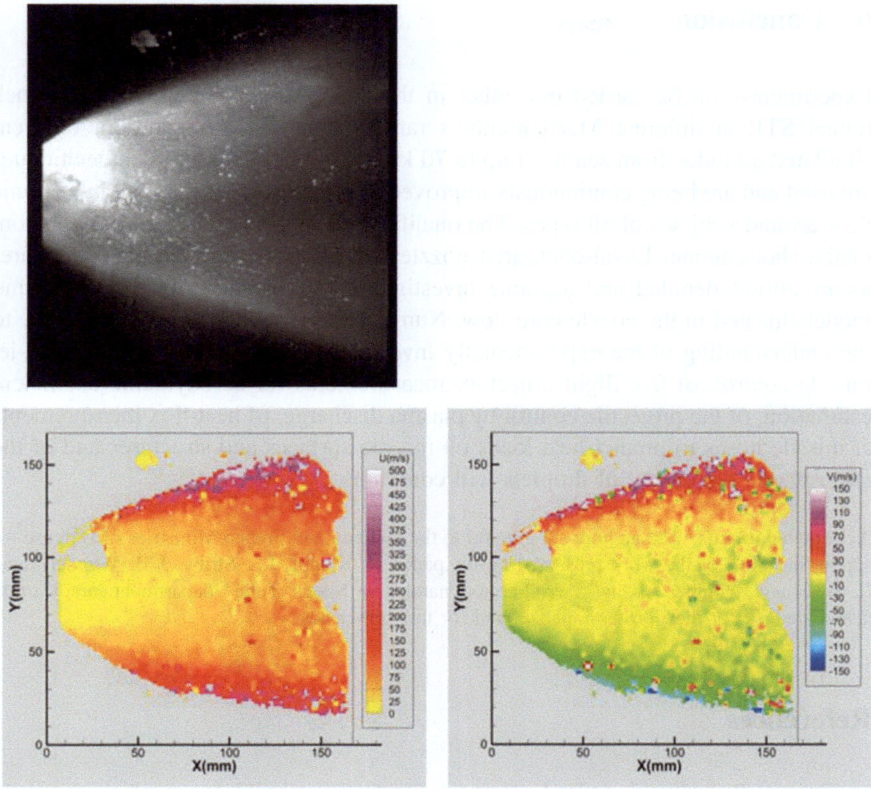

Fig. 46 Fragmentation of water at $M = 4.5$, $h = 40$ km (*top*), velocity measurement by PIV (*bottom*)

front of the camera to observe the field of view. The camera spatial resolution was of 2048 pixels × 2048 pixels, the optical magnification was 82.3 μm/pixel, the field of view was 168 mm × 168 mm and the delay between subsequent laser pulses was 2 μs. The laser and the camera were synchronized and triggered by a heat-flux sensor flush mounted in the shock-tube wall. The synchronizer separation time was checked by a fast-response photodiode and timing errors were found to be less than 1 %. The PIV images were analyzed after the experiment by an inter-correlation algorithm. The width and height of the correlation windows were 48 pixels × 48 pixels and the grid spacing step was 16 pixels. The intercorrelation function was calculated by a fast Fourier transform. The upper picture of Fig. 46 shows a raw image of the drops illuminated by the laser-light sheet for an experiment conducted with water at Mach number 4.5 at 40-km altitude. The lower images of Fig. 46 present the corresponding horizontal and vertical components of the velocity.

Future experiments will focus on the implementation and adaptation of appropriate and sophisticated optical measurement techniques in order to get a better insight of the physics of the process.

6 Conclusion

Experiments can be carried out either in the shock tunnel STA or in the shock tunnel STB at different Mach numbers ranging from 3 to 14 and for different simulated altitudes from sea level up to 70 km. Advanced measurement techniques are used and are being continuously improved in order to investigate the hypersonic flow around vehicles of all types. The qualification of the free-stream flow in front of the shock-tunnel Laval-contoured nozzles by Pitot-pressure and PIV measurements allows detailed and accurate investigations of the flow developed around models located in the nozzle-core flow. Numerical investigations also contribute to the understanding of the experimentally investigated flows. Studies of the side-jet missile control, of free-flight trajectory measurements for aerodynamic coefficient estimation, of the projectile control by plasma discharge, of heat-flux measurements at missile noses to predict heat loads on missile surfaces and structures and of the atmospheric dispersion of droplets will continue.

Acknowledgments The authors are grateful to the German and French Ministries of Defense for supporting most of the work presented in this paper. R. Kempf, J.L. Striby, J.M. Wunderly and R. Bernhard, member of the "Aerothermodynamics & Shock Tube" department, are warmly acknowledged for their professionalism and for their efficiency.

References

1. Gnemmi, P., Srulijes, J., Seiler, F.: Overview of activities at the ISL hypersonic shock tunnels. Int. J. Eng. Syst. Model. Simul. **3**(1/2), 74–86 (2011)
2. Patz, G.: Das Hyperschallstossrohrlabor des ISL, 3. Teil: Stossrohr B, ISL Report N 30/70 (1970)
3. Patz, G.: *Das* Hyperschallstossrohrlabor des ISL, 2. Teil: Stossrohr A, ISL Report N 27/71 (1971)
4. Oertel, H.: Stossrohre. Springer, New York (1966)
5. Smeets, G., Patz, G., Srulijes, J., Seiler, F., Havermann, M.: Software Codes Developed in the Shock Tube Laboratory of ISL (1980–2009)
6. Sauerwein, B., Hruschka, R., Gnemmi, P., Rey, C., Bastide, M.: ISL shock tunnel operation with contoured Mach-4.5-6-8-10-nozzles for hypersonic test applications. In: 29th International Symposium on Shock Waves (ISSW29), Madisson/WI, USA, 15–19 July 2013
7. Smeets, G., George, A.: Instantaneous laser doppler velocimeter using a fast wavelength tracking Michelson interferometer. Rev. Sci. Instrum. **49**(11), 1589–1596 (1978)
8. G. Smeets: *Interferometry*, ISL Report CO 214/90 (1990)
9. Jessen, C.: Messungen von Druck, Temperatur und Kraft an Modellen im Stroßwellenkanal, PhD. Thesis Technische Hochschule Aachen (1993)
10. Olivier, H.: Available at http://www.swt-aachen.de/products/thin-film-gauges/ (2014)
11. Haertig, J., Havermann, M., Rey, C., George, A.: Application of particle image velocimetry to high-speed supersonic flows. In: Proceedings of the PIVNET II International Workshop on the Application of PIV in Compressible Flows (2005)
12. Haertig, J., Rey, C., Sourgen, F.: Etude expérimentale de l'écoulement supersonique confiné dans un conteneur de missile à lancement vertical. In: 11ème Congrès Francophone de Technique Laser, Poitiers, France, 11–16 Sept 2008

13. Havermann, M., Haertig, J., Rey, C., George, A.: PIV Measurements in Shock Tunnels and Shock Tubes, Topics in Applied Physics 112, Particle Image Velocimetry, New Developments and Recent Applications, pp. 429–443. Springer, Berlin (2008)
14. Gauthier, T., Comet, M., Pichot, V., Sourgen, F., Spitzer, D., Leopold, F., Martinez, B.: Nouvelles possibilités des Peintures Sensibles à la Pression Nanostructurées (NANOPSP). In: 13ème Congrès Francophone de Visualisation et de Traitement d'Images en Mécanique des Fluides, Reims, France, 16–20 Nov 2009
15. Gauthier, T., Sourgen, F., Leopold, F., Martinez, B., Comet, M., Pichot, V., Spitzer, D., Piazzon, N.: Improvement in pressure measurements using nanostructured paints (NanoPSP). In: 27th AIAA Aerodynamic Measurement Technology and Ground Testing Conference, Chicago/IL, USA (2010)
16. Chaplin, R., MacManus, D., Leopold, F., Gauthier, T., Martinez, B., Birch, T.: Experimental investigation into the interference aerodynamics of two slender bodies in close proximity. Exp. Fluids **50**, 2 (2011)
17. Igra, O., Takayama, K.: Shock-tube study of the drag coefficient of a sphere in a non-stationary flow. Proc. R. Soc. Lond. A **442**(1915), 231–247 (1993)
18. Seiler, F., Mathieu, G., George, F., Srulijes, J., Sauerwein, B., Havermann, M.: Development of a free flight force measuring (FFM) technique at the ISL shock tube laboratory. In: 25th International Symposium on Shock Waves (ISSW25), Bangalore, India, 17–22 July 2005
19. Jaeggy, B.C.: Calcul d'une tuyère supersonique plane ou de révolution, ISL Report R110 (1975)
20. Shope, F.L.: Contour design techniques for super/hypersonic wind tunnel nozzles. In AIAA 2006-3665 (2006)
21. Joly, M., Grumbach, J.P.: Correction de couche limite pour le calcul de l'écoulement dans une tuyère supersonique. ISL Report N34 (1973)
22. Smeets, G., Gatau, F., Srulijes, J., Seiler, F., Havermann, M.: Software Codes Developed in the Shock Tube Laboratory of ISL (1980–2009)
23. Zierep, J.: Grundzüge der Strömungslehre. Springer, Berlin (1997)
24. Srulijes, J., Sauerwein, B., Seiler, F., Bastide, M., Rey, C., Hennig, P., Gleich, P.: Wärmestrommessungen an Flugkörperspitzen im Stossrohr-Windkanal mit einer Mach-8-Düse, ISL Report RV 231/2009 (2009)
25. Landau, L., Teller, E.: Zur Theorie der Schalldispersion. Phys. Z. Sowjetunion **10**, 1 (1936)
26. Champigny, P., Lacau, R.G.: Lateral jet control for tactical missiles, special course on missile aerodynamics. AGARD-FDP Von Karman Institute, Brussels, Belgium, 3.1-3.57, 6–10 June 1994
27. Gnemmi, P., Eichhorn, A., Emunds, H., Esch, H., Gülhan, A., Leopold, F., Schäfer, H.J.: Experimental and computational study of the interaction between a lateral jet and the supersonic external flow on a generic missile body. In: NATO RTO-AMP-AVT-135 Symposium on "Innovative Missile Systems", Amsterdam, NL, 15–18 May 2006
28. Zukoski, E.E., Spaid, F.W.: Secondary injection of gases into a supersonic flow. AIAA J. **2** (10), 1689–1696 (1964)
29. Seiler, F., Sauerwein, B., Bastide, M., Schacherer, M., Weinand, K., Hennig, P.: Untersuchungen zur Seitenstrahlsteuerung einer aktuellen Flugkörper-konfiguration im Strossrohr-Windkanal bei realen Konditionen, ISL Report RV 250/2008 (2008)
30. Seiler, F., Sauerwein, B., Bastide, M., Schacherer, M., Srulijes, J., Leopold, F., Weinand, K., Hennig, P.: Wirkung des Seitenstrahlsteuerung an einem zylindrischen Flugkörper im Strossrohr-Windkanal STB des ISL. ISL Report RV 226/2009 (2009)
31. Seiler, F., Srulijes, J., Gnemmi, P., Sauerwein, B., Bastide, M., Dahlem, K.J., Weinand, K.: Untersuchungen zur Seitenstrahlsteuerung an dem zylindrischen KV-Flugkörper. ISL Report RV 235/2011 (2011)
32. Sauerwein, B., Gnemmi, P., Srulijes, J., Bastide, M., Seiler, F., Dahlem, K.J., Weinand, K.:: Untersuchungen zur Seitenstrahlsteuerung an dem zylinrischen KV-Flugkörper, ISL Report RV 229/2012 (2012)

33. Sauerwein, B., Bastide, M., Kempf, R.: Untersuchungen zur Seitenstrahlsteuerung an dem zylinrischen KV-Flugkörper, ISL Report RV 256/2012 (2012)
34. Wey, P., Bastide, M., Martinez, B., Srulijes, J., Gnemmi, P.: Determination of the aerodynamic coefficients from shock tunnel free flight trajectories. In: 28th Aerodynamic Measurement Technology, Ground Testing and Flight Techniques Conference. AIAA Paper 2012-3321, New Orleans/LO, USA, 25–28 June 2012
35. Rondot, F., Berner, C.: Performance of aerodynamically optimized EFP simulants. In 17th International Symposium on Ballistics, Midrand, South Africa, 23–27 Mar 1998
36. Winchenbach, G., Krieger, J., Hathaway, W., Whyte, R.: EFP Free Flight Test Aerodynamic Results, Air Force Research Laboratory, AT-98-08-02 (1998)
37. Fleck, V., Berner, C., Winchenbach, G., Hathaway, W.: Aerodynamic testing for long range explosively formed penetrators. In 18th International Symposium on Ballistics, San Antonio/TX, USA, Nov 1999
38. Harris, C.G., Stephens, M.J.: A combined corner and edge detector, pp. 147–151. In: Fourth Alvey Vision Conference, Manchester, UK (1988)
39. Gnemmi, P., Charon, R., Dupéroux, J.P., George, A.: Feasibility study for steering a supersonic projectile by a plasma actuator. AIAA J. **46**(6), 1308–1317 (2008)
40. Gnemmi, P.: Concept of the Steering of a Supersonic Projectile by an Electric Discharge Generating a Plasma, Research Management Accreditation Thesis, ISL Report R 101/2013 (2013)
41. Gnemmi, P., Rey, C.: Plasma actuation for the control of a supersonic projectile. J. Spacecraft Rockets **46**(5), 989–998 (2009)
42. Seiler, F., Werner, U., Patz, G.: Theoretical and experimental modelling of real projectile flight heating. J. Spacecraft Rockets **38**(4), 497–503 (2001)
43. Gnemmi, P., Sommer, E., Junod, E.: Aerothermal study of a projectile nose in flight. In AIAA Atmospheric Flight Mechanics Conference and Exhibit, Austin/TX, USA, 11–14 Aug 2003
44. Gnemmi, P., Sommer, E., Mortel, P., Bredif, M., Emunds, H.: Aerothermal study of a supersonic missile forebody. In: NATO RTO-AMP-AVT-135 Symposium on "Innovative Missile Systems", Amsterdam, NL, 15–18 May 2006
45. Srulijes, J., Seiler, F., Hennig, P., Gleich, P.: Heat transfer at the nose of a high-speed missile. In: 15. STAB/DGLR Symposium, Aachen, Germany, 3-4 Nov 2008
46. Schlöffel, G., Sauerwein, B., Bastide, M., Kempf, R.: Abschlussbericht zum Thema Aerothermodynamik von Gefechtsköpfen beim Eintritt in die Atmosphäre mit hoher Machzahl. ISL Report RV 257/2012 (2012)
47. Srulijes, J., Hruschka, R., Seiler, F., Sauerwein, B., Bastide-Luquet, M., Dahlem, K.J.: Shock tube experiments on heat transfer at generic re-entry bodies. In: 28th International Symposium on Shock Waves, Manchester, UK, 17–22 July 2011
48. Srulijes, J., Sauerwein, B., Bastide, M., Seiler, F., Hruschka, R., Gnemmi, P., Dahlem, K.J., Weinand, K.: Abschlussbericht zum Thema "Experimentelle und Numerische Untersuchung der Umströmung von Gefechtsköpfen beim Eintritt in die Atmosphäre". ISL Report RV 228/2012 (2012)
49. Bird, G.A.: Molecular Gas Dynamics and the Direct Simulation of Gas Flows. Clarendon Press, Oxford (1994)
50. Schlöffel, G., Srulijes, J., Albers, H., Bastide, M., Rey, C.: Preliminary shock-tunnel experiments on liquid fragmentation and atomization in hypersonic flows. In ICLASS 2012, 12th Triennial International Conference on Liquid Atomization and Spray Systems, Heidelberg, Germany, 2–6 Sept 2012
51. Smeets, G., Patz, G.: Zerstauben einer bei hoher Fluggeschwindigkeit austretenden Fluessigkeit, ISL Report R 107/98 (1998)
52. Srulijes, J., Seiler, F.: Fragmentation of liquid in hypersonic cross flow, lecture series 2009-04: liquid fragmentation in high-speed flow, von Karman Institute for Fluid Dynamics, Rhode-Saint-Genèse, Belgium, 16–18 Mar 2009

Author Biographies

Patrick Gnemmi earned his accreditation on research management in mechanics-energetics in 2012 at the University Paris Nanterre La Défense in Paris, France. The subject was on the concept of the steering of a supersonic projectile by an electric discharge generating plasma. He earned his Ph.D. degree in mechanical engineering in 1993 at the Conservatoire National des Arts et Métiers (CNAM) in Paris, France on the prediction of the main rotor noise of helicopters at subsonic speed. He worked during about twelve years on the generation and propagation of such a noise at ISL.

He is a senior scientist presently involved in the management of scientific research projects at ISL. Since 1995, he specializes in the field of the aerodynamics and aerothermodynamics of super- and hypersonic projectiles and missiles; in particular he studies guidance systems for such vehicles. In 2007, he joined the "Aerothermodynamics & Shock Tube" department and in 2011 he became head of this laboratory. In parallel since 2009, he is also involved in the research project of a Gun Launched Micro Air Vehicle (GLMAV). He published more than sixty archival journals or conference proceedings papers and book chapters among more than hundred-eighty publications.

Julio Srulijes was born in Argentina in 1947. He received a degree in Mechanical Engineering from the Technological University, Buenos Aires, Argentina in 1972. After several years in the industry, in 1975 he took up a Ph.D. work at the Institute of Fluid Mechanics of the University of Karlsruhe, Germany where in 1979 he obtained his Dr.-Ing. degree with his work on cellular convection in containers with horizontal temperature gradients. From 1980 to 1983, Chief Engineer at the Institute for Theoretical Mechanics of the University of Essen, Germany, he was involved among other in planning a space experiment on thermocapillary migration of bubbles and drops flown successfully in the D-1 Space Shuttle mission. From 1983 to 2012, Scientist at the French-German Research Institute of Saint-Louis ISL, France, he worked on gasdynamics of high enthalpy flows with and without reactive combustion at the Shock Tube Laboratory. He participated in the development of numerous optical measuring techniques.

Friedrich Seiler received his "Dipl.-Phys." degree and his "Dr.-Ing." degree from the University of Karlsruhe, Germany, now Karlsruhe Institute of Technology (KIT). His doctoral study dealt with the experimental investigation of the structure of a shock wave traveling along a wall, carried out in a low pressure shock tube, and compared with theoretical calculations, using the direct Monte-Carlo Simulation Technique DSMC. As scientist at the French-German Research Institute of Saint-Louis (ISL), France, from 1980 on, he worked with shock tubes in the field of gas-dynamics, hypersonics and measuring techniques. Additionally, he did a basic research in the field of ram acceleration. At the ISL he became head of the Aerothermodynamics & Shock Tube Laboratory until his retirement in the year 2011. Since his "Habilitation" in 1992 and the upgrade to be "Professor for Fluid Mechanics" at the University of Karlsruhe, his student courses are dealing with the aerothermodynamic behavior of gases and with optical measuring techniques. Associated at KIT as "Lecturing Professor", he is in the editorial board of the "International Journal of Shock Waves" and member of the advisory committees of the "International Symposium on Shock Waves (ISSW)", the "International Shock Interaction Symposium (ISIS)" and the "International Symposium on Flow Visualization (ISFV)". More than 300 publications are available.

Berthold Sauerwein received his graduate mechanical design engineer (Dipl. Ing. (Fh)) in 1988 in Friedberg, Germany.

At ISL, he was first working as engineer in planning and design of scientific installations for 10 years. After changing to the "Aerothermodynamics & Shock Tube" department of ISL, he was in charge of designing shock-tunnel models, of planning shock-tunnel operation and of the manufacturing and calibration of heat-flux probes used in shock-tunnel measurements. He mainly carries out shock-tunnel experiments with pressure and heat-flux measurements. Since 2008 he is responsible for operating the shock-tunnel facilities of ISL.

Myriam Bastide is a graduate Engineer (DEST) of the Conservatoire National des Arts et Métiers (CNAM); she obtained her diploma on Physics Instrumentation in 1995, in Paris, France.

During around 20 years, she was involved in holography, microholography, focometry, interferometry, holographic PIV, flow visualisations, optical simulations, etc. As an expert Engineer in Optics, she joined the "Aerothermodynamics & Shock Tube" department in 2007 in which she is responsible for all optical measurements developed for high-speed flow experiments.

Christian Rey received his graduate Engineer diploma (DEST) in Measurement and Instrumentation Engineering at the Conservatoire National des Arts et Métiers in Paris in 1999.

He is a Senior Engineer in the "Aerodynamics, Measurements and Simulations" department at ISL. He is in charge of PIV measurements in super- and hypersonic flows. He also develops an embedded generator for projectile steering by electric discharge generating plasma and he is responsible for measurements and visualizations of the plasma interaction with a supersonic flow. He is also in charge of flight tests and measurements on a Gun Launched Micro Air Vehicle (GLMAV) demonstrator.

Pierre Wey received a Master of Science in Engineering with a major in Computational Science from the Ecole des Mines de Saint-Etienne, France, in 1984.

He is a senior research scientist at ISL. His main interests include the design and development of simulation models in exterior ballistics and the analysis of weapon systems performance. He is currently in charge of the Electromagnetic Artillery System project which combines railgun launchers and hypervelocity projectiles. Since 2009, he has been leading the NATO STANREC 4618 Team of Experts, the goal of which is to define a standard 6 and 7 degrees-of-freedom trajectory model for guided projectiles and to develop a common exterior ballistics code. Previously, he was in charge of the Guided Supersonic Projectile project which was focused on the control of fin-stabilized projectiles using sequences of lateral impulses.

Bastien Martinez received his Engineer Diploma in Aeronautics Engineering and his Master of Science in Fluid Mechanics at the Ecole Nationale Supérieure de Mécanique et d'Aérotechnique (ENSMA) in Poitier, France, in 2000. He joined the ISL in 2001 and he contributed to the use of Pressure Sensitive Paints by writing the image processing software. He did his graduate work at ISL and earned his Ph.D. from Université de Franche-Comté in 2009 after developing an optical setup for Doppler Global Velocimetry applied to supersonic flows. Since then, he works in the field of experimental and numerical aerodynamics dedicated to new steering devices in relation with projectiles. Those activities led him to develop the processing algorithm for the Free-Flight Method used at the ISL shock tunnel.

Hermann Albers is technical officer from the BAAINBw Germany since 1982. His academic specialization is "Arms and Ammunition".

He works as Engineer at ISL since 1993. His previous project was on investigations of the effect of a shock wave on the ignition behavior of aluminum particles in a shock tube. His current project is on investigations of the mechanisms of the dispersion of combat agents at high velocities and altitudes.

Gunther Schlöffel received his Master Degree in Mathematics and Scientific Computing at the University of Heidelberg in 2005. He earned his Ph.D. in Aerospace Science at the University of the Bundeswehr of Munich in 2009. In 2006 he started his research in the "Aerothermodynamics & Shock Tube" department at ISL. Since 2009, he is scientist in the field of reactive flow simulations.

Robert Hruschka received his Master of Science in Aerospace Engineering from the Technical University of Munich in Germany in 2005. He earned his Ph.D. degree from the University of New South Wales in Australia in 2010. He has worked in the field of experimental and numerical analysis of super- and hypersonic flows. Since 2011, he is a junior scientist in the department for "Aerodynamics, Measurements and Simulations" at the ISL.

ThibautGauthier received his graduate Engineer diploma in Measurement and Instrumentation Engineering at the Conservatoire National des Arts et Métiers (CNAM) in Paris in 2007.

He is Engineer since 13 years in the "Aerodynamics, Measurements and Simulations" department at ISL. He is particularly responsible of the Pressure Sensitive Paint (PSP) method for steady and unsteady pressure measurements in flows. He is also involved in the development of nanoPSPs for improving their response time in high-speed flows.

Free Piston Shock Tunnels HEG, HIEST, T4 and T5

Klaus Hannemann, Katsuhiro Itoh, David J. Mee and Hans G. Hornung

1 Introduction

In hypervelocity flows the speed of the considered fluid is much larger than the speed of sound. Commonly the hypersonic flow regime is considered to start above a Mach number, M, of 5. As outlined, e.g., by Hornung [41], ground based testing of such flows is performed in many different types of facilities. The reason for this is the large range of flow conditions and phenomena encountered in hypersonic flight and the fact that no single facility can simulate all relevant flow parameters simultaneously. Therefore, in hypersonic testing, partial simulation of different flow phenomena is performed in different types of facilities. Examples are Mach-Reynolds number simulation in cold hypersonic ground based test facilities, verification and qualification of hot structures of space vehicles in arc–heated test facilities or the investigation of the influence of the chemically reacting flow past an entry or re-entry vehicle on its aerodynamic behaviour in shock tunnels or shock

K. Hannemann (✉)
Spacecraft Department, German Aerospace Center, DLR, Institute of Aerodynamics and Flow Technology, 37073 Göttingen, Germany
e-mail: Klaus.Hannemann@dlr.de

K. Itoh
Japan Aerospace Exploration Agency, JAXA, Kakuda Space Center, Kakuda Miyagi 981-1525, Japan

D.J. Mee
Centre for Hypersonics, School of Mechanical and Mining Engineering, Brisbane, The University of Queensland, Saint Lucia, QLD 4072, Australia
e-mail: d.mee@uq.edu.auProfessor

H.G. Hornung
Graduate Aerospace Laboratories, California Institute of Technology, Pasadena, CA 91125, USA
e-mail: hans@caltech.edu

O. Igra and F. Seiler (eds.), *Experimental Methods of Shock Wave Research*, Shock Wave Science and Technology Reference Library 9,
DOI 10.1007/978-3-319-23745-9_7

expansion tunnels. Comprehensive overviews of ground based testing of hypersonic flows are given by e.g. Lukasiewicz [61] and Lu and Marren [60].

One possibility to increase the Mach number in ground based facilities is by reducing the free stream temperature, i.e. the free stream speed of sound. In this case, high Mach numbers can be achieved while the free stream velocity is significantly lower than the actual flight velocity. However, characteristic of high Mach number hypersonic flight with $M \approx 10$ and higher is that the kinetic energy of the flow is large enough that high temperature effects such as vibrational excitation or dissociation of the fluid molecules occur in the flow past hypersonic vehicles. For such hypersonic, high velocity flows the term hypervelocity flow is used. The high flow velocities and subsequently the high temperature effects are not duplicated in cold hypersonic ground based testing facilities. During the re-entry flight of a space vehicle in the earth's atmosphere or the interplanetary atmospheric entry of space vehicles or meteorites, speeds in excess of 6 km/s are achieved. Considering a flow with this speed in a test section with an area of 1 m^2 and a density of 0.003 kg/m^3, the flow power itself is already over 300 MW (see also [41]). Therefore, continuous flow facilities are not a practical way to generate such high enthalpy, hypersonic flows. Additionally, the correct simulation in ground based testing of the chemical relaxation length of the dissociation reactions of the fluid molecules occurring, for example behind the strong bow shock in front of the nose of a re-entry vehicle, requires the duplication of the flight binary scaling parameter, the product of the free stream density ρ times a characteristic flow length L [96]. Consequently, the smaller the scale of the wind tunnel model chosen, the higher the free stream density or pressure needs to be. Considering the flight trajectory range of a re-entry vehicle at about 70 km altitude, where typically the highest heat flux occurs, the atmospheric density is approximately 10^{-4} kg/m^3. Using a geometrical scaling factor of 30, a free stream density in the ground based facility of 0.003 kg/m^3 is required. If a flow with this free stream density and a velocity of 6 km/s is generated by expansion in a convergent—divergent hypersonic nozzle from a reservoir at rest without adding energy, a total specific enthalpy of about 23 MJ/kg and a nozzle reservoir pressure in the order of 90 MPa is required. This results in a nozzle reservoir temperature of about 10,000 K. It is clear that such conditions can only be achieved in impulse facilities with short flow duration [41]. The most successful types of facility which are able to generate high enthalpy and high pressure hypersonic flows are shock tunnels and shock expansion tunnels with typical test times of approximately 5 ms and less. The principle of operation of these facilities is to store the energy over a long period of time, therefore reducing the necessary power requirement and subsequently releasing the stored energy rapidly. Due to the high flow speeds, test times in the order of a few milliseconds are still sufficient for the development of a steady flow over a model. According to Hornung [41], a reasonable, conservative correlation of the necessary test time to establish a steady flow is

$$\tau = 20\frac{L}{V_\infty},$$

where L is the model length and V_∞ is the free stream velocity. For a test using the above mentioned flow condition and a 0.3 m long wind tunnel model, the required test time would be 1 ms. The high pressure, high velocity flows which can be generated in shock tunnels and shock expansion tunnels makes these facilities not only suitable for the investigation of space vehicle aerothermodynamics but also for studying complete air breathing propulsion systems, particularly supersonic combustion ramjets (scramjets) at flight Mach numbers of $M = 8$ and above. In this framework it is important that, in addition to the free stream Mach and Reynolds number, the correct static pressure and temperature are established at the combustor inflow. Further, if hypersonic flight configurations are considered which can be tested at 1:1 scale, the flight free stream conditions can be duplicated in these facilities generating the same pressure and similar heat flux loads as experienced in flight. Laminar to turbulent hypersonic boundary layer transition has been studied successfully in shock tunnels. In addition to fundamental studies of the development of the so called second mode instabilities, the influence of high temperature effects on the transition process or the influence of ultrasonic coatings on the transition process were investigated (see e.g., [1, 58, 69, 78, 109, 116]).

2 Working Principle of Free Piston Driven Shock Tunnels

2.1 Reflected Shock Tunnel Operation

The principal layout and wave diagram of a reflected shock tunnel is shown in Fig. 1. The facility is characterised by a convergent—divergent nozzle which is attached to the end of the shock tube. A thin secondary diaphragm is placed at the nozzle entrance in order to allow evacuation of the nozzle, test section and vacuum tank before the run. The nozzle entrance diameter is chosen sufficiently small such that the incident shock wave is almost completely reflected, thus compressing and raising the temperature of the test gas. The stagnant slug of test gas, generated by the shock reflection, is subsequently expanded through the hypersonic nozzle. The facility sketched in Fig. 1 has different driver section and driven section diameters. This area change at the driver-driven tube intersection generates a steady expansion until the condition of $M = 1$ is reached at the throat. This is followed by an unsteady expansion wave. For a given pressure difference between the driver and driven section, a steady expansion provides a higher velocity increase in subsonic flow than an unsteady expansion and vice versa (see e.g., [61]). Therefore, the area change is a means of improving the shock tube and shock tunnel performance.

Even for ideal shock tube performance, i.e. not considering effects such as viscosity or multi-dimensional flows, the test flow quality and resulting test time

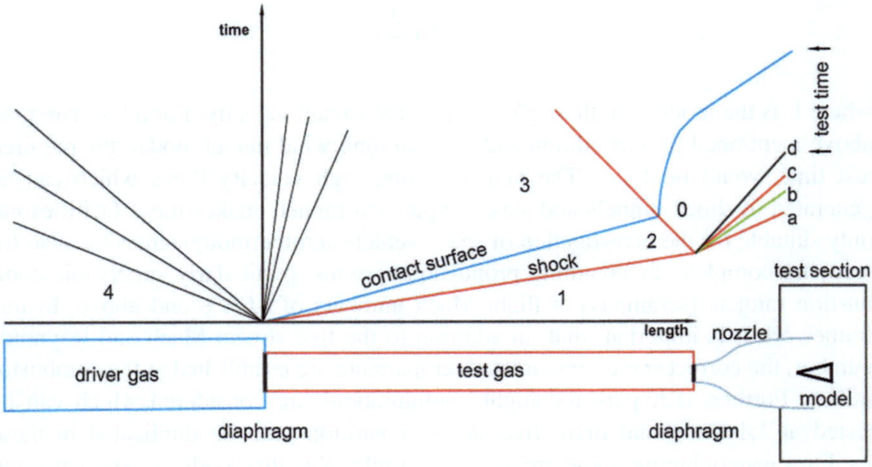

Fig. 1 Schematic and wave (x-t) diagram of a reflected shock tunnel

depends on several aspects. The first one is the wave pattern which is obtained when the reflected shock interacts with the contact surface. The shock tunnel operation should be tuned in such a way that a so called tailored interface condition is obtained, resulting in a nozzle reservoir with constant conditions (see e.g., [61]). Further, the shock tunnel geometry should be chosen such that the head of the left running expansion wave, which reflects at the left end of the driver tube and subsequently travels downstream, does not reach the nozzle entrance before the contact surface. An additional flow feature which influences the available test time is the nozzle starting process. A wave system has to pass through the nozzle before a steady flow is established. This process was studied using a quasi-one dimensional analysis by Smith [91]. The waves resulting from this analysis are shown in Fig. 1. The incident shock wave (a) is followed by a contact surface (b), an upstream facing secondary shock wave (c) and the upstream head of an unsteady expansion (d). Compared to the straight through shock tunnel mode, significantly larger test times, of the order of several milliseconds, can be achieved with a reflected shock tunnel.

2.2 *Techniques for Total Specific Enthalpy Augmentation in Reflected Shock Tunnels*

For reflected shock tunnels, a good approximation of the total specific enthalpy of the nozzle reservoir condition is given by $h_0 = u_s^2$, where u_s denotes the speed of the incident shock wave (see e.g., [41]). Introducing the shock Mach number M_s and still assuming that the gases are thermally and calorically perfect, the relation

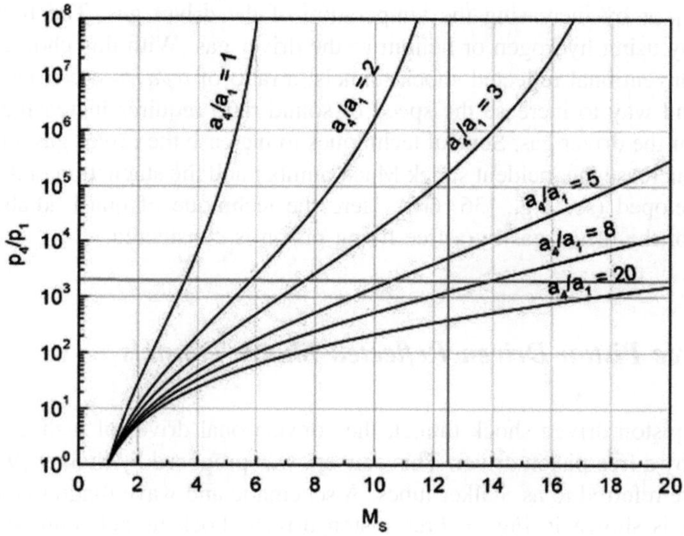

Fig. 2 Shock Mach number M_S as function of p_4/p_1 and a_4/a_1 for perfect monatomic driver ($\gamma_4 = 5/3$) and diatomic test gases ($\gamma_1 = 7/5$)

between the initial pressure and speed of sound ratio between the driver and driven section reads:

$$\frac{p_4}{p_1} = \frac{2 \cdot \gamma_1 \cdot M_S^2 - (\gamma_1 - 1)}{\gamma_1 + 1} \left[1 - \frac{\gamma_4 - 1}{\gamma_1 + 1} \cdot \frac{a_1}{a_4} \cdot \left(M_S - \frac{1}{M_S} \right) \right]^{-\frac{2\gamma_4}{\gamma_4 - 1}}. \tag{1}$$

This correlation is plotted in Fig. 2. Considering high enthalpy flows, as a first approximation, the total specific enthalpy is approximately equal to the kinetic energy of the flow. Assuming the case that an air flow in the test section with a velocity of $u_\infty = 6$ km/s should be generated, a total specific enthalpy of 18 MJ/kg ($h_0 \approx 0.5 \cdot u_\infty^2$) would be required. Due to practical reasons, air at room temperature is used in the driven section of reflected shock tunnels and consequently an approximate shock wave speed of 4.2 km/s and a shock Mach number of $M_S = 12.5$ results. According to Fig. 2, using a pressure ratio of $p_4 / p_1 = 2000$, this shock Mach number requires a ratio of speed of sounds of about $a_4 / a_1 = 8$. When increasing the shock Mach number in air such that the post shock temperature exceeds an approximate value of 800 K, inner degrees of freedom of the test gas molecules are excited and at temperatures above approximately 2500 K, oxygen molecules start to dissociate. Therefore, the assumption of a constant specific heat ratio is no longer valid. The influences of these high temperature effects must be considered for the design and performance prediction of shock tunnels. Here the discussion is restricted to reveal basic dependencies.

A high ratio of a_4 / a_1, i.e. a high value of a_4 (a_1 is fixed by the choice of using air at room temperature as test gas) can be obtained by reducing the molecular weight m_4 or by increasing the temperature of the driver gas. The first point is fulfilled by using hydrogen or helium as the driver gas. With this choice of driver gas, in conventional reflected shock tunnels, a ratio of $a_4/a_1 = 3$ can be achieved. The second way to increase the speed of sound ratio requires increasing the temperature of the driver gas. Several techniques to increase the driver-gas temperature and thus increase the incident shock Mach number and the stagnation enthalpy have been developed (see e.g., [36, 60]). Here the technique of quasi adiabatic compressing of the driver gas by a free flying piston is considered.

2.3 Free Piston Driven Reflected Shock Tunnels

In a free piston driven shock tunnel, the conventional driver of a shock tunnel is replaced by a free piston driver. This concept was proposed by Stalker [97] and the tunnels are referred to as Stalker tubes. A schematic and wave diagram of this type of facility is shown in Fig. 3. Free piston driven shock tunnels consist of an air buffer, a compression tube, separated from an adjoining shock tube via the primary diaphragm, and a subsequent nozzle, test section and vacuum tank. The high pressure air stored in the air buffer is utilized to accelerate a heavy piston down the compression tube. During this quasi-adiabatic compression of the light driver gas (typically helium or a helium/argon mixture), the piston reaches a maximum speed

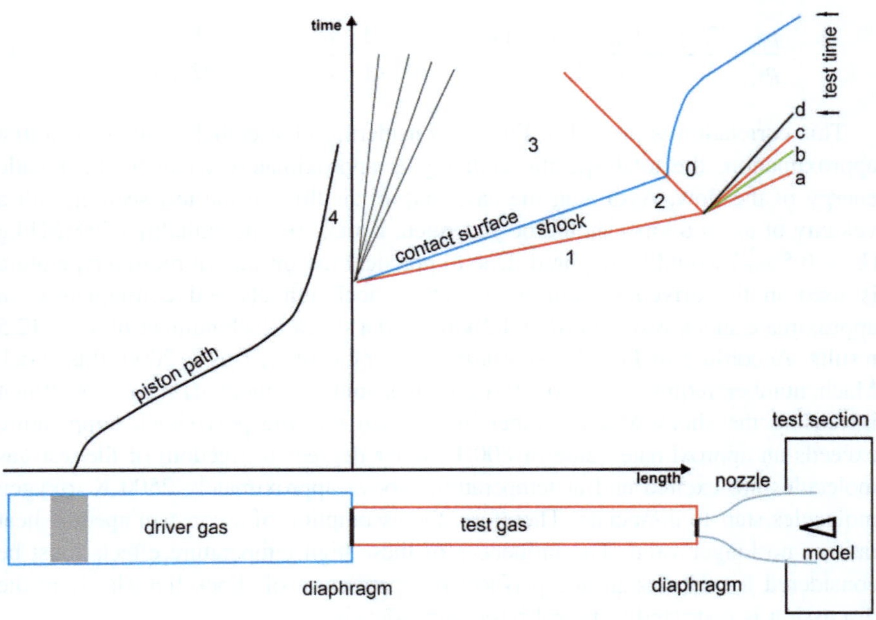

Fig. 3 Schematic and wave (x-t) diagram of a free piston driven shock tunnel; waves *a*, *b*, *c*, and *d* are explained in Sect. 2.1

in the order of 300 m/s. The driver gas temperature increases with the driver gas volumetric compression ratio. In principle, there is no limit to the used compression ratio and therefore, there is also—in principle—no limit to the stagnation enthalpy which can be generated [92].

The numbers used in Fig. 3 (see also Fig. 1) denote distinct regions of the flow. Region 1 contains the test gas at the initial shock tube filling conditions and region 4 contains the hot, compressed driver gas after piston compression. Region 2 contains the shock compressed test gas, while in region 3, the driver gas processed by the unsteady expansion wave is contained. The test and driver gas are separated by a contact surface. When the main diaphragm burst pressure is reached it ruptures and the wave process as in a conventional reflected shock tunnel is initiated. The trajectory of the piston is chosen in a way that, after main diaphragm rupture, the pressure and temperature of the driver gas is maintained approximately constant. This is achieved by selecting the velocity of the piston at diaphragm rupture, and, therefore, the subsequent movement of the piston such that it compensates for the loss of the driver gas flowing into the shock tube. For that reason, in contrast to the constant volume driver of conventional shock tunnels, the free piston driver is a constant pressure driver. Due to the large forces occurring during the operation of the free piston driver, the compression tube, shock tube, nozzle assembly is allowed to move freely in the axial direction. After reflection of the incident shock wave at the right end wall of the shock tube, the test gas is brought to rest in region 0. Subsequently, the secondary diaphragm breaks and the nozzle flow is initiated. Further, the reflected shock wave penetrates the contact surface.

An inertial mass placed at the compression tube/shock tube junction can significantly reduce the recoil motion of the facility during operation. The test section and the vacuum tank remain stationary. A sliding seal is used at the nozzle/test section interface. While, in principle, there is no limit on the achievable stagnation enthalpy for free piston driven shock tunnels, in real facility operation, effects such as driver gas contamination of the test gas impose upper limits. Tailored interface operation at shock Mach numbers of M_s = 18.5 in air is reported [92, 97]. This results in a total specific enthalpy of about 40 MJ/kg.

Considering the different types of high performance shock tunnel drivers, the free piston technique appears to be the most developed and most distributed. After a series of free piston driven shock tunnels, T1–T3 (Australian National University, Canberra) and T4 (The University of Queensland, Brisbane) were built in Australia between the early 1960s and the late 1980s [92, 96], this type of facility was implemented in a number of different institutions worldwide. The largest ones of these are the T5 at the Graduate Aeronautical Laboratories, California Institute of Technology, United States [40], the High Enthalpy Shock Tunnel Göttingen, HEG, of the German Aerospace Center [23] and the High Enthalpy Shock Tunnel, HIEST, of the Japan Aerospace Exploration Agency (JAXA) at Kakuda [43]. The latter uses the largest nozzles with an exit diameter of 1200 mm. The maximum stagnation enthalpy used is 25 MJ/kg at 1500 bar nozzle reservoir pressure resulting in a test time of 2 ms.

3 High Enthalpy Shock Tunnel Göttingen (HEG)

The HEG is a free piston driven shock tunnel [33, 35, 37] at Göttingen, Germany which was developed and constructed in the framework of the European HERMES program over the period 1989–1991. It was commissioned for use in 1991, at that time being the largest facility of its type worldwide. Since then it was extensively used in a large number of national and international space and hypersonic flight projects. The research activities which were always strongly linked with CFD investigations range from the calibration process of the facility and the study of basic aerodynamic configurations, which are well suited to investigate fundamental aspects of high enthalpy flows, to the investigation of complex entry, re-entry, hypersonic flight and integrated scramjet configurations. A schematic and photos of HEG are given in Figs. 4 and 5, respectively.

The overall length of HEG is 62 m and it weighs 280 t. Approximately a third of its weight is contributed by the inert mass (see Fig. 4 and the left picture of Fig. 5). The compression tube is closed by a hydraulic oil system (quick disk connect) at the main diaphragm station. The shock tube is connected to the nozzle of the tunnel at the downstream closure, which is also driven by oil hydraulics to close and seal the tunnel. The compression tube has a length of 33 m and a diameter of 0.55 m. The shock tube is 17 m long with a diameter of 0.15 m. The HEG was designed to provide a pulse of gas to a hypersonic convergent—divergent nozzle at stagnation pressures of up to 200 MPa, and stagnation enthalpies of up to 23 MJ/kg. Regarding

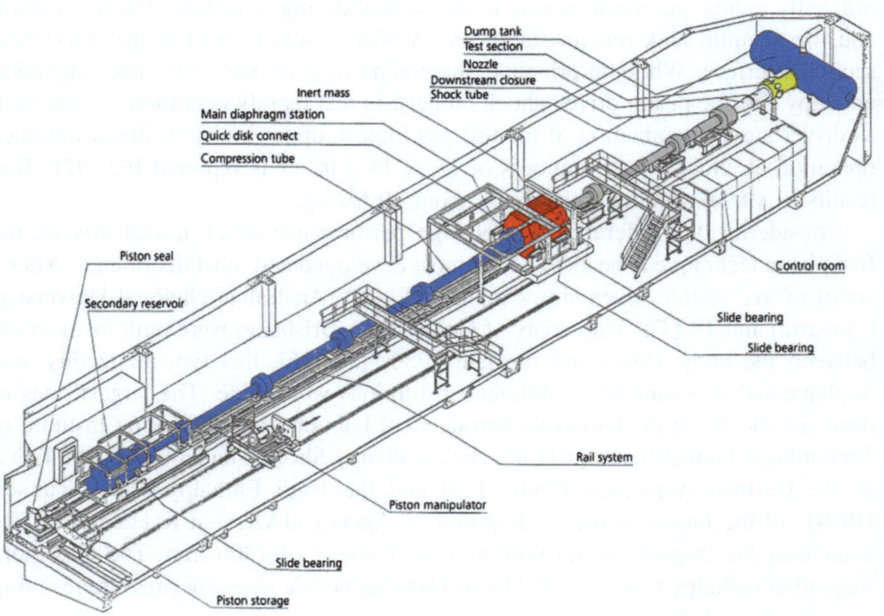

Fig. 4 Schematic of the High Enthalpy Shock Tunnel Göttingen (HEG)

the test gas, no basic limitation exist. The operating conditions given here are related to the test gas air. Additionally, operating conditions using nitrogen and carbon dioxide exist.

Originally, HEG was designed for the investigation of the influence of high temperature effects such as chemical and thermal relaxation on the aerothermodynamics of entry or re-entry space vehicles. In order to correctly model the chemical relaxation occurring behind the bow shock of a re-entry vehicle, the flight binary scaling parameter must be reproduced in ground based testing. Further, for high enthalpy testing an additional driving parameter which must be reproduced is the flow velocity. Therefore, the operating conditions of HEG are shown in Fig. 6 in terms of the binary scaling parameter ρL and the flow velocity u. Here L represents a characteristic length of the considered configuration. In addition to the HEG

Fig. 5 Photographic views of the High Enthalpy Shock Tunnel Göttingen (HEG)

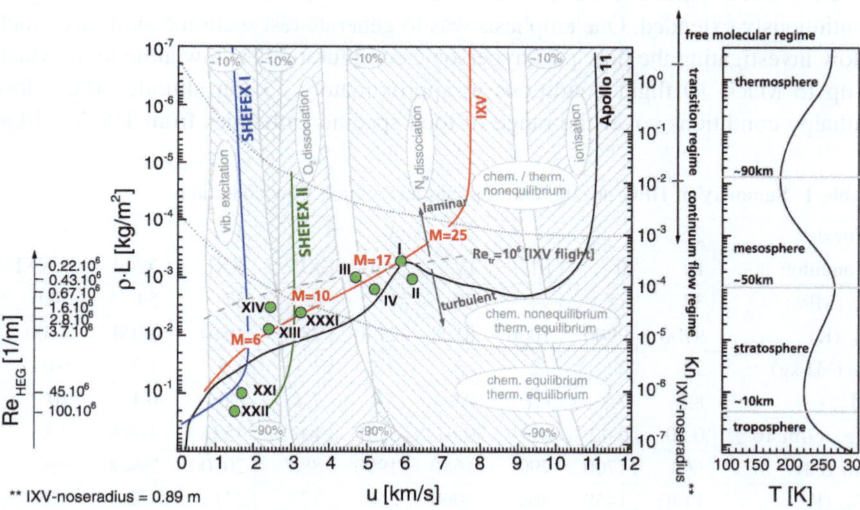

Fig. 6 HEG operating conditions in terms of the binary scaling parameter ρL and the flow velocity u

operating conditions, the most important fluid mechanical and chemical processes occurring during re-entry of a spacecraft in the Earth's atmosphere are depicted in Fig. 6. Further, as reference, the flight trajectories of a lifting body re-entry from low Earth orbit (ESA intermediate experimental vehicle (IXV)), a ballistic super orbital re-entry (Apollo 11) and two hypersonic flight experiments (DLR flight test SHEFEX I and SHEFEX II) are provided.

An indication of the corresponding flight altitudes is given in the right diagram of Fig. 6 showing the temperature variation of the Earth's atmosphere. The transitions between regimes of different physical and chemical properties shown in Fig. 6 depend on the chosen reference length and vary when different configurations are considered. Further, the boundaries shown have only symbolic character. In reality, no clear-cut dividing lines between the different regimes exist. The Knudsen number given in Fig. 6 shows that the HEG operating conditions are located in the continuum flow regime. Depending on the ratio of the relaxation time to a characteristic timescale of the flow, the chemical and thermal relaxation processes can be either in non-equilibrium or in equilibrium. Further, along a re-entry trajectory, the Reynolds number varies over several orders of magnitude. In high altitude flight the wall boundary layer of a re-entry vehicle is initially laminar. After exceeding a critical Reynolds number (shown exemplarily for the IXV configuration in Fig. 6), the transition from a laminar to a turbulent boundary layer takes place. This process is linked with an increase of the skin friction and the wall heat flux.

A summary of the main HEG operating conditions is provided in Table 1. The HEG operating conditions I–IV are the high enthalpy conditions covering a total specific enthalpy range from 12–23 MJ/kg. A detailed discussion of these conditions is given in Hannemann et al. [32]. They have been used, for example, for the investigation of several re-entry configurations including the Atmospheric Re-entry Demonstrator, ARD [35, 51], and X-38 [35]. The HEG operating range is continuously extended. One emphasis was to generate test section conditions which allow investigating the flow past hypersonic configurations at low altitude for Mach 6 up to Mach 10 flight conditions at approximately 33 km altitude. These low enthalpy conditions cover the range of total specific enthalpies from 1.5–6 MJ/kg.

Table 1 Summary of HEG nozzle reservoir and test section flow conditions

Nozzle	2				3		4		5
Condition	I	II	III	IV	XIII	XIV	XXI	XXII	XXXI
p_0 (MPa)	35	85	44	90	17	8	37	54	70.0
T_0 (K)	9100	9900	7000	8100	2740	2810	1640	1200	4400
h_0 (MJ/kg)	22	23	12	15	3.3	3.4	1.5	1.3	6.0
M_∞ (–)	8.2	7.8	8.1	7.9	7.4	7.4	6.0	6.1	10.3
Re_m (1/m 10^6)	0.20	0.42	0.39	0.67	3.70	1.60	45.0	100.0	2.8
p_∞ (Pa)	660	1700	790	1680	1990	880	20100	29400	930
T_∞ (K)	1140	1450	800	1060	266	277	221	152	253
ρ_∞ (g/m^3)	1.7	3.5	3.3	5.3	25.9	11.0	327.0	682.0	12.6
u_∞ (m/s)	5900	6200	4700	5200	2410	2450	1750	1510	3270

For 1:1 scale wind tunnel models, conditions XIII and XIV duplicate M = 7.4 flight conditions at 28 and 33 km, respectively. They were used for the ground based testing of the HyShot II [38, 59] and IV [66] supersonic combustion flight experiment configurations. Condition XXI is related to Mach 6 flight conditions at 15 km flight altitude and condition XXII is the HEG operating condition with the highest unit Reynolds number. Condition XXI was used, for example, in the framework of the SHEFEX I post flight analysis [67]. Condition XXXI focuses on Mach 10 flight conditions at 33 km altitude. It was used in the framework of the DLR SHEFEX II flight experiment [6].

With reducing flight velocity (i.e. reducing total specific enthalpy), the binary scaling parameter becomes less important. However, for 1:1 scale models of hypersonic flight configurations, the ρL versus u diagram gives an indication of the dynamic pressure range which can be reproduced in HEG. In particular for air-breathing propulsion driven hypersonic vehicles this quantity is of interest. Typically they fly at dynamic pressures between 50 and 100 kPa, and, in order to model the combustion process in e.g. a scramjet engine correctly, the dynamic pressure is one of the quantities which must be duplicated. In Fig. 7, the low enthalpy HEG operating conditions are also given in terms of Mach and Reynolds number. As reference length, a fixed wind tunnel model length of 0.3 m and the lengths of the considered flight configuration were used.

The test time for the high enthalpy conditions I–IV is up to 1 ms. For the low enthalpy conditions, the test time ranges from 3–6 ms.

In order to realise the operating conditions of Table 1, a series of different Laval nozzles had to be designed, constructed and implemented in HEG. The nozzles used to generate the corresponding test conditions are also given in Table 1. Further, different pistons are utilised on HEG for generating different operating conditions. In order to allow a large flexibility in tuning new operating conditions, four pistons (without brakes) with different weight (275, 481, 700 and 815 kg) are available. One additional 848 kg piston with brakes is utilised.

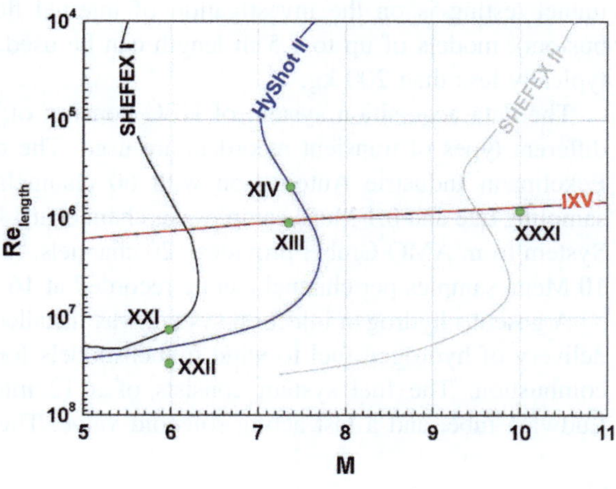

Fig. 7 HEG operating conditions in terms of Mach and Reynolds number

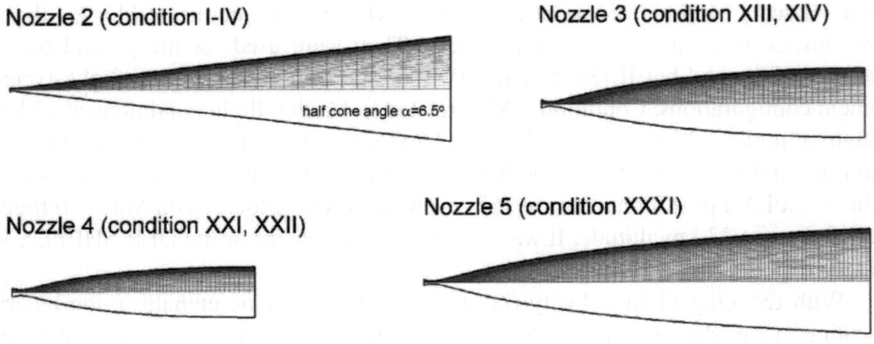

Nozzle	2	3	4	5
nominal M	8	8	6	10
d*[mm]	22	40	50	26
d_{Exit}[mm]	880	590	430	880
A_{Exit}/A^*	1600	218	74	1146
L [mm]	3780	2710	2015	3780

Fig. 8 Geometry of operational HEG nozzles

The set of HEG nozzles comprises of a conical nozzle used for the high enthalpy conditions I-IV and three contoured nozzles for the low enthalpy conditions. Their nominal design Mach number, area ratio and length are given in Fig. 8. Note that for nozzle 2 the Mach number is lower than the corresponding flight Mach number (see also Fig. 6) due to chemical and thermal freezing effects during the nozzle expansion [32]. However, for high enthalpy testing the Mach number is of less importance and the flight velocity must be reproduced correctly.

Depending on the chosen operating condition and the angle of attack, model configurations with a typical length between 0.4 and 1.0 m and a width of up to 0.4 m can be mounted in the test sections. In case where the major emphasis of the tunnel testing is on the investigation of internal flow paths (e.g. scramjet combustors), models of up to 1.5 m length can be used. The weight of the models is typically less than 200 kg.

The data acquisition system of HEG consists of a total of 180 channels. Two different types of transient recorders are used. The original KRENZ System from Eckelmann Industrie Automation with 60 channels is characterised by 1 MHz sampling rate and 0.1 Mega samples per channel at 14 Bit resolution. The SATURN System from AMO GmbH provides 120 channels. Its sampling rate is 50 MHz and 10 Mega samples per channel can be recorded at 16 Bit resolution.

A gaseous hydrogen injection system was installed at HEG in order to allow the delivery of hydrogen fuel to wind tunnel models for the investigation of scramjet combustion. The fuel system consists of a 12 mm diameter and 38.4 m long Ludwieg tube, and a fast acting solenoid valve. The maximum filling pressure of

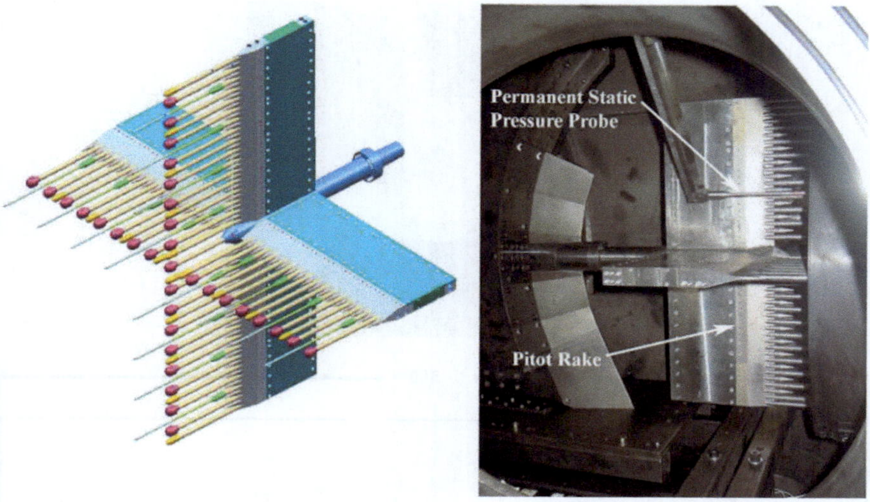

Fig. 9 HEG calibration rake—design drawing (*left*) and rake installed in the test section (*right*)

the Ludwieg tube is 15 MPa and it can deliver a pulse of fuel with constant pressure for up to 50 ms.

A modular cross arm calibration rake was designed and manufactured. It can be used for the detailed calibration of the test section free stream flows generated by nozzles 2–5. Probes to measure Pitot pressure, static pressure and stagnation point heat transfer can be mounted on this rake (see Fig. 9).

3.1 Selected Applications of HEG

3.1.1 High Enthalpy Cylinder Shock Layer Investigations

When considering the re-entry trajectory of winged space vehicles returning from low earth orbit into the atmosphere, the most critical point concerning the heating loads on the vehicle is found in the continuum flow region in approximately 70 km altitude. In this region of the re-entry path, the velocity of the vehicle is approximately 6 km/s and the flow past the vehicle is accompanied by strong shock waves, leading to high temperatures and ensuing dissociation reactions. The fundamental influence of the thermal and chemical relaxation processes caused by these high temperature effects on the external aerodynamics, i.e. the pressure distribution, flap efficiency, shock/shock and shock/boundary layer interactions and on the heating loads can be investigated by looking at the flow past basic generic flow configurations which are especially designed in order to focus on one of these effects. Additionally, these studies are well suited to validate the ground based facility performance, measurement techniques and computational fluid dynamics (CFD) codes.

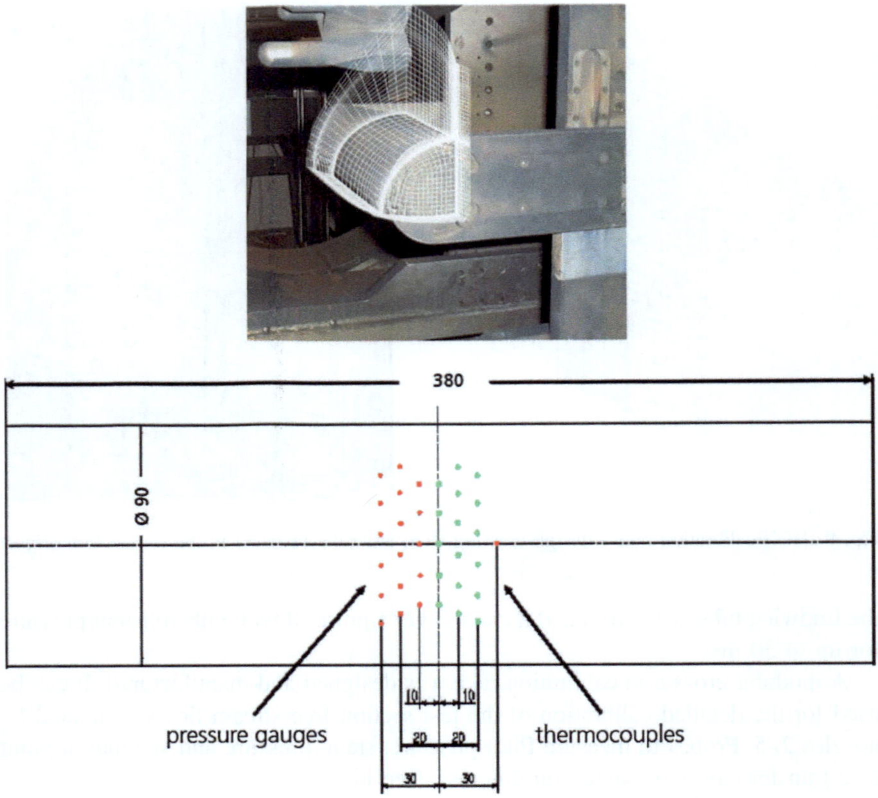

Fig. 10 Cylinder model in the HEG test section including the grid used for the three-dimensional flow field computations (*top*) and model dimensions as well as gauge positions in mm (*bottom*)

A test campaign was performed in HEG to study the relaxation processes in the shock layer of a cylinder placed with its axis transverse to the flow [34, 53] . This configuration was chosen because of the large shock stand-off distance that permits optical measurement techniques to investigate gas properties in the shock layer. Surface pressure and surface heat flux measurements and the determination of two dimensional shock layer density distributions by phase step holographic interferometry were performed. The diameter of the cylinder is 90 mm and its length amounts to 380 mm. The cylinder model was equipped with pressure transducers to measure surface pressure distributions and thermocouples to measure surface heat flux distributions, and the model was mounted on the nozzle centreline (Fig. 10).

The results presented here were obtained using HEG operating condition III (h_0 = 12 MJ/kg,) and condition I (h_0 = 22 MJ/kg, see Table 1). Two and three-dimensional computations of the flow past the cylinder model were performed assuming laminar flow, chemical nonequilibrium, thermal equilibrium and a fully catalytic cylinder wall using the DLR CEVCATS-N code [34]. The computational

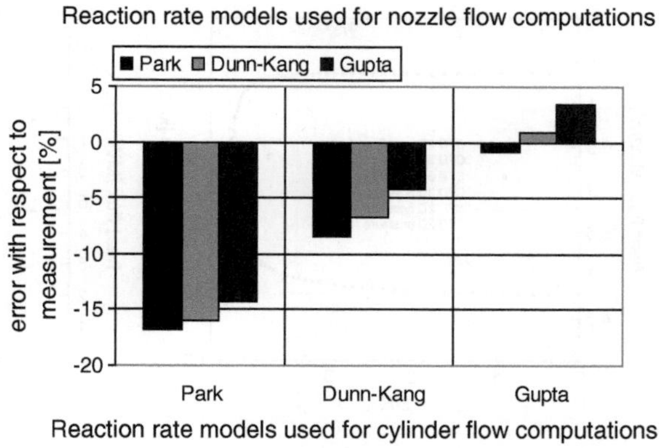

Fig. 11 Comparison of the computed shock stand-off distance resulting from 2D flow field analyses with the values obtained in HEG

grid used for the 3D simulations is shown in the upper part of Fig. 10. The symmetry plane of the 3D grid was also used for the 2D reference computations.

Two dimensional simulations were first carried out using the reaction rate models by Park [73], Dunn and Kang [20] and Gupta et al. [31]. For the latter the third body efficiencies were set to unity. Hereby the dissociation reactions are delayed. The different free stream conditions used for this study resulted from nozzle computations, each performed with one of the reaction rate models mentioned above. The resulting shock standoff distances were compared with the experimental values and a summary of the results is given in Fig. 11. It can be seen that using the modified reaction rate model of Gupta et al. results in the smallest deviation between computed and measured shock stand-off distance. For that reason this chemistry model was also chosen for further 2D and 3D cylinder flow field computations. In order to investigate the influence of the flow past the edges of the cylinder on the flow in the central part of the model, 3D flow field computations were performed. A comparison between 2D and 3D results for surface pressure and heat flux along the stagnation line in the span wise direction is shown in Fig. 12 (left). The centre plane of the cylinder is located at $z/R = 0$. Two different sets of inflow conditions were used for the 3D simulations. The first set is referred to as "parallel" and represents a parallel inflow with averaged free stream conditions in the plane of the cylinder position. The second set is referred to as "conical". Here the diverging flow resulting from the conical nozzle computation is used to interpolate inflow conditions on the CFD grid around the cylinder. A comparison of the computed surface pressure and heat flux distributions in the centre plane is given in Fig. 12 (right). It is clear from these plots that the surface data in the centre plane resulting from 3D computations compares well with the data obtained from the 2D computation. The heat flux and pressure distributions along the stagnation line of the cylinder obtained with the 3D computations show the influence of the cylinder edge effects on the surface pressure and heat flux. From the centre plane up to about $z/R = 2.5$, no significant impact of the edge effects exists. It can also be seen from Fig. 12 that compared to the "parallel" inflow condition, the utilisation of the

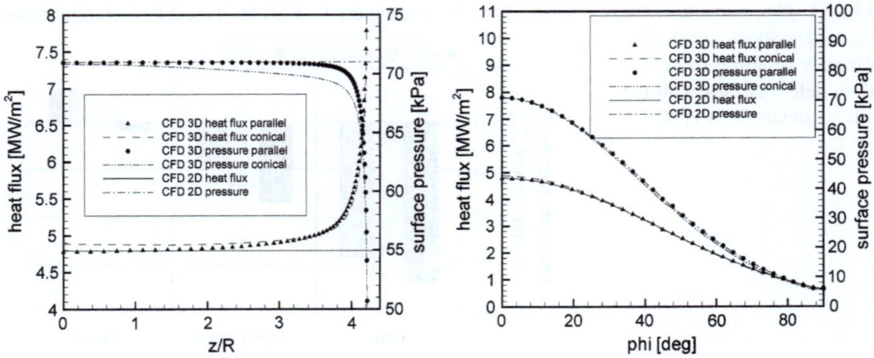

Fig. 12 Comparison of surface pressure and heat flux along the stagnation line on the cylinder in span wise direction resulting from 2D and 3D computations (*left*); comparison of surface pressure and heat flux in the centre-plane resulting from 2D and 3D computations (*right*)

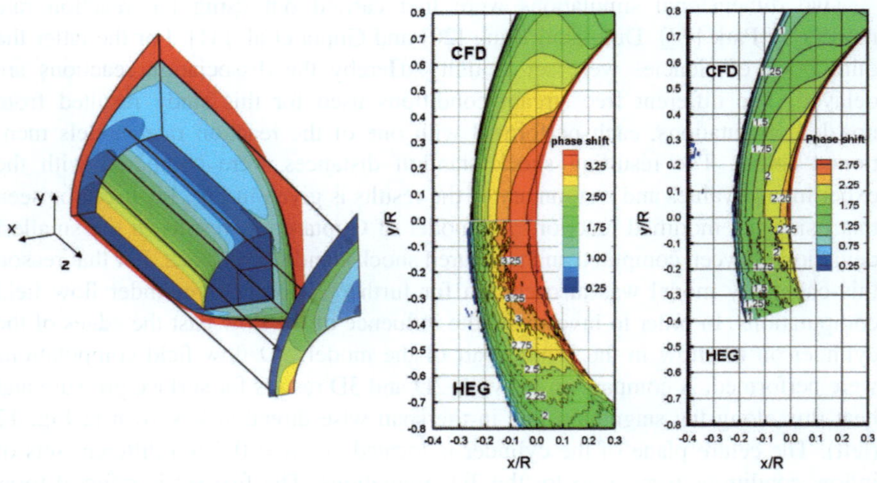

Fig. 13 Sketch of the line of sight phase shift distribution reconstruction based on the computed 3D flow fields (*left*) and comparison of computed and measured phase shift distribution (*right*); HEG operating condition III (run 627) and operating condition I (run 619); the phase shift values are normalized by 2π

"conical" condition results in only a small difference of the obtained surface pressure and heat flux in the centre plane.

It was shown through the comparison of the results of the two and three dimensional computations, that the flow in the symmetry plane of the cylinder can be regarded as two dimensional. However 3D effects become important for optical line of sight methods. Therefore, numerical phase shift distributions were obtained from the CFD solutions by using a procedure as described by Hannemann et al. [34]. A computational ray tracing algorithm was implemented (see Fig. 13) and numerical phase shifts were obtained for the 2D computations and the 3D

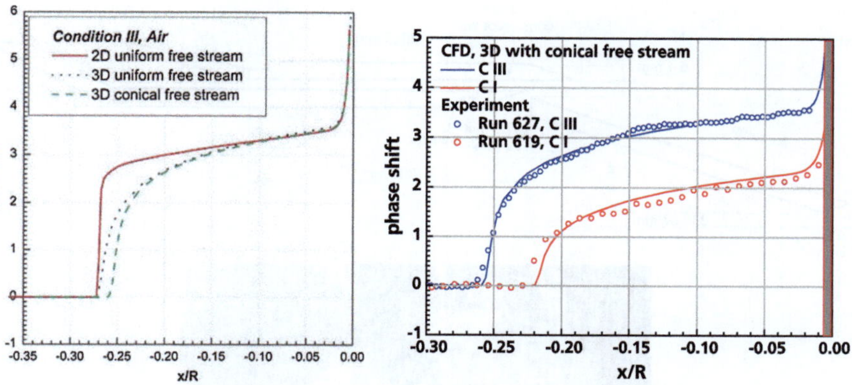

Fig. 14 Comparison of computed phase shift distributions along the stagnation streamline (*left*); comparison of computed and measured phase shift distributions along the stagnation streamline for HEG operating condition III and condition I [65] (*right*); the phase shift values are normalized by 2π

computations with both "parallel" and "conical" free stream conditions. The comparison of the computed phase shift distributions along the stagnation line is given in Fig. 14 (left). This figure shows clearly that the result obtained from the 3D computations by ray tracing along a set of lines of sight differ from the 2D CFD results. It should be emphasized that these differences are only due to the contributions of the outer flow regions past the cylinder edges. The use of the "conical" free stream condition for the 3D computation results in a shock stand-off distance which is reduced by about 3 %.

Figure 14 (right) shows the computed and measured phase shift distributions along the stagnation streamline. Using the result of the 3D computation with "conical" free stream condition, very good agreement was obtained with the measured phase shift distribution for both conditions [65]. The corresponding two-dimensional phase shift distributions are plotted in the right part of Fig. 13.

3.1.2 Free Jet Testing and Numerical Analysis of Scramjet Flow Paths

Experimental investigations of supersonic combustion ramjet (scramjet) flow paths in HEG are combined with numerical investigations using the DLR TAU code [27, 52, 79]. The applied methodology is outlined in Hanneman et al. [38]. Its capability has been demonstrated by applying it to the flow path of the generic HyShot II scramjet flight experiment configuration [75, 88]. This generic scramjet consisting of intake, combustor and nozzle was designed such that it contains all necessary components of a scramjet but still generates an overall flow field which is as simple as possible with well-defined combustor inflow. Therefore, it is considered to be a configuration well suited for basic investigations and tool validation purposes.

A cut along the symmetry plane of the HyShot II model configuration assembly is shown in the upper part of Fig. 15. The intake is a simple 18° wedge. Between the

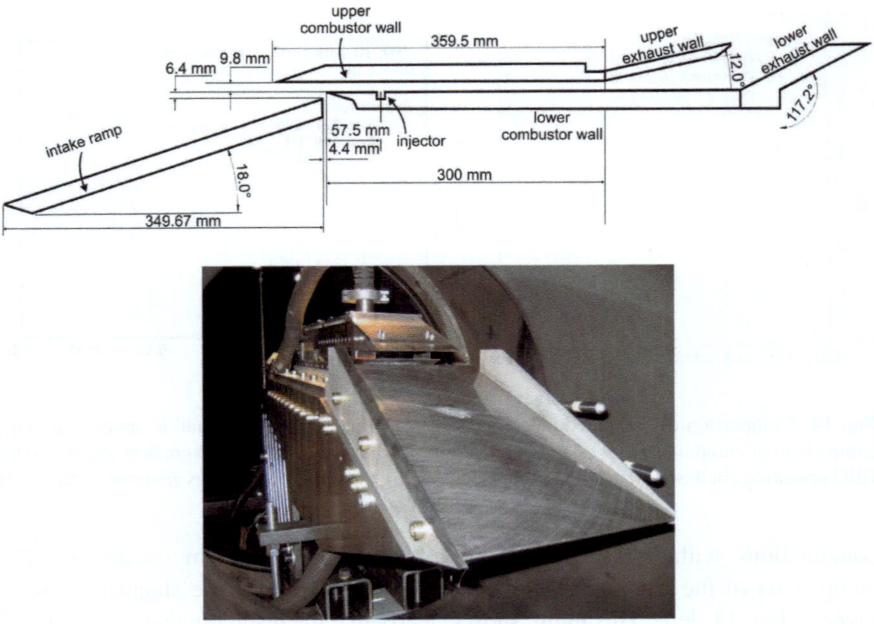

Fig. 15 HyShot II model configuration assembly (*upper*) and model installed in the HEG test section (*lower*) (from Hanneman et al. [38])

intake and the combustion chamber, a boundary bleed channel is used such that the boundary layer and entropy layer of the intake ramp do not enter the combustor. Further, the bleed allows the shock generated by the leading edge of the cowl to pass outside the combustion chamber. The combustion chamber is parallel and the cross sectional dimensions are 9.8 × 75 mm. The length of the constant area combustor is 308.5 mm. Fuel is injected into the combustion chamber by a series of four holes, with radius of 1 mm, equally spaced along one line in cross flow direction in the lower combustor wall, 57.5 mm downstream of its leading edge. The HyShot II wind tunnel model installed in the HEG test section is shown in the lower part of Fig. 15.

The experimental and numerical treatment of scramjet flow path generated in ground based test facilities closely depends on the utilisation of a suitable and correct set of free stream data. In addition to the free stream Mach number and the unit Reynolds number, the knowledge of the absolute values of the free stream pressure and temperature are of particular importance since they determine via the intake flow the combustor inflow condition which in turn have a strong influence on the combustion process. Therefore, the present approach is split up in three main steps, namely: (1) Determination of the HEG free stream flow, (2) Analysis of the intake flow, and (3) Analysis of the combustor and exhaust flow [38].

The evaluation of the free stream conditions in the test section of the HEG free piston driven shock tunnel is performed by numerical analysis using a suitable set

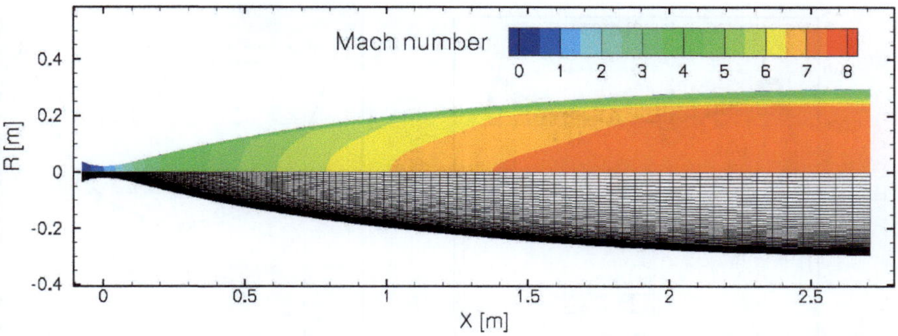

Fig. 16 CFD grid and Mach number contours for the HEG nozzle flow (operating condition XIII)

of measured input and reference parameters. First, the nozzle reservoir temperature is computed with a one-dimensional simulation of the shock tube. The relevant input parameters are the measured values of the initial shock tube filling pressure and temperature, the shock speed and the nozzle reservoir pressure. Based on the nozzle reservoir conditions (see Table 1), the free stream is subsequently determined by numerical computation of the nozzle flow utilising the DLR TAU code. The grid and Mach number contours, resulting from a computation using the assumption that a thermal non-equilibrium nozzle expansion is established, are shown in Fig. 16.

It should be noted that for the present operating condition, the influence of chemical reactions is considered to be small due to the low reservoir temperature. However, the temperatures are sufficiently high that molecular species are vibrationally excited. Due to the fact that the thermal relaxation behaviour in the nozzle expansion is not known a priori, assumptions of both thermal equilibrium and thermal non-equilibrium were investigated separately in the computations.

The accuracy of this CFD based approach for the determination of the free stream conditions was assessed by comparison with available experimental calibration data at the nozzle exit plane. The data was obtained with the calibration rake shown in Fig. 9. In Fig. 17, the comparison of measured and computed normalised Pitot pressure and static pressure at the nozzle exit plane are shown. The data are normalised using the nozzle reservoir pressure. Regarding the Pitot pressure, the computed data resulting from computations assuming either a thermal equilibrium or a thermal non-equilibrium nozzle expansion lie within the experimental scatter bars. For the present operating condition, the wall boundary layer is assumed to be fully turbulent and the difference due to the application of different turbulence models is negligible. The excellent reproduction of the measured Pitot pressure profile shape confirms the assumption regarding the state of the boundary layer. The best agreement between computed and average measured Pitot pressure profiles is obtained with the thermal equilibrium assumption. The computed static pressure profiles reveal pronounced deviations resulting from the application of different thermal relaxation models. Excellent agreement between the numerical and

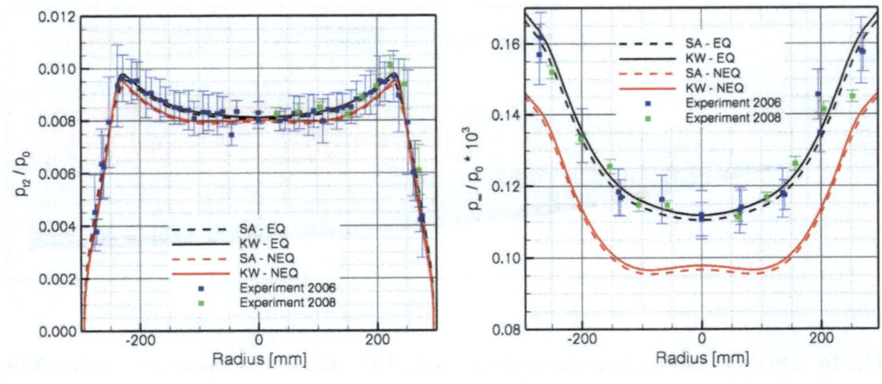

Fig. 17 Comparison of measured and computed normalised Pitot pressure (*left*) and static pressure (*right*) at the nozzle exit plane; *NEQ* thermal non-equilibrium, *EQ* thermal equilibrium, *SA* Spalart-Allmaras turbulence model, *KW* Wilcox k-ω turbulence model

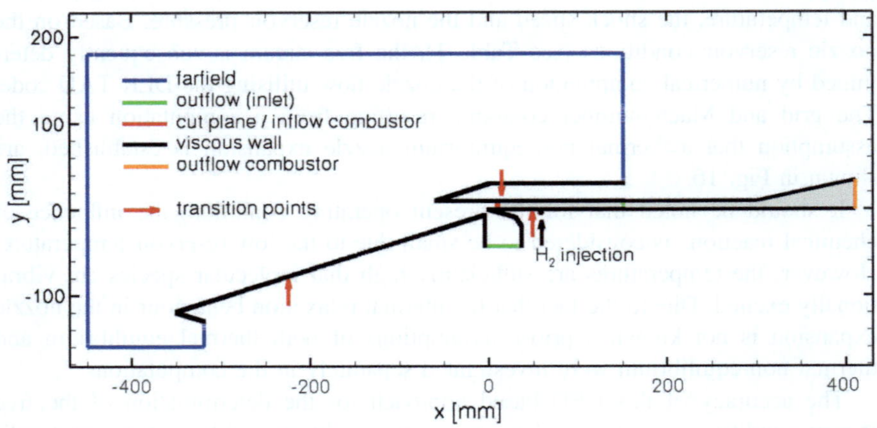

Fig. 18 Computational domains and boundary conditions used for the intake and combustor/exhaust investigations [38]

experimental results was achieved using the thermal equilibrium assumption. These results highlight the importance of the static pressure measurements since they help to select the most suitable numerical model to determine the free stream pressure and consequently the free stream temperature.

The computational domains and the applied boundary conditions used to study the intake and the combustor/nozzle flow fields of the HyShot II configuration are sketched in Fig. 18 in a plane cutting through an injector. The HEG free stream used for the present studies is related to a flight trajectory point of HyShot II in an altitude of 28 km [38] (condition XIII in Table 1). From the flight data [75], an angle of attack of $\alpha = 3.6^{\circ}$ was derived and used for the present studies.

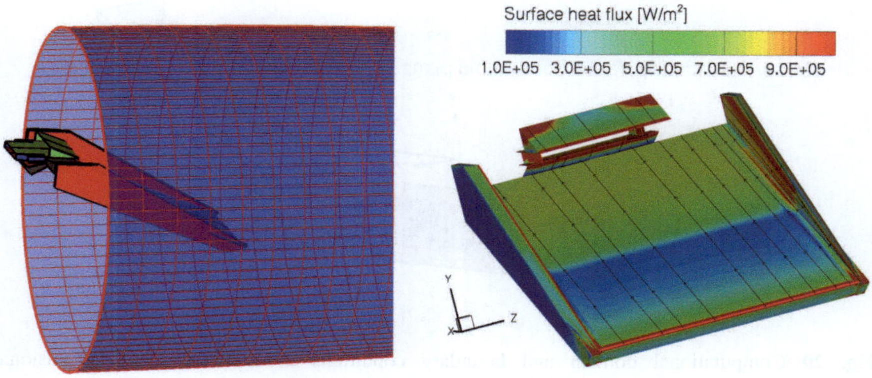

Fig. 19 Position of the HyShot II intake relative to the exit section of the HEG nozzle (*left*); Skin friction lines and surface heat flux contours on the HyShot II intake ramp (*right*)

The intake flow was first investigated by a three-dimensional flow field computation. The inflow conditions were determined by interpolation from nozzle flow computations. The position of the intake relative to the HEG nozzle exit section is shown in Fig. 19 (left).

Computed skin friction lines and the surface heat flux distribution on the intake are plotted in Fig. 19 (right). The laminar to turbulent boundary layer transition occurring on the intake ramp was determined with an empirical correlation based on $[Re_\theta/M_e]_{crit} = 200$ [5]. Here, $Re_\theta = \rho_e u_e \theta/\mu_e$ denotes the Reynolds number based on the boundary layer momentum thickness and M_e the Mach number at the boundary layer edge. The computational result indicates that due to the side walls, a highly two-dimensional flow is produced on the intake ramp and hence at the inlet to the combustion chamber.

The computational domain and boundary conditions used for the combustor computations is shown in Fig. 20. In order to minimise the computational cost all existing flow symmetries were exploited. The combustor inflow conditions result from intake computations. Transition from laminar to turbulent boundary layer flow at the top and bottom combustor wall is modelled using the same criterion as for the intake.

The normalised wall pressure and heat flux distributions generated by hydrogen combustion in the HyShot II combustor on the lower and the upper wall are plotted in Figs. 21 and 22, respectively. The fuel equivalence ratio was chosen such that a steady state combustor flow was obtained. According to the design of the HyShot II flight experiment [76], the combustor is expected to show a steady state supersonic combustion flow up to an equivalence ratio of $\phi \approx 0.35$. The fuel equivalence ratio is defined as the ratio of the fuel-to-oxidizer ratio to the stoichiometric fuel-to-oxidizer ratio:$\phi = 8 \cdot \dot{m}_{H_2}/\dot{m}_{O_2}$. For clarity, the experimental data are subdivided into three groups with different ranges of fuel equivalence ratio, $0.266 < \phi < 0.288$, $0.315 < \phi < 0.329$ and $0.341 < \phi < 0.351$. The computed normalised pressure and heat flux distributions were obtained with the DLR TAU

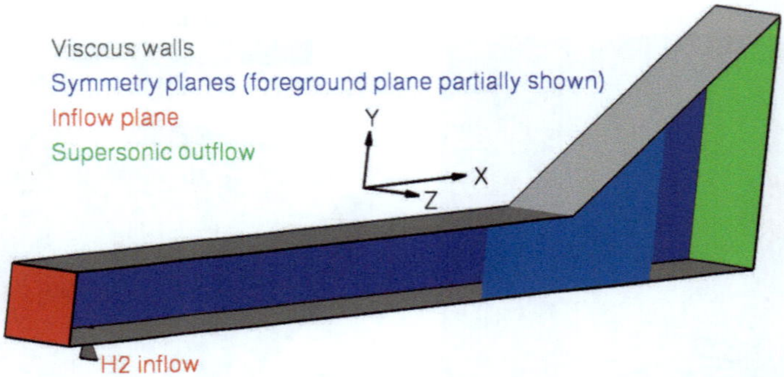

Fig. 20 Computational domain and boundary conditions used for the three-dimensional combustor fuel on computations

code using the Spalart-Allmaras and the Wilcox *k*-ω turbulence model and an equivalence ratio of $\phi = 0.295$. For all equivalence ratios, the measured wall pressure distributions are characterized by a continuous increase along the combustor indicating supersonic combustion. Good agreement between the measured and the computed pressure distributions is achieved. Further, the computed normalised pressure distributions depend only weakly on the applied turbulence model.

Regarding the comparison of measured and computed wall heat flux distributions, good agreement was obtained on the upper combustor wall using the Spalart-Allmaras turbulence model (Fig. 22). However, on the lower combustor wall significant differences occur between experimental and numerical results. While the computed heat flux increases continuously downstream of the fuel injection location, the measured values remain approximately constant after an initial stepwise increase (Fig. 21). A possible reason for this discrepancy between measured and computed heat flux values could be the strong span wise heat flux gradients obtained downstream of the hydrogen injectors. The heat flux gauges are positioned downstream of a fuel injector and even a small flow misalignment could cause the observed discrepancy.

In general, the comparison of measured and computed pressure and heat flux distributions shows that the pressure generated by supersonic combustion can be predicted numerically with satisfactory accuracy. However, apart from the qualitative difference between the measured and computed heat flux distributions on the lower combustor wall, the CFD data depends strongly on the applied turbulence model. The computed peak values on the lower combustor wall differ by a factor of almost two and on the upper wall the heat flux levels differ by approximately 30 %.

In Fig. 23 schlieren and OH* chemiluminescence visualizations of the section of the combustion chamber immediately downstream of the injection (56–136 mm) are shown. The injected hydrogen has penetrated to approximately half the duct height by the downstream end of the visualized section. The OH* visualization shows the onset of combustion to be clearly linked to the shock structure in the duct, namely, the first reflection of the injection shock appears to initiate combustion close to the injector-side wall; the second reflection then "kicks" the flame further out into the duct and increases the intensity of combustion [59].

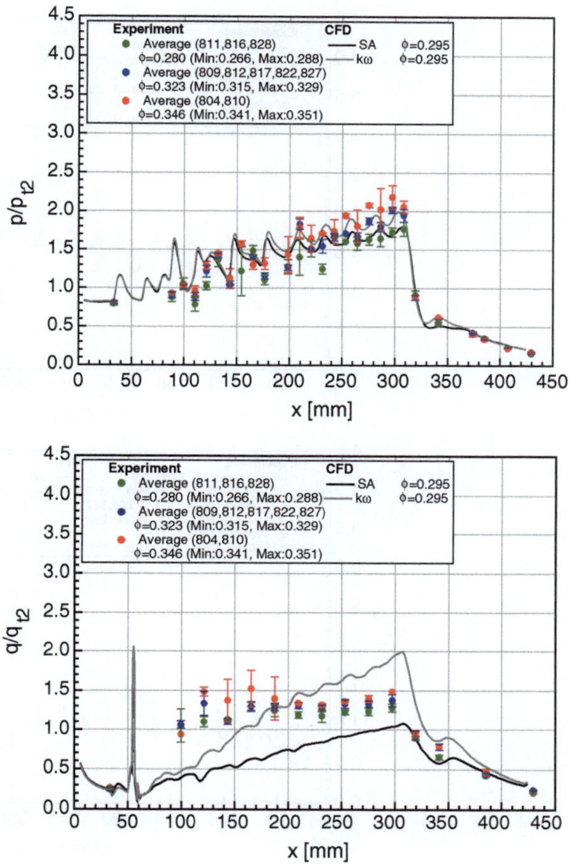

Fig. 21 Normalised computed and measured surface pressure (*top*) and wall heat flux (*bottom*) distributions on the lower HyShot II combustor wall; fuel on conditions; the numbers in parenthesis denote the HEG run numbers; p_{t2} and q_{t2} are the Pitot pressure and the stagnation point heat transfer measured on spherical reference probes

In addition to the investigation of steady combustor flows, an experimental and numerical investigation of the unsteady phenomena induced in the HyShot II combustor under high-equivalence-ratio conditions was carried out, focusing on the processes leading up to unstart [59]. The experimental results were compared with those obtained from unsteady numerical simulations. High-speed schlieren and OH* chemiluminescence visualization, together with time-resolved surface pressure measurements, allowed links to be drawn between the experimentally observed flow and combustion features. The transient flow structures signalling the onset of unstart are observed to take the form of an upstream-propagating shock train. Both the speed of propagation and the downstream location at which the shock train originates depend strongly on the equivalence ratio. The physical nature of the incipient shock system, however, appears to be similar for different equivalence ratios. Both experiments and computations indicate that the primary mechanism responsible for the transient behaviour is thermal choking, though localized boundary-layer separation is observed to accompany the shock system as it moves upstream. In the numerical simulations, the global choking behaviour is dictated by the limited region

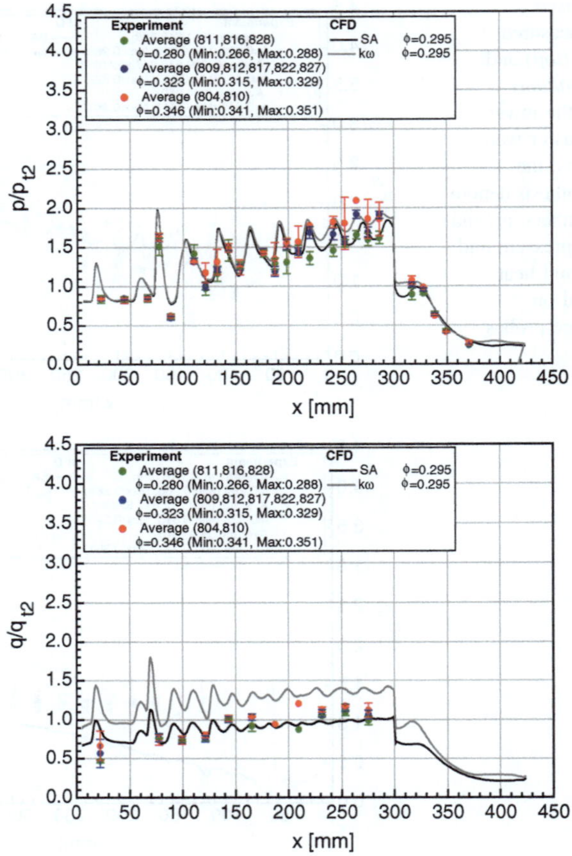

Fig. 22 Normalised computed and measured surface pressure (*top*) and wall heat flux (*bottom*) distributions on the upper HyShot II combustor wall; fuel on conditions; the numbers in parenthesis denote the HEG run numbers; p_{t2} and q_{t2} are the Pitot pressure and the stagnation point heat transfer measured on spherical reference probes

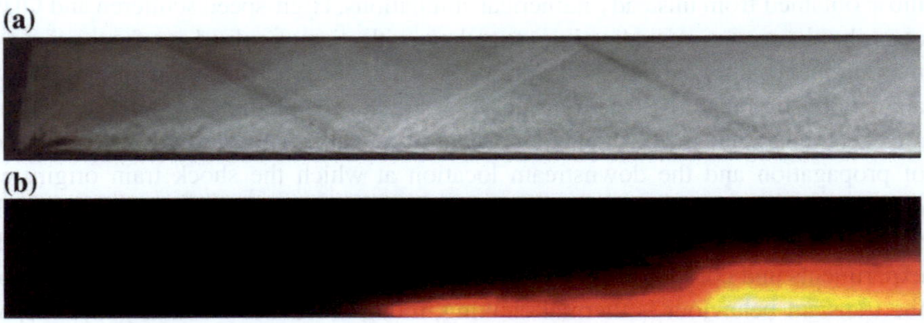

Fig. 23 Schlieren **a** and OH* chemiluminescence **b** visualizations of the flow downstream of the injector (x = 56–136 mm) for an equivalence ratio of 0.33 (from Laurence et al. [59])

of maximum heat release around the shear layer between the injected hydrogen and the incoming air flow. This leads to the concept of "local" thermal choking and results in a lower choking limit than is predicted by a simple integral analysis [59].

In Fig. 24, sequences of schlieren and OH* visualizations of the flow region immediately downstream of the injector (x = 56–136 mm) is shown for an equivalence ratio of $\phi \approx 0.66$. The picture is initially similar to that for $\phi = 0.33$, except

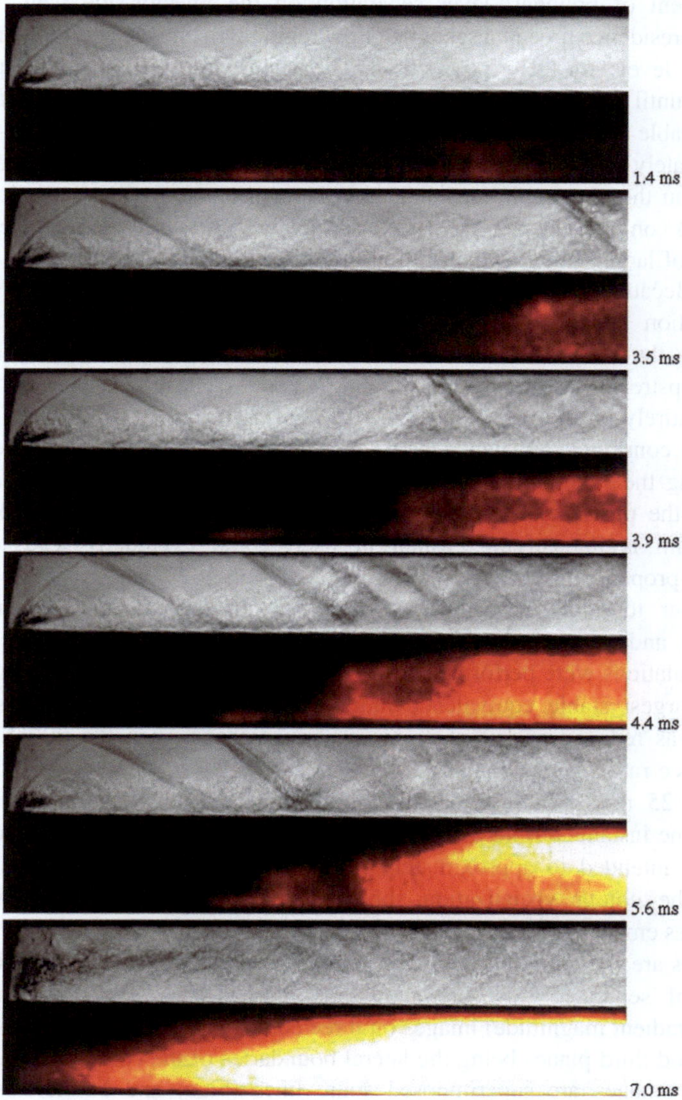

Fig. 24 Sequences of quasi-synchronous schlieren and OH*chemiluminescence images of the flow near the injector (x = 56–136 mm) for an equivalence ratio of $\phi \approx 0.66$ (from Laurence et al. [59])

with a higher degree of injectant penetration and stronger injection-related shocks caused by the increased mass flow of hydrogen. Weak combustion is again visible in the OH* image in areas close to the injector-side wall, where reflections of the injection barrel shock impinge on the hydrogen jet. By t = 3.5 ms, the arrival of the upstream-propagating shock system is visible in the schlieren image. Corresponding to this, a bulging structure appears in the OH* visualization, slightly preceding the shock visible on the cowl-side wall. This structure suggests the development of boundary-layer separation on the injector-side wall, since the increased residence time in a separated flow region will enhance ignition and lead to increased levels of OH* production. The shock train continues to propagate upstream until approximately 4.4 ms, whereafter the combustor flow appears to find a quasi-stable configuration with an oblique shock lodged on the cowl-side wall approximately three duct heights downstream of the injector. A region of intense combustion then begins immediately downstream of the impingement location of this shock on the injector-side wall. Unsteady flow structures, suggesting the presence of large-scale flow separation in this region, are observed in the schlieren images. Because of the limited test time, it is difficult to conclude whether this configuration is truly stable or simply a transient quasi-stable flow topology. Following the conclusion of the test time (6.0 ms), the shock system continues moving upstream past the injection point. It is not clear, however, whether this is caused entirely by the increasing effective equivalence ratio or if, given sufficient time at a constant equivalence ratio, such behaviour would develop regardless. Comparing the schlieren and OH* images at 4.4 and 5.6 ms, it is apparent that, although the position of the cowl-side shock is relatively fixed, the combustion downstream is intensifying, which in itself could eventually lead to further upstream propagation of the shock train.

In order to gain further insight into the transient unstart phenomena just described, and, in particular, to help isolate the mechanism responsible, unsteady CFD simulations were performed. A steady-state initial solution was first obtained for the largest stable equivalence ratio. Subsequently the inflow pressure and density was reduced such that a smooth transition between the initial and final equivalence ratios was obtained [59].

In Fig. 25, numerical results highlighting the flow features within the combustor at four time instants during the unsteady development for a final equivalence ratio of 0.715, intended to approximately match the ϕ = 0.66 experiments. At each instant, the colour visualization shows contours of OH mass fraction, with streamlines emanating from the injector indicated in black. On the lateral cut planes, sonic lines are shown in red; boundaries of negative streamwise velocity indicating regions of separated flow are shown as red surfaces. Pseudo-schlieren (i.e. density-gradient magnitude) images on three equally spaced planes are also shown; the first and third planes being the lateral boundaries of the computational domain. On these images are superimposed sonic lines (blue) and boundaries of flow reversal (red). At t = 0.01, the expected shock pattern extends down the duct. A significant subsonic region develops over a narrow zone downstream of the injection port-hole, but subsonic flow is limited to the wall boundary layers

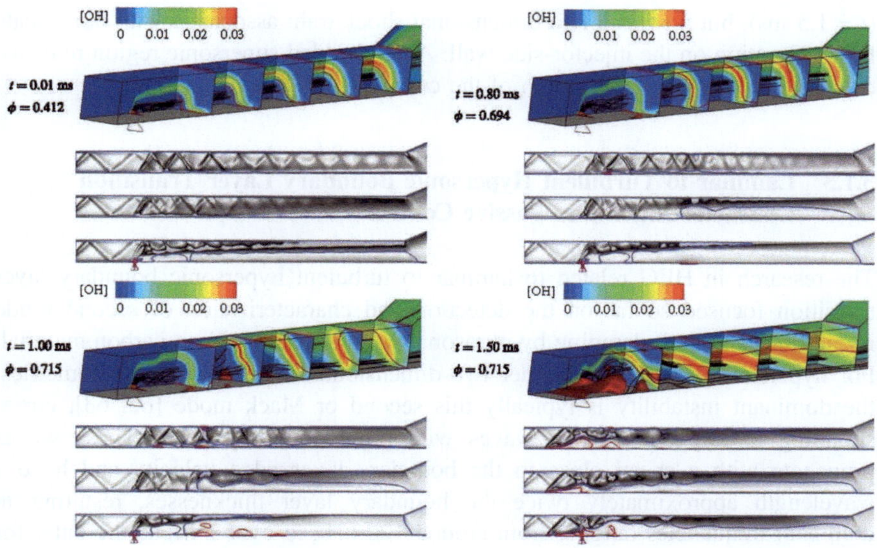

Fig. 25 Numerical results of the HyShot II combustor choking process at different time instants for a final equivalence ratio of 0.715; the axial coordinate in all images has been compressed by a factor of two (from Laurence et al. [59])

throughout the remaining combustor domain. In the OH image, we see that the sonic line on each lateral cut plane lies below the main combustion region centred at the shear layer between the incoming air and the injected hydrogen. Thus, combustion is taking place under principally supersonic conditions at this time. No flow separation is visible. At the second instant shown (t = 0.8 ms), the subsonic region on the injector plane has grown considerably to encompass the entire combustor height and has also expanded laterally. The shock system near x = 150 cm has strengthened significantly and already begun to propagate upstream. A small localized separation bubble has just begun to develop on the injector-side wall, visible midway down the combustor on the injector plane, but this is clearly not responsible for the formation of the propagating shock train. The OH visualization shows that flow conditions in the main combustion region are now almost completely subsonic by the fourth cut plane downstream, or, more correctly, that the heat release in this region has driven the flow to subsonic conditions by this point downstream. In other words, the conditions for thermal choking have been attained locally in the stream tube containing the main combustion region. At t = 1.0 ms, additional separation bubbles are visible both on the cowl-side wall and further downstream on the injector-side wall. Of further note is the embedded region of flow reversal in the middle of the combustor: this feature would explain the "two-tiered" appearance of the experimental OH* structures as the shock system approaches the injector in Fig. 24 (t = 3.5 and 3.9 ms). In contrast to the corresponding experiments, no significant pausing of the shock train motion is observed near the injection location. The shock motion continues upstream of the injectors

(t = 1.5 ms), but now as a two-dimensional shock train associated with large-scale flow separation on the injector-side wall. An embedded supersonic region remains, extending down the entire length of the combustor.

3.1.3 Laminar to Turbulent Hypersonic Boundary Layer Transition Characterization and Passive Control

The research in HEG related to laminar to turbulent hypersonic boundary layer transition focused so far on the detection and characterization of second mode instabilities and their damping by ultrasonically absorptive carbon-carbon material. For hypersonic flows over slender two-dimensional or axisymmetric geometries, the dominant instability is typically this second or Mack mode [62, 63], corresponding to trapped acoustic waves within the boundary layer. These waves propagate with a speed close to the boundary layer edge velocity and have a wavelength approximately twice the boundary layer thicknesses, resulting in dominant frequencies ranging from around 100 kHz to over 1 MHz, the latter for hypervelocity boundary layers in impulse facilities. The HEG conditions used for the boundary layer transition experiments reported here are based on conditions XIII and XIV (see Table 1) and are characterized by a free stream Mach number of M = 7.4, free-stream unit Reynolds number varying over a range of $Re_m = 1.5 \times 10^6$ m^{-1} to $Re_m = 6.4 \times 10^6$ m^{-1} at a stagnation enthalpy of $h_0 \approx 3.2$ MJ/kg and a wall temperature ratio of $T_w/T_0 \approx 0.1$.

The possibility of employing time-resolved schlieren visualizations in HEG to determine the structural and propagation characteristics of second mode instability waves within a hypersonic boundary layer was demonstrated by Laurence et al. [57] (Fig. 26). The model was a 1.1 m long, 7° half-angle cone, having an interchangeable nose with a blunted-tip radius of either 2.25 or 5 mm. Visualization was by means of a dual-field-lens schlieren setup employing a 1000 W short-arc Xe lamp together with a Shimadzu HPV-1 camera, capable of recording 102 10-bit images at a fixed resolution of 312 × 260 pixels and at frame rates of up to 1 MHz. In addition to this technique using high frame rate (i.e., at least as high as the dominant instability frequency) schlieren visualization with a continuous light source, a modified version of the technique was introduced whereby a pulsed light source allows the use of a higher-resolution camera with a lower frame rate. This provides significant benefits in terms of spatial resolution and total recording time. A detailed picture of the surface-normal intensity distribution for individual wave packets was obtained, and the images provide comprehensive insight into the unsteady flow structures within the boundary layer. In Fig. 27, power spectra at different heights above the cone surface and time-developing power spectra at the height above the cone surface of maximum disturbance intensity, evaluated from schlieren visualization such as the ones included in Fig. 26, are shown. Power spectra in terms of inverse wavelength were first derived from the schlieren images by taking the Fourier transform of the row-wise intensity profiles. In order to

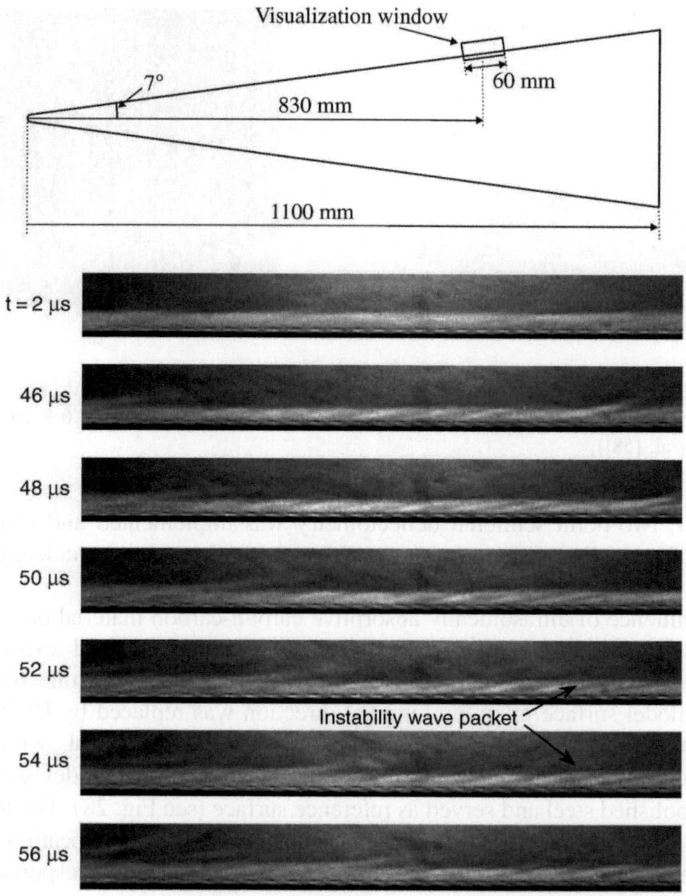

Fig. 26 Schematic of the cone model (top) and sequence showing the development of an instability wave packet (46–56 μs) in the boundary layer (bottom) for $Re_m = 6.2 \times 10^6\ m^{-1}$. The regular structures visible in the two rows of pixels immediately above the cone surface are artefacts of the image-processing techniques employed (from Laurence et al. [58])

convert these to power spectra in terms of frequency (Fig. 27), image correlation techniques to determine the propagation speed of the wave packet were utilized [57].

Compared with the power spectra obtained with the continuous light source schlieren set-up [57, 58] (not shown here), the benefit of using a camera with a higher pixel count is immediately obvious. The second-mode frequency here is approximately 260 kHz; and a double-extremum profile is observed, with one local maximum inside the boundary layer and the other at the cone surface. Such a detailed picture of the evolution of an individual wave packet is not possible with any other currently available technique.

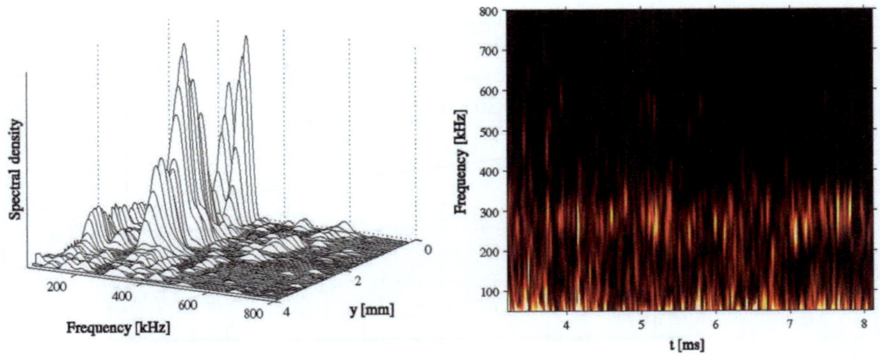

Fig. 27 Example of power spectra at different heights above the cone surface (*left*) and time-developing power spectra 1.5 mm above the surface (*right*) for $Re_m = 2.6 \times 10^6\ m^{-1}$ (from Laurence et al. [58])

Further, two-point schlieren deflectometry was implemented and shown to be capable of providing second-mode growth information in the challenging shock tunnel environment [58].

The influence of ultrasonically absorptive carbon-carbon material on hypersonic laminar to turbulent boundary-layer transition was investigated experimentally using the same cone configuration as shown in Fig. 26. Here, one third of the metallic model surface in circumferential direction was replaced by DLR in-house manufactured ultrasonically absorptive carbon-carbon material with random microstructure for passive transition control. The remaining model surface consisted of polished steel and served as reference surface (see Fig. 28). The model was equipped with co-axial thermocouples to determine the transition location by means of surface heat-flux distribution. Flush mounted piezoelectric fast-response pressure

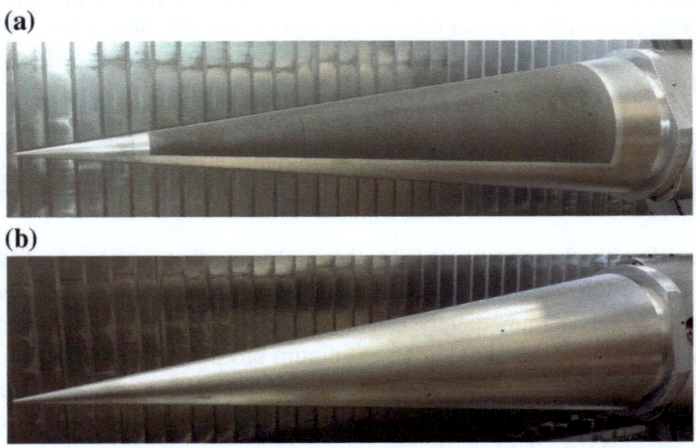

Fig. 28 Integrated cone model (from Wagner et al. [116]). **a** Porous insert. **b** Stainless steel reference surface

transducers were used to measure the pressure fluctuations in the boundary-layer associated with second mode instabilities [116].

The concept to passively control transition on hypersonic vehicles using ultrasonically absorptive surfaces was developed by Fedorov et al. [25] and Malmuth et al. [64]. Rasheed et al. [78] experimentally confirmed the stabilizing effect by showing a delay of the transition onset using a porous surface with regularly spaced cylindrical blind holes on a sharp 5° half-angle cone at Mach 5 in the T-5 high-enthalpy shock tunnel of the California Institute of Technology (see also Sect. 6.1.1). Further studies on ultrasonically absorptive surfaces with regular microstructure were, e.g., conducted by Fedorov et al. [24] in the ITAM T-326 Mach 6 hypersonic wind tunnel on a 7° half-angle sharp cone.

Based on the promising results obtained on porous surfaces with regular and random microstructures and based on the fact that the majority of thermal protection system (TPS) materials have random microstructures, a material with practical TPS potential was used for passive transition control. Hence, carbon fibre reinforced carbon (C/C) was chosen as model material to assess the potential of ceramic matrix composites. It represents an intermediate state of C/C-SiC which has been successfully used as TPS on hypersonic flight test vehicles [111, 117]. C/C material is lightweight and offers high temperature stability in absence of oxygen, but might need to be adapted for real TPS applications to improve its oxidation resistance.

A series of tests at unit Reynolds numbers between $Re_m = 1.5 \times 10^6\ \mathrm{m}^{-1}$ and $Re_m = 6.4 \times 10^6\ \mathrm{m}^{-1}$ were conducted. Figure 29 shows the Stanton number distribution versus the free-stream Reynolds number based on the model x-coordinate, originating at the model tip and pointing downstream along the centre line. A clear shift of the transition onset towards higher Reynolds numbers is observed on the porous surface for $Re_m = 4.0 \times 10^6\ \mathrm{m}^{-1}$ and $Re_m = 6.4 \times 10^6\ \mathrm{m}^{-1}$. At a unit Reynolds number of $Re_m = 4.0 \times 10^6\ \mathrm{m}^{-1}$ the transition Reynolds number increases from $Re_t = 2.7 \times 10^6$ to about $Re_t = 3.0 \times 10^6$. The relative transition delay appears to be higher at a unit Reynolds number of $Re_m = 6.4 \times 10^6\ \mathrm{m}^{-1}$ with a transition Reynolds number of $Re_t = 3.0 \times 10^6$ on the smooth surface and $Re_t = 3.9 \times 10^6$ on the porous surface.

The result that the efficiency of the C/C material depends on the Reynolds number, i.e., the transition delay increases with increasing unit Reynolds number, indicates that the C/C properties are not optimal for the test conditions applied. The surface porosity and the average pore size are too small. A further development of the material to improve the ultrasonically absorptive properties is conceivable by adapting the phenolic resin composition to optimize the pore size distribution. The assessment of the ultrasonic absorption properties of the C/C material performed by Wagner et al. [115] represents a first necessary step towards such an optimization process.

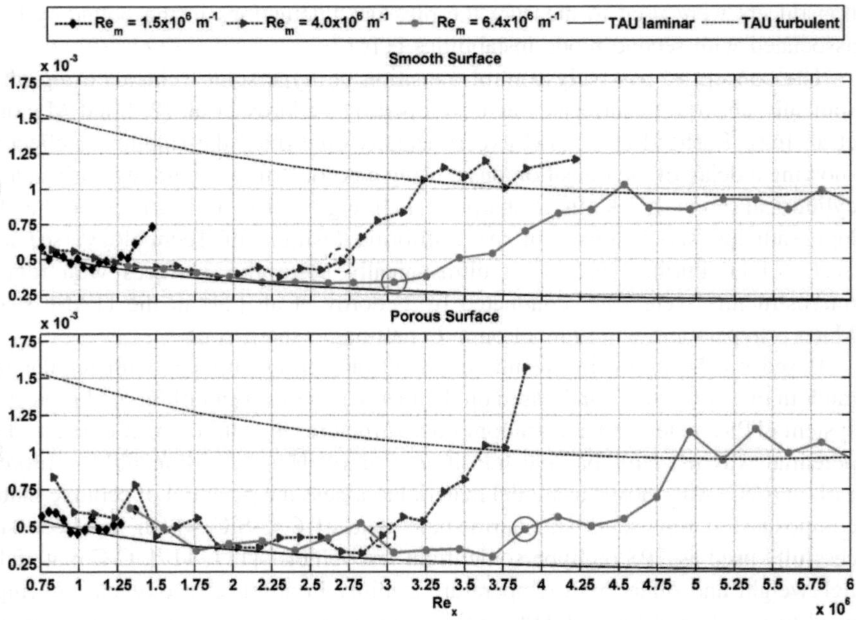

Fig. 29 Normalized heat flux distribution (Stanton number) as function of Reynolds number based on the distance from the cone tip (Re_x) on the porous and the smooth surface for different unit Reynolds numbers (transition Reynolds numbers marked by *circles*) (from Wagner et al. [116])

4 JAXA High Enthalpy Shock Tunnel (HIEST)

In Japan in the latter half of the 1980s, research focusing on a space vehicle to be launched by the H-II rocket, named HOPE (H-II Orbiting Plane) began. In parallel, research into hypersonic air-breathing engines also started for a prospective SSTO space plane. To ensure progress in the aerothermodynamic and propulsion research for these space transportation system programs, the need for hypervelocity ground testing became apparent. A plan for building a large free-piston shock tunnel was made in 1990. To build a knowledge base for the design of the large free-piston shock tunnel, experimental studies on the characteristics of the free-piston shock tunnels were performed with small- and medium-sized pilot facilities, which were supported by numerical simulations. The most remarkable result from the preliminary studies was the establishment of the tuned-operation method that safely improves the performance of a free-piston driver [44]. The large facility, named High Enthalpy Shock Tunnel, HIEST, located at the JAXA Kakuda Space Center was designed to accomplish the tuned operation throughout the requirements for the HOPE aerothermodynamics and the scramjet tests [45]. The stagnation enthalpy and the free-stream density were principal factors to reproduce chemical nonequilibrium effects on the HOPE reentry aerothermodynamics. The velocity range in the flight path to be tested was 3–7 km/s, corresponding to total specific stagnation enthalpies of approximately 4–25 MJ/kg. The maximum test model size, required to realize a binary scaling factor corresponding to that of the HOPE flight path, was 500 mm. Considering a time needed for the flow establishment and the force measurement with a test model of 500 mm, the test flow duration should be longer than 2 ms at maximum enthalpy conditions. For scramjet testing, the flight Mach number range to be tested was 8–14. The corresponding stagnation enthalpy range was at the lower range of the HOPE tests, but higher dynamic pressures of 50–100 kPa were required. The model engine size to be tested was 2 m. Therefore, longer test flow duration in lower enthalpy conditions and a larger test section were required. To meet the testing requirements, the nozzle reservoir condition and geometries were decided first. To reduce the time spent for nozzle starting in the higher enthalpy condition and to obtain a large exit diameter for tests at high angles of attack, a conical nozzle 3 m long and 12-degree half angle was designed for the HOPE tests. On the other hand, higher dynamic pressure and parallel flow were of higher priority than the size of the core flow for the scramjet tests. Thus, a contoured nozzle of smaller area ratio was also designed. Next, the shock tube diameter and length were determined to fulfill the test time requirements, considering the effect of the boundary layer growing behind the incident shock wave on the reduction of the test slug length, as well as the early arrival of the driver gas. Then, the diameter and length of the compression tube and the piston masses were determined using the tuned operation theory [44]. Photographic views of HIEST are shown in Fig. 30 and some of its characteristics are listed in Table 2.

Fig. 30 Photographic views of the HIEST; overall view (*left*), nozzle and test section (*right*)

Table 2 Characteristic dimensions and operating parameters of HIEST

Compression tube	42 m in length	600 mm in diameter
Shock tube	17 m in length	180 mm in diameter
Piston mass	220, 290, 440, 580, 780 kg	
Nozzle (conical)	1.2 m exit diameter	24–50 mm throat diameter
Nozzle (contoured)	0.8 m exit diameter	50 mm throat diameter
Maximum stagnation enthalpy	25 MJ/kg	
Maximum stagnation pressure	150 MPa	
Test time	>2 ms	

Construction of HIEST began in 1995 and was completed in 1997 [43]. The HIEST has been operated about 2400 times up to 2013. In the following, selected applications of the HIEST for the modern development of the impulsive measurement techniques and the high enthalpy aerothermodynamic studies are presented.

4.1 Selected Applications of HIEST

4.1.1 Development of Force Measurement Technique and Investigation of Real-Gas Effect on Aerodynamic Characteristics

An innovative force-measurement technique has been developed and implemented at HIEST, in which the test model is completely unrestrained during the test and thus experiences free-flight conditions for a period on the order of milliseconds [105]. Advantages over conventional free-flight techniques include the complete absence of aerodynamic interference from a model support system and less variation in model position and attitude during the test itself. A miniature on-board data recorder, which was a key technology for this technique, was also developed in order to acquire and store the measured data [107]. The technique was demonstrated in the HIEST test campaign in which three-component aerodynamic force measurements were performed using a blunted cone of length 316 mm, and total mass of 19.75 kg (Fig. 31) [103].

During the test campaign, axial force, normal force and pitching moment coefficients were obtained at angles of attack from 14 to 32° under two conditions: h_O = 4 MJ/kg, p_O = 14 MPa, and h_O = 16 MJ/kg, p_O = 16 MPa. For the first, low-enthalpy condition, the test flow was considered to be nearly perfect gas;

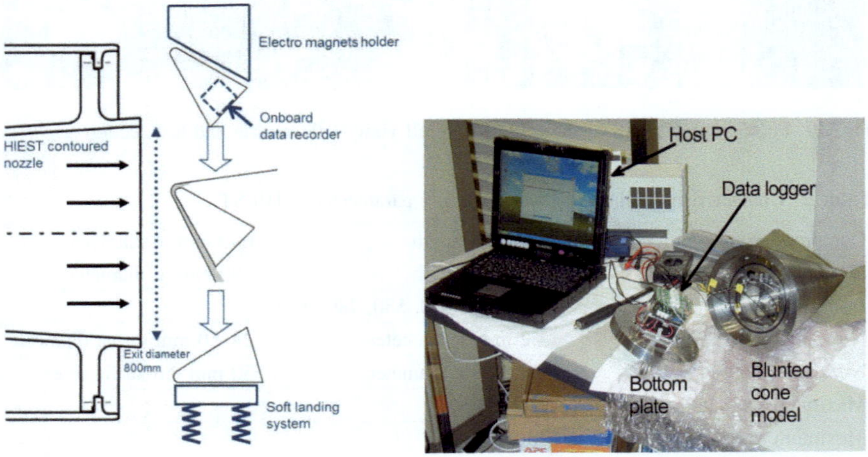

Fig. 31 Free-flight force measurement test in HIEST

measurements were thus directly compared with those obtained in a conventional blow-down wind tunnel (JAXA-HWT2) to evaluate the accuracy of the present free-flight force measurement technique. The second test condition was a high-enthalpy condition in which more than 80 % of oxygen molecules were expected to be dissociated; high-temperature effects were, therefore, evaluated by comparison with results obtained in the low enthalpy conditions. In Fig. 32 sequential high-speed video images of the falling model taken with a monitoring video camera are shown. In Fig. 33, examples of the time-resolved axial force coefficient, C_A, normal force coefficient, C_N, and pitching moment coefficient, C_M, for condition 1 (low enthalpy condition) at an angle of attack of 16.7° are presented.

In Fig. 34, all the present measurements in HIEST, depicted as open symbols, show the relationship between aerodynamic coefficients and the angle of attack as the latter is varied from 14 to 32°. The coefficients are the average values during each shot. Circles and triangles in the diagrams show the results at condition 1 (low enthalpy condition) and condition 2 (high enthalpy condition), respectively. Aerodynamic coefficients measured in the blow-down hypersonic wind tunnel JAXA-HWT2 are also plotted to estimate the present measurement accuracy under condition 1 that can be considered as nearly perfect gas flow. As shown in Fig. 34, these two measurement groups agree to within their 95 % prediction interval, and we can be confident that the accuracy of the present free-flight force measurements in HIEST is sufficiently high for the investigation of the aerodynamic characteristics.

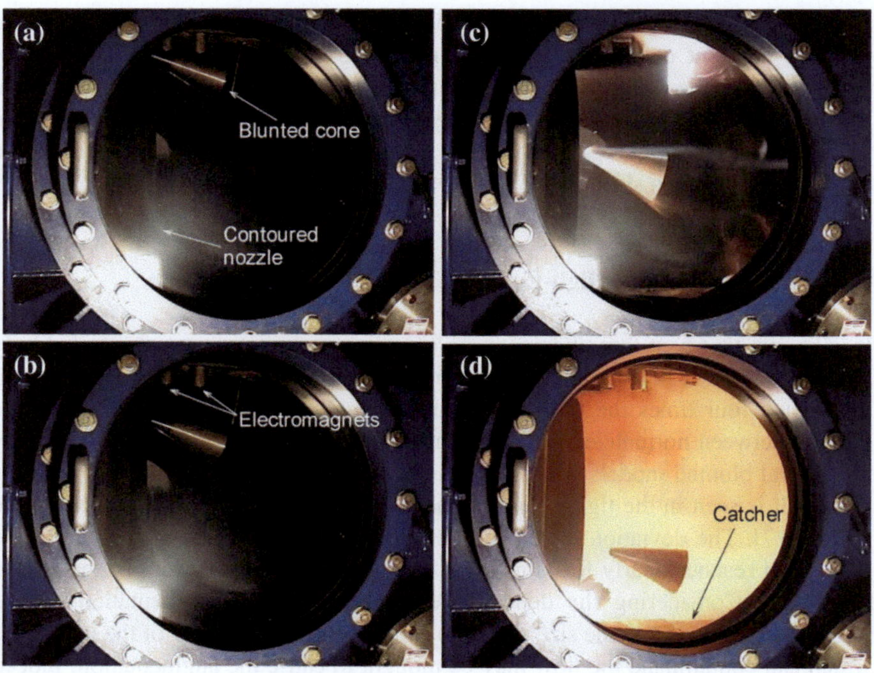

Fig. 32 Sequential images of model dropping

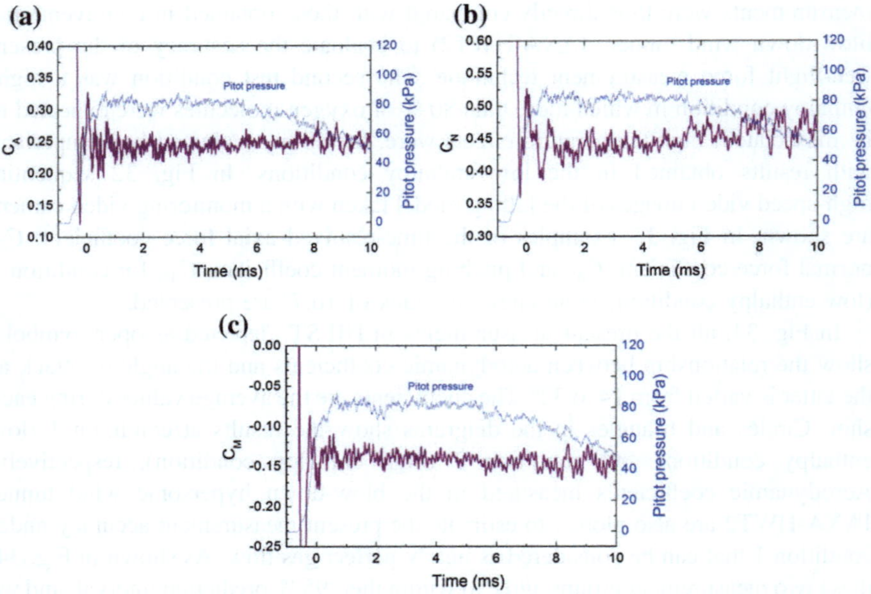

Fig. 33 Time-resolved aerodynamic coefficients; C_A, axial force coefficient **a**; C_N, normal force coefficient **b**; C_M, pitching moment coefficient **c**

Comparing the aerodynamic characteristics obtained at condition 1 and 2, the pitching moment coefficient C_M for condition 2 lies consistently above that of condition 1, while no significant differences are observed in the axial and normal force coefficients C_A and C_N. The aerodynamic characteristics for condition 2 can be considered to show high-temperature effects which cause a forward shift of the centre of pressure as shown in Fig. 35.

4.1.2 Investigation of High-Enthalpy Heat Flux Augmentation

It is known that heat flux augmentation of unknown origin occurs for particularly high enthalpy and high pressure conditions in a reflected type shock tunnel, which may reach four times the theoretical value at the maximum [39]. In Fig. 36 the relation between normalized stagnation heat flux ($St \times R_N$) and stagnation enthalpy for several blunted models with different nose radii R_N obtained in HIEST [108] is shown. It is seen in the figure that the measured heat flux becomes larger than the theory [17]. The deviation increases with increasing stagnation enthalpy and nose radius and reaches nearly four times the theoretical value in the case of the largest nose radii. Considering the dependence of the heat flux augmentation on the stagnation enthalpy and the size of the nose radii, radiation heating of the impurities which can contaminate the flow may be thought to cause the augmentation. Indeed the heat flux augmentation can be characterized by the Stefan-Boltzmann law with

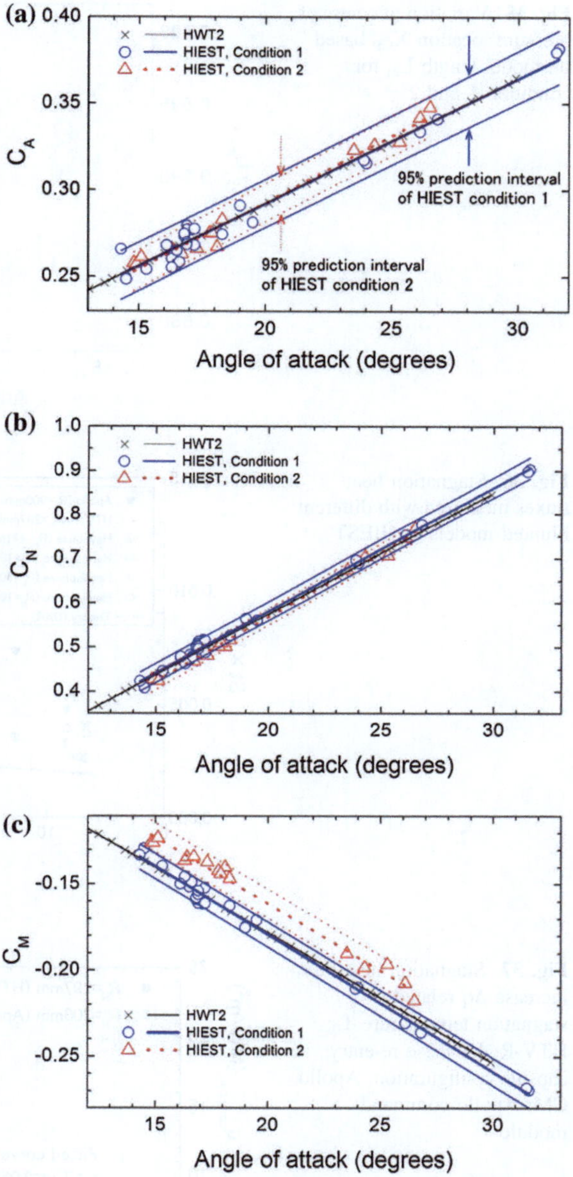

Fig. 34 Aerodynamic characteristics obtained for both HIEST conditions (1—low enthalpy, 2—high enthalpy) compared with JAXA-HWT2 hypersonic blow-down wind tunnel results

an emissivity of $\varepsilon = 0.055$, as shown in Fig. 37, suggesting that the radiation may cause the heat flux augmentation.

To investigate the relation between the radiation and the heat flux augmentation, a simple flat plate was specially designed and manufactured with four clusters of miniature thermocouples on the surface (Fig. 38 left) [104]. Each cluster includes five co-axial chromel-constantan thermocouples. The plate has glass windows in

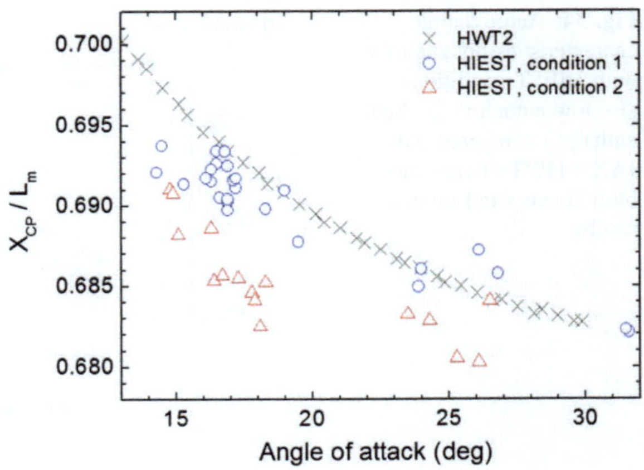

Fig. 35 Variation of centre of pressure location X_{CP} based on model length L_m for condition 1 and 2

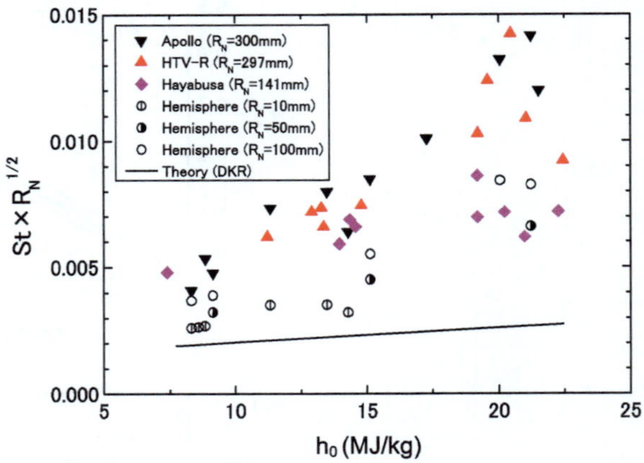

Fig. 36 Stagnation heat fluxes measured with different blunted models in HIEST

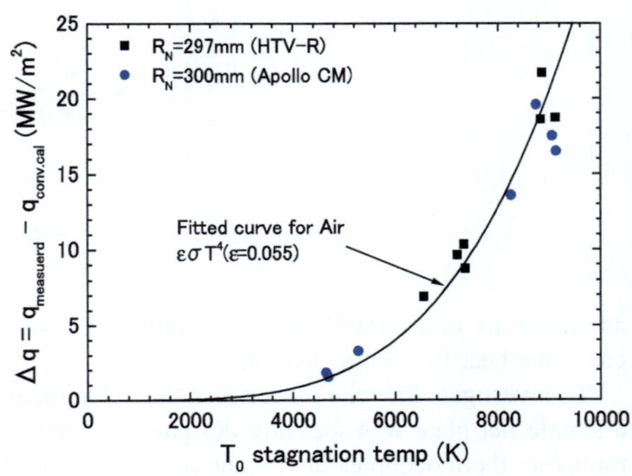

Fig. 37 Stagnation heat flux increase Δq related to stagnation temperature T_0; HTV-R: Japanese re-entry capsule configuration, Apollo CM: Apollo command module

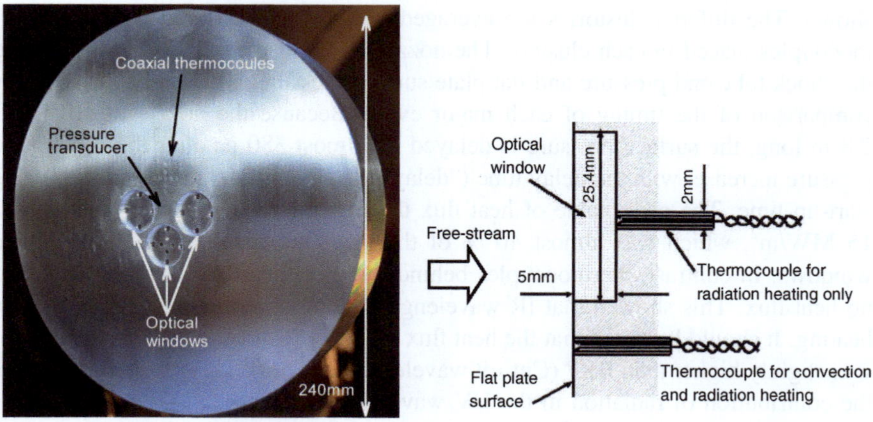

Fig. 38 Flat plate test model for radiation heating measurement (*left*); Schematic of thermocouple mounts with and without optical window (*right*)

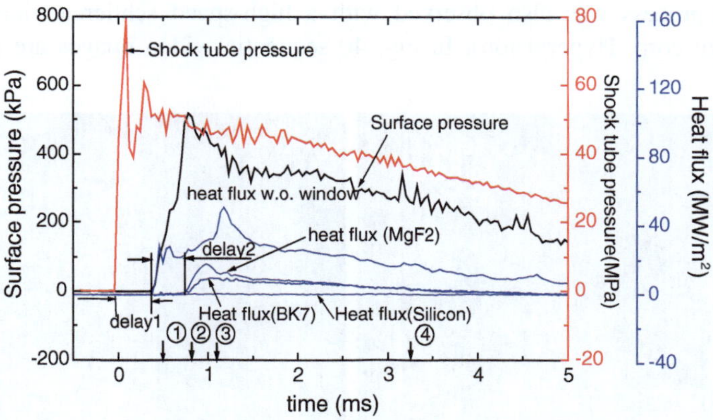

Fig. 39 Example of surface heat flux records measured on the flat plate model

front of three of the four thermocouple clusters (Fig. 38 right), to isolate them from surface convection heating. These optical windows can be fitted with different window materials. In this test campaign, BK7, MgF_2 and metallic silicon windows were tested, with different transparency wavelengths. Because metallic silicon is transparent only to infrared wavelengths, it was used to estimate their contribution to the heating. The remaining thermocouple cluster had no glass windows, so it could register the total surface heat flux for reference, that is, the heat flux from both radiation heating and convection heating.

The test campaign was conducted in HIEST at stagnation conditions varied from 8 to 23 MJ/kg in total specific enthalpy while the stagnation pressure was kept constant at 50 MPa. In Fig. 39 an example of heat flux histories on the flat plate surface with and without optical windows at $h_0 = 21$ MJ/kg and $p_0 = 50$ MPa is shown. The different histories are averaged heat fluxes measured with five thermocouples placed in each clusters. The nozzle reservoir pressure record, including the shock tube end pressure and flat plate surface pressure are also shown, to allow comparison of the timing of each major event. Because the nozzle of HIEST is 2.8 m long, the surface pressure is delayed by almost 380 μs after the shock tube pressure increase, with the delay time ("delay1" in Fig. 39) defined from the nozzle start-up time. The peak value of heat flux through the MgF_2 window was almost 15 MW/m^2, which was almost 40 % of the heat flux measured without optical windows. In contrast, thermocouples behind the metallic silicon window detected no heat flux. This showed that IR wavelength has no contribution to the radiation heating. It should be noted that the heat flux of MgF_2 (Cut-off wavelength 200 nm) was slightly higher than BK7 (Cut-off wavelength 350 nm). This result may explain the contribution of radiation in the UV wave length range.

It should also be noted that the heat flux rise measured with the optical windows (MgF_2 and BK7) showed a 340 μs delay ("delay2" in Fig. 39) compared to the heat flux rise measured without windows. This means that the radiation was emitted 340 μs after the test flow arrived. To analyse the time interval, the flow establishment process was also observed with a high-speed schlieren video camera (Shimadzu corp. Hypervision). In Fig. 40 sequential video images are presented,

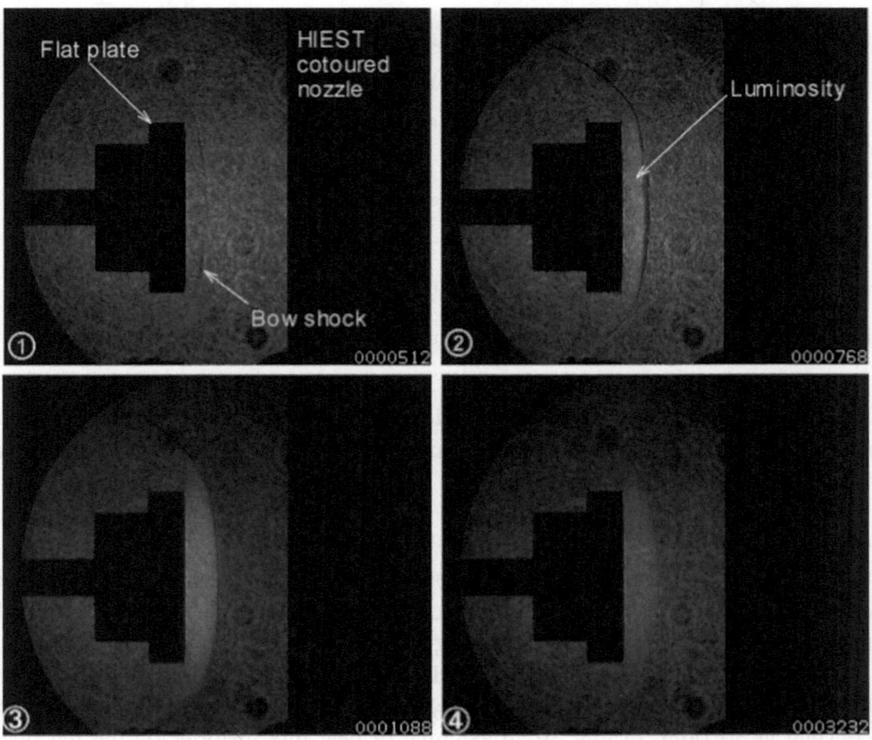

Fig. 40 Sequential high-speed schlieren video images

where the numbers on the bottom-left of each picture match the numbers at the bottom of Fig. 39 are presented. In picture 1 ($t = 526$ μs), the flow around the plate is not yet fully established, and the bow shock in front of the plate has not yet completely formed. Consistent with the video image, heat flux was not observed through the windows, but heat flux was seen with no window. Picture 2 ($t = 768$ μs) shows that the bow shock had formed and luminosity can be observed in the shock layer. Heat flux through the optical windows began to increase at that time. Almost 300 μs later, in picture 3 ($t = 1088$ μs), both the luminosity and the radiation heat flux increased remarkably. Finally in picture 4 ($t = 3232$ μs), the test flow had already terminated, so the radiation decreased. Since the characteristics of the heat flux records can be correlated to the events shown in the sequential video images, it was concluded that the radiation comes from the shock layer ahead of the model.

4.1.3 Investigation of Hypersonic Boundary Layer Transition

If a free-piston shock tunnel is large enough to produce a flow of sufficiently high Reynolds number, it is useful for an investigation on laminar to turbulent boundary layer transition phenomena at hypersonic flow conditions [1, 85]. Since 2008 hypersonic boundary layer transition has been investigated on a cone model (7-degree half cone angle, 1.1 m long) in HIEST (Fig. 41) [106, 109]. With this configuration, several conventional ground experiments and also a flight experiments were conducted [54]. In the present investigation the particular interest was to resolve the second mode instability which appears in the laminar to turbulent transition process [10, 26]. To observe the second mode instability of which the characteristic frequency may exceed several hundred kHz at high enthalpy conditions, the surface pressure measurement was improved by noise reduction utilizing a special technique for mounting the pressure transducers as shown in Fig. 42.

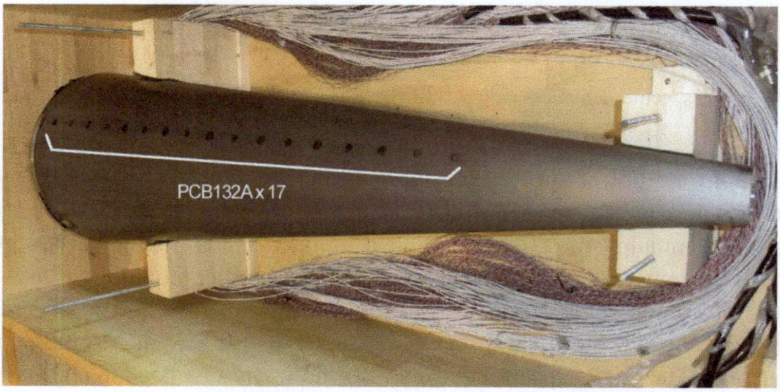

Fig. 41 7-degree half cone angle model equipped with 17 PCB pressure transducers for the second mode instability investigations

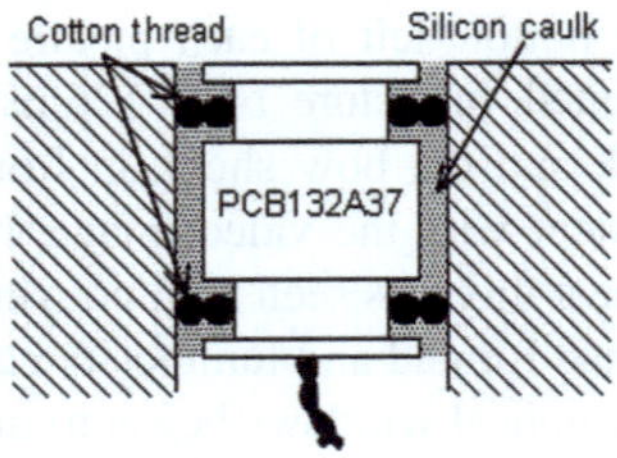

Fig. 42 Schematic of the pressure transducer installation with cotton threads and silicone caulk

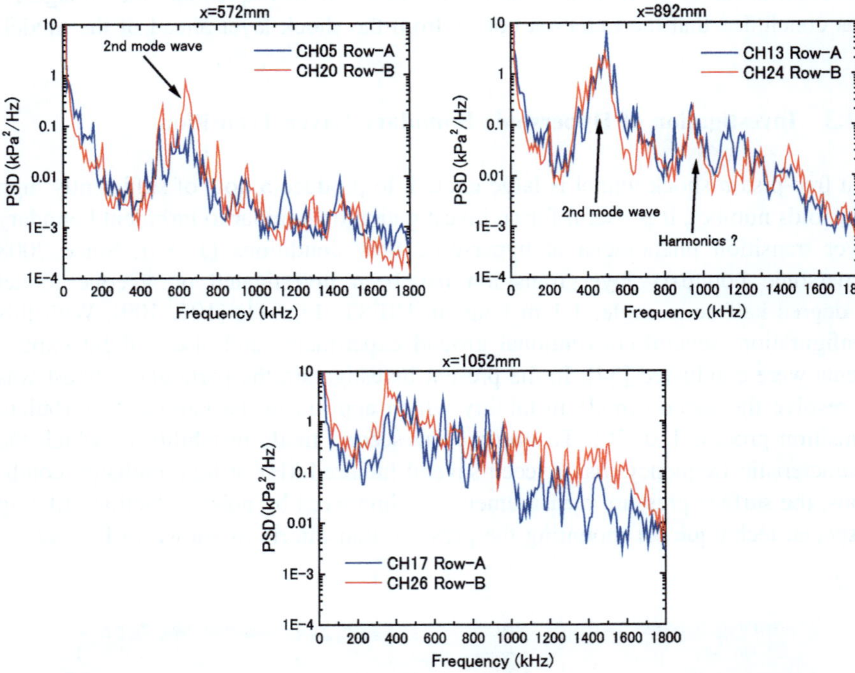

Fig. 43 Power spectral densities (h_0 = 7.0 MJ/kg, Re = 2.1 × 10^6)

In Fig. 43 the power spectral densities (PSD) calculated from surface pressure fluctuations measured at $x = 572$ mm, 892 mm and 1052 mm from the cone apex for a stagnation enthalpy of 7.0 MJ/kg and a unit Reynolds number 2.1 × 10^6 are shown. It is remarked that the second mode instability could be successfully observed at around 600 kHz at x = 572 mm. Consequently at x = 892 mm the frequency of the second mode instability reduced to 500 kHz due to the boundary layer growth. The power spectral density also shows another peak at approximately 1 MHz, which is probably the higher harmonic wave of the second mode instability. All the peaks became unclear at $x = 1052$ mm, at which boundary layer break-down seems to occur. Thus, the boundary layer transition process with the change of the

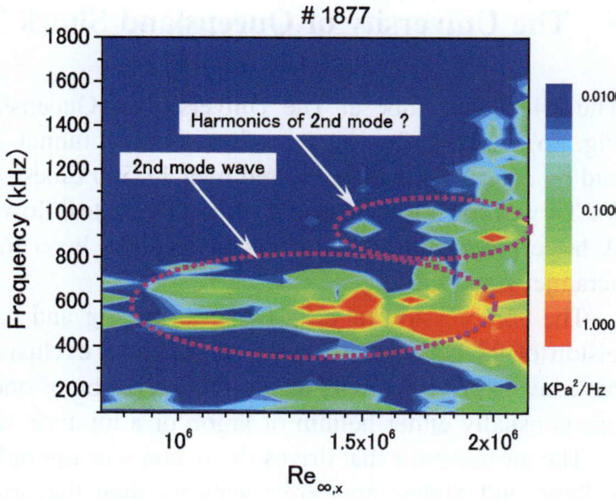

Fig. 44 Distribution of the power spectral density at a unit Reynolds number of $Re_m = 2.1 \times 10^6\ m^{-1}$

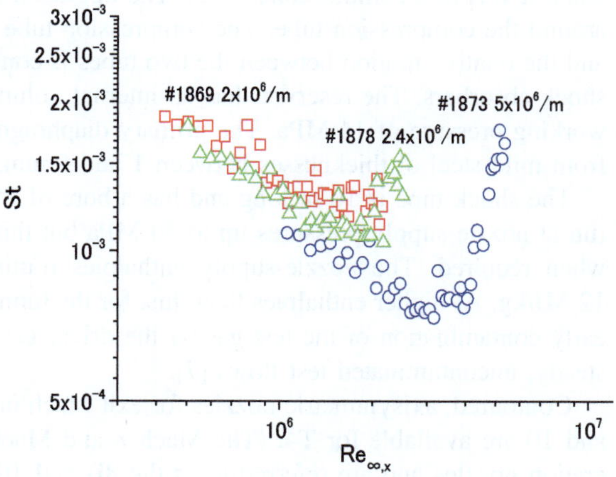

Fig. 45 Comparison of Stanton number distributions for different unit Reynolds numbers

second mode instability characteristic could be successfully resolved by the present surface pressure measurements.

In Fig. 44 the distribution of the power spectral density versus the local Reynolds number $Re_{\infty,x}$ is plotted for run number 1877. It is clearly seen that the second mode instability appears at a Reynolds number of around 1×10^6 and breaks down at around 2×10^6 showing the boundary layer transition process.

In Fig. 45 the comparison of the normalized heat flux distributions for three different unit Reynolds numbers $2.0 \times 10^6\ m^{-1}$ (run 1869), $2.4 \times 10^6\ m^{-1}$ (run 1878) and $5.0 \times 10^6\ m^{-1}$ (run 1873) obtained by the tests at three different free stream pressures is shown. In the case of $Re_m = 2.0 \times 10^6\ m^{-1}$, the heat flux slightly increased at the end of the model. The increase of the heat flux at the end of the model became remarkable with increasing Reynolds number and showed that the turbulent boundary layer was fully established eventually. Thus, the results of the heat flux measurement were quite consistent with the boundary layer transition process obtained by the observation of the second mode instability.

5 The University of Queensland Shock Tunnel T4

The T4 Stalker tube at The University of Queensland, Brisbane, Australia (see Fig. 46) is also a free-piston reflected shock tunnel. It was commissioned in 1987 and by June 2014 had been fired over 11,500 times. An important design objective for T4 was for it to be capable of producing test flows suitable for scramjet testing. A large proportion of the shots in T4 have been for research into scramjets and scramjet subsystems.

The compression tube of T4 is 26 m long and has a bore of 229 mm. A steel piston of 92 kg mass has been used almost exclusively. A small number of tests have been performed with an aluminium piston of one-third of the mass. The driver gas is usually either helium or argon or a mixture of the two gases [48].

The air reservoir that drives the piston was upgraded in 2000 to provide a larger volume and higher pressure capability than the original reservoir for the tunnel when it was first commissioned [95]. The upgraded reservoir wraps concentrically around the compression tube. The compression tube can slide within the reservoir and the relative motion between the two tubes is controlled by a set of springs and shock absorbers. The reservoir has an internal volume of 1.2 m^3 and a maximum working pressure of 14 MPa. The primary diaphragms are unscored and are made from mild steel of thicknesses between 1 and 6 mm.

The shock tube is 10 m long and has a bore of 76 mm. The tunnel is routinely run at nozzle-supply pressures up to 50 MPa but this can be increased to 90 MPa when required. The nozzle-supply enthalpies routinely used range from 2.5 to 12 MJ/kg. At higher enthalpies than this, for the tunnel in its present configuration, early contamination of the test gas by the driver gas compromises the duration of steady, uncontaminated test flows [7].

Contoured, axisymmetric nozzles for exit Mach numbers of nominally 4, 6, 7, 8 and 10 are available for T4. (The Mach 8 and Mach 10 nozzles are second generation nozzles and are referred to as the 8B and 10B nozzles, respectively.) The nozzles are designed for a particular nozzle-supply condition (enthalpy and pressure) and the Mach number at exit of the nozzle varies, particularly with nozzle-supply enthalpy due to high-temperature effects on the gas. For example, the mean exit Mach number in the test core of the Mach 6 nozzle decreases from 6.7 at 3 MJ/kg to 5.5 at 12 MJ/kg.

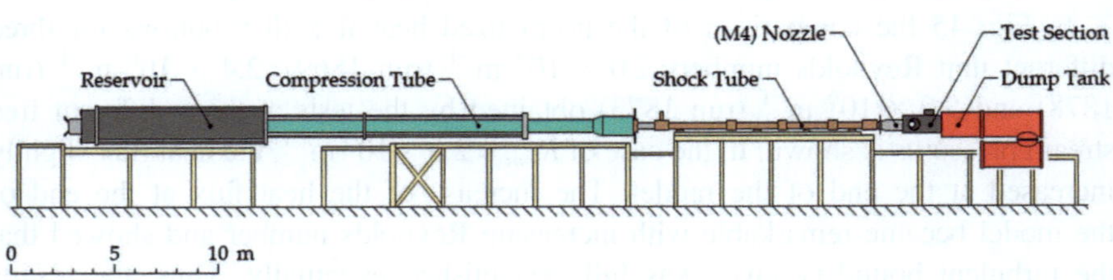

Fig. 46 The T4 Stalker Tube [18]

Fig. 47 The T4 shock tunnel nozzle and test section. **a** The end of the shock tube, Mach 8B nozzle and test section. **b** Mach 8B nozzle protruding into test section. Rake of Pitot and conical probes shown

The Mach 4 nozzle has an exit diameter of 135 mm, the Mach 6, 7 and 8B nozzles have exit diameters of 262, 273 and 270 mm, respectively, and the Mach 10B nozzle has an exit diameter of 375 mm. Figure 47a shows the end of the shock tube and the Mach 8B nozzle protruding into the test section. The test section and dump tank are held stationary when the tunnel is fired and the rest of the tunnel is free to move horizontally on rollers. A sliding seal between the nozzle and the test section allows the tunnel to recoil while the test section and model are held stationary. Figure 47b shows the test section with the side-door open. The nozzle can be seen extending into the test section. A rake of probes for calibration of the flow downstream of the nozzle exit is also shown in the figure. When the tunnel fires, allowance must be made for positioning the model such that it is at the correct location in the core flow of the nozzle when the recoil has occurred. Therefore, upon recoil of the tunnel, the nozzle in Fig. 47b will protrude into the test section by a smaller amount.

Some of the higher Mach number nozzles used in T4 were manufactured to be shorter than their full design length. This was done to reduce the time it takes for the boundary layers formed on the walls of the nozzles to become fully established, noting that boundary layers will take longer to establish on longer nozzles. This allows the establishment of flows over models in the tunnel to occur more quickly than if the full nozzle length was used. The trade-off for the reduced nozzle-stating time is a reduction in the size of the uniform test core available for locating models. The last part of the expansion of a high Mach number contoured nozzle is typically quite gradual, so a relatively larger length is required for a relatively small area increase. For example, for the design of T4's Mach 10B nozzle, the last 5 % of area ratio increase in the divergent part of the nozzle occurs over the last 14 % of the nozzle length. Also, for nozzles designed using a "method of characteristics" approach (e.g. [49]), there is little to gain in extending the nozzle length beyond the point where the final characteristic intersects the edge of the nozzle boundary layer. The Mach 4 and Mach 6 nozzles were designed using a "method of characteristics" approach [49]. The Mach 7, 8B and 10B nozzles were designed using viscous computational fluid dynamics and an optimizer [15]. Either equilibrium or finite-rate chemistry was used, depending on the total enthalpy of the design condition. Optimization was used to adapt the nozzle contour to produce minimal variations in the flow uniformity leaving the nozzle. The starting profiles for the iterative processes were based on either a previous nozzle design or a profile given by the method of characteristics.

T4's Mach 6 nozzle is one of the truncated nozzles. Figure 48 shows the location of a model in the uniform test flow core. The core is shown by the dashed lines. Shown also are Pitot pressure surveys at two streamwise locations. The Pitot pressure, p_{Pitot}, is normalised by the nozzle-supply pressure, p_s, in Fig. 48. It can be seen that the maximum diameter of the test flow core is located approximately 0.23 m downstream of the exit plane of the nozzle.

The speed of the primary shock wave as it travels down the shock tube in T4 is measured by three piezoelectric pressure transducers at 2-m spacing. A fourth timing is indicated by two nozzle-supply pressure sensors located 65 mm from the nozzle-end of the tube. Typical instrumentation used in the test section of T4 includes pressure measurement (static and Pitot) made using piezoelectric and/or piezoresistive sensors. Heat transfer is measured using surface-mounted thin-film gauges (vacuum-deposited nickel) and thermocouple gauges. Both types of sensors are manufactured in-house and operate on a semi-infinite substrate principle [86]. Aerodynamic forces are measured using stress-wave force balances [70, 84, 112]. High-speed video schlieren is made at frame rates up to 1 MHz. Various qualitative and quantitative optical measurements are made using techniques including chemiluminescence [9] and laser-induced fluorescence (e.g. [8]). Measurements of species concentrations have been made using time-of-flight mass spectrometers (e.g. [87]).

The data-acquisition used on T4 is based around a 128-channel 14-bit National Instruments system running on an in-house developed LabView code. The maximum sampling rate is 2.5 MS/s per channel.

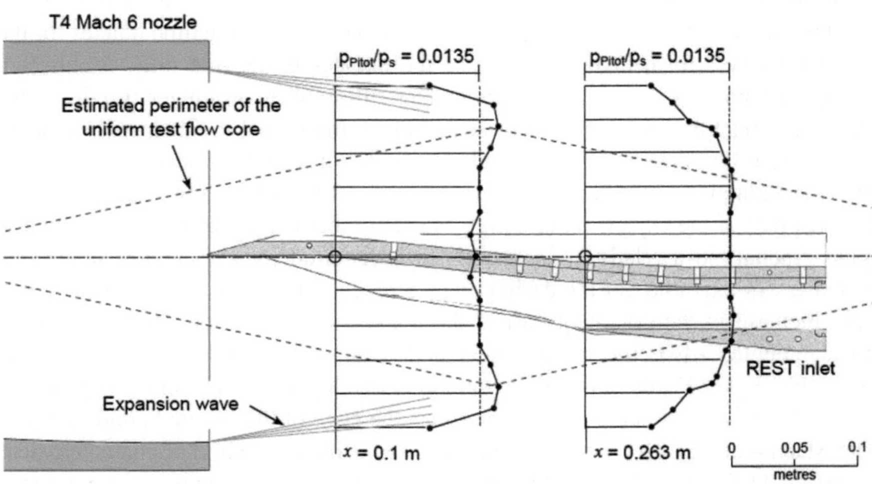

Fig. 48 Model located in the uniform test core of the T4 Mach 6 nozzle [12]

The conditions of the flow in the test section for each shot are obtained through a combination of measurements and computational simulation. Measurements are made of the initial filling pressure and temperature of the gas in the shock tube, the speed of the primary shock wave in the shock tube, and the measured pressure in the nozzle-supply region after reflection of the primary shock from the end of the shock tube. The overall procedure for using these measurements to calculate the flow conditions in the test section is implemented in an in-house code called NENZFR [19]. The first step in the calculation is to model the compression of the test gas by the primary and reflected shock waves in the shock tube, assuming one-dimensional flow with chemical and thermal equilibrium. The conditions behind the reflected shock are then expanded isentropically to the measured nozzle-supply pressure. These calculations are done using a code called ESTCJ [47]. This gives the conditions of the gas that supply the nozzle flow. These conditions are then further expanded isentropically to Mach 1 to provide inflow properties for the next step in the calculation—the expansion of the test gas through the nozzle.

The nozzle flow is modelled within NENZFR using another in-house code called EILMER [46]. The nozzle flow can be simulated using the code running in viscous, axisymmetric mode using different chemical reaction schemes and turbulence models. For example, Doherty [18] performed axisymmetric Mach 10 nozzle flow calculations with air as the test gas using a thermally-perfect mixture of N_2, O_2, N, O and NO, with the reaction scheme of Gupta et al. [31] and the 2006 Wilcox *k*-*ω* turbulence model as detailed in Chan et al. [12]. The location of transition from laminar to turbulent flow in the boundary layer formed on the wall of the nozzle is set to best match measured profiles in the test section. An example of the result from NENZFR is shown in Fig. 49 in the form of Mach number contours for two different pressure levels at a nominal nozzle-supply enthalpy of 5 MJ/kg [18]. The results from the NENZFR calculations can be averaged over the location of the model in the test section to obtain the average flow conditions over the model.

An example of a nozzle Pitot pressure survey is shown for the Mach 7 nozzle in Fig. 50. Plots are shown of the measured ratio of Pitot pressure, p_{Pit}, to nozzle-supply pressure, p_s, at three planes downstream of the exit plane of the nozzle. Shown also are computations of the ratio made using a viscous, axisymmetric simulation using

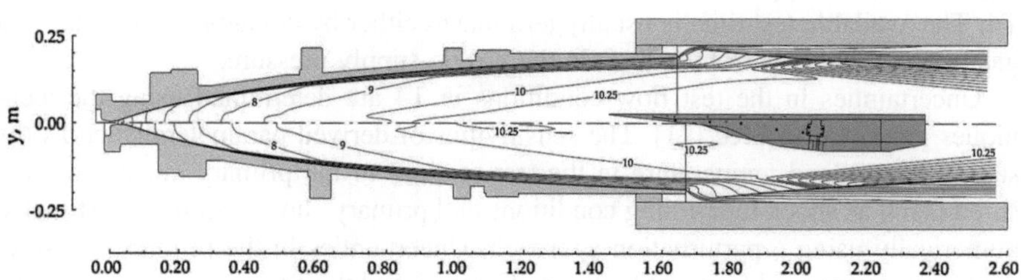

Fig. 49 Mach 10B nozzle flow simulation. Contours of Mach number for 5 MJ/kg flow conditions. Top half shows contours for a high-Re flow condition (test section unit Reynolds number 2.2×10^6 m^{-1}) and lower half shows contours for low-Re flow condition (test section unit Reynolds number 1.2×10^6 m^{-1}). The typical location of a model in the core flow is shown (from Doherty [18])

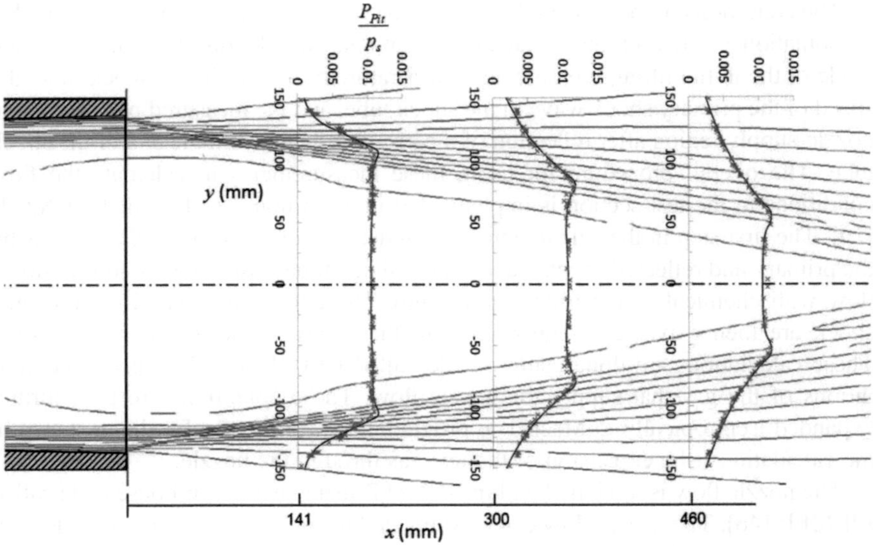

Fig. 50 Mach 7 nozzle calibration showing the ratio of Pitot to nozzle-supply pressure. Measurements from the nozzle Pitot survey are shown at three survey planes located 141, 300 and 460 mm from the nozzle exit plane. Measurements (indicated by crosses) are from five experiments. Axisymmetry is assumed in presenting experimental results. Solid lines in the plots at the three survey planes and contour lines in the background are from the EILMER simulation. Nozzle supply enthalpy of 2.4 MJ/kg. Data and simulation results provided by Dr W.Y.K. Chan

EILMER with the *k*-*ω* turbulence model. Transition from laminar to turbulent flow in the nozzle boundary layer was set to occur 50 mm downstream of the nozzle throat. The results show that the size of the test core, the level of Pitot pressure in the core and the shear layer at the edge of the core are well reproduced using the code.

T4 is smaller than HEG, HIEST and T5 and can be fired routinely up to five times a day. However, the smaller scale also means that the steady test time is usually shorter than in the larger facilities. For example, for a nozzle-supply enthalpy of 7.6 MJ/kg, the available steady test time (after nozzle starting time) is around 1.5 ms. Figure 51 shows typical nozzle-supply and Pitot pressure traces for such a flow in T4. The available test time is usually terminated either by contamination of the test gas by the driver gas or by a drop in the nozzle-supply pressure.

Uncertainties in the test flow conditions in T4 are determined using the techniques presented in Mee [71]. The sensitivities of derived parameters (such as the static pressure and temperature in the test section) to the primary quantities measured (such as shock tube filling conditions and primary shock speed) are estimated numerically using a perturbation approach. Uncertainties in the primary measurements are determined from calibrations of sensors and the processing methods used. Typical uncertainties in conditions in the test section are ±6 % in static pressure and static density, ±3 % in flow speed and ±2 % in Mach number.

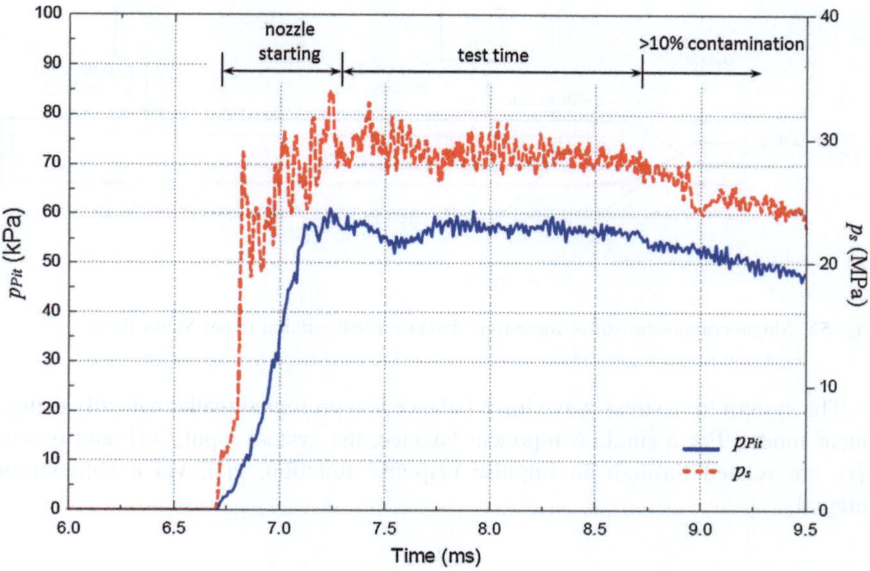

Fig. 51 Typical nozzle-supply pressure, p_s, and Pitot pressure, p_{Pit}, for T4 (adapted from Suraweera [99]); Nozzle-supply enthalpy of 7.6 MJ/kg

5.1 Selected Applications of T4

5.1.1 Measurement of Forces on Scramjets

A method for measurement of aerodynamic forces in the short-duration flows typical of shock tunnels, referred to as the stress-wave force balance technique, was developed at The University of Queensland and validated via experiments in the T4 Stalker tube [70, 84, 112]. The technique has been applied to measure the performance of scramjets and their components in T4.

The generic arrangement for a single-component stress-wave force balance is shown in Fig. 52. The test model, shown at the left of the figure as a cone, is exposed to the aerodynamic flow. Attached to the base of the model is a stress bar. A strain gauge on the stress bar measures strains caused by the stress waves generated by the aerodynamic loads applied to the model. The stress bar and model are supported by fine wires that allow the model/stress bar arrangement to swing freely in the direction of the oncoming test flow. The wires are attached to the top surface of the test section. Everything apart from the test model is shielded aerodynamically from the oncoming flow with only a small gap at the base of the model to permit the model/stress bar to move a small amount horizontally during the test period.

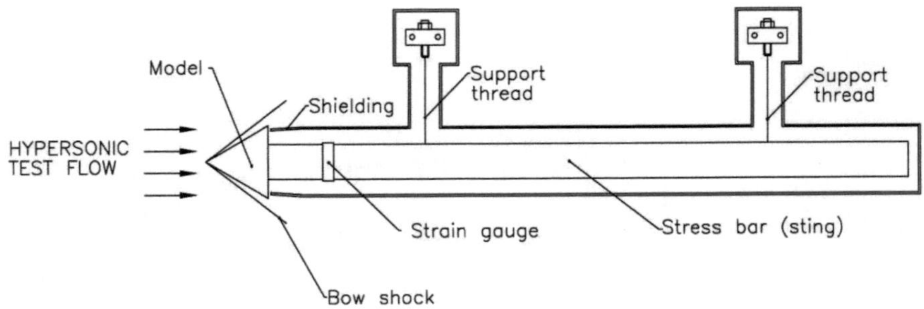

Fig. 52 Single-component stress-wave force-balance configuration (from Smith [90])

The system for a stress-wave force balance is represented mathematically using a linear model. For a single-component balance, the system input, *u(t)*, and output, *y(t)*, are related through an impulse response function, *g(t)*, via a convolution integral,

$$y(t) = \int_0^t g(t - \tau)u(\tau)d\tau \tag{2}$$

The input is the force applied to the model and the output is the strain measured in the stress bar. The impulse response is determined through a dynamic calibration process [68]. In an experiment, the strain is measured and a deconvolution process is used to determine the force on the model as a function of time. The system is capable of measuring not just the mean aerodynamic force during the test time, but also how the overall force varies with time during the test.

Single-component (thrust/drag) measurements have been made on scramjet models in T4 (e.g. [77, 94, 101]). The general arrangement used for such tests is shown in Fig. 53. A fuel tank is located between the base of the model and the stress bar. Stress waves initiated by forces on the scramjet model travel through the tank to the stress bar for measurement. A fast-acting valve located in the fuel tank is used to initiate the fuel flow. Typical measurements from an experiment for an unfuelled test are shown in. The Pitot pressure measured in the test section and the measured drag force on the model have similar time histories. The time history of the coefficient of drag for the model (based on intake capture area) during the test can be inferred as described in Tanimizu et al. [101]. A typical uncertainty in the average measured drag coefficient during the period of steady test flow is ±12.5 % [101]. Measurements made using this stress-wave force balance arrangement, reported in Paull et al. [77], demonstrated for the first time that a complete model scramjet vehicle could produce more thrust than drag (Fig. 54).

Thrust/drag, lift and pitching moments were measured on a fuelled scramjet model in T4 by Robinson et al. [80], see Fig. 55. For these measurements, four stress bars were used to connect the scramjet model to a base plate and an inertial

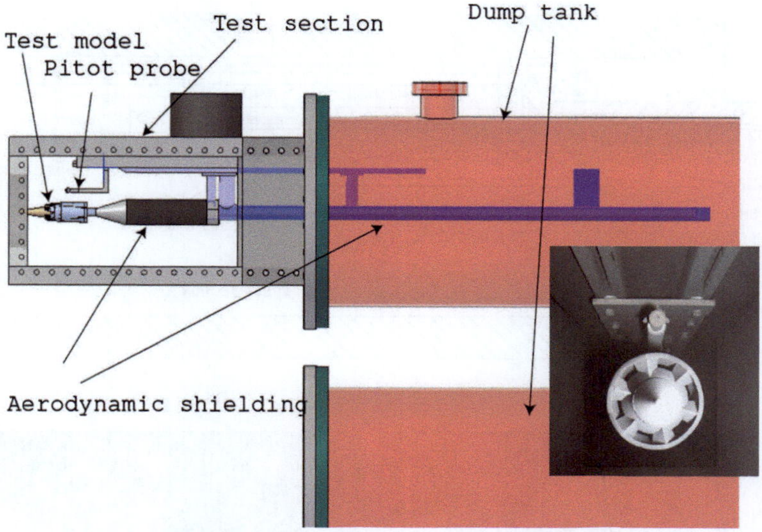

Fig. 53 General arrangement for a thrust/drag stress-wave force balance for a scramjet model in T4. Insert shows a photograph of the model looking downstream from the nozzle centre line (from Tanimizu [102])

Fig. 54 Typical drag force measurements for a scramjet test without fuel injection (from Tanimizu [102]) **a** Force and Pitot pressure measurements **b** Drag coefficient

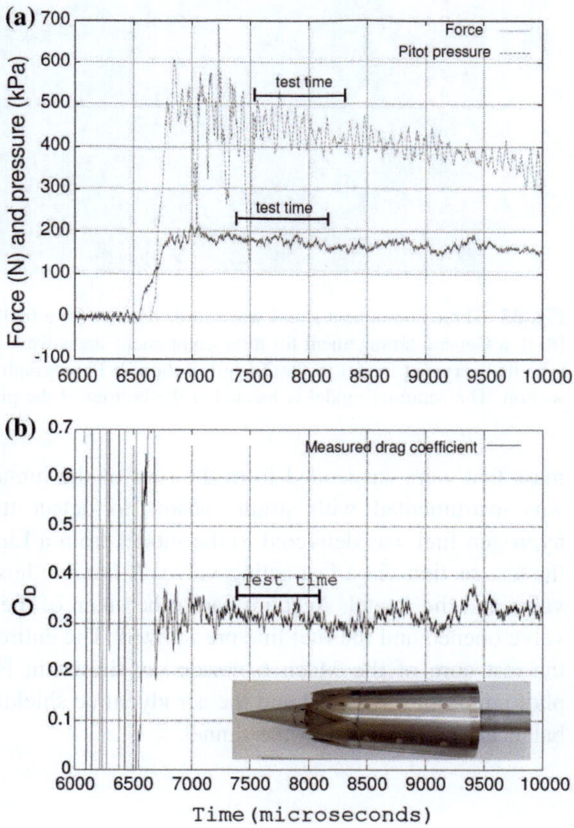

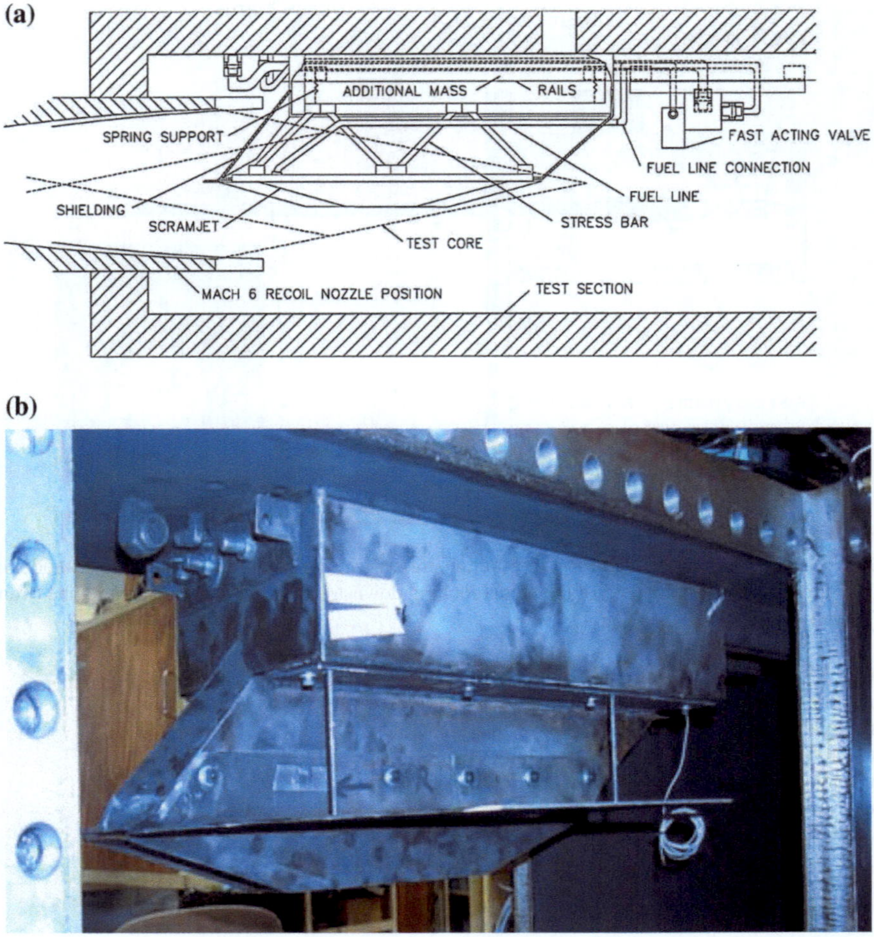

Fig. 55 Three-component stress-wave force balance for a fuelled scramjet model (from Robinson [81]). **a** General arrangement for three component stress-wave force balance for measurements on a fuelled scramjet model in the T4 test section. **b** Photograph of model and shielding in the test section. The scramjet model is located at the bottom of the photograph

mass that were suspended from the roof of the tunnel by springs. Each stress bar was instrumented with strain sensors to detect the stress wave activity. The hydrogen fuel was delivered to the model from a Ludwieg tube located outside of the test section via a fast-acting valve. A flexible hose was connected between the valve and the model. Account had to be taken of the force on the model when the valve opened and the fuel line pressurized. The entire model was positioned within the test core of the Mach 6 nozzle, as shown in Fig. 55a. Figure 55b shows a photograph of the model and the aerodynamic shielding when the model and force balance were installed in the tunnel.

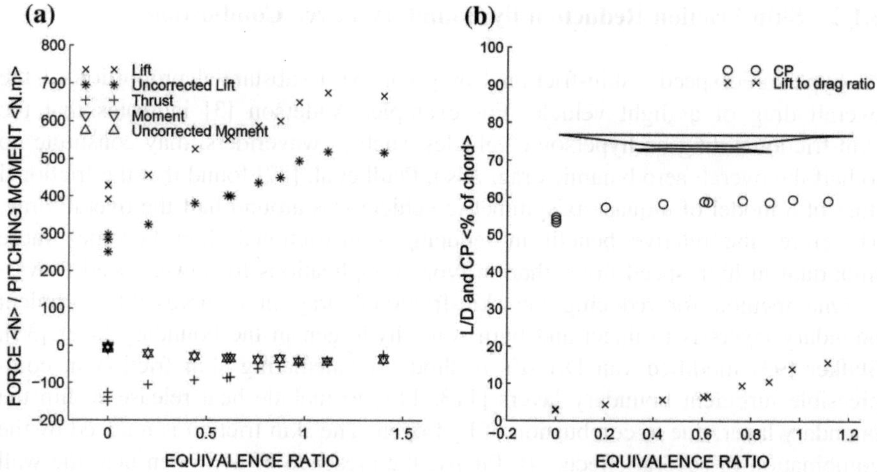

Fig. 56 Results from three-component stress wave force balance for fuelled scramjet model; 3.3 MJ/kg flow condition (from Robinson [81]). **a** Lift, thrust and pitching moment measurements. **b** Centre of pressure (CP) (shown as a percentage of the chord-length of the model) and ratio of lift to drag (L/D)

Coupled, multidimensional deconvolution was used to extract the drag/thrust, lift and pitching moment from the strain signals, as detailed in Robinson et al. [80]. Tests were performed over a range of fuel equivalence ratios at a fixed flow condition and at a fixed fuel equivalence ratio for a range of flow conditions. Example results are shown in Fig. 56. This shows the forces on the scramjet model for variations in the rate at which fuel was injected. Figure 56a shows results for both "Lift" and "Uncorrected Lift". Before the tunnel is fired, the test section and dump tank are evacuated. When the flow arrives, pressure builds up inside the shielding and this provides a pressure force that acts to reduce the measured lift force on the model. This pressure rise inside the shielding was measured and the magnitude of the force this pressure applied to the top of the model was inferred and corrected for in determining the aerodynamic lift force on the model. The "Uncorrected Lift" does not account for this force build-up but the "Lift" shown in Fig. 56 does account for the build-up. A similar effect was obtained for pitching moment and a correction for this was also applied.

The results clearly show how the drag force on the model (a negative thrust force in Fig. 56) decreases as the rate of fuel injection increases and the pressure in the nozzle increases. Note that the lift force increases as more fuel is injected and burns. The increased pressure in the nozzle produces a nose-down pitching moment, as indicated by a rearward movement of the centre of pressure for the model. The combination of increasing lift force and reducing drag force leads to a large increase in the lift-to-drag ratio with fuelling.

5.1.2 Skin-Friction Reduction by Boundary Layer Combustion

At hypersonic speeds, skin-frictional drag can be a substantial proportion of the overall drag of a flight vehicle. For example, Anderson [3] indicates that the skin-frictional drag on hypersonic vehicles, such as waveriders, may constitute up to half the overall aerodynamic drag. Also, Paull et al. [77] found that the frictional drag of a model of a quasi-axisymmetric vehicle was around half the overall drag. Therefore, the relative benefit in reducing skin-frictional drag becomes more important in high-speed flows than in typical applications for lower-speed flows.

One method for reducing the skin-frictional drag in compressible turbulent boundary layers is to inject and burn some hydrogen in the boundary layer [30]. Stalker [93] modified van Driest's method for calculating skin friction in compressible turbulent boundary layers [113, 114] to include heat release within the boundary layer, due to combustion of hydrogen. The skin friction is reduced by the combination of several effects [4]. Firstly, the presence of hydrogen near the wall reduces the viscosity of the gas near the wall. Secondly, the low speed and low momentum of the injected gas relative to that of the freestream reduces the velocity gradient near the wall. Both of these effects will occur whether or not the hydrogen burns. The third effect is due to combustion and heat release. The layer of hot, low-density gas beneath the flame front dampens the transfer of turbulent momentum to the wall, thus reducing the Reynolds stresses. The T4 Stalker tube has been used to investigate the range of conditions over which boundary-layer combustion can be used to reduce skin friction and how it could be implemented in a scramjet combustor to reduce the overall drag of a scramjet engine.

Goyne et al. [30] measured skin friction at discrete points in a duct of rectangular cross section in T4. The arrangement allowed for hydrogen to be ejected into the boundary layer from an injector located on the wall of the duct. The skin friction was measured using flush mounted, floating-element skin-friction gauges that had been designed specifically for operation in shock tunnels [29]. Skin friction was measured with and without injection of hydrogen. When hydrogen was injected and it burned in the boundary layer for a Mach 4.5 flow at a nozzle-supply enthalpy of 7.8 MJ/kg, reductions in the skin friction coefficient of as much as 70–80 % were observed. Corresponding numerical calculations indicate that boundary layer combustion caused the skin friction reduction to persist for at least 5 m downstream. An investigation into the range of conditions over which the skin friction could be reduced by boundary-layer combustion was conducted by Suraweera et al. [100] using a similar arrangement. The nozzle-supply enthalpy was varied from 4.8 to 9.5 MJ/kg. The largest reduction in skin friction coefficient was obtained at the 7.6 MJ/kg condition with smaller reductions at lower and higher enthalpies.

A good demonstration of the effect on skin friction of releasing heat in a compressible turbulent boundary layer is provided by Chan [14]. He used EILMER [46] to simulate the viscous flow in a circular duct for three different thermal situations. One was an adiabatic flow, another was a flow in which heat was released uniformly throughout the duct and a third was when the same amount of heat was released but it was all released within the boundary layer (see Fig. 57). The flow conditions

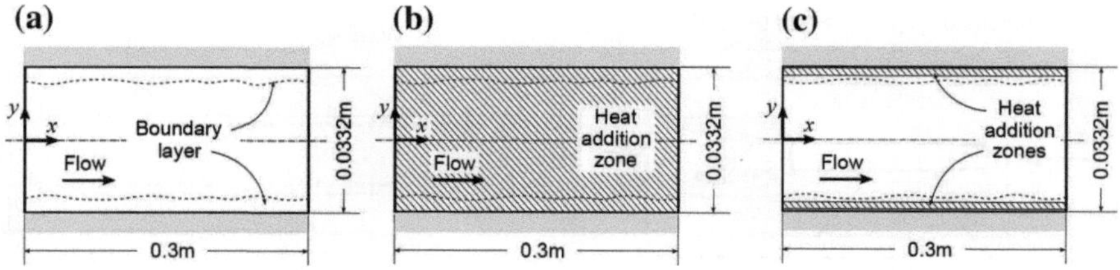

Fig. 57 Arrangements for heat addition in a constant area duct (from Chan [14]). **a** Adiabatic. **b** Uniform heat addition. **c** Boundary-layer heat addition

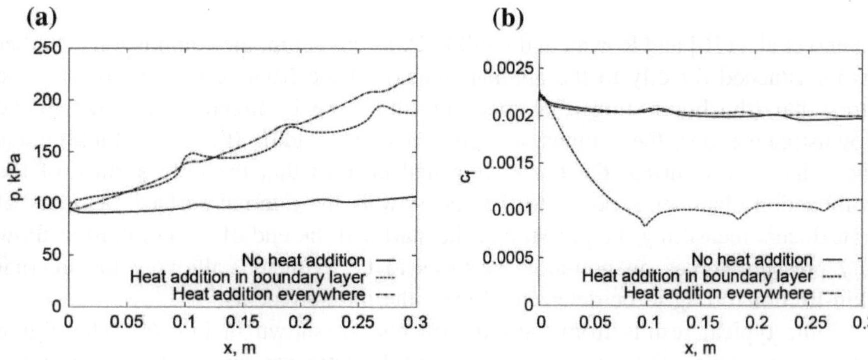

Fig. 58 Effects on skin-friction coefficient of heat addition in a compressible turbulent boundary layer (from Chan [14]). **a** Surface static pressures. **b** Skin friction coefficient

correspond to those typical of experiments that Chan conducted in T4 (duct diameter 33.2 mm; duct length 300 mm; incoming velocity, temperature and pressure of 2576 m/s, 1126 K and 88.5 kPa, respectively). The amount of heat released corresponded to the heat release expected for hydrogen combustion at an equivalence ratio of 0.5. The results are shown in Fig. 58. For the adiabatic case, the pressure rises slowly and the skin-friction coefficient decreases slowly as the boundary layer grows on the walls of the cylinder. The small bumps in the pressure and skin friction distributions are associated with weak shocks generated by the boundary layer growth and subsequently reflect across the duct. When heat is added uniformly, the pressure increases in the duct as the Mach number reduces. The corresponding skin-friction coefficient remains essentially unchanged. When the same amount of heat is released in the boundary layer, the pressure again increases in the duct but the skin friction coefficient is reduced by approximately 50 %.

Kirchhartz et al. [55] performed experiments in T4 to measure the overall skin-frictional drag on a scramjet duct of circular cross section. The duct was equipped with the capability to inject hydrogen into the boundary layer as it entered the combustor. They used a stress-wave force balance to measure the skin-frictional drag on the internal surface of the combustor. A complex experimental arrangement was required to isolate the internal skin-frictional drag from other forces acting on the model (see Fig. 59). This arrangement was developed from previous studies by

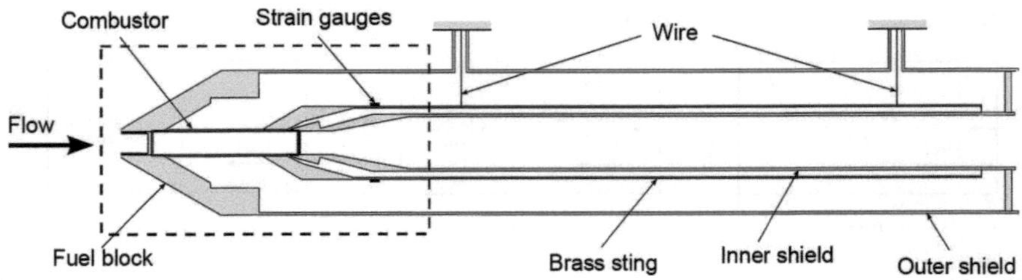

Fig. 59 Experimental arrangement of stress-wave force balance for measurement of the internal skin-frictional drag on a scramjet combustor (from Chan [14])

Tanno et al. [110] and Rowan and Paull [82]. In this setup, the combustion chamber is not attached directly to the annular injector at the front of the fuel block. The stress bar (the brass sting) that measures the force is directly connected to the downstream end of the combustor. Aerodynamic shielding (the inner shield) ducts the exhaust flow inside the brass sting and ensures that the only surface of the combustion chamber exposed to the test flow is the internal surface. A series of transducers measuring the pressure at the start and the end of the combustor allow any pressure forces on non-axial surfaces to be extracted, allowing the internal skin-frictional drag to be determined from the measurements.

Some typical results from the experiments are shown in Fig. 60. This figure shows the nozzle-supply pressure and a sample static pressure measurement in the combustor duct for a fuel-off experiment (Fig. 60a). The measured voltage signal from the strain gauge, from which the drag force on the combustor is inferred, is shown in Fig. 60b. The voltage initially decreases when the flow arrives due to the compressive stress generated in the stress bar by the skin-frictional drag force on the combustor. After about 1 ms, the strain is relieved as the primary stress wave reflects from the free end of the bar. This primary wave continues to reflect up and down the stress bar. The corresponding drag coefficient, CD, inferred from the measurements using deconvolution, is shown in Fig. 60c.

The results of the study indicate that the viscous drag reduced when hydrogen was injected into the boundary layer but did not burn. A test gas of nitrogen was used in these "suppressed-combustion" experiments. This is a "film-cooling" effect, in which a cooler foreign gas is injected close to the surface (e.g. [11]) and leads to reductions in heat transfer and skin friction. Larger reductions in skin friction were obtained when the hydrogen burned in the boundary layer. The leading edges of forebodies and intakes for scramjets are likely to have blunted leading edges. This will result in relatively thick entropy layers entering scramjet combustors. The influence of this on skin friction with boundary layer combustion is also of interest. Therefore, Kirchhartz et al. [55] also conducted tests with blunted leading edges on the intake to the combustor (at the left of the model in Fig. 59). Larger reductions in skin friction were obtained with the blunted leading edges than with sharp leading edges. This was attributed to more vigorous burning in the boundary layer. Also, when combustion occurred, larger reductions in skin friction were obtained at lower nozzle-supply enthalpies.

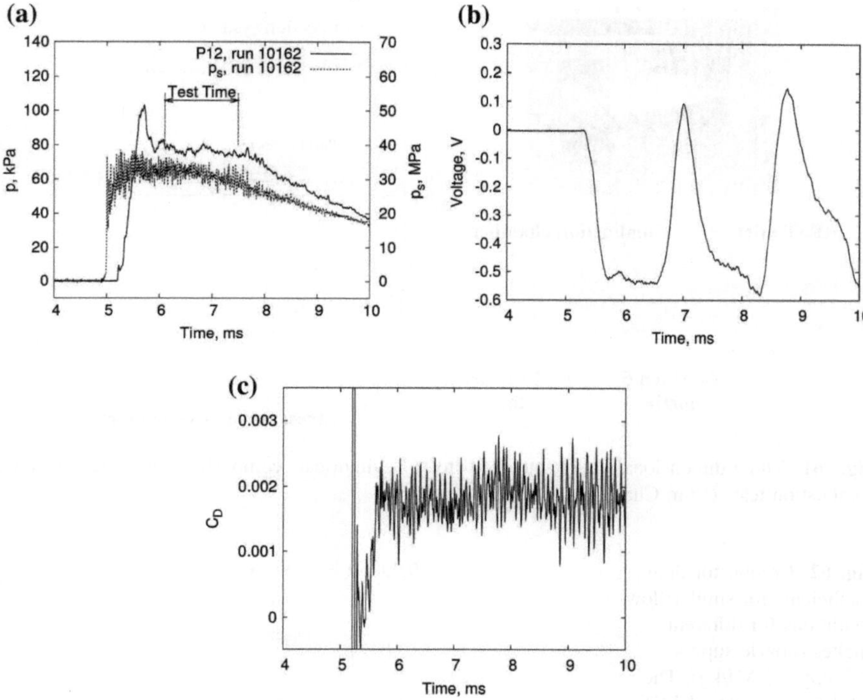

Fig. 60 Typical signals obtained from experiment to measure skin-frictional drag force on a scramjet combustor. Nozzle-supply enthalpy of 7.3 MJ/kg; oncoming flow Mach number 4.3 (from Kirchhartz [56]). **a** Nozzle-supply pressure, p_s, and static pressure in the duct, p_{12}. **b** Raw signal from strain-gauge amplifier. **c** Processed combustor Drag Coefficient, C_D

The tests conducted in the study by Goyne et al. [30], Suraweera et al. [100] and Kirchhartz et al. [55] had quite uniform oncoming flows entering the combustors. In a scramjet intake, there may be relatively large non-uniformities in the flow entering the combustor. For example, a Rectangular-to-Elliptical-Shape-Transition (REST) inlet will generate non-uniform flow and highly three-dimensional boundary layers [89]. A series of tests in T4 was conducted by Chan et al. [13] to identify the influence of such non-uniformities on skin-friction reduction due to boundary layer combustion. A REST inlet was attached in place of the axisymmetric intake used by Kirchhartz et al. [55]. The arrangement is shown in Fig. 61. The cylindrical intake used by Kirchhartz et al. [55] captured the flow exiting T4's Mach 4 nozzle. The REST inlet used in the study of Chan et al. [13] captured the flow exiting T4's Mach 6 nozzle. This flow was compressed by the inlet to conditions similar to those entering the combustor in the experiment of Kirchhartz et al. [55].This allows a direct comparison of the influence of the intake on the skin friction in the combustor for both combusting and non-combusting flows. The inlet used by Chan et al. [13] had a modest geometrical contraction ratio of 3.84 to achieve similar conditions to those of Kirchhartz et al. [55] in the combustor.

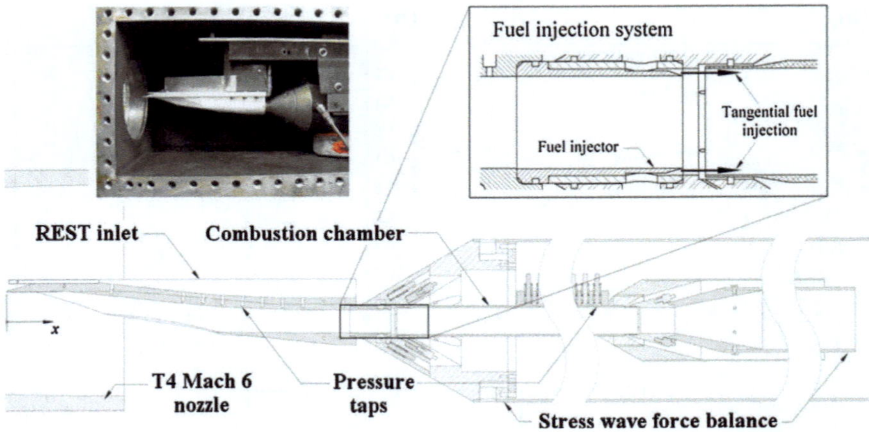

Fig. 61 Three-dimensional inlet attached to a cylindrical combustor for boundary-layer combustion tests (from Chan [14])

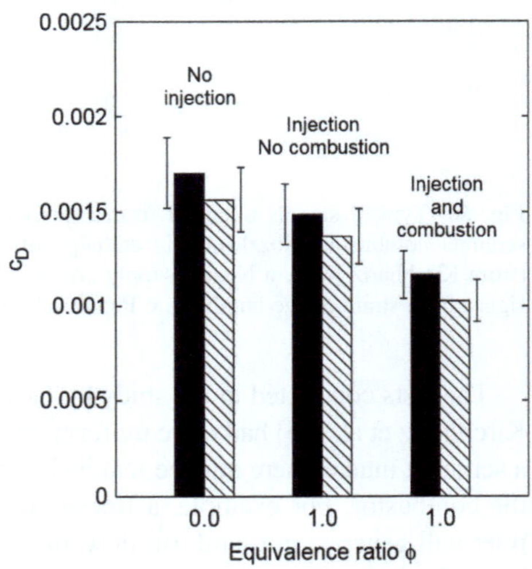

Fig. 62 Combustor drag coefficients for similar flow conditions for different intakes (nozzle-supply enthalpy 4.8 MJ/kg). The solid bars are for a REST intake and the hashed bars are for an axisymmetric intake (adapted from Chan [14])

The major results from the experiments are shown in Fig. 62. For the tests, in which no fuel was injected, it can be seen that the drag coefficient for Chan's test with the combustor downstream of the REST inlet (the solid-filled bar to the far left in Fig. 62) was approximately 5 % higher than for Kirchhartz's test with an axisymmetric, constant area intake (the cross-hatched bar in Fig. 62). The difference is within the accuracy with which the drag coefficient could be resolved but the higher drag coefficient for the more disturbed flow coming from the REST inlet is expected. In the suppressed combustion experiments (fuel injected into a flow of nitrogen), the drag coefficients for both experiments reduced by a similar amount, indicating that there is some "film-cooling" reduction in skin friction for both cases.

For the combustion cases, the drag coefficients further reduced, again by similar amounts for both intakes. Chan et al. [13] conclude that the effect of boundary-layer combustion reducing skin friction in the combustor is maintained for this REST inlet with its modest contraction ratio.

6 California Institute of Technology Shock Tunnel T5

The T5 hypervelocity shock tunnel at GALCIT (Graduate Aerospace Laboratories, California Institute of Technology, Pasadena, USA) is one in a series of free-piston driven facilities of the type pioneered by Stalker [97]. Its operation started in December 1990. The facility is capable of producing flows of air or nitrogen up to a specific reservoir enthalpy of 21 MJ/kg, a reservoir pressure of 80 MPa, and a reservoir temperature of 10,000 K. It achieves this by using a free piston to adiabatically compress the driver gas of the shock tunnel to a pressure up to 130 MPa. A schematic of the facility with enlargements of the important components is shown in Fig. 63

A test starts with the 120 kg piston loaded into the launch manifold between the secondary reservoir (2R) and the compression tube (CT). The secondary reservoir is filled with compressed air (typically at 8 MPa, but sometimes as high as 15 MPa) used to drive the piston, and the compression tube (30 m long and 0.3 m in diameter) is filled with the driver gas, a mixture of helium and argon. The piston is held in place by evacuating the space behind it. A stainless steel diaphragm (typically 7 mm thick) separates the compression tube from the shock tube (ST). The shock tube (12 m long and 0.09 m in diameter) is filled with the test gas. A thin

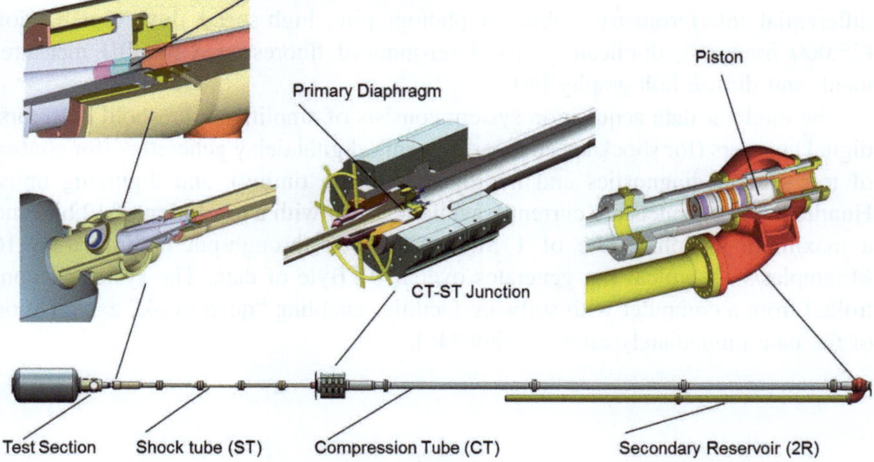

Fig. 63 Schematic of the T5 facility with enlargements of the important components

mylar diaphragm just downstream of the nozzle throat isolates the test gas from the test section and dump tank, which are both evacuated before a test.

The facility is fired by opening a valve to let high pressure air into the space behind the piston. Then the piston moves forward, uncovering large slots open to the high pressure air in the secondary reservoir. The piston can reach maximum speeds in excess of 300 m/s. When the diaphragm bursts, the piston continues to move, keeping the pressure in front of it approximately constant. This makes the device a constant-pressure driver as distinct from the constant-volume driver of conventional shock tubes. The compression tube, shock tube, and nozzle are attached together as one unit, free to move horizontally and recoil with the shift in center of mass due to the piston motion. An inertial mass of 14 tons at the compression tube—shock tube junction greatly reduces this recoil, but the facility still moves about 10 cm during a test. The test section and dump tank remain stationary while the nozzle moves in a sliding seal. The contoured T5 nozzle has an exit diameter of 314 mm and an interchangeable throat. It is designed to operate at Mach 5.2 with high temperature air and a 30 mm diameter nozzle throat.

Pressure transducers mounted at points along the shock tube and at the reservoir allow measurements of the shock speed and the stagnation pressure to be made. These are used to determine the free stream conditions in the test section. T5 is capable of producing stagnation pressures of up to 80 MPa. The free stream conditions at the nozzle exit are typically on the order of 30 kPa pressure and 2000 K temperature, and the flow speeds are typically in the range 3–6 km/s, sufficient to reproduce many of the high temperature effects that are present in hypervelocity flows. The useful test time, before any significant loss of reservoir pressure, is typically 1–2 ms. A more detailed description of the T5 facility and its performance envelope is given by Hornung [40].

The instrumentation available at the facility includes thermocouple type surface heat transfer sensors, piezoelectric surface pressure transducers, and Pitot pressure probes. In addition, non-intrusive visualization is routinely used in the form of differential interferometry, schlieren photography, high-speed flow visualization (25,000 frames/s), duplicate planar laser-induced fluorescence for OH measurement, and diffuse holography [40].

The modular data acquisition system consists of amplifiers, threshold detectors, digital counters (for shock speed measurement), digital delay generators (for control of test section diagnostics and hydrogen injection timing), and digitizing units. Hundred A/D channels are currently available, each with a resolution of 12 bits and a maximum sampling rate of 1 MHz. The total through-put is limited to 16 MSamples/s. A typical run generates over 250 kByte of data. The system is controlled from a computer with software facilities enabling "quick-look" examination of the data immediately after the shot [40].

6.1 Selected Applications of T5

6.1.1 Enthalpy Effects on Hypervelocity Boundary Layer Transition and Passive Control of Transition

The knowledge of the transition from laminar to turbulent boundary layer flow on a entry or re-entry vehicle is an important phenomenon in the aerothermodynamic design of such vehicles, because the prediction of heat flux and drag depends critically on this process. As pointed out in Sect. 3.1.3, for hypersonic flows over slender two-dimensional or axisymmetric geometries, the dominant instability is typically the second or Mack mode [62, 63]. Here the boundary layer acts as a wave guide for acoustic noise, where selected frequencies are trapped and amplified, eventually leading to transition. Another important difference, discussed by Hornung [42], is that in high enthalpy hypersonic flows the relaxation processes associated with vibrational excitation and dissociation provide mechanisms for damping acoustic waves and may therefore be expected to affect the second mode.

An extensive experimental program on transition in hypervelocity flows has been conducted at GALCIT in the T5 facility in which the focus is specifically on the regime where relaxation processes associated with vibrational excitation and dissociation are important [2, 28]. Since the frequencies of the most strongly amplified modes are typically 0.8-2 MHz, the short test time in the order of 1–2 ms is not a serious limitation. The configuration used for the transition experiments in T5 was a slender cone (see Fig. 64). The flow over this configuration has the advantages that the pressure gradient is zero, and that it is free of side effects. A 5° half-angle cone was selected in order to be able to compare the new high enthalpy results with those from cold hypersonic testing. In hypervelocity flow simulation it is important to reproduce the actual speed of the flow, so that the vibrational excitation and dissociation are reproduced correctly. This is often done at the expense of reproducing the Mach number. This is the case in the T5 experiments also, where the free-stream Mach number is typically 5.5, but the speed ranges up to

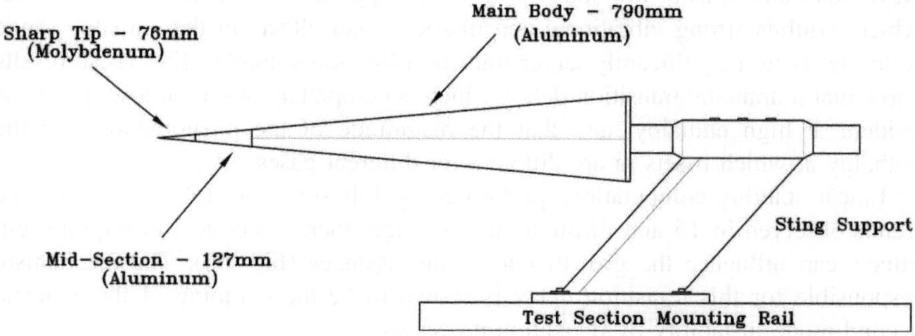

Fig. 64 Slender cone model used for the transition experiments in T5; the main body is hollow and instrumented with thermocouple surface heat flux gauges (from Adam [1])

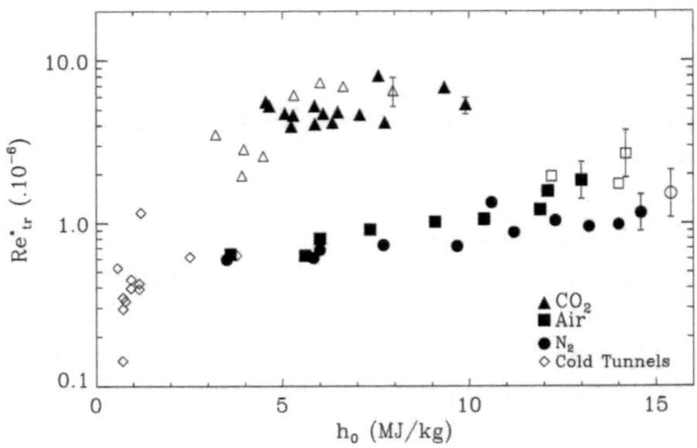

Fig. 65 Transition Reynolds number evaluated at reference conditions as function of total specific enthalpy for different test gases. Open symbols correspond to cases where the flow was laminar to the end of the cone. The references for the cold tunnel data are given in Hornung [42]; the carbon dioxide results show a large transition delay relative to the nitrogen and air results (from Hornung [42])

6 km/s. Thus, the boundary layer edge temperature in a free-flight situation is very different than in the T5 experiments, but the temperature profile in the inner part of the boundary layer is almost the same in both cases. It is therefore more meaningful to compare any two flows in terms of the Reynolds number evaluated at the reference condition rather than at that based on the edge conditions [42]. The transition location on the cone as a function of the total specific enthalpy was determined from the distinct rise in heat flux which coincides with the transition process. The Reynolds number at transition, evaluated at the reference temperature and based on the distance from the cone tip to the transition location, versus the total specific enthalpy of the flow is shown for different test gases in Fig. 65.

It is obvious that a significant increase in transition Reynolds number with increasing total specific enthalpy is obtained. This increase is slightly larger in air than in nitrogen. Tests with helium, which behaves like a perfect gas in the T5 total specific enthalpy conditions, revealed that even at 15 MJ/kg, the same transition Reynolds number as at low total specific enthalpy was obtained. Carbon dioxide, which exhibits strong vibrational and dissociational effects in the enthalpy range tested leads to a significantly larger transition Reynolds number [2]. These results show that a dramatic transition delay, which is completely absent at low speeds, is evident at high enthalpy, and that the magnitude of the phenomenon and the enthalpy at which it sets in are different for different gases.

Linear stability computations performed by Johnson et al. [50] agree with the trends observed in T5 and illustrate how strongly thermochemical nonequilibrium effects can influence the growth rate of disturbances (Fig. 66). The mechanism responsible for this transition delay is shown to be the damping of the acoustic second mode instability by relaxation processes.

In additional tests in T5 it was demonstrated that transition could be further delayed by up to a factor of two by suitable blind porosity of the surface [42, 78].

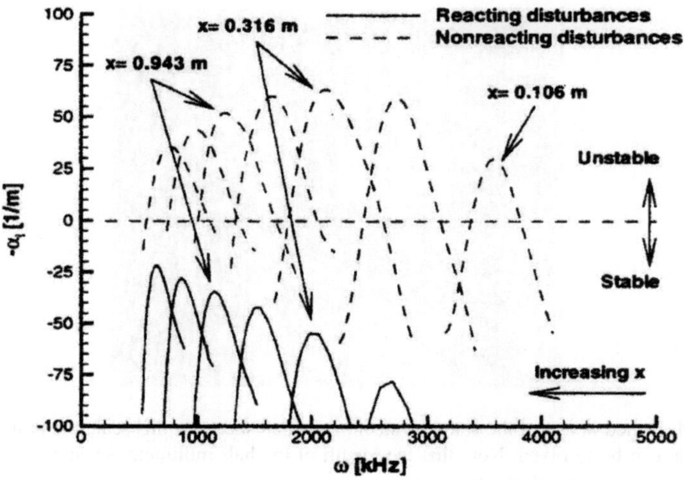

Fig. 66 Results of linear stability calculations with thermochemical nonequilibrium at the conditions of T5 (4.0 MJ/kg in carbon dioxide); the diagram shows growth rate of disturbances as function of disturbance frequency at several distances along the cone; the dotted curves show the results at the same conditions as for the continuous curves but with the rate processes turned off [50]

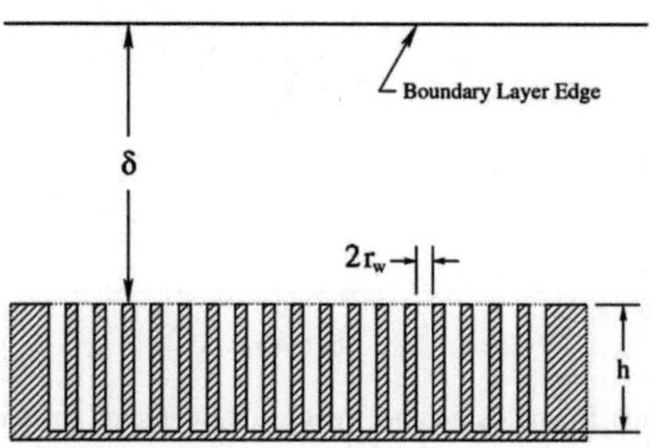

Fig. 67 Approximate proportions of the hole diameter, spacing and depth in relation to the laminar boundary layer thickness. With a typical boundary layer thickness of one mm, this makes the desirable hole depth 0.5 mm and the hole diameter and spacing 0.05 mm and 0.1 mm, respectively (from Hornung [42])

This work was motivated by the results of Fedorov et al. [25] who had shown theoretically that the acoustic mode could also be damped by wall porosity. At the conditions of the cone experiments in T5, small-diameter, deep, blind holes that are closely spaced were predicted to produce suitable damping. The proportions of the configurations chosen are related to the boundary layer thickness in the schematic sketch of Fig. 67.

A magnified image of a perforated sheet which was wrapped around the aluminium cone model is shown in Fig. 68. In order to provide a control experiment in every shot by making half of the cone surface porous and half non-porous, the

Fig. 68 Magnified image of the stainless steel perforated sheet. At this scale the grain boundaries of the metal can be resolved. Note that the length of the half-millimeter scale bar is equal to the depth of the holes (from Hornung [42])

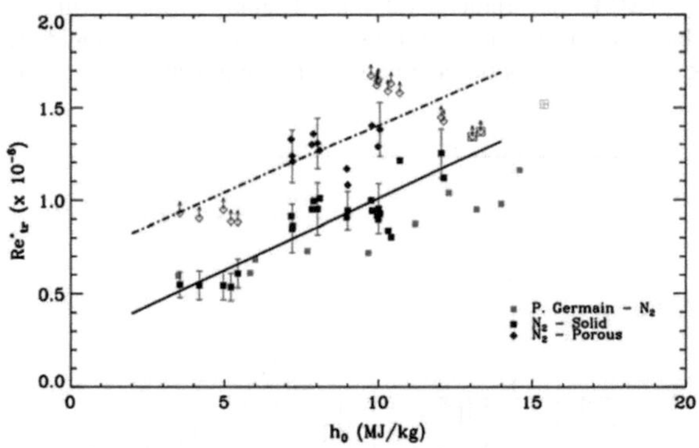

Fig. 69 Plot of transition Reynolds number versus total enthalpy for the N2 data. Dark squares show the results from the non-porous side of the cone. *Gray squares* show the nitrogen data from Fig. 65 for comparison. The filled diamonds show the values from the porous side of the cone. The open diamonds symbolize situations in which the boundary layer was laminar on the porous side all the way to the end of the cone. In these cases, the Reynolds number plotted is that based on the length of the cone. The lines are linear fits to the points to guide the eye (from Rasheed et al. [78])

porous sheet was formed into a half cone and welded to a similar half cone sheet without holes. This hollow cone was mounted on the original model such that a flush transition from the non-porous tip to the porous surface was realized [42].

The results obtained in nitrogen flow are shown in Fig. 69. They confirm approximately the results of the previous experiments and exhibit a dramatic transition delay on the porous side of the cone. The increase of the transition Reynolds number is typically 400,000 which is as much as 80 % at the low-enthalpy end of the range. Both at the low and at the high end of the range, transition could not even be achieved on the porous side, the boundary layer remaining laminar all the way to the end of the cone.

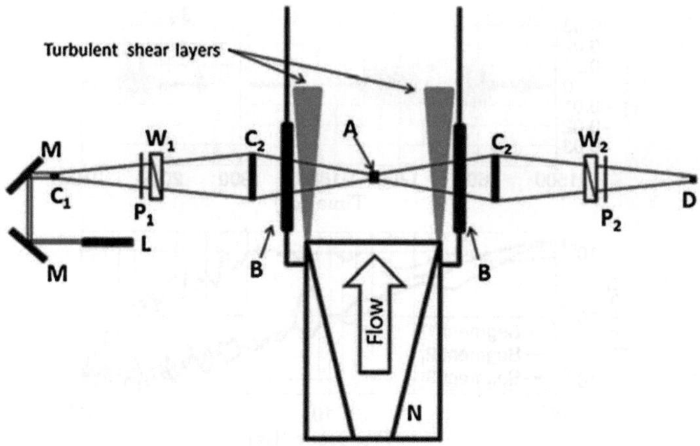

Fig. 70 Annotated schematic of the FLDI, showing the laser (*L*), mirror (*M*), lens (*C*), polarizer (*P*), Wollaston prism (*W*), window (*B*), probe volume (*A*), photo detector (*D*), and nozzle (*N*) (from Parziale et al. [74])

For the quantitative characterization of the second modes within a hypersonic boundary layer, traditionally, fast-response piezoelectric pressure transducers, heat-flux gauges, or hot-wire anemometry techniques are used. However, the high frequency and small wavelength of the disturbances render these techniques inadequate above 1 MHz for conditions in T5 [74]. Here optical techniques have the potential to overcome these limitations (see also Sect. 3.1.3). In T5, focused laser differential interferometry (FLDI) was used to investigate disturbances in a hypervelocity boundary layer on a 5° half-angle [74]. Quantitative measurements of density perturbations with high temporal (20 MHz) and spatial (700 μm) resolution were performed.

To measure the acoustic instability on a slender body in a large-scale reflected shock tunnel such as T5, five requirements of the diagnostic are clear: (1) high temporal resolution of the measurement technique (>10 MHz), (2) high spatial resolution to capture the small wavelength of the disturbance (<1 mm), (3) insensitivity to mechanical vibration, (4) the capability to have a small focal volume near the surface of the cone, and (5) a straightforward and repeatable means of extracting quantitative data from the technique. These requirements are met with focused laser differential interferometer, FLDI [74]. The laser used in the experiments is a Spectra-Physics Excelsior diode-pumped solid-state continuous-wave laser (532 nm wavelength, 200 mW power). The high-quality beam (Transverse Electromagnetic Mode 00, TEM00) does not require additional beam conditioning for use as an interferometer. Following the optical path in Fig. 70, starting from the laser, the beam is turned by a periscope arrangement for precise directional control. The beam is expanded by a lens, C1, and linearly polarized by P1 at 45° to the plane of separation of the first Wollaston prism,W1. The plane of separation of W1 is chosen to be parallel to streamlines in the boundary layer of the 5° half-angle cone. The prism splits the light by a narrow angle (2 arc min) into orthogonally polarized

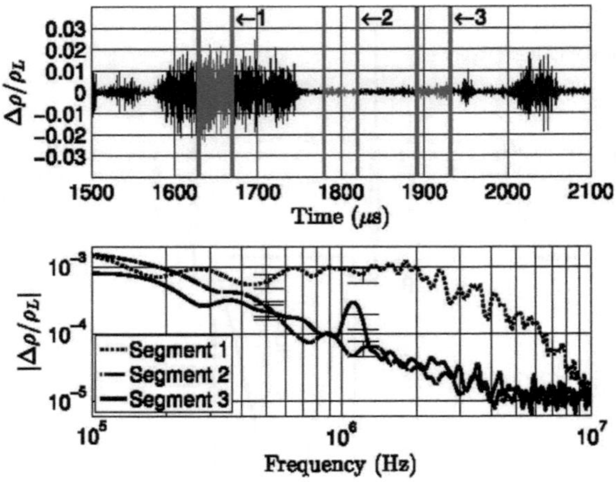

Fig. 71 FLDI results (density difference $\Delta\rho$ normalized by the mean local density ρ_L) from T5 shot 2695 (7.15 MJ/kg, $Re_\infty^{unit} = 7.5 \times 10^6$ 1/m): the processed response (*top*) and spectral response from the three chosen segments (*bottom*) (from Parziale et al. [74])

beams. The separation of the beams is fixed at 350 μm by a lens, C2, while the diameter of the beams is reduced to small values in the center of the test section. This arrangement creates two beams with orthogonal polarization that share much of the same optical path. The orthogonally polarized beams do not share the same optical path within ± 10 mm of the focal point (along the beam direction, centered at A in Fig. 70). In this region, the beams are calculated to be less than 100 μm in diameter and traverse separate but very closely spaced volumes; they are 350 μm apart (assuming $1/e^2$ Gaussian beam propagation). It is primarily within this small focal region that the diagnostic is sensitive to changes in optical path length. The spatial resolution of the technique (700 μm) is set by doubling the beam spacing to satisfy the Nyquist sampling theorem. Beyond the beam focus, the optical paths are again common and an additional lens, C2, refocuses the beams. The Wollaston prism, W2, and polarizer, P2, recombine and then mix the orthogonally polarized beams so that the interference will be registered as irradiance fluctuations by the photo detector. The response of the photo detector is amplified at a gain of 5 and digitized at 100 MHz by a 14-bit Ethernet oscilloscope and a 20 MHz antialiasing filter [74].

A relation between the fluctuations in density and output voltage from the photo detector is used for post processing. The technique detects differences in phase, primarily due to the density differences at the two spatially separated focal regions, thus making the interferometer sensitive to spatial density differences in the streamwise direction. The FLDI technique was used to try to measure the disturbances in the boundary layer. Two examples (Figs. 71 and 72) are presented where both turbulent bursts and wave packets are detected; the spectral content estimation in these examples is obtained using Welch's method, with 50 % overlapping 20 μs Hann windows [74].

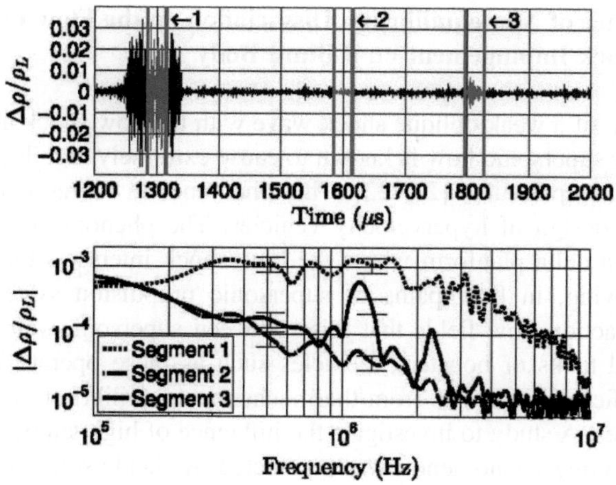

Fig. 72 FLDI results (density difference Δρ normalized by the mean local density ρ_L) from T5 shot 2702 (8.77 MJ/kg, $Re_{\infty}^{unit} = 5.7 \times 10^6$ 1/m): the processed response (*top*) and spectral response from the three chosen segments (*bottom*) (from Parziale et al. [74])

The FLDI response for shot 2695 (Fig. 71) reveals interesting phenomena at 1650 and 1915 μs; 40 μs segments centered at 1650, 1800, and 1915 μs are highlighted. This shows the spectral content of the interrogated point of the boundary layer when minimal disturbances are detected (segment 2), when a turbulent spot passes (segment 1), and when a wave packet passes (segment 3). The spectral content of the turbulent spot (segment 1) shows broadband response; the wave packet (segment 3) has a strong peak in response at 1.11 MHz. The FLDI response for shot 2702 (Fig. 72) reveals interesting phenomena at 1300 and 1810 μs; 30 μs segments centered at 1300, 1600, and 1810 μs are highlighted. This shows the spectral content of the interrogated point of the boundary layer when minimal disturbances are detected (segment 2), when a turbulent spot passes (segment 1), and when a wave packet passes (segment 3). The spectral content of the turbulent spot (segment 1) shows broadband response; the wave packet (segment 3) has a strong peak in response at 1.17 MHz with a harmonic at 2.29 MHz.

As discussed in Parziale et al. [74], the most-unstable frequency was also derived from numerical computations of the multicomponent, nonequilibrium boundary layer flow based on the correlation $f \approx 0.8\, u_{edge}/2\delta_{99}$. Very good agreement with the peak frequencies shown in Figs. 71 and 72 was found.

The error and noise floor associated with the FLDI technique and the facility were quantified and found to be sufficiently low such that acoustic instability growth rate measurements seem to be feasible by placing an additional FLDI downstream of the first FLDI [74].

The results of this work and the associated FLDI measurement of the free-stream noise in T5 led to the important conclusion that, as far as the second-mode instability is concerned, the high frequency of the most unstable mode causes free-piston shock tunnels to be quiet facilities.

6.1.2 Influence of Non-equilibrium Dissociation on the Flow Produced by Shock Impingement on a Blunt Body

The interaction of a weak oblique shock wave with the bow shock ahead of a blunt body in steady supersonic flow is known to cause extremely high local heat transfer rates and surface pressures [21, 22]. This phenomenon represents a severe constraint on the design of hypervelocity vehicles. The phenomenon is observed in vehicles with a delta planform where the bow shock intersects the swept leading edge of the wing, in flow paths of supersonic propulsion systems and in the complex interaction flow fields that arise between supersonic vehicles and externally mounted tanks or boosters. Vehicles such as these operate in regimes that produce significant departures from thermochemical equilibrium in the flow about vehicle surfaces. A study to investigate the influence of high temperature effects on this design-limiting phenomenon was conducted by Sanderson et al. [83].

The most complete description of the shock impingement phenomenon is due to Edney [21] who observed and classified six distinct interaction regimes known as types I–VI. Severe heating occurs when the incident shock wave impinges in the vicinity of the geometrical stagnation point and a type IV flow results (Fig. 73). The remaining interaction regimes are less intense and therefore not design limiting.

As outlined by Sanderson et al. [83],the type IV interaction is characterized by a three-shock λ-pattern occurring at the point where the impinging shock wave, 1 (numbers refer to indices in Fig. 73 left) interacts with the bow shock, 2. A strong vortex sheet emanates from the λ-point. The shock layer flow above the vortex sheet is subsonic whereas the flow between the vortex sheet and reflected shock wave, 3, remains supersonic. A second, less distinct, inverted λ-pattern is observed at the intersection between the reflected shock, 3, and the continuation of the strong bow shock below the interaction region, 4. An additional oblique wave, 5, is reflected upwards at the secondary λ-point and a second vortex sheet is produced. Observe that the two vortex sheets bound a supersonic jet that is embedded in the surrounding subsonic shock layer. The jet is formed from the fluid passing between the two λ-points. The oblique wave, 5, crossing the jet is guided along the supersonic jet by successive reflections from the two shear layers. A strong terminating

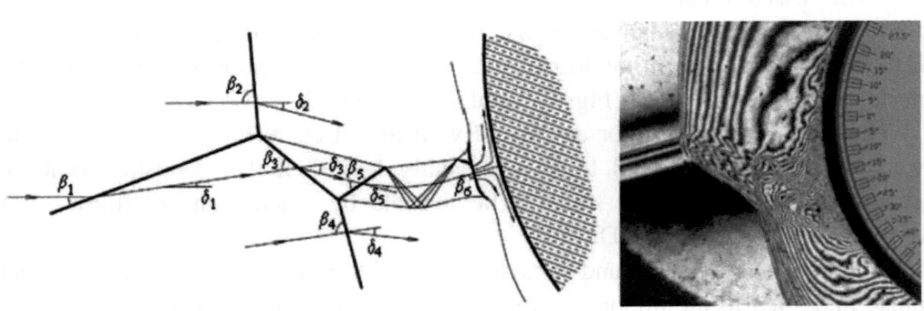

Fig. 73 Schematic (*left*) and holographic interferogram (*right*) for a type IV interaction taken in T5 (from Sanderson et al. [83])

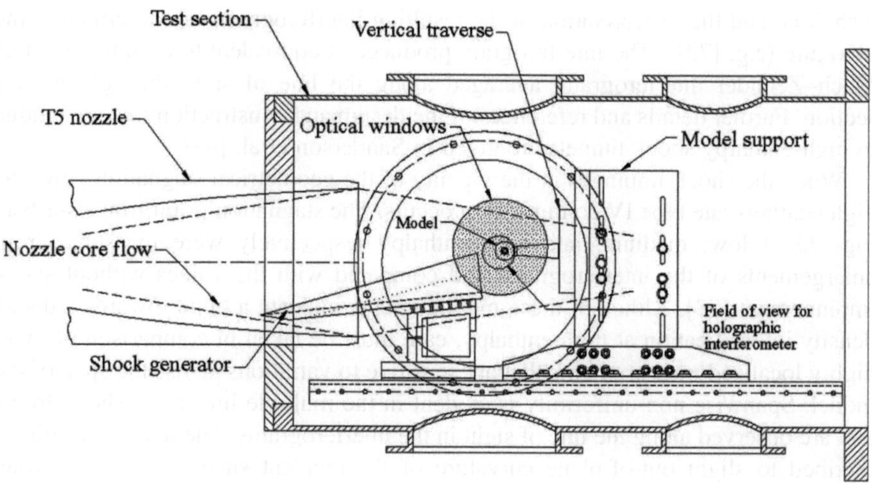

Fig. 74 Layout of apparatus in the test section of T5 (to scale) (from Sanderson et al. [83])

shock, 6, is observed at the base of the jet. The two shear layers turn rapidly upward and downward respectively with the strong pressure differential across the terminating shock being balanced by extreme streamline curvature.

The experimental arrangement in T5 is shown in Fig. 74. The gas was expanded into the test section using a contoured axisymmetric nozzle with an area ratio of 109 and an exit diameter of 314 mm. A cylinder with aspect ratio 4.5 (length over diameter) provided the largest model that adequately approximated a two-dimensional flow and optimization of the geometrical arrangement led to a model diameter of 40.6 mm. The surface temperature rise of the model during the test time was sufficiently small to approximate an isothermal wall boundary condition. A large shovel-shaped shock generator was partially inserted into the uniform nozzle core flow that extends upstream of the nozzle exit plane. Shock generators of this type produce a planar oblique shock wave, with uniform downstream flow, and are closely related to a class of supersonic lifting body shapes known as waveriders. Variations in the shock impingement flow field were produced by vertically translating the model with respect to the exit of the shock generator. The shock generator was inclined at an angle of 6° for all experiments discussed here.

Surface junction thermocouples were used to record the time-resolved surface temperature at 24 circumferential locations around the model forebody. The sensitive area of the gauges subtended an angle of 2° at the centre of the cylinder and were spaced at 5° intervals in the most densely instrumented portion. Surface heat fluxes may be inferred from the time histories of the surface temperatures via a semi-infinite body model of transient heat conduction into the model [86]. A 75 mm field of view holographic interferometer was constructed and used to visualize the flow field and to provide quantitative measurements of the density field. The fundamentals of the technique and the interpretation of the resulting interferograms are described in the literature (e.g. [72]). The interferogram produced is equivalent to an infinite-fringe Mach–Zehnder interferogram, averaged along the line of sight through the test section. Further details and references of the design and construction that are unique to high-enthalpy shock tunnels are given in Sanderson et al. [83].

When the shock impinges in the vicinity of the geometrical stagnation point the high heating- rate type IV configuration occurs. The stagnation-point fringe shifts in Fig. 75 at low, medium, and high enthalpy respectively were measured from enlargements of the interferograms and compared with the values without shock impingement [83]. Although these measurements indicate a trend towards reduced density intensification at high enthalpy, care must be taken in comparison of these highly localized measurements that are sensitive to variations across the span of the model. Spanwise non-uniformity is evident in the multiple images of shock fronts that are observed along the line of sight in the interferograms. The non-uniformity is ascribed to slight out-of-plane curvature of the incident shock wave, the strong sensitivity of the flow field to impingement location, and potential unsteady oscillations of the jet structure that are not synchronous along the span of the cylinder. Comparison of the type IV thermocouple data shown in Fig. 75 indicates a similar trend of reduced heat transfer intensification at the higher enthalpy conditions, and importantly these data are insensitive to variations across the span of the model. In all cases the maxima of fringe shift and instantaneous heat transfer occur at the same angular location. The standard deviation of the heat transfer data increases in the vicinity of the jet impingement point, indicative of flow unsteadiness.

In summary, the experiments in T5 performed at three different test conditions with varying specific total enthalpy indicated that existing predictions of greatly increased stagnation density at high enthalpy were not realized. Time-resolved heat transfer measurements confirmed this observation and were correlated, along with data from previous studies. It could be demonstrated that 85 % of the variability in the heat transfer data was correlated by variations in free-stream Mach number and impinging shock wave angle. No statistically significant trends with Reynolds number or stagnation enthalpy were resolvable.

In addition to the experimental campaign in T5, an improved model was developed to elucidate the finite-rate thermochemical processes occurring in the interaction region. It was shown that severe heat transfer intensification is a result of a jet shock structure that minimizes the entropy rise of the supersonic jet fluid whereas strong thermochemical effects are promoted by conditions that maximize the entropy rise (and hence temperature). This dichotomy underlies the smaller than anticipated influence of high temperature effects on the heat transfer intensification. The model could accurately predict the measured heat transfer rates. The improved understanding of the influence of high temperature effects on the shock interaction phenomenon reduces a significant element of risk in the design of hypersonic vehicles.

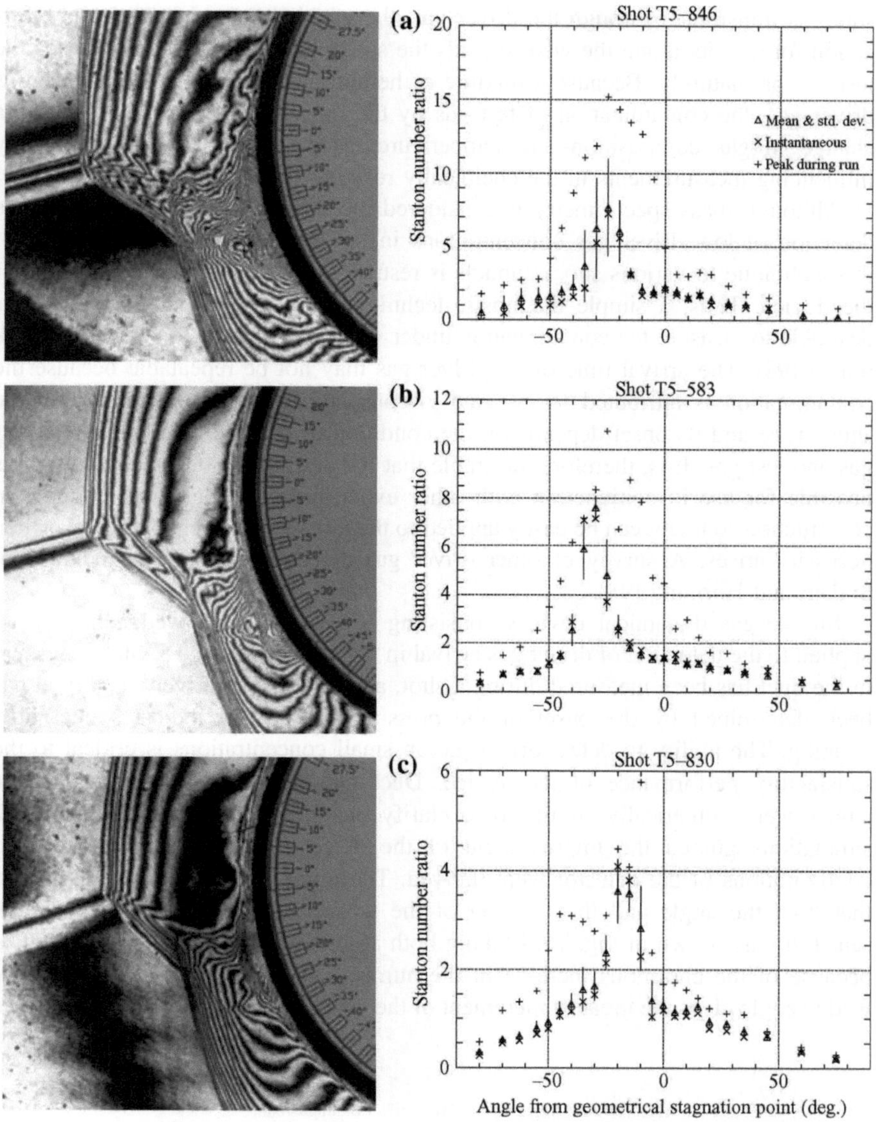

Fig. 75 Holographic interferograms and heat transfer distributions for type IV interactions at conditions **a** 3.88 MJ/kg, **b** 12 MJ/kg and **c** 19.1 MJ/kg (from Sanderson et al. [83])

6.1.3 Gas Dynamical Detection of Driver Gas Contamination

In reflected shock tunnels, driver gas contamination is one of the most serious problems that limit the test time, especially for high-enthalpy flow conditions. The interaction between the reflected shock wave and the boundary layer on a shock tube wall causes bifurcation at the foot of the reflected shock. When the reflected shock is transmitted through the driver gas, the gas that passes through the bifurcation forms a jet along the wall towards the shock tube end and contaminates the test gas prematurely. Because a mixture of helium and argon is normally used as driver gas, the contamination of test gas by the monatomic driver gas leads to a change in gas composition and temperature in the test section flow, thereby influencing measurements in the chemically reacting flow field.

Although mass spectrometry is considered the most suitable technique for the detection of low driver gas concentrations in the test section, the application of this technique to various shock tunnels is restricted by the size and complexity of the device. Thus, a simple diagnostic technique using gas dynamical effects is desirable to measure the contamination under various test conditions in each shock tunnel [98]. The arrival time of the driver gas may not be repeatable because the contamination is attributed to intricate viscous and turbulent phenomena in the shock tube and its onset depends on the condition of the interface between driver gas and test gas. It is, therefore, desirable that the detector be designed as small as possible for use in conjunction with other experimental models or be simple in construction so that it can be easily applied to the test section of shock tunnels when necessity arises. A survey of other driver gas detection methods is included in Sudani and Hornung [98].

Simple gas dynamical devices consisting of a duct and a wedge have been applied to the detection of driver gas arrival in the test section of T5. Static pressure in the duct has been measured during a shot, and the time of driver gas arrival has been determined by the onset of the pressure rise, which indicates duct flow choking. The ability to detect driver gas in small concentrations is critical to the satisfactory performance of the device. Duct internal flows for various wedge angles were numerically simulated to clarify the flow choking mechanism. The simulations guided the improvement of the detector sensitivity, and modified configurations of the detector were derived. The numerical simulations suggested that both the angle and the location of the wedge be adjustable to attain high sensitivity as shown in Fig. 76. Making both adjustments, however, is impractical because of the enormous increase in the number of shots required. The variable wedge angle gives the greatest increment of the shock angle close to the detachment

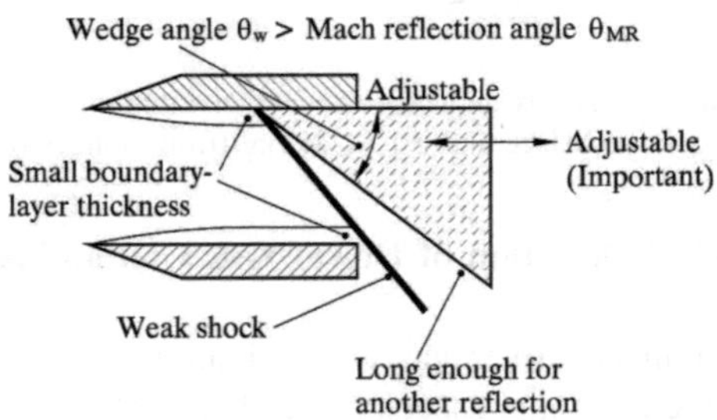

Fig. 76 Optimization of the duct detector for high sensitivity to a small concentration of driver gas (from Sudani and Hornung [98])

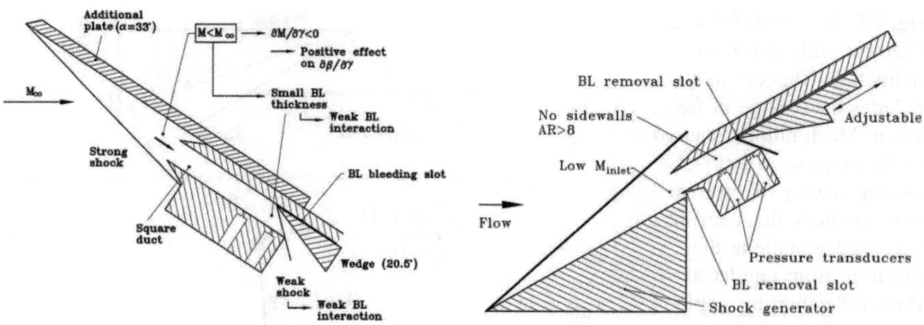

Fig. 77 *Left* Modified configuration of the square-duct detector (γ: ratio of specific heats; β: shock angle) (Type III); *Right* Two-dimensional duct detector (Type IV) allowing flow visualization of duct internal flow (AR: duct width over height aspect ratio) (adapted from Sudani and Hornung [98])

point. However, in relation to the choking criterion, the shock detachment from the wedge tip is not critical. The adjustment of the wedge location is more essential and directly relates to sensitivity because the shock should impinge on the duct wall at the driver gas arrival time. Thus, the wedge angle does not need to be variable if it is greater than the wedge angle at the transition from regular to Mach reflection (for more details see Sudani and Hornung [98]).

Figure 77 left illustrates a modified configuration of the square-duct detector for higher sensitivity and for weak shock/boundary-layer interaction. A flat plate is attached to the outer wall of the square duct, and both are installed at a relatively high angle of attack to the free stream. A simple calculation shows that, when the angle of attack of the flat plate is greater than approximately 15°, driver gas contamination causes the Mach number behind the oblique shock to decrease despite the fact that it increases the free stream Mach number. The change in duct inlet Mach number yields a positive effect on sensitivity in contrast to the case of the detector without inclination. The inlet Mach number is approximately 2 under typical conditions in T5 ($M_\infty \approx 5$), and the unit Reynolds number of the duct flow is almost the same as the free stream Reynolds number. Because of this low Mach number at the same Reynolds number, the boundary layers in the duct should be thinner and the wedge shock weaker. In addition, the wedge is placed a short distance away from the inside wall to avoid the boundary-layer influence. Much weaker interaction between the shock and the boundary layers is, therefore, expected in this modified configuration. A numerical simulation at a duct inlet Mach number of 2 shows that the duct flow chokes when a wedge shock with a large angle impinges on the duct wall, as in the case of a duct inlet Mach number of 5. This is because the duct exit is too small to let all the subsonic flow behind the Mach stem and the first reflected shock to pass through.

Another modified detector (Type IV) was designed as illustrated in Fig. 77 right to visualize the duct internal flow for a better understanding of the recorded pressure traces. A couple of pressure transducers were installed in the spanwise direction to

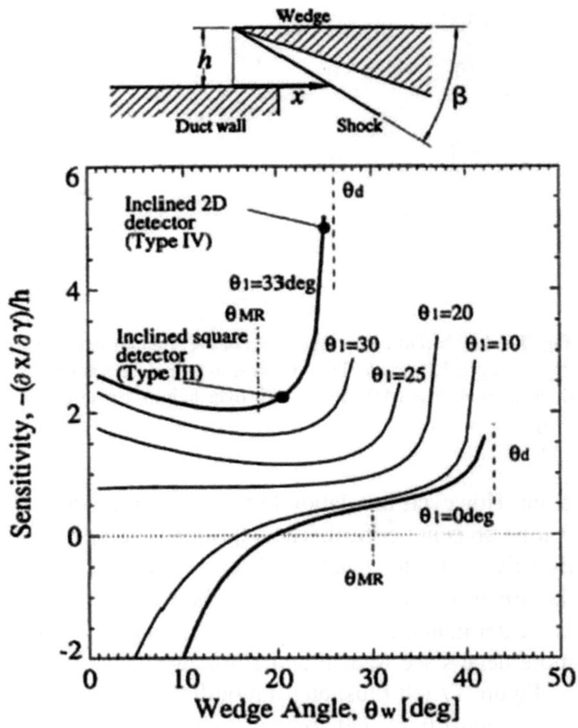

Fig. 78 Sensitivity of duct detectors with and without inclination. This calculation includes the change in free stream Mach number caused by the increase in ratio of specific heats γ (θ_d: wedge angle at shock detachment; θ_{MR}: wedge angle at the transition from regular to Mach reflection; θ_1: angle of attack of the duct) (from Sudani and Hornung [98])

check the two-dimensionality. This detector is also expected to yield a decrease in the number of shots required for the adjustment of the wedge location. The chosen experimental design points are also plotted in Fig. 78, where the type IV detector appears to be very close to the best configuration for test conditions in T5.

Figure 78 shows results of a sensitivity analysis with a fixed duct height *h* for a calorically perfect gas. The calculation includes the change in free stream Mach number caused by the increase in the ratio of specific heats γ. The lowest thick line is for the duct without inclination ($\theta_1 = 0^o$). As the duct is inclined to the free stream, the sensitivity improves markedly.

An example of the pressure traces obtained with the type IV detector using unseeded and seeded free streams is shown in Fig. 79. Holographic interferograms of the duct internal flow for an unseeded test gas are shown in Fig. 80. The pictures were taken in different shots, for which the detector setup was not exactly the same, but each picture qualitatively corresponds to the part of the pressure trace with the same label. For uncontaminated and unchoked flow (period "a" in Figs. 79 and 80a), the wedge shock does not impinge on the duct inner wall, and hence the flow in the duct is unchoked and everywhere supersonic. The pressure in the unseeded case is not affected by the presence of the wedge shock, thus being nearly equal to the static pressure of the outer flow. This period is considered useful test time with uncontaminated free stream. For slightly contaminated but unchoked flow (period

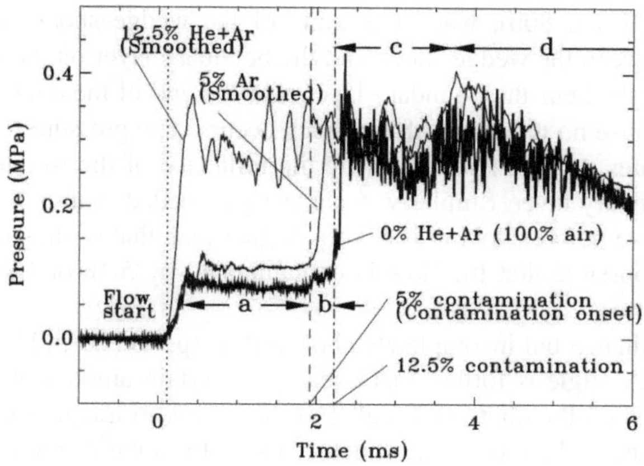

Fig. 79 Duct pressure traces in the two-dimensional duct detector (Type IV) at a moderate specific reservoir enthalpy for an unseeded test gas and driver gas seeded test gases: $h_0 \approx 14$ MJ/kg and $p_0 \approx 15$ MPa (adapted from Sudani and Hornung [98])

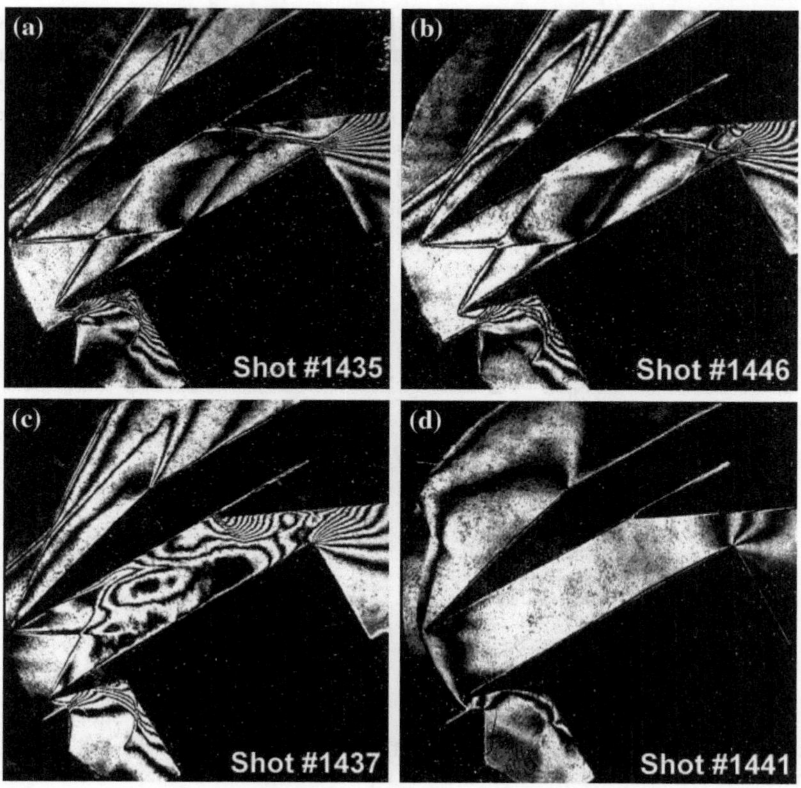

Fig. 80 Holographic interferograms of duct internal flows. Each picture qualitatively corresponds to the part of the pressure trace with the same label in Fig. 79: $h_0 \approx 14$ MJ/kg and $p_0 \approx 15$ MPa (adapted from Sudani and Hornung [98])

"b" in Figs. 79 and 80b), when the angle of the wedge shock is increased, an interaction between the wedge shock and the boundary layer on the duct inner wall is observed. It thickens the boundary layer near the end of the duct, but the flow is unchoked because no reflection of the shock occurs. The pressure for the unseeded test gas indicates a slight rise caused by the influence of the wedge shock on the duct wall boundary layer. Similarly, for a test gas seeded with 5 % of rnonatornic gas, the pressure before choking is slightly higher than that of the unseeded case. It is, therefore, possible that the flow is contaminated by 5 % or less of driver gas during this period.

For contaminated but incompletely choked flow (period "c"in Figs. 79 and 80c), when the shock angle is further increased, by a certain amount of driver gas, the shock impinges on the duct inner wall and then starts to travel upstream. Because subsonic flow behind the traveling shock can spill from the duct (no sidewalls), the shock stays in the duct. It was observed that the shock moves very fast to the point where it comes to rest, so that the pressure traces show an abrupt rise. For a test gas seeded with 12.5 % of monatomic gas, this type of flow is observed immediately after the flow reaches the duct. This means that 12.5 % or less of driver gas contaminates the test gas at the time of the abrupt pressure rise in the case of the unseeded test gas. For contaminated and fully choked flow (period "d" in Figs. 79 and 80d), when a high amount of driver gas reaches the duct, the shock travels farther upstream and is ultimately detached in front of the duct. The duct internal flow is everywhere subsonic and fully choked. Because no shock waves exist in the duct, the pressure trace is not as noisy as the trace during period "c".

All of the experimental data measured with the duct detectors in T5 are plotted as functions of specific reservoir enthalpy h_0 in Fig. 81. The origin of the time axis

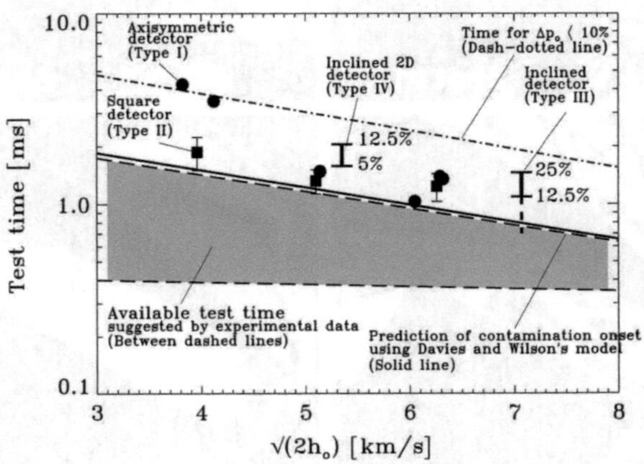

Fig. 81 Useful test time with uncontaminated free stream determined by the experimental data with duct detectors in T5 and a comparison with a prediction based on a simple shock-bifurcation flow model. Note the late choking of the type IV detector at intermediate enthalpy. This is thought to be related to slightly off-tailored interface operation. Type I: axisymmetric duct detector, type II: square duct detector (adapted from Sudani and Hornung [98])

in this figure corresponds to the time when flow reaches the test section, i.e., when the pressure in the duct starts to rise. The time, therefore, includes a period until the steady flow is established in the duct and is represented by the lower dashed line. The upper dashed line is derived from the earliest choking time for each enthalpy ($h_0 \approx 8$, 13, 20, or 25 MJ/kg). The contamination occurs earlier than the time at which p_0 drops by 10 % of its nearly constant value irrespective of h_0 (denoted by a dash-dotted line). Thus, the shaded region between the two dashed lines is defined as the useful test time with uncontaminated free stream.

The contamination onset is predicted by the method of Davies and Wilson [16] based on a simple shock-bifurcation model (solid line) using measured values for the incident and reflected shock speeds. This is in excellent agreement with the upper dashed line. Davies and Wilson's method gives a reliable prediction of the early contamination onset for a wide range of enthalpies in T5.

References

1. Adam, P.H.: Enthalpy effects on hypervelocity boundary layers. PhD Thesis, California Institute of Technology (1997)
2. Adam, P.H, Hornung, H.G.: Enthalpy effects on hypervelocity boundary-layer transition: Ground test and flight test data. J. Spacecr. Rockets **34**(5), 614–619 (1997)
3. Anderson, J.D.: Hypersonic and high-temperature gas dynamics, 2nd edn. American Institute of Aeronautics and Astronautics, Reston (2006)
4. Barth, J.E., Wheatley, V., Smart, M.K.: Hypersonic turbulent boundary-layer fuel injection and combustion: skin-friction reduction mechanisms. AIAA J. **51**(9), 2147–2157 (2013). doi:10.2514/1.J052041, 2013
5. Bertin, J.J., Stetson, K.F., Bouslog, S.A., Caram, J.M.: Effect of isolated roughness elements on boundary-layer transition for shuttle orbiter. J. Spacec. Rocket **34**, 4 (1997)
6. Böhrk, H., Wartemann, V., Eggers, T., Martinez Schramm, J., Wagner, A., Hannemann, K.: Shock tunnel testing of the transpiration-cooled heat shield experiment AKTiV, AIAA 2012-5935. In: Proceedings 18th AIAA/3AF International Space Planes and Hypersonic Systems And Technology Conference, Tours, France, 24–28 Sept 2012
7. Boyce, R.R., Takahashi, M., Stalker, R.J.: Mass spectrometric measurements of driver gas arrival in the T4 free-piston shock-tunnel. Shock Waves **14**(5/6), 371–378 (2005). doi:10.1007/s00193-005-0276-3
8. Brieschenk, S., Gehre, R., Wheatley, V., Boyce, R.: Fluorescence studies of jet mixing in a hypersonic flow. In: Proceedings of the 29th International Symposium on Shock Waves, 14–19 July 2013
9. Brieschenk, S., Lorrain, P., McIntyre, T.J., Boyce, R.R.: Chemiluminescence imaging in a radical-farming scramjet. In: Proceedings of the XXI International Symposium on Air Breathing Engines (ISABE 2013), pp. 1403–1408 (2013)
10. Capser, K.M., Beresh, S.J., Henfling, J.F., Spillers, R.W., Pruett, B., Schneider, S.P.: Hypersonic wind-tunnel measurements of boundary layer pressure fluctuations. AIAA 2009-4054, 39th AIAA Fluid Dynamics Conference, San Antonio, Texas, 22–25 June 2009
11. Cary Jr, A.M., Hefner, J.N.: Film-cooling effectiveness and skin friction in hypersonic turbulent flow. AIAA J. **10**(9), 1188–1193 (1972)
12. Chan, W.Y.K., Jacobs, P.A., Mee, D.J.: Suitability of the k-ω turbulence model for scramjet flowfield simulations. Int. J. Numer. Meth. Fluids **70**(4), 493–514 (2012). doi:10.1002/fld.2699

13. Chan, W.Y.K., Mee, D.J., Smart, M.K., Turner, J.C.: Effects of flow disturbances from cross-stream fuel injection on the drag reduction by boundary layer combustion. In: AIAA 2012-5889, presented at the 18th AIAA/3AF International Space Planes and Hypersonic Systems and Technologies Conference, Tours, France (2012), 24–28 Sept 2012
14. Chan, W.Y.K.: Effects of flow non-uniformities on the drag reduction by boundary layer combustion. PhD Thesis, The University of Queensland, Brisbane. Australia (2012)
15. Craddock, C.S.: Design of the axisymmetric HyShot nozzle for T4, Research Report 02/2000. Department of Mechanical Engineering, The University of Queensland (2000)
16. Davies, L., Wilson, J.L.: Influence of reflected shock and boundary-layer interaction on shock-tube flows. Phys. Fluids (Supplement I), 1-37–1-43 (1969)
17. Detra, R.W., Kemp, N.H., Riddell, F.R.: Addendum to heat transfer to satellite vehicles reentering the atmosphere. Jet Propuls. **27**(12), 1256–1257 (1957)
18. Doherty, L.J.: Experimental investigation of an airframe integrated 3-D scramjet at a Mach 10 flight condition. PhD Thesis, The University of Queensland (2014)
19. Doherty, L.J., Chan, W.Y.K., Zander, F., Jacobs, P.A., Gollan, R.J., Kirchhartz, R.M.: NENZFR: Non-Equilibrium Nozzle Flow, Reloaded, Division of Mechanical Engineering Report 2012/08, Brisbane. School of Mechanical and Mining Engineering, The University of Queensland, Australia (2012)
20. Dunn, M.G., Kang, S.W.: Theoretical and experimental studies of reentry plasmas. NASA CR-2232 (1973)
21. Edney, B.E.: Anomalous heat transfer and pressure distributions on blunt bodies at hypersonic speeds in the presence of an impinging shock. FFA Rep. 115 (1968)
22. Edney, B.E.: Effects of shock impingement on the heat transfer around blunt bodies. AIAA J. **6**, 15–21 (1968)
23. Eitelberg, G., McIntyre, T.J., Beck,W.H., Lacey, J.: High Enthalpy Shock Tunnel in Göttingen. AIAA 92-3955 (1992)
24. Fedorov, A., Kozlov, V., Shiplyuk, A., Maslov, A., Malmuth, N.: Stability of hypersonic boundary layer on porous wall with regular microstructure. AIAA J. **44**(8), 1866–1871 (2006)
25. Fedorov, A.V., Malmuth, N.D., Rasheed, A., Hornung, H.G.: Stabilization of hypersonic boundary layers by porous coatings. AIAA J. **39**(4), 605–610 (2001)
26. Fujii, K.: Experiment of the two-dimensional roughness effect on hypersonic boundary-layer transition. J. Spacecr. Rocket **43**(4), (2006)
27. Gerhold, T., Friedrich, O., Evans, J., Galle, M.: Calculation of complex three-dimensional configurations employing the DLR-TAU-code. AIAA 1997-0167 (1997)
28. Germain, P., Hornung, H.G.: Transition on a slender cone in hypervelocity flow. Exp. Fluids **22**, 183–190 (1997)
29. Goyne, C.P., Stalker, R.J., Paull, A.: Transducer for direct measurement of skin friction in hypervelocity impulse facilities. AIAA J. **40**(1), 42–49 (2002)
30. Goyne, C.P., Stalker, R.J., Paull, A., Brescianini, C.P.: Hypervelocity skin-friction reduction by boundary-layer combustion of hydrogen. J. Spacecr. Rocket. **37**(6), 740–746 (2000)
31. Gupta, R.N., Yos, J.M., Thompson, R.A., Lee, K.P.: A Review of Reaction Rates and Thermodynamic and Transport Properties for an 11-Species Air Model for Chemical and Thermal Nonequilibrium Calculations to 30000 K, NASA Reference Publication, No. 1232 (1990)
32. Hannemann, K., Krek, R., Eitelberg, G., Latest Calibration Results of the HEG Contoured Nozzle. In: Sturtevant, B., Sheperd, J.E., Hornung, H.G. (eds.) Proceedings of the 20th International Symposium on Shock Waves, Pasadena, CA, USA, July 1995, pp. 1575–1580, World Scientific (1996)
33. Hannemann, K., Schnieder, M., Reimann, B., Martinez Schramm, J.: The influence and delay of driver gas contamination in HEG, AIAA 2000-2593, 21st AIAA Aerodynamic Measurement Technology and Ground Testing Conference, Denver, CO, 19–22 June 2000
34. Hannemann, K., Martinez Schramm, J., Karl, S., Beck, W.H.: Cylinder Shock Layer Density Profiles Measured in High Enthalpy Flows in HEG, AIAA 2002-2913. 22nd AIAA

Aerodynamic Measurement Technology and Ground Testing Conference, St. Louis, MO, June 24–28 2002

35. Hannemann, K.: High Enthalpy Flows in the HEG Shock Tunnel: Experiment and Numerical Rebuilding, AIAA 2003-0978, 41st AIAA Aerospace Sciences Meeting and Exhibit, 6-9 Jan. Reno, Nevada (2003)
36. Hannemann, K., Martinez Schramm, J.: High enthalpy, high pressure short duration testing of hypersonic flows. In: Tropea, C., Foss, J., Yarin, A. (eds.) Springer Handbook of Experimental Fluid Mechanics, pp. 1081–1125. Springer, Berlin (2007)
37. Hannemann, K., Martinez Schramm, J., Karl, S.: Recent extensions to the High Enthalpy Shock Tunnel Göttingen (HEG). In: Proceedings of the 2nd International ARA Days "Ten Years after ARD", Arcachon, France, 21–23 Oct 2008
38. Hannemann, K., Karl, S., Martinez Schramm, J., Steelant, J.: Methodology of a Combined Ground Based Testing and Numerical Modelling Analysis of Supersonic Combustion Flow Paths, Shock Waves, Vol. 20, No. 5, pp. 353–366. Springer (2010)
39. Hollis, B.R., Prabhu, D.K.: Assessment of Laminar, Convective Aeroheating Prediction Uncertainties for Mars Entry Vehicles, AIAA 2011-3144. 42nd AIAA Thermophysics Conference. Honolulu, Hawaii, 27–30 June 2011
40. Hornung, H.G.: Performance Data of the New Free-Piston Shock Tunnel at GALCIT, AIAA 92-3943, AIAA 17th Aerospace Ground Testing Conference, Julv 6-8, Nashville, TN (1992)
41. Hornung, H.G.: Experimental hypervelocity flow simulation, needs, achievements and limitations. First Pacific International Conference on Aerospace Science and Technology, PICAST'1, Tainan, Taiwan (1993)
42. Hornung, H.G.: Hypersonic real-gas effects on transition. In: IUTAM Symposium on One Hundred Years of Boundary Layer Research, Solid mechanics and its applications, Vol. 129, pp. 335–344 (2006)
43. Itoh, K., Ueda, S., Komuro, T., Saito, K., Takahashi, M., Miyajima, H., Koga, K.: Design and Construction of HIEST (High Enthalpy Shock Tunnel). In: Proceedings of the International Conference on Fluid Engineering, Vol. 1. JSME Press, Tokyo, pp. 353–358 (1997)
44. Itoh, K., Ueda, S., Tanno, H., Komuro, T., Sato, K., Takahashi, M., Miyajima, H., Muramoto, H.: Improvement of free piston driver for high enthalpy shock tunnel. Shock Waves **8**(4), 215–233 (1998)
45. Itoh, K., Ueda, S., Tanno, H., Komuro, T., Sato, K.: Hypersonic aerothermodynamic and scramjet research using high enthalpy shock tunnel. Shock Waves **12**, 93–98 (2002)
46. Jacobs, P.A., Gollan, R.J., Potter, D.F.: The Eilmer3 Code: User Guide and Example Book 2014 Edition, Mechanical Engineering Report 2014/05. School of Mechanical and Mining Engineering, The University of Queensland, Brisbane, Australia (2014)
47. Jacobs, P.A., Gollan, R.J., Potter, D.F., Zander, F., Gildfind, D.E., Blyton, P., Chan, W.Y.K., Doherty, L.J.: Estimation of High-Enthalpy Flow Conditions for Simple Shock and Expansion Processes Using the ESTCj Program and Library, Division of Mechanical Engineering Report 2011/02, Brisbane. School of Mechanical and Mining Engineering, The University of Queensland, Australia (2011)
48. Jacobs, P.A., Morgan, R.G., Stalker, R.J., Mee, D.J.: Use of argon-helium driver-gas mixtures in the T4 shock tube. In: Brun, R., Dimitrescu, L.Z. (eds.) Shock Waves at Marseille. Springer, Berlin (1995)
49. Jacobs, P.A., Stalker, R.J.: Mach-4 and Mach-8 axisymmetrical nozzles for a high-enthalpy shock tunnel. Aeronaut. J. **95**(949), 324–334 (1991)
50. Johnson, H.B., Seipp, T., Candler, G.V.: Numerical study of hypersonic reacting boundary layer transition on cones. Phys. Fluids **10**, 2676–2685 (1998)
51. Johnston, I.A., Weiland, M., Martinez Schramm, J., Hannemann, K., Longo, J.: Aerothermodynamics of the ARD: Postflight Numerics and Shock-Tunnel Experiments, AIAA 2002–0407, 40th AIAA Aerospace Sciences Meeting and Exhibit, Reno, NV, 14–17 Jan 2002

52. Karl, S.: Numerical Investigation of a Generic Scramjet Configuration. PhD Thesis, TU Dresden, Dresden, Germany (2011)
53. Karl, S., Martinez Schramm, J., Hannemann, K.: High enthalpy shock tunnel flow past a cylinder: a basis for CFD validation. New Results in Numerical and Experimental Fluid Mechanics IV, Vol. 87. Springer, Berlin (2004)
54. Kimmel, R.L., Adamczak, D., Gaitonde. D., Rougeux, A., Hayes, J.R.: HIFiRE-1 Boundary layer transition experiment design, AIAA 2007-534, 45th AIAA Aerospace Sciences Meeting and Exhibit, 2007, Reno, Nevada, 8–11 Jan 2007
55. Kirchhartz, R.M., Mee, D.J., Stalker, R.J.: Supersonic skin-friction drag with tangential wall slot fuel injection and combustion. AIAA J. **50**(2), 313–324 (2012). doi:10.2514/1.J05107
56. Kirchhartz, R.: Upstream wall layer effects on drag reduction with boundary layer combustion. PhD Thesis, The University of Queensland, Brisbane, Australia (2009)
57. Laurence, S., Wagner, A., Hannemann, K., Wartemann, V., Lüdeke, H., Tanno, H., Ito, K.: Time-resolved visualization of instability waves in a hypersonic boundary layer. AIAA J. **50** (1), 243–246 (2012)
58. Laurence, S.J., Wagner, Hannemann, K.: Schlieren-based techniques for investigating instability development and transition in a hypersonic boundary layer. In: Experiments in Fluids, Vol. 55, p. 1782. Springer, Berlin (2014). doi:10.1007/s00348-014-1782-9
59. Laurence, S., Karl, S., Martinez Schramm, J., Hannemann, K.: Transient fluid-combustion phenomena in a model scramjet. J. Fluid Mech. **722**, 85–120 (2013)
60. Lu, F.K., Marren, D.E. (eds.): Advanced Hypersonic Test Facilities. Progress in Astronautics and Aeronautics, Vol. 198. AIAA, USA (2002)
61. Lukasiewicz, J.: Experimental methods of hypersonics. Marcel Dekker Inc, New York (1973)
62. Mack, L.: Linear stability theory and the problem of supersonic boundary-layer transition. AIAA J **13**(3), 278–289 (1975)
63. Malik, M.R.: Prediction and control of transition in supersonic and hypersonic boundary layers. AIAA J. **27**(11), 1487–1493 (1989)
64. Malmuth, N., Fedorov, A., Shalaev, V., Cole, J., Khokhlov, A., Hites, M., Williams, D.: Problems in high speed flow prediction relevant to control. In: 2nd AIAA, Theoretical Fluid Mechanics Meeting, AIAA 98-2695 (1998). doi: 10.2514/6.1998-2695
65. Martinez Schramm, J.: Aerothermodynamische Untersuchung einer Wiedereintrittskonfiguration und ihrer Komponenten in einem impulsbetriebenen Hochenthalpie-Stoßkanal, Dissertation Universität Göttingen (2008)
66. Martinez Schramm, J., Sunami, T., Ito, K., Hannemann, K.: Experimental Investigation of Supersonic Combustion in the HIEST and HEG Free Piston Driven Shock Tunnels, AIAA 2010-7122, 46th AIAA/ASME/SAE/ASEE Joint Propulsion Conference and Exhibit, Nashville, TN, 25–28 July 2010
67. Martinez Schramm, J., Barth, T., Wagner, A., Hannemann, K.: Post Flight Analysis of SHEFEX I: shock tunnel testing and related CFD analysis. In: Proceedings of the 7th European Symposium on Aerothermodynamics for Space Vehicles. Brugge, Belgium, , 9–12 May 2011
68. Mee, D.J.: Dynamic calibration of force balances for impulse hypersonic facilities. Shock Waves **12**(6), 443–455 (2003). doi: 10.1007/s00193-003-0181-6
69. Mee, D.J.: Boundary-layer transition measurements in hypervelocity flows in a shock tunnel. AIAA J. **40**(8), 1542–1548 (2002)
70. Mee, D.J., Daniel, W.J.T., Simmons, J.M.: Three-component force balance for flows of millisecond duration. AIAA J. **34**(3), 590–595 (1996). doi:10.2514/3.13108
71. Mee, D.J.: Uncertainty analysis of conditions in the test section of the T4 shock tunnel, Research Report 4/93. Department of Mechanical Engineering, The University of Queensland (1993)
72. Merzkirch, W.: Flow Visualization. Academic Press, Waltham (1974)
73. Park, C.: On convergence of computation of chemically reacting flows. In: AIAA-85-0247, AIAA 23rd Aerospace Sciences Meeting, Reno, NV (1985)

74. Parziale, N.J., Shepherd, J.E., Hornung, H.G.: Differential interferometric measurement of instability in a hypervelocity boundary layer. AIAA J. **51**(3) (2013)
75. Owen, R. Cain, T.: Reconstruction of the Hyshot-2 Flight from onboard sensors. In: Proceedings of the Fifth European Symposium on Aerothermodynamics for Space Vehicles, Cologne, Germany, 8–11 Nov 2004
76. Paull, A., Alesi, H., Anderson, S.: The HyShot flight program and how it was developed. AIAA 2002-5248, AIAA/AAAF 11th International Space Planes and Hypersonic Systems and Technologies Conference. Orleans, France (2002)
77. Paull, A., Stalker, R.J., Mee, D.J.: Experiments on supersonic combustion ramjet propulsion in a shock tunnel. J. Fluid Mech. **296**, 159–183 (1995)
78. Rasheed, A., Hornung, H.G., Fedorov, A.V., Malmuth, N.D.: Experiments on passive hypervelocity boundary-layer control using an ultrasonically absorptive surface. AIAA **40**(3), 481–489 (2002)
79. Reimann B., Johnston I., Hannemann V.: The DLR TAU—code for high enthalpy flows, notes on num. Fluid Mech. Multidisc. Design **87**, (2004)
80. Robinson, M.J., Mee, D.J., Paull, A.: Scramjet lift, thrust and pitching-moment characteristics measured in a shock tunnel. J. Propul. Power **22**(1), 85–95 (2006). doi:10.2514/1.15978
81. Robinson, M.: Simultaneous lift, moment and thrust measurements on a scramjet in hypervelocity flow. PhD Thesis, The University of Queensland, Australia (2003)
82. Rowan, S.A., Paull, A.: Performance of a scramjet combustor with combined normal and tangential fuel injection. J. Propul. Power **22**(6), 1334–1338 (2006). doi:10.2514/1.18744
83. Sanderson, S.R., Hornung, H.G., Sturtevant, B.: The influence of non-equilibrium dissociation on the flow produced by shock impingement on a blunt body. J. Fluid Mech. **516**, 1–37 (2004)
84. Sanderson, J.R., Simmons, J.M.: Drag balance for hypervelocity impulse facilities. AIAA J. **29**(12), 2185–2191 (1991). doi:10.2514/3.10858
85. Schneider, S.P.: Hypersonic laminar-turbulent transition on circular cones and scramjet forebodies. Prog. Aerosp. Sci. **40**, 1–50 (2004)
86. Schultz, D.L., Jones, T.V.: Heat-Transfer Measurements in Short-Duration Hypersonic Facilities. AGARDograph Report No. 165 (1973)
87. Skinner, K.A., Stalker, R.J.: Species measurements in a hypersonic, hydrogen-air, combustion wake. Combust. Flame **106**(4), 478–486 (1996). doi:10.1016/0010-2180(96)00018-1
88. Smart, M.K., Hass, N.E., Paull, A.: Flight data analysis of the HyShot 2 scramjet flight experiment. AIAA J. **44**(10), 2366–2375 (2006)
89. Smart, M.K.: Design of three-dimensional hypersonic inlets with rectangular-to-elliptical shape transition. J. Propul. Power **15**(3), 408–416 (1999). doi:10.2514/2.5459
90. Smith, A.L.: Multiple component force measurement in short duration test flows. PhD Thesis, The University of Queensland, Brisbane, Australia (1999)
91. Smith, C.E.: The starting Process in a Hypersonic Nozzle. J. Fluid Mech. **24**(part 4), 625–640 (1966)
92. Stalker, R.J.: Modern development in hypersonic wind tunnels. Aeronaut. J. 21–39 (2006)
93. Stalker, R.J.: Control of hypersonic turbulent skin friction by boundary-layer combustion of hydrogen. J. Spacecr. Rockets **42**(4), 577–587 (2005). doi:10.2514/1.8699
94. Stalker, R.J., Paull, A.: Experiments on cruise propulsion with a hydrogen scramjet. Aeronaut. J. **102**(1011), 37–43 (1998)
95. Stalker, R.J., Morgan, R.G.: The University of Queensland free piston shock tunnel T4—Initial operation and preliminary calibration. In: Proceedings of the 4th National Space Engineering Symposium, Adelaide, pp. 182-198. Barton, ACT: Institution of Engineers, Australia, 12–14 July 1988 (1988)
96. Stalker R.J.: Shock tunnel for real-gas hypersonics. AGARD CP 428 (1987)
97. Stalker, R.J.: A study of the free-piston shock tunnel. AIAA J. **5**(12), 2160–2165 (1967)

98. Sudani, N., Hornung, H.G.: Gasdynamical detectors of driver gas contamination in a high-enthalpy shock tunnel. AIAA J. **36**(3), 313–319 (1998)
99. Suraweera, M.V.: Reduction of skin friction drag in hypersonic flow by boundary layer combustion. PhD thesis, The University of Queensland. Brisbane. Australia (2006)
100. Suraweera, M., Mee, D.J., Stalker, R.J.: Skin friction reduction in hypersonic turbulent flow by boundary layer combustion. AIAA Paper 2005-613. Presented at the 43rd Aerospace Sciences Meeting and Exhibit, Reno, Nevada, 10–13 Jan 2005
101. Tanimizu, K., Mee, D.J., Stalker, R.J., Jacobs, P.A.: Drag force on quasi-axisymmetric scramjets at various flight Mach numbers: theory and experiment. Shock Waves **19**(2), 83–93 (2009). doi:10.1007/s00193-009-0194-x
102. Tanimizu, K.: Nozzle optimization study and measurements for a quasi-axisymmetric scramjet model. PhD Thesis, The University of Queensland, Brisbane. Australia. August (2008)
103. Tanno, H., Sato, K.,Komuro, T., Itoh. K.: Free-flight Aerodynamic Tests of Reentry Vehicles in High-temperature Real-gas Flow. AIAA 2014-3109, 19th AIAA International Space Planes and Hypersonic Systems and Technologies Conference, Atlanta, Georgia, 16–20 June 2014
104. Tanno, H., Komuro, T., Ohnishi, N., Ishihara, T., Ogino, Y., Sawada, K.: Experimental study on heat flux augmentation in high-enthalpy shock tunnels. AIAA 2014-2548, 11th AIAA/ASME Joint Thermophysics and Heat Transfer Conference, Atlanta, Georgia, 16–20 June 2014
105. Tanno, H., Sato, K., Komuro, T., Itoh, K., Takahashi, M., Fujita, K., Laurence, S., Hannemann, K.: Free-flight force measurement technique in shock tunnel. AIAA 2012-1241, 50th AIAA Aerospace Sciences Meeting including the New Horizons Forum and Aerospace Exposition, Nashville, Tennessee, 9–12 Jan 2012
106. Tanno, H., Komuro, T., Sato, K., Itoh, K., Takahashi, M., Fulii, K.: Measurement of hypersonic high-enthalpy boundary layer transition on a 7° cone model. AIAA 2010-310, 48th AIAA Aerospace Sciences Meeting Including the New Horizons Forum and Aerospace Exposition, Orlando, Florida, 4–7 Jan 2010
107. Tanno, H., Komuro, T., Sato, K., Itoh, K., Takahashi, M.: Miniature data-logger for aerodynamic force measurement in impulsive facility, AIAA 2010-4204, 27th AIAA Aerodynamic Measurement Technology and Ground Testing Conference, Chicago, Illinois, 28 June–1 July (2010)
108. Tanno, H., Komuro, T., Sato, K., Itoh, K., Yamada, T., Sato, N., Nakano, E.: Heat flux measurement of Apollo capsule model in the free-piston shock tunnel HIEST, AIAA 2009-7304. 16th AIAA/DLR/DGLR International Space Planes and Hypersonic Systems and Technologies Conference, Bremen, Germany, 19–22 Oct 2009
109. Tanno, H., Komuro, T., Sato, K., Itoh, K., Takahashi, M., Fujii, K.: Measurement of hypersonic boundary layer transition on cone models in the free-piston shock tunnel HIEST. AIAA 2009-781, 47th AIAA Aerospace Sciences Meeting Including The New Horizons Forum and Aerospace Exposition, Orlando, Florida, 5–8 Jan 2009
110. Tanno, H., Paull, A., Stalker, R.J.: Skin-friction measurements in a supersonic combustor with crossflow fuel injection. J. Propul. Power **17**(6), 1333–1338 (2001). doi:10.2514/2.5883
111. Turner, J., Hörschgen, M., Jung, W., Stamminger, A., Turner, P.: SHEFEX Hypersonic re-entry flight experiment; vehicle and subsystem design, flight performance and prospects. AIAA 2006-8115, 14th AIAA/AHI Space Planes and Hypersonic Systems and Technologies Conference (2006)
112. Tuttle, S.L., Mee, D.J., Simmons, J.M.: Drag measurements at Mach 5 using a stress wave force balance. Exp. Fluids **19**(5), 336–341 (1995). doi:10.1007/BF00203418
113. van Driest, E.R.: Turbulent boundary layer in compressible fluids. J. Spacecr. Rockets **40**(6), 1012–1028 (2003). doi:10.2514/1.10862
114. van Driest, E.R.: The Problem of Aerodynamic Heating. Aeronaut. Eng. Rev. **15**(10), 26–41 (1956)
115. Wagner, A., Kuhn, M., Hannemann, K.: Ultrasonic absorption characteristics of porous carbon-carbon ceramics with random microstructure for passive hypersonic boundary layer transition control. Exp. Fluids **55**(6), 1–9 (2014). doi 10.1007/s00348-014-1750-4

116. Wagner A., Kuhn M., Martinez Schramm J., Hannemann K.: Experiments on passive hypersonic boundary layer control using ultrasonically absorptive carbon-carbon material with random microstructure. Exp. Fluids **54**(10), 1–10 (2013). doi:10.1007/s00348-013-1606-3
117. Weihs, H., Longo, J., Turner, J.: Key experiments within the SHEFEX II mission. In: IAC 2008, Glasgow, Scotland UK, IAC-08.D2.6.4 (2008)

Author Biographies

Klaus Hannemann is head of the Spacecraft Department at the German Aerospace Center (DLR) in Göttingen and Professor for Spacecraft at the Justus-Liebig-Universität Gießen, Germany. He obtained his diploma and doctoral degree in Mechanical Engineering from the Universität Karlsruhe, Germany. His research interests are in the field of numerical and experimental aerothermodynamics of space vehicles including re-entry flows, rocket propulsion, chemical and electrical small- and micro-thrusters, high speed airbreathing propulsion and hypersonic boundary layer transition. He is member of a number of scientific committees including the editorial board of Shock Waves and he serves as field editor aerothermodynamics for the CEAS Space Journal.

Katsuhiro Itoh is Senior Researcher at the Japan Aerospace Exploration Agency, Department of High Enthalpy Shock Tunnel Laboratory, Kakuda Space Center. He received his Diploma in 1983 at the Faculty of Mechanical Engineering, Tohoku University and his Master Eng. in 1985 at the Institute of Fluid Science, Tohoku University. In 1988 he received his Ph.D. from Tohoku University, Institute of Fluid Science, under the supervision of Prof. K. Takayama. The titel of his thesis is: "Unsteady Transonic Flows in Shock Tube". Before joining JAXA, Dr. Itho conducted research as DAAD Research Fellow at the Shock Wave Laboratory of RWTH Aachen University, Germany from April 1988 until September 1989. From October 1989 until March 1990 he worked as Research Associate at the Institute of Fluid Science, Tohoku University, Sendai, Japan. His main research interests are: Characteristics of high enthalpy shock tunnels, high enthalpy nozzle flows, reentry aerothermodynamics and measurement techniques for impulse facilities.

David J. Mee has been conducting research in hypersonic flows for more than 20 years. He has a BE in Mechanical Engineering and a PhD from The University of Queensland (UQ). After five years as a Research Fellow in the Osney Turbomachinery Group at Oxford University, he returned to UQ to join the Centre for Hypersonics. His research interests are in force measurements in hypersonic impulse facilities, boundary layer transition and hypersonic aerothermodynamics. Along with Professors Ray Stalker and Allan Paull, he was a member of the team that demonstrated for the first time that a scramjet-powered vehicle could produce more than enough thrust to overcome its drag. Professor Mee is currently the Head of the School of Mechanical and Mining Engineering at UQ.

Hans G. Hornung received his bachelor (1960) and master (1962) degrees from the University of Melbourne and his Ph. D. (1965) in Aeronautics from Imperial College, London. He worked in the Aeronautical Research Laboratories, Melbourne (1962–63, and 1965–67), and on the faculty of the Physics Department of the Australian National University (1967–80), with a sabbatical year as a Humboldt Fellow in Darmstadt, Germany, 1974. In 1980 he accepted an offer to head the Institute for Experimental Fluid Mechanics of the DLR in Goettingen, Germany. He was appointed director of the Graduate Aeronautical Laboratories at Caltech in 1987. He is Emeritus since 2005. He made contributions in gasdynamics, notably in Mach reflection and effects of dissociation, in separated flows, and in wind tunnel technology.

Honors and awards include Royal Swedish Academy of Engineering Science, US National Academy of Engineering, Ludwig Prandtl Ring of the German Aerospace Society (DGLR), ICAS von Karman Award for international cooperation in aeronautics, AIAA Fluid Dynamics Award, D.Sc. h. c. ETH Zurich. Fellow of the Royal Aeronautical Society, of the American Institute of Aeronautics and Astronautics, of the American Association for the Advancement of Science and of the Australasian Fluid Mechanics Society.

Development of the New Piston-Driven Shock-Tunnel HELM

Christian Mundt

1 Introduction

The simulation of high enthalpy, hypersonic flows is a field of continued research efforts and provides a difficult task despite all progress made in the past years. This is the case for both experimental and numerical simulations of the real flight of a high speed vehicle, typically either winged or un-winged re-entry configurations, high speed aircraft or rockets [1]. A lot of different phenomena which are often of coupled nature govern the flow and the characteristics of the geometry studied, such as thermochemistry, compressible shock-layers, high temperature boundary-shear-layers, expansions/rarefaction, thermochemical freezing etc. Due to these complex interactions a lot of different similarity parameters have to be considered [2, 3]. Conducting experiments with a reduced scale model leads to contradictions concerning the full consideration of all of the similarity parameters. Thus, different facilities are used which are able to simulate at least a few of the parameters. Numerical methods and the computers they run on have become very powerful tools in the last decades. The challenge here is in developing an accurate and stable approach describing all parameters affecting the considered flow and is suitable for modelling the physicochemical phenomena.

It has to be stated, that both branches of the simulation have fertilized each other and the present status of capabilities would not have been reached without the constant effort to improve both, see e.g. [4, 5]. Experimental campaigns are necessary e.g. to validate and verify the numerical results. In addition the latter can be used for understanding the facilities better and to characterize the specific conditions of the test object. It is generally accepted that a numerical method validated

C. Mundt (✉)
Faculty of Aerospace Sciences, Institute for Thermodynamics, University of the Federal Armed Forces München (UniBw Munich), 85577 Neubiberg, Germany
e-mail: christian.mundt@unibw.de

O. Igra and F. Seiler (eds.), *Experimental Methods of Shock Wave Research*,
Shock Wave Science and Technology Reference Library 9,
DOI 10.1007/978-3-319-23745-9_8

with respect to experimental results of a suitable wind-tunnel test is able to predict the phenomena seen during the flight of a vehicle (extrapolation to flight).

For the rationale described above a new high enthalpy facility was created recently. In its design, construction and operation the emphasis was laid on the fact, that the facility is operated at a university. Therefore, a small to medium size tunnel is appropriate. It is dedicated to basic research related to the flow in the tunnel itself, and on high speed/high enthalpy flows in the test section. Typical examples for the latter are flows around re-entry-bodies or inside of high-speed propulsion systems.

To complement an existing plasma-driven wind-tunnel [6] with typical features of a long duration high enthalpy experiment at densities too low for simulating the Reynolds- or Damköhler numbers, for example, a shock-tunnel was chosen due to its high density capabilities together with the possibility to generate high enthalpies. To achieve the latter, it relies on a powerful driver in order to simulate both the density- and enthalpy- range aimed for. Of the different options (detonation driven [7], electrical [8] or piston driven) the free-piston driver as pioneered by Stalker [9] was selected. Thus the new facility is in its principle similar to T4 at the University of Queensland, Australia [10], and T5 at Caltech, California, USA [11]. Larger shock-tunnels operated by national research establishments are e.g. HEG, "Hoch-Enthalpiekanal Göttingen/High Enthalpy shock tunnel Göttingen", Germany [12] and HIEST "High Enthalpy Shock Tunnel", Japan [13].

2 Description of the Facility and Capabilities

2.1 Working Principle

For a general overview on shock tubes, see e.g. [14]. We restrict ourselves to the working principle of a piston-driven shock-tunnel as it was pioneered by Stalker [9], which is briefly described here.

Before the test the piston is located at the upstream end of the compression tube. A pressure reservoir has to be filled for the piston acceleration. Usually air is used for this purpose. The driver tube is filled with the driver gas, for high enthalpies usually Helium is used due to its high speed of sound, which is separated from the shock tube gas by a membrane. The latter is made from a ductile material and often consists of a plate of milled steel. Depending on the pressure conditions aimed for its opening, it is 2–6 mm in thickness and can be grooved for minimization of particle debris. The shock tube is filled with the test gas, in many cases with air, but almost any other gas or gas mixture to be studied can be used. The test gas is contained by the strong membrane upstream at the intersection to the driver tube and by a thin membrane (usually made of a polymer) downstream at the beginning of the nozzle and the test chamber. The latter are evacuated to very low pressures to facilitate the flow start-up process in the nozzle.

Once the experiment is started and the piston is released, it is accelerated by the pressure difference between the high pressure air in the reservoir and the lower pressure in the driver gas. The driver gas is compressed very fast and therefore quasi isentropically. When the pressure is high enough the membrane opens relatively quickly. Thus a very high pressure gas at high temperature is now available to power the shock into the shock tube. The high temperature is very essential for achieving high enthalpy conditions due to the high speed of sound. It is required that the pressure in the driver tube increases for some time after rupture, or at least stays approximately constant. This helps for the strong deceleration of the piston with the aim of a vanishing velocity at the end of the driver tube thereby ensuring high pressure driving the shock into the shock tube. Reducing the shock-tube diameter relative to that of the driver is required for achieving this. The corresponding condition is called tuned operation. For further details, see e.g. [15, 16].

When the high temperature, high pressure driver gas enters into the low temperature and low pressure test gas after rupture of the membrane, a very strong shock is building up after some initial forming processes which involve pressure losses. This shock is then travelling downstream the shock-tube and compresses the test gas and heats it, again almost isentropically due to the short times involved. This compression and heating can already be used for investigating the flow state behind the incident shock. However, for achieving the required high pressures and high enthalpies for the operation as a reflected shock-tunnel, the shock is reflected at the initially closed end wall of the shock-tube. Thus it acts on the test gas again, which is now already in a state of high pressure and high temperature, and therefore increases the absolute levels by more than what it experienced during the first pass of the shock. Behind the reflected shock, the flow is close to stagnant and available to be expanded through the nozzle. The corresponding condition is called tailored operation. In a one-dimensional consideration the tailored operation also stops the interface between driver-gas and the test-gas thus optimizing the test time. The conditions of the test gas are so severe, that the plastic membrane is destroyed thermally and mechanically immediately after the shock reflection and opens the way for the expansion flow through the convergent-divergent nozzle into the test section where it interacts with the test model or for further investigation of the flow field.

It is useful to support the understanding of the processes mentioned above with a x-t-diagram, Fig. 1. Depicted is the end of the driver-tube with the piston (white stripe due to its extended length), which is at the end of its deceleration phase due to the high pressure in front of it. The piston stops and recoils at a later time. At the (arbitrary) time 0.24 s the membrane opens and a strong shock compresses the test gas in the shock tube (x-coordinates 21.5–32.5 m) and is finally reflected at the end of the shock tube, compressing the test gas again. The second, thin membrane is destroyed and the high enthalpy test gas is released into the nozzle. The black line following the shock is the contact surface between test- and driver gas. It is almost brought to rest by the reflected shock. The time until this interface is accelerated again while entering the nozzle is an (optimistic) estimate of the available test time.

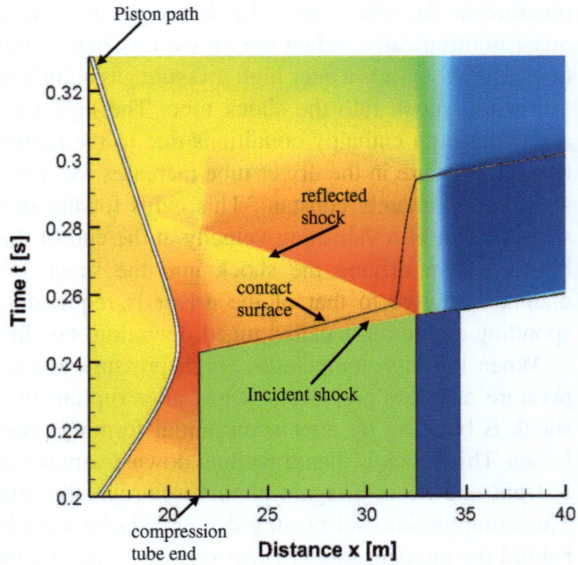

Fig. 1 x-t-diagram of the end of the driver tube, the shock tube and the front end of the nozzle

2.2 *Geometric Design and Capabilities*

The process described above is realized at the UniBw, Munich in the facility called high enthalpy laboratory Munich (HELM) [17, 18]. It consists of a 21 m long driver tube (inner, aerodynamically usable length) having an inner diameter of 286 mm which is enclosed by the compressed air reservoir in approximately the first 25 % of its length. The reservoir has a volume of about 1.3 m^3. Subsequent to the junction from driver tube to shock tube (driven tube) with the membrane the shock tube has an inner diameter of 95 mm. The lengths can be changed easily due to a modular design from 8 to 11 m in steps of 1 m. The following nozzle has a variable throat which ensures a smooth transition (second order continuous) to the conical nozzle with an end diameter of 685 mm. Thus its area ratios (exit/throat) can vary from 400 to 2000.

The geometric parameters of this facility are listed in Table 1 [17, 18] and compared with other piston-driven reflected shock-tunnels. HELM is thus similar to T4 and T5; HEG is bigger and more powerful. Figure 2 shows a schematic description of the facility with the nozzle truncated and the test-section omitted. Some details are enlarged.

Two pistons are available presently. They are of identical design. One is fabricated of aluminium and weighs some 53 kg; the second is heavier, it is made of steel and weighs almost two times more. Both can be equipped with a sensor measuring the acceleration (and changing the mass, of course). In the future it is planned to use pistons with brakes and a design is analysed and in fabrication, but up to now the pistons are just decelerated by the pressure differential (and friction).

Table 1 Parameters of selected free-piston shock tunnels

Facility name	T4[10]	T5[11]	HELM	HEG[12]
Location	UQ Brisbane Australia	Caltech Pasadena USA	UniBw Munich Germany	DLR Göttingen Germany
In operation since	1987	1990	2010	1991
Total enthalpy h_o (MJ/kg)	2.5–15	20	20–25	25
Total pressure p_o (MPa)	10–50	100	100	150
Flow velocity U (km/s)	8.9	6.3	5–7	7
Compression tube (ct)				
Length L (m)	26	30	21	33
Diameter D (mm)	229	300	286	550
Ratio L/D (−)	113	100	73	60
Shock tube (st)				
Length l (m)	10	12	8, 9, 10, 11	17
Diameter d (mm)	76	90	95	150
Ratio l/d (−)	132	133	84, 95, 105, 116	113
Ratio D/d (−)	3.01	3.33	3.01	3.67
Piston mass m_p (kg)	92	120–150	53, 150	250–760
Compression ratio l (−)	60–120	40–60	64–124	60
Nozzle exit diameter (mm)	135–388	314	684	880
Measuring time t (ms)	>1	1–2	<2	1.8

At the high pressure end of the driver and throughout the shock tube the facility is designed for a pressure of 200 MPa. During the certification test in 2010 the corresponding level was almost reached due to limitations of the hydraulic pump and thus the high pressure parts are now rated at 185 MPa. The air reservoir driving the piston is designed for 30 MPa but rated presently to 22 MPa due to compressor limitations. The HELM thus provides a good pressure capability. Figures 3 and 4 show photos of the main components; the driver-tube (yellow, in Fig. 3) with air reservoir (silver) and the shock-tube (red, in Fig. 4).

With the geometric dimensions and the maximum pressures achievable two parameters can be defined: The compression ratio λ is the volume ratio of the driver gas before and after compression. An important boundary condition for the evaluation of its maximum value is, of course, that the piston has to be brought to rest at the end of the tube. The parameter P [19] defines the operation conditions and thus the performance of the driver. Although defined as a function of pressure and speed of sound (amongst others) it can be reformulated to be just a function of the compression ratio and ratio of the diameters of driver- and shock-tube. With the latter ratio being 3, the facility is similar to T4 with respect to the driver and a P of close to 2 should be possible.

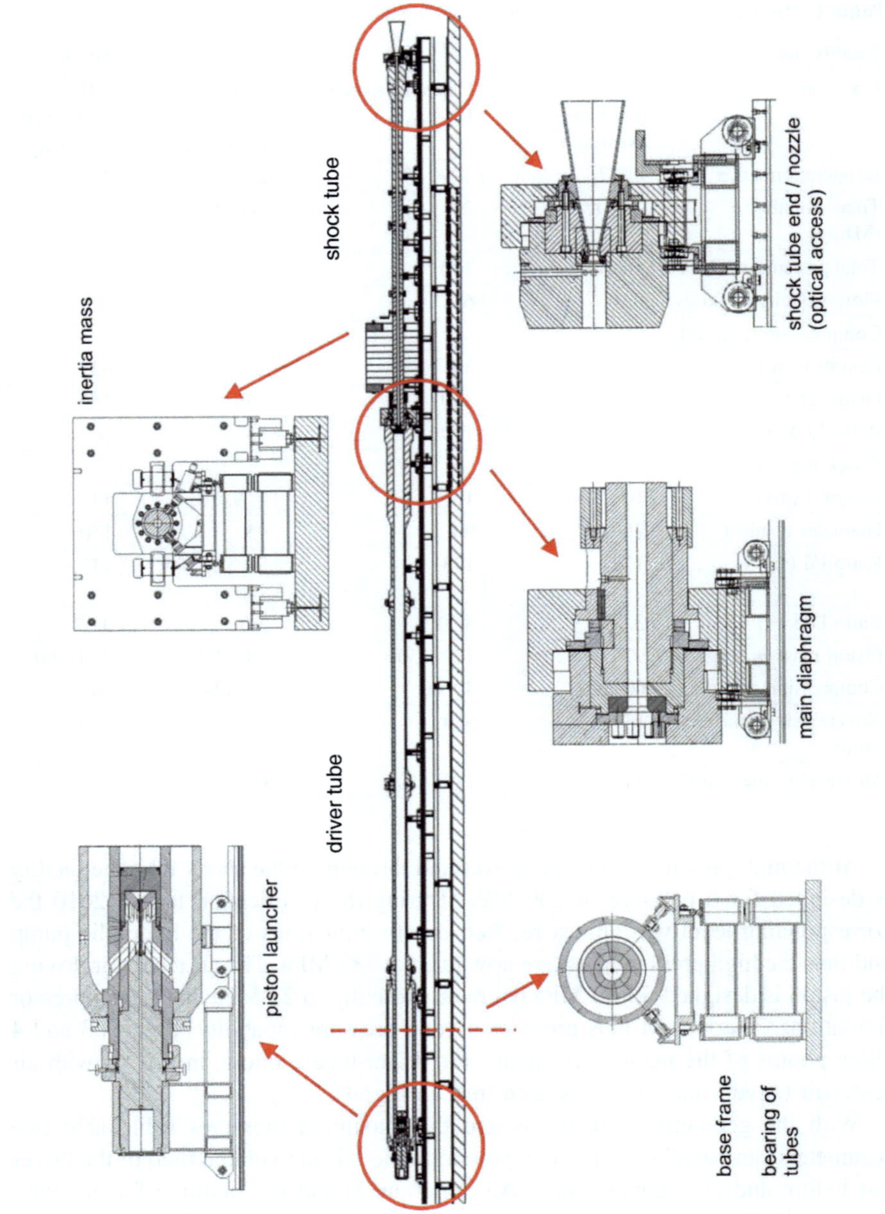

Fig. 2 Side view of the tunnel with enlargement of important regions

Fig. 3 Driver tube of HELM

Fig. 4 Shock tube of HELM

2.3 *General Simulation Capabilities*

The operating principle of the reflected shock tunnel provides a very high enthalpy reservoir of the test gas to be expanded through the nozzle. Due to the total enthalpy conservation of this quasi-steady flow this is converted into high velocities. Depending on the nozzle reservoir conditions velocities up to typically 5–7 km/s can be reached. Thus the speed of re-entry vehicles coming from low earth orbit can be simulated in the middle of their trajectory where maximum thermal and mechanical loads occur. Of course, lower velocities can be generated for the simulation of high speed range aircraft or air breathing high speed propulsion components.

A model of reduced scale is installed in the test section exposed to the generated flow. If conditions were chosen to directly duplicate the density of the free stream in flight at a corresponding flight altitude, the chemical reactions would not occur fast enough (with respect to time considerations) or at representative lengths (steady flow at a constant speed). However, for binary reactions like dissociation processes in the frontal part of most configurations of interest this can be easily corrected by increasing the flow density. This is mostly known as $\rho \times L$ scaling or binary scaling [2], but is essentially a similarity of the first Damköhler number, and means, for example, that a model reduced to 1/10 of the flight vehicle has to be exposed to a free stream with a density increased by ten. The big advantage is that this is achievable with the pressure capabilities of shock-tunnels (in contrast e.g. to plasma facilities).

Thus the main parameters which are duplicated are velocity and, through the $\rho \times L$ scaling, the chemical effects caused by the strong shocks in hypervelocity flows. Since the test gas is the same as the gas to be simulated, the Reynolds-number is fulfilled as well as the parameters related to physical and chemical characteristics of the test gas [2]. To be accurate, this is an approximation since in the nozzle freezing of chemical and thermodynamic effects occur and thus the free-stream is slightly different from conditions prevailing in real flight. Behind the usually strong shock (for hypervelocity flows) the deviation is adjusted. However, due to the very short experimental time, the model surface remains cold and all effects connected with the thermal state of the model cannot be simulated. A possible solution for overcoming this difficulty will be discussed in the next chapter.

Due to the freezing of thermodynamic properties in the nozzle (i.e. vibrational freezing) and the nozzle's limited area ratio the Mach number is not correctly reproduced despite the fact that the velocity is. Due to the Oswatitsch Mach number independence principle [20] many parameters expressed in their limit case when the flow Mach number approaches infinity can be extrapolated from medium to higher Mach numbers, especially for those experienced in re-entry (Mach number 20 and more). The minimum Mach number for the independence principle to hold depends on the body shape and the amount of its bluntness, and how low the minimum Mach number is [21, 22] (contrary to some different opinions). Thus most of the compressibility effects can be studied in a shock-tunnel, too.

3 Features of the Facility

3.1 Variable Tube Lengths

The HELM is designed in such a way that both the driver and the shock-tube length can be varied easily. This is achieved by a modular design and the fact that the tubes are installed on small moveable chariots. Also the additional mass of inertia can be moved by some 5 m since it is mounted on some kind of rail. For the driver tube this is just a provision for a future increase in performance by adding length and thus the compression ratio.

The length of the shock tube can be altered easily in small steps of one meter. The minimum length is 8 m, and the maximum, presently, is 11 m. The change only involves dismantling the locks of either the 1 m piece or the 2 m piece and removing it (or both of them if 8 m total length is sought for). Keeping the downstream parts in the same position the upstream parts are now pushed to a new position and the junction locked again. The scientific importance of these modifications lies in the fact that the ratio length to diameter can be varied from 85 to 115. This has an influence on the growth of the boundary layer at the inner side of the walls. The dissipative effects lead to a deceleration of the incident shock wave which is worth for studying and quantifying. For measurements at least 4 pressure transducer locations are available at different axial locations and an optical access is possible at the end of the shock-tube. Using this possibility it is planned to conduct research on the boundary layer state and the shock attenuation as a function of length or length/diameter, respectively.

3.2 Optical Access at the End of the Shock Tube

Limited information is published in the literature on the state of the test gases at the end of the shock tube, after passing of the incident and the reflected shock waves that generated high pressure and high enthalpy conditions in shock tunnels. For determination of chemical kinetics and thermodynamic relaxation processes in shock tubes many literature data are available, but these involve mostly low enthalpy and low density conditions. Often the test gases are trace components diluted in noble gases in these experiments.

The gas details at the end of the shock tube are of interest for their own sake, but also for the conditions in the test section for which they form the "initial/boundary condition" before and during the expansion and acceleration process through the nozzle. Thus the HELM features three sites for optical probing located at the same axial position, close to the shock tube end, separated circumferentially by 90°; see in Fig. 2. Therefore through-light methods and an observation of signals perpendicular to the beam are possible. The scientific objective is to quantify the thermodynamic and chemical state of the gas (together with the pressure measurement possibilities in the same plane). Also this access can be used for characterisation of the boundary layer developed in the tube, as well as for studying the behaviour of the interface separating the test gas from the driver gas (driver gas contamination).

Several laser-optical measurement methods exist for temperature determination of gases. Although the institute has experience in using Raman- [23, 24] and LIF-methods [25, 26] for measurements in high enthalpy flow, they don't meet the requirements for gathering data in a shock tunnel. The requirements are (a) high pressure in addition to the high enthalpy (difficult for e.g. LIF due to quenching), (b) fast measurement (of the order of a fraction of a ms, which makes the methods requiring collection and averaging unattractive), and (c) triggering by the shock and by the facility movement (just a practical implication, approximately the same for all methods). A thorough review of available non-intrusive methods was conducted. It was concluded that the use of Laser induced thermal acoustics (LITA), also called Laser induced grating spectroscopy, is the preferred option. Also, the analysis of the self-luminescence of the hot gas seems to be promising.

The measurement relies on the following principle [27, 28], see Fig. 5: two coherent laser beams of equivalent frequency cross each other at a shallow angle. At the intersection, an interference pattern is created (therefore the name laser induced grating spectroscopy). The laser beams are pulsed, and thus this interference pattern decays with time after being formed. The decay is a function of the local speed of sound (therefore the name laser induced thermal acoustics) which in turn is a function of temperature. The interference pattern is read out by diffraction of a third laser beam and recorded for post-processing.

Two possibilities for the generation of the interference pattern can be distinguished: the electrostrictive (LIEGS) and the thermostrictive (LITGS) one. With the non-resonant electrostrictive process polarizable molecules orientate themselves according to the electromagnetic field. Thus more molecules tend toward the regions of higher electromagnetic fields which cause a density gradient. This changes the refraction index. In the case of a thermostrictive process the frequency of the laser has to be chosen such that it excites an absorption line of a specific molecule in the gas. More molecules will be exited in the regions of a higher electromagnetic field. By collisional quenching the energy is transferred to the surroundings, resulting in a locally higher temperature. Again the refraction index of the gas is changed, but in this case the density is lower in regions of high field strengths.

Both strategies have been demonstrated in principle in the laboratory [29], and even in parallel with the same optical set-up [30], which is very beneficial for arranging and aligning the measurement system before conducting a shock-tunnel test. For lower temperatures and densities good results are reached when using electrostrictive method. However, for higher temperatures and densities the thermostrictive method is preferable.

In the present case the NO_2-molecule will be used for excitation. Concentrations of some 10 ppm are sufficient for generating signals with a LITA process [30], which is exceeded easily for most of the medium to high enthalpy conditions in shock-tunnels. Significant advantage of the LITGS process is the improvement of the signal with higher pressures [31]. However, some areas are still of concern: First, the evaluation of the speed of sound (which is the transfer function for the

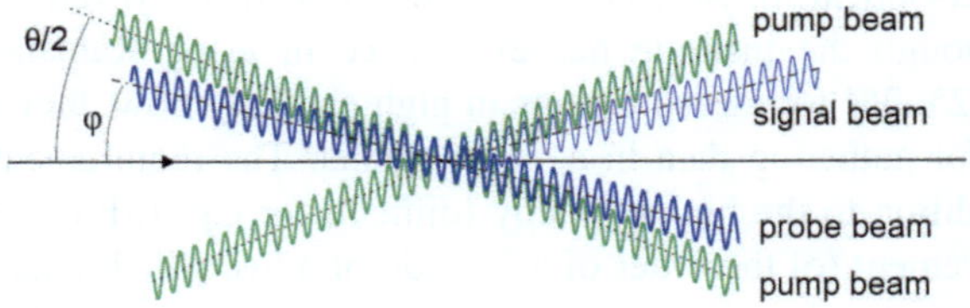

Fig. 5 Principle of laser induced gratings

temperature) for thermally perfect gas is trivial and for a gas at equilibrium conditions straightforward. However, for chemical non-equilibrium conditions, which we want to explore, the evaluation of the relation between temperature and speed of sound not only becomes complicated, but is also non-unique [32]. Second, we have to rely on single "shot" measurements. This has been done before [31] and also been demonstrated in a test-cell in the laboratory [30] but has to be proven to result in a meaningful data-set for the conditions aimed for here.

3.3 Integration of Plasma-Windtunnel and Hot Model Technique

As mentioned earlier, despite the fact that the gases in the facility are at very high temperature all parts of the wind tunnel (except the nozzle throat) as well as the model and the equipment in the test-section are thermally loaded to a small degree only. This is due to the short duration of the test-time in this impulse facility. This offers a lot of advantages as it make construction and handling of a shock tunnel easier in considering the huge amount of available power (order of GW/m^2 [2]). However, this includes also a problem, since in real flight the characteristic times are much longer and therefore the surface of vehicles become very hot. In fact, the control and management of the high thermal loads is a task of first importance in hypersonic/high speed flight and research on high temperature materials and heat shields/thermal protection systems is important.

The unrealistic thermal boundary condition at the model has implications on the flow in the vicinity of the model surface. For example, the boundary layer thickness is altered and the skin friction is changing as a function of temperature. The boundary condition concerning chemical species conservation is also changed, and catalytic effects may become very important contributors to the total heat flux due to recombination processes at the wall (catalycity). Without going into details, it is clear that investigation of these phenomena is worth for pursuing. Research was and is done using electrical heating of models (e.g. [33, 34]). However it is often difficult to apply the energy source at the locations of highest thermal loads experienced in real flight, i.e. stagnation point and leading edges, and thus the results are useful as CFD, but still very generic, test cases.

Therefore the integration of a plasma wind tunnel has been a requirement during the development of the shock tunnel. The plasma facility, though being a small one, is in operation for decades and was dedicated to research ranging from supersonic combustion of hydrogen to material testing of thermal barrier coatings/heat shields. Recently, it has been adapted to provide testing for high temperature, high pressure (atmospheric) erosion investigations [35]. Also, the development of non-intrusive laser measurement methods is easier in this continuous facility and a research focus in recent years.

Once the requirements of infrastructure are fulfilled, the plasma facility can be directly plugged into HELM. The plan is to heat the models aerothermodynamically. Thus meaningful distributions of the thermal loads are generated by the high enthalpy flow. Since the plasma heated flow will be perpendicular to the flow direction in the shock-tunnel, the model has to be turned by 90° shortly before or during the time of flow establishment in the nozzle. Thus another similarity parameter, i.e. flow-enthalpy/wall-enthalpy (temperature) can be varied and the related effects can be studied.

It is not denied that a few hurdles have to be overcome in order to generate meaningful experimental results. First, the model has to be made of a material capable to withstand the high temperature. This is seen as an advantage, since the material should be the same as the flight body for catalycity reasons. However, it makes the model more expensive and difficult to fabricate. Second, the instrumentation has to be adapted and possibly changed to be used in a high temperature. It is anticipated that most measurements will be made optically, and measurement equipment in the model has to fulfil the same requirements as flight hardware with respect to thermal loads. Again, this is not necessarily a disadvantage, but it is more costly and complex.

4 Present Status of Obtained Results

During the aerothermodynamic design of the shock-tunnel extensive use of Jacobs' [36] method (called L1d) proposed for quasi-one-dimensional modelling of a free-piston shock tube was made. It solves the Euler equations for a shock tube, shock tunnel, or expansion tube including the piston and membrane dynamics using a Lagrange approach. Dissipative effects are considered by adding empirical engineering expressions, thus it is called a quasi 1-D-method. The method was validated by comparing its predictions to results obtained from various tunnels [37]. Also during operation of the HELM the method is used for prediction and analysis and is shown in the following figures.

In the following a few details are given from experiments conducted in the HELM so far. Initially the facility was operated as a shock tube at very moderate pressure and enthalpy levels [38]. Air was thus used as driver gas. The aim was to use the experience gained with a small shock tube and operated for basic experiments [39], to develop and provide some low enthalpy test conditions for the LITA-measurements mentioned above, and to gain experience with the facility. All experiments were done using the full length driver and the full length shock-tube. It is acknowledged, that the shock is attenuated unnecessarily because of the high length to diameter ratio of the shock-tube.

Most of tested conditions in the shock-tube mode involved two different conditions [38]. By increasing the reservoir pressure needed for accelerating the piston, two rupture pressures of the main membrane can be chosen: 8.5 or 12.5 MPa. They result in nozzle reservoir conditions of about 3.5 MPa of total pressure and 2 MJ/kg of total enthalpy or 4.5 MPa and 2.5 MJ/kg, respectively. These are very mild conditions for a facility of this kind.

As described above, the L1d scheme is used for the prediction and analysis of the tunnel performance. In Fig. 6 the pressure history in the driver tube is shown for the first operating condition. Following initial continuous compression at low pressure levels higher pressure compression is governed by wave processes and is thus exhibiting a step-wise behaviour. Overall, the agreement between simulations and experiment is good. In the next diagram (Fig. 7) the pressure history at the shock-tube end wall is shown and compared with the L1d-results. For this over-tailored condition the general agreement is also good. Results of basic solution of the one dimensional shock tube equations assuming equilibrium gas by ESTCj [40] are introduced in this and in Figs. 6 and 7.

For a higher pressure in the nozzle reservoir at the end of the shock-tube (6 MPa), and in the shock tunnel mode, the flow conditions in the nozzle have been analysed. At such low pressures the Reynolds-number is also small. Even then, the flow in the nozzle and especially in the boundary layer shows sound flow behaviour. In Fig. 8 the distribution of the flow Mach number inside the nozzle is shown; it was evaluated using the CFD++ scheme [41]. The method predicts a Mach number of 9 at the nozzle exit. Although the boundary layer is rather thick for such a Reynolds-number, a core of inviscid flow having a 0.45 m width is obtained. The wall pressure was measured and agrees reasonably well with the numerical results. However, since the measurements were done with sensors not suitable to the range of prevailing pressures at the four measurement locations, between 1.5 and 2 m in direction of the flow, the quality of the signal was low.

The test condition having the highest total enthalpy used so far is the one with 15 MJ/kg enthalpy condition, based on an ESTCj computation [40] using the shock speed in the middle of the shock tube. One special feature of this experiment is the

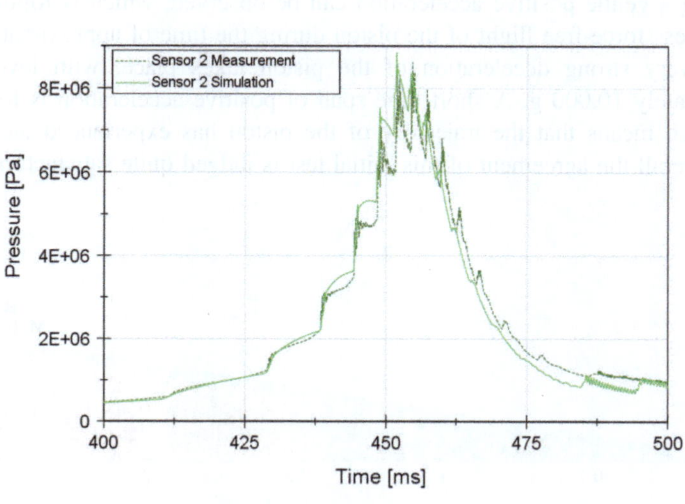

Fig. 6 Pressure history at the end of the driver-tube

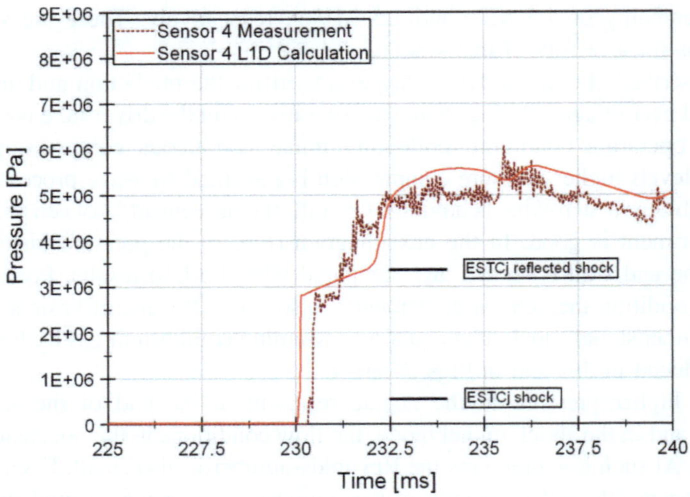

Fig. 7 Pressure history at the end of the shock tube

use of a piston equipped with an acceleration sensor. It was provided on loan from the German-French Research Institute at Saint Louis, France, (ISL) and was mounted at the back side of the piston. The piston is then placed inside the piston-launcher. After starting the experiment and the launch of the piston the sensor required a short delay time before recording started. After the test, the piston with the sensor is removed from the driver-tube and the recorded signals can be read out.

The acceleration curve obtained in this experiment is shown in Fig. 9, together with the acceleration result obtained numerically with the L1d method. In the beginning a gentle positive acceleration can be observed, which is followed by a more or less force-free flight of the piston during the time of approximately 0.1 s. Then a very strong deceleration of the piston takes place, with levels up to approximately 10.000 g. A short time span of positive acceleration is found after this, which means that the trajectory of the piston has experienced an inflexion point. Overall the agreement of this initial test is judged quite satisfactorily. Some

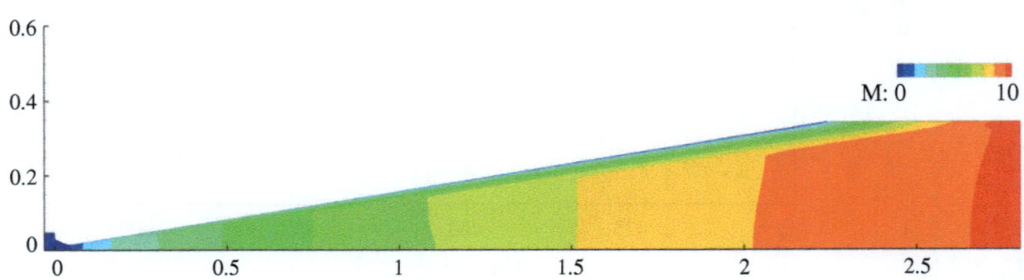

Fig. 8 Mach-number distribution in the conical nozzle of HELM

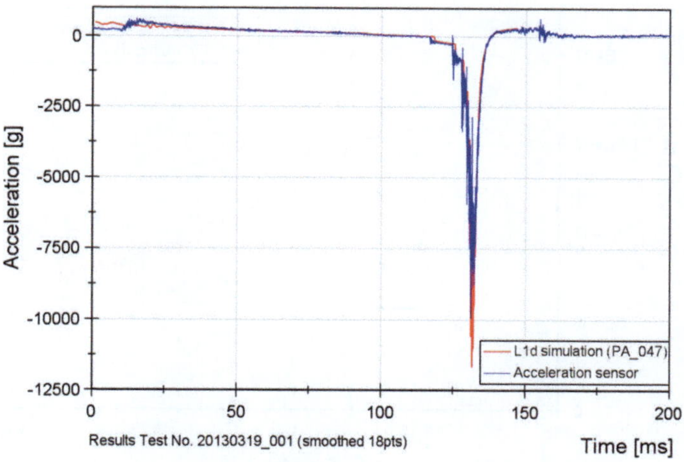

Fig. 9 Comparison between measured and simulated piston acceleration

deviations occur at the beginning of the recording process, which might be connected to the initiation process.

A close-up of the maximum g-loads acting on the piston is depicted in Fig. 10. It is apparent from this figure that a discrepancy exists between measured and simulated results during the 2 ms time of maximum deceleration. While the acceleration sensor measured during this time values in the region of 7.500 g with only one peak reaching 10.000 g, the simulation suggests a maximum peak of about 10.700 g. Anyway the mechanical loads are very high and the suitability of a piston-brake is obvious for removing the observed maxima.

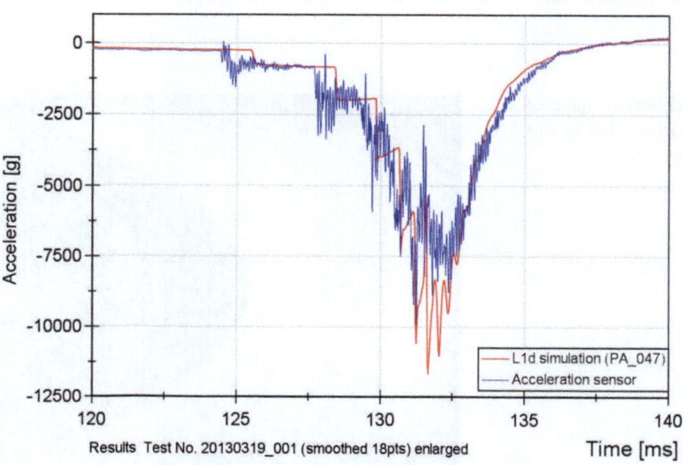

Fig. 10 Comparison between measured and simulated piston acceleration, enlargement of Fig. 9

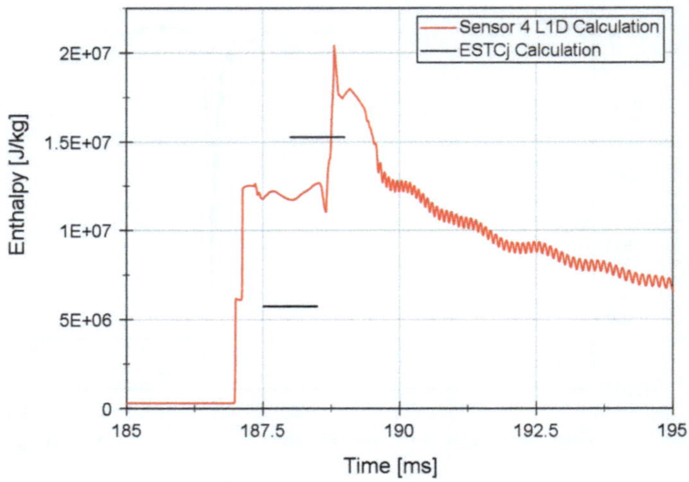

Fig. 11 Numerical results of total enthalpy

For the considered condition, Fig. 11 shows the total enthalpy as function of time, computed with the L1d method and the ESTCj schemes. When using the shock speed at the middle of the tube, approximately 15 MJ/kg are reached. However, due to further damping effects only a total enthalpy of ~12.5 MJ/kg is observed during the first 2 ms. The condition is not tailored though, and it is expected that by a fine tuning the constant flow conditions can be improved. The observed oscillations show the arrival of the driver gas. Due to the very high speed of sound the discretization chosen is not good enough and leads to instabilities.

As a typical body for configurations in hypersonic conditions, a hyperboloid is installed in the test section (Fig. 12). The body has been used for extensive testing in the T3 tunnel, and a lot of results were produced for validation purposes

Fig. 12 Schlieren image of a hyperboloid installed in the test-section

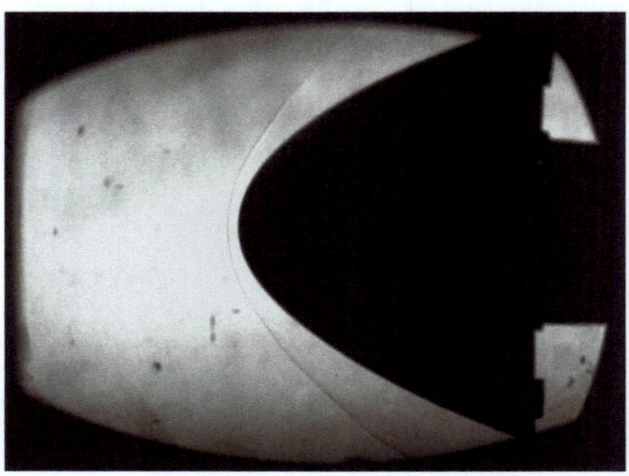

[4, 42–44]. It is considered a benefit to conduct experiments with this configuration in the future that will hopefully enable a cross-facility check, even though T3 is dismantled now.

5 Summary

A new piston-driven shock-tunnel, the HELM has been developed during the recent years. The rationale behind designing and building the tunnel is given, together with an overview of its characteristics, its special features and the first possible directions of research. Results from operations whose main objective were to widen the operational envelope (conditions) of the HELM are provided and compared with numerical predictions.

Acknowledgements This worked wouldn't have been realized without the support and collaboration of several colleagues, especially K. Schemperg, P. Altenhöfer and T. Sander. This is gratefully acknowledged.

References

1. Hirschel, E.H.: Basics of Aerothermodynamics; Springer-Verlag and AIAA, vol. 204 of Progress in Astronautics and Aeronautics (2005)
2. Hornung, H.G.: Experimental real-gas hypersonics, ZFW 12 (1988)
3. Mundt, Ch., Hirschel, E.H.: Modelling of chemical and physical effects with respect to flows around reentry bodies, 3. Aerospace Symp. Braunschweig, 1991, Örtel, H., Körner, H. (eds.) Orbital Transport—Technical, Meteorological, and Technical Aspects, pp. 233–244., Springer, Berlin (1993)
4. Mundt, Ch., Boyce, R.R., Hirschel, E.H.: Simulation of hypersonic, reacting flow around reentry bodies—Comparisons of numerical and experimental results. Z. Flugwiss. Weltraumforsch. **19**, 492–501 (1995)
5. Nguyen, T., Behr, M., Reinartz, B., Hohn, O., Gülhan, A.: Effects of sidewall compression and relaminarization in a scramjet inlet. J. Propul. Power **29**, 628–638 (2013)
6. Langkau, R.: Eine neue Forschungsanlage zur Untersuchung zur Untersuchung chemischer Reaktionen in Gasströmungen hoher Enthalpie, Dissertation Universität der Bundeswehr München (1981)
7. Olivier, H.: Transonic and hypersonic flow research using different shock tube technologies. In: Interdisciplinary Shock Wave Research. Proceedings of the International Symposium on Interdisciplinary Shock Wave Research, pp. 391–408. Sendai, Japan (2004)
8. Holden, M.S., Parker, R.A.: LENS hypervelocity tunnels and application to vehicle testing at duplicated flight conditions. In: Progress in Astronautics and Aeronautics, vol. 198. Advanced Hypersonic test facilities (2002)
9. Stalker, R.J.: Development of a hypervelocity wind tunnel. Aeronaut. Journ. **76**, 374–384 (1972)
10. Stalker, R.J., Morgan, R.G.: The University of Queensland free piston shock tunnel T4—initial operation and preliminary calibration. In: Proceedings of the 4th Nature Aerospace Engineering Symposium, Inst. Eng. Australia, 1988

11. Hornung, H., Sturtevant, B., Bélanger, J., Sanderson, S., Brouilette, M., Jenkins, M.: Performance data of the new free-piston shock tunnel at GALCIT. In: Proceedings of the 18th International Symposium on Shock Waves, Sendai (1991)
12. Eitelberg, G.: First results of calibration and use of HEG, AIAA, pp. 94–2525 (1994)
13. Itoh, K.: Characteristics of the HIEST and its capability for hypersonic aerothermodynamic and scramjet research. In: Progress in Astronautics and Aeronautics, vol. 198, Advanced Hypersonic test facilities (2002)
14. Oertel, H.: Stoßrohre. Springer Verlag (1966)
15. Hornung, H.: The piston motion in a free-piston driver for shock tubes and tunnels, GALCIT report FM, pp. 1–88 (1988)
16. Itoh, K.: Tuned operation of a free piston shock tunnel. In: Proceedings of the 20th International Symposium on Shock Waves, Pasadena (1995)
17. Schemperg, K., Mundt, Ch.: On the free piston shock tunnel at UniBwM (HELM), 26. In: Hannemann, K., Seiler, F. (ed.) International Symposium on Shock Waves. Springer, Berlin, pp. 477–482 (2007)
18. Schemperg, K., Mundt, Ch.: Study of numerical simulations for optimized operation of the free piston shock tunnel HELM, AIAA-2008–2653, 15. In: AIAA International Space Planes and Hypersonic Systems and Technologies Conference (2008)
19. Hornung, H., Bélanger, J.: Role and techniques of ground testing for simulation of flows up to orbital speed, 3rd Joint Europe/US short course in hypersonics (1990)
20. Oswatitsch, K.: Ähnlichkeitsgesetze für Hyperschallströmungen, ZAMP, also: similarity laws for hypersonic flow, KTH-AERO TN 16, vol. II, 1951, pp. 249–264 (1950)
21. Kliche, D., Mundt, Ch., Hirschel, E.H.: The hypersonic Mach number independence principle in the case of viscous flow. Shock Waves **21**, 307–314 (2011)
22. Lorenz, V., Mundt, Ch.: Numerische Untersuchung des Oswatitsch'schen Machzahlunabhängigkeitsprinzips, 16. STAB-Tagung, Göttingen (2013)
23. Hatzl, S., Sander, T., Mundt, Ch.: Measurements of high enthalpy flow temperature using a one-dimensional spontaneous Raman spectroscopy procedure, AIAA-2010–4809, 27. In: AIAA Aero-Dynamic Measurement and Ground Testing Conference (2010)
24. Hatzl, S., Sander, T., Mundt, Ch.: One-dimensional measurements of high enthalpy flow temperature using spontaneous Raman spectroscopy, AIAA-2011–2212, 17. In: AIAA International Space Planes and Hypersonic Systems and Technologies Conference (2011)
25. Kirschner, M., Sander, T., Mundt, Ch.: Laser induces fluorescence of Nitric Oxide A-X(0,0) in high enthalpy flow, pp. 11-1–11-8, GALA Conference (2013)
26. Kirschner, M., Sander, T., Mundt, Ch.: Rotational temperature measurement in an arc-heated wind tunnel by Laser induced fluorescence of nitric oxic A-X(0,0), AIAA-2014-2528, AIAA Ground Testing Conference (2014)
27. Cummings, E.: Laser-induced thermal acoustics: simple accurate gas measurements. Opt. Lett **19**(17), 1361–1363 (1994)
28. Dillmann, M.: Entwicklung laserspektroskopischer Verfahren für den quantitativen Nachweis von Stickoxid und die Bestimmung der Temperatur in Hochdruckflammen. Ph.D. Dissertation, Institut für Verbrennungstechnik, DLR e.V. (2004)
29. Sander, T., Altenhöfer, P., Mundt, Ch.: Experimental study on the applicability of laser-induced grating spectroscopy for temperature measurements in shock tunnels and shock tubes, AIAA-2012–5955, 18. In: AIAA International Space Planes and Hypersonic Systems and Technologies Conference (2012)
30. Sander, T., Altenhöfer, P., Mundt, Ch.: Development of laser-induced grating spectroscopy for the application in shock tunnels. AIAA J. Thermophys. Heat Transfer **28**(1), 27–31 (2014)
31. Stampanoni-Panariello, A.: Laser-induced gratings in the gas phase: formation mechanisms and applications for diagnostics. Dissertation, Swiss Federal Institute of Technology Zürich, ETH Zürich (2003)
32. Vincenti, W., Kruger, Ch.: Introduction to physical gas dynamics. R.E. Krieger Publ. Comp, Malabar (1965)

33. Bleilebens, M., Olivier, H.: Druckmessungen hoher räumlicher Auflösung an einem aufgeheizten Rampenmodell im Stoßwellenkanal, DGLR-2003-129 (2003)
34. Zander, F., Morgan, R., Sheikh, U., Buttsworth, D., Teakle, P.: Hot-wall reentry testing in hypersonic impulse facilities. AIAA J. **51**(2), 476–484 (2013)
35. Kirschner, M., Wobst, T., Rittmester, B., Mundt, Ch.: Erosion testing of thermal barrier coatings in a high enthalpy wind tunnel, ASME Turbo expo GT2014-25523 (2014)
36. Jacobs, P.: Quasi-one-dimensional modelling of a free-piston shock tunnel. AIAA J. **32**(1), 137–145 (1994)
37. Mundt, Ch., Boyce, R., Jacobs, P., Hannemann, K.: Validation study of numerical simulations by comparison to measurements in piston driven shock-tunnels. Aerosp. Sci. Technol. **11**(2–3), 100–109 (2007)
38. Mundt, Ch., Altenhöfer, P.: Results of HELM in initial operation as a shock tube. In: 11th International workshop on shock tube technology, Brisbane (2011)
39. Altenhöfer, P., Mundt, Ch.: Comparison of L1d simulations with measurements on a double-diaphragm shock tube, EFMC8 (2010)
40. McIntosh, M.: Computer program for the numerical calculation of frozen and equilibrium conditions in shock-tunnels, Internal report Australian National University (1968)
41. Kimmerl, J.: Analysis of the unsteady flow establishing in the nozzle and over a compression corner model in the shock tunnel HELM. Master Thesis, Universität der Bundeswehr München (2013)
42. Boyce, R., Mundt, Ch., Houwing, F.: Computational fluid dynamics validation using surface heat transfer measurements of hypersonic flowfields. AIAA-paper 96–351 (1996)
43. Boyce, R.R., Morton, J.W., Houwing, A.F.P., Mundt, Ch., Bone, D.J.: Computational fluid dynamics validation using multiple interferometric views of a hypersonic flowfield. J. Spacecraft Rockets **33**(3), 319–325 (1996)
44. Boyce, R.R., Pulford, D.R.N., Houwing, A.F.P., Mundt, Ch.: Rotational and vibrational temperature measurements using CARS in a hypervelocity shock layer flow and comparison with CFD calculations. Shock Waves J. **6**, 41–51 (1996)

Author Biography

Christian Mundt is the head of the institute of Thermodynamics in the faculty of aerospace sciences at the "Universität der Bundeswehr München", in Germany. His research is focused on high enthalpy / high velocity flows and their interaction with surfaces. This comprises experimental and numerical investigations. Further, optical laser diagnostics form an important filed of research. He received his Dipl.-Ing. degree from the "Technische Universität München" and his Dr.-Ing. dealing with the numerical simulation of reentry flows from the same institution. He has held industry positions with MBB (now Airbus) developing high performance aircraft and Rolls-Royce Deutschland developing jet-engines.

Development and Calibration of Detonation-Driven High-Enthalpy and Hypersonic Test Facilities

Zonglin Jiang and Hongru Yu

1 Introduction

Hypersonics for the foreseeable future is the domain of space access and hypersonic flight. A hypersonic vehicle encounters an aero-thermodynamic environment characterized by strong shocks and high temperatures, and the flow in the hypersonic regime behaves quite differently from that at supersonic Mach numbers [1–3]. The kinetic energy associated with hypersonic flight is converted into internal energy which increases the temperature of the air and gives rise to endothermic reactions, such as vibration, dissociation and ionization of the air near the vehicle surface. Thus, the air becomes a gaseous reacting flow. The mechanisms behind this conversion include adiabatic compression and viscous energy dissipation. Heat is transferred from the surrounding high temperature air to the vehicle surface. The rate at which heat is transferred to the surface depends upon many factors, including the free-stream conditions, the configuration of the vehicle and its orientation to the flow, the temperature difference between the air and surface, and the surface catalysis. The aforementioned phenomena are different from those in supersonic flows and so called the real gas effects, which vary with the flight Mach number and altitude. Hypersonic reflects the close coupling of the aerodynamic forces with the heating environment and its research is the frontier of gas dynamics [4, 5]. Actually, it represents the fullest integration of both scientific fundamentals and engineering applications into "aerospace" systems full of challenges.

Hypersonic vehicles have been designed and built since the 1950s. In many of these programs, flight tests have revealed problems that had not been predicted either through analysis/computation or through ground-based testing. These problems exposed hypersonic flow phenomena not previously identified, what we shall

Z. Jiang (✉) · H. Yu
State Key Laboratory of High Temperature Gas Dynamics (LHD), Institute of Mechanics, Chinese Academy of Sciences (CAS), 100190 Beijing, China
e-mail: zljiang@imech.ac.cn

O. Igra and F. Seiler (eds.), *Experimental Methods of Shock Wave Research*,
Shock Wave Science and Technology Reference Library 9,
DOI 10.1007/978-3-319-23745-9_9

call “unknown unknowns” [4]. The severity of the hypersonic flight environment strongly depends on the mission profile and the vehicle configuration. Figure 1 shows typical trajectories of various spacecrafts and the regions of different physical and chemical processes in the shock layer near such vehicles [6]. It is important to note that the hypersonic flight regime covers a large range of speed and altitude. As an example, an air-breathing hypersonic vehicle must operate at relatively low altitudes in order to maintain the relatively high dynamic pressures required for maximum engine performance, therefore it often flies at Mach numbers 5–10 and altitudes from 30 to 50 km, as indicated in Fig. 1 where the aero-thermal-environment is dominated by fluid dynamics effects. Although the real gas effect is weak in these cases, the vehicle heating is high at leading edges, in the regions with shock interactions, and in downstream locations where the flow transitions to turbulence. Since there is considerable uncertainty in the boundary-layer transition prediction and since the convective heating from a turbulent boundary layer is much greater than that from a laminar boundary layer, this transition uncertainty presents critical challenges to the researchers. On the other hand, a high speed planetary entry blunt vehicle, such as the Galileo probe entering Jupiter or a high speed Earth return capsule, encounters an aero-thermal-environment dominated by high enthalpy effects. A high enthalpy flow in planetary entry is characterized by elevated levels of ionization and may result in significant radiation and heating. The strong real gas effects and low density effects also play an important role in the flight regime. As depicted in Fig. 1, the middle part of the typical trajectories of various spacecraft is for TSTO, and ascent and re-entry vehicles. Strong real gas effects dominate the flow around the vehicles.

In order to design new hypersonic vehicles, it is important to develop hyper-velocity test facilities for ground-based experimental research. No single ground-test facility can fully simulate all aspects of hypersonic flight. The flow duration,

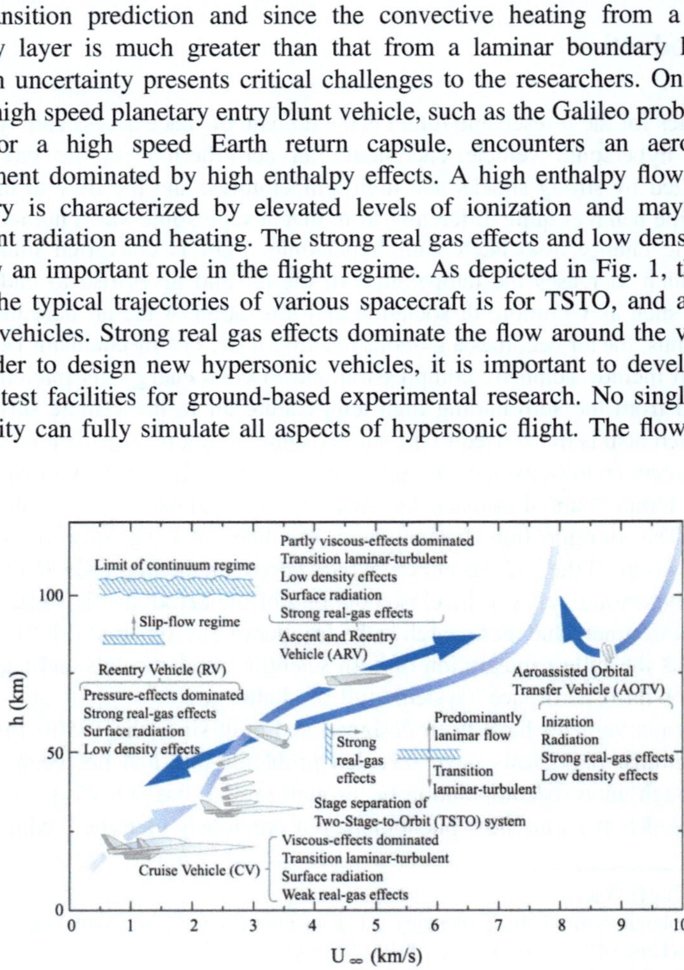

Fig. 1 Typical trajectories of various spacecraft and major aero-thermodynamic effects [6]

velocity, gas chemistry effects, Mach number, altitude or Reynolds number, model surface temperature, ablation effects, and the quality of the free-stream flow cannot be controlled simultaneously in any single facility, if at all. Nearly all tunnels suffer from noise levels much higher than in flight, and all tunnels simulating gas chemistry effects also have free-stream chemistry contaminants. Ground test simulations are an exercise in the art of combining partial simulations, each having unique advantages and disadvantages. After more than 60 years developments, high-enthalpy and hypervelocity test facilities suitable for studying aero-thermochemistry are still based on shock tunnels [7]. Three main kinds of high enthalpy shock tunnels have been developed over the world for several decades. The first one is the heated-light-gas driven shock tunnel [8–10]. The second one is the free-piston driven high-enthalpy shock tunnel [11–15], and the third is the detonation-driven high-enthalpy shock tunnel [16–19]. During the last decades, many such hypervelocity test facilities have been built over the world and a lot of valuable experimental data have been produced, which are very precious for hypersonic study.

For hypersonic shock tunnels, one of the great challenges is to create a true hypersonic environment at a reasonable scale on the ground due to the high energy requirement. The detonation driver was first proposed to solve this problem by Bird in 1957 [16]. Later, a 13.3 m long detonation driver was built by Yu [17]. Grönig et al. developed TH2 in Aachen, Germany and the TH2 shock tunnel is operated in the backward-running detonation mode [18, 19]. Then in 1994, the bigger detonation-driven shock tunnel, so-called HYPULSE was setup in GASL, USA. It could be also operated as an expansion tube [7]. These detonation-driven shock tunnels can be operated either in the backward-running detonation mode to achieve stable test flows with long driving time at a relatively low enthalpy level or in the forward-running detonation mode to obtain high-enthalpy flows, but with a short test duration [20–22]. The recent rapid progress shows that the detonation-driven hypersonic test facilities become a promising method for hypersonic vehicle ground tests.

In the following sections, the research progress in the detonation-driven hypersonic test facilities developed in the State Key Laboratory of High Temperature Gas Dynamics (LHD), Institute of Mechanics, Chinese Academy of Sciences (CAS) is summarized, and the report includes the detonation-driven high-enthalpy shock tunnel (JF10), the long-test-duration hypervelocity detonation-driven shock tunnel (JF12) [24, 28] and the detonation-driven orbital-speed expansion tube (JF16). JF10 shock tunnel is operated for Mach numbers from M10 to 20, JF12 shock tunnel performance covers M = 5–9, and JF16 expansion tube is capable of producing the near orbital speed flow at M = 20–25. With these facilities, the typical trajectories of various spacecrafts could be covered so that research on aero-thermo-dynamics and hypersonic propulsion could be carried out in the laboratory to gain fundamental understanding on hypersonics.

2 Detonation-Driven High-Enthalpy Shock Tunnel JF10

The JF10 shock tunnel was developed by Yu in 1998 [17] for exploring re-entry physics and it could be operated both in forward- and backward-running detonation modes. In order to achieve high quality test flows at high enthalpy level, the JF10 shock tunnel was improved by replacing the forward-running detonation driver with a new FDC (Forward-running Detonation Cavity) driver [21] in 2003 and its performance tests were completed in 2006. Tunnel calibration results presented in this section include the improvement work on the JF10 shock tunnel, P5 pressure variations, the experimental data on nozzle flow uniformity and the incident shock decay in the driven section [23]. The experimental data have shown that the improvement on the JF10 shock tunnel is successful and future research work will benefit from the thus-obtained high quality hypersonic test flows.

2.1 *Layout of the JF10 Shock Tunnels*

The original and the improved JF10 shock tunnels are schematically shown in Fig. 2. The improved JF10 shock tunnel, as shown in Fig. 2b, consists of three main parts. The first part is the FDC driver being about 6.225 m in length and 150 mm in diameter, the second part is a driven section being 12.5 m in length and 100 mm in diameter, and the last part is a conical nozzle with a 500 mm diameter exit. For the original JF10 shock tunnel, as shown in Fig. 2a, its driver section is 10 m in length, and it is about 4 m longer than the FDC driver. Other parts are the same as the improved JF10 shock tunnel. The cavity of the FDC driver is 300 mm in diameter and 2 m in length. The detailed research work on the FDC driver was reported by Jiang et al. [20, 21]. For high-enthalpy shock tunnels, the shorter driver section is quite effective in reduction of experimental costs and nozzle throat ablation.

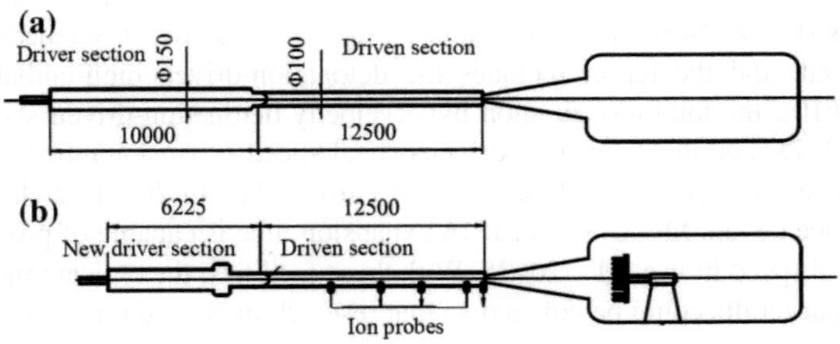

Fig. 2 Schematic of the improved and the original JF10 shock tunnels (length unit: millimeter)

2.2 Pressure Variations at the End of the Driven Section

For shock tunnels, P_5 pressure variation at the end of the driven section is a key parameter that indicates the test flow quality and stands for one of the most important performance parameter of shock tunnels. To obtain P_5, the calibration experiments were carried out in the condition of P_{4i} = 3.0 MPa and P_1 = 11 kPa where P_{4i} and P_1 are initial pressures of the driver section and driven section, respectively. The incident shock Mach number is found to be 11.8 and the P_5 pressure history is plotted in Fig. 3.

Figure 3 shows the pressure history measured at the end of the driven section. This pressure profile acts as a measure of the reservoir pressure during nozzle working duration. The more stable the reservoir pressure is, the more uniform the nozzle flows becomes. From this figure, it is observable that the pressure level maintains approximately constant for more than 6 ms. By comparing it with the 2 ms test time achieved with the free piston driven shock tunnel, it is apparent that this pressure variation is very promising for extending effective test duration and demonstrated well the performance of the improved JF10 shock tunnel.

Figure 4 shows pitot pressure variations measured in the nozzle flow in the same experiment. The pressure profile looks similar to the reservoir pressure variation as shown in Fig. 3. The similarity agreement between these two pressure profiles indicates not only experimental data certification, but also effective test time possibly available from the improved JF10 shock tunnel. This time duration is not only long enough for both flow pressure and heat transfer measurements, but also provides good conditions for aerodynamic force tests with existing measurement instrumentations. As it is well known, stress force balances are very difficult for shock tunnel force measurements because of the short test duration. The results of aerodynamic force measurement will be more reliable if the shock tunnels could provide longer test duration. Therefore, a 6 ms long test duration is very useful for conducting aerodynamic force experiments.

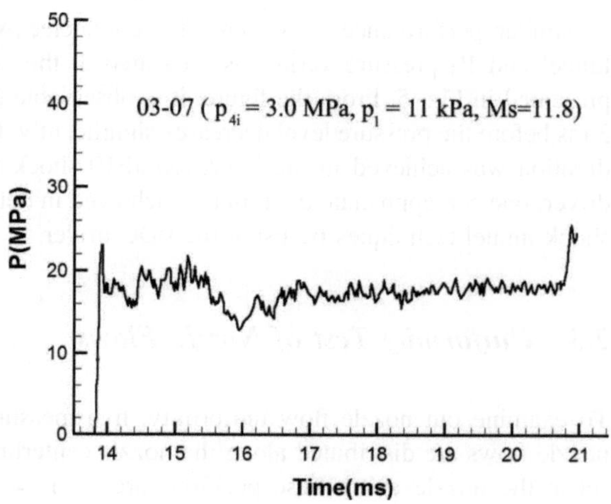

Fig. 3 Pressure history measured at the end of the improved JF10 shock tunnel

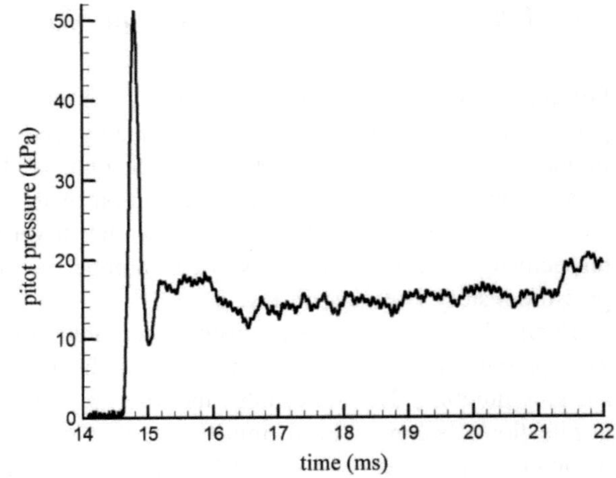

Fig. 4 Pitot pressure variations measured in the nozzle flow

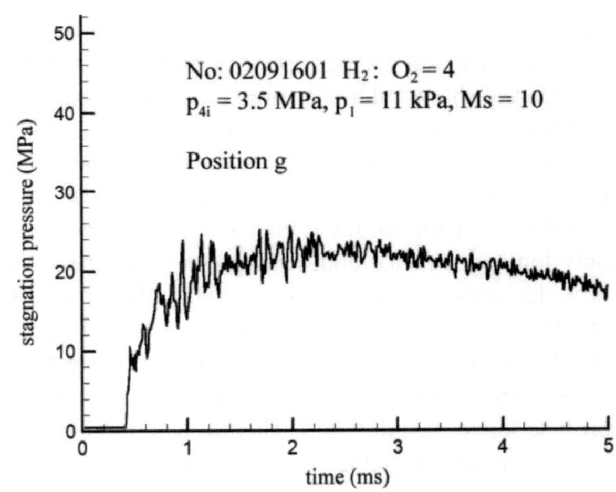

Fig. 5 Pressure variations measured at the end of the original JF10 shock tunnel

Similar performance tests were also conducted with the original JF10 shock tunnel and P_5 pressure variations measured at the end of the driven section are presented in Fig. 5. From the figure, it is observable that the test duration is about 3 ms before the pressure level decreases significantly. Considering that the 6 ms test duration was achieved in the improved JF10 shock tunnel with the shorter FDC driver, one can appreciate the progress achieved in detonation-driven high-enthalpy shock tunnel techniques by using the FDC driver.

2.3 Uniformity Test of Nozzle Flows

To examine out nozzle flow uniformity, five measurement positions in the JF10 nozzle flows are distributed along the nozzle centerline and the starting position is set at the nozzle exit. These positions are (a) L = −50 mm; (b) L = 100 mm;

(c) L = 265 mm; (d) L = 400 mm and (e) L = 600 mm, respectively. The experimental setup of pitot pressure measurements is shown in Fig. 6. Measurements were carried out at each measurement position both in the vertical and the horizontal directions. The measurement data were plotted together to demonstrate the flow uniformity of the nozzle.

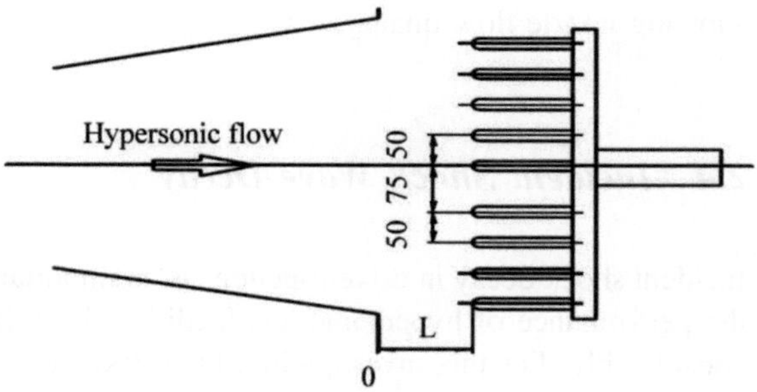

Fig. 6 Experimental setup of pitot pressure measurements in the improved JF10 nozzle flow

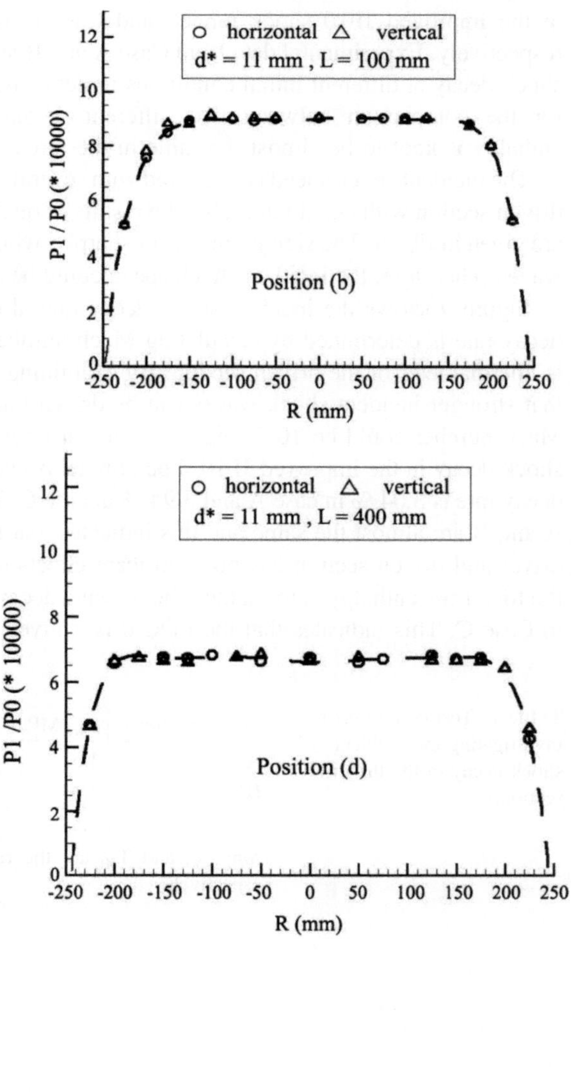

Fig. 7 Pitot pressure distributions in nozzle flows at two measurement positions in two directions

Figure 7 shows pitot pressure distributions in nozzle flows at two positions, i.e., (b) and (d). It is observable that the hypersonic flow in the nozzle central area appears quite uniform if evaluated from a view point of pitot pressure distribution. This uniform flow area is found to be approximately 700 mm in length and 400 mm in diameter. Although the pressure uniformity does not fully confirm hypersonic flow uniformity, these results are quite promising and very encouraging for showing nozzle flow quality.

2.4 *Incident Shock Wave Decay*

Incident shock decay in driven sections is an important parameter for demonstrating the performance of hypersonic test facilities since the test flow quality is affected considerably. For this investigation, three test cases were selected for comparison and the relevant test conditions are listed in Table 1. Cases A and B were carried out in the improved JF10 shock tunnel, and Case C was done in the original one, respectively. Experimental data from Case A and B are used to examine the incident shock decay at different initial conditions in detonation drivers, and Case C is used for the comparison between two different detonation drivers. The total flow enthalpy is kept to be almost the same in the three cases.

The incident shock speed is measured with several ion probes distributed along the driven section with equal intervals. The results of qualified tests for these ion probes are given in Fig. 8. The sign jumps occur sharply, which indicate the arrival of shock waves. Therefore, the incident shock speed could be calculated with high accuracy.

Figure 9 shows the incident shock decay rate along the driven section and the decay rate is determined by calculating Mach number drop per tube diameter, that is, the diameter of the driven section. By examining cases A and C, it can be seen that stronger incident shock waves can be driven out with the FDC driver and the Mach number could be 10 % higher in similar operating conditions. The incident shock decay in the improved JF10 shock tunnel is about 30 % less, and actually the decay rate is 3.04 % in case A and 3.94 % in case C. The experimental data of cases A and B are almost the same and this indicates that the initial pressure ratio of the driver and driven section has no significant effects on the incident shock decay if the total flow enthalpy is the same. Shock wave decay is weaker in Case A than that in Case C. This indicates that the FDC driver gives good performance.

Table 1 Test conditions for investigating the incident shock decay in the driven section

Test conditions	P_{4i} (MPa)	P_1 (KPa)	P_{50} (MPa)	T_{50} (K)
A	1.5	4.5	8.0	7480
B	3.0	11	19.4	7920
C	1.5	4.7	8.3	7200

Note P_{50} and T_{50} are the reservoir pressure and temperature, respectively

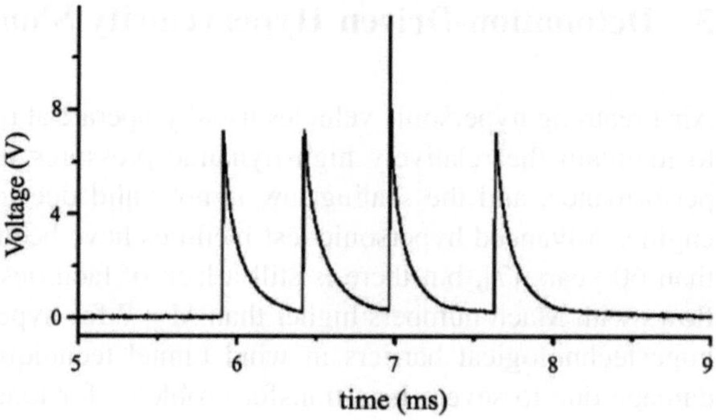

Fig. 8 Ion probe signs for incident shock speed measurement

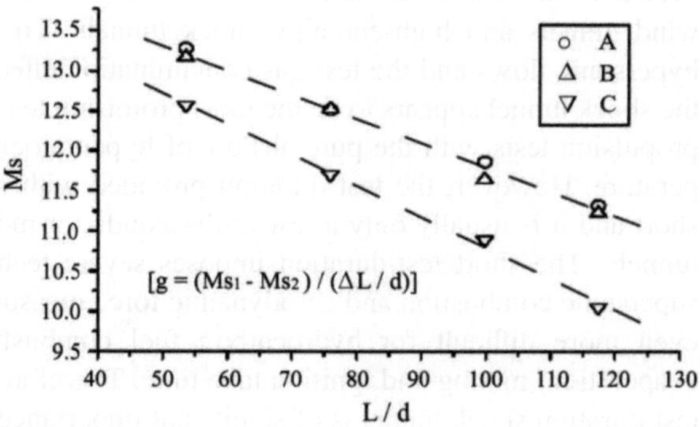

Fig. 9 Incident shock wave decay for three test cases

2.5 *Concluding Remarks*

From experimental data and the above discussions, the research progress on the detonation-driven high-enthalpy shock tunnel can be summarized as follows:

(1) The improvement on JF10 shock tunnel with a new FDC driver was demonstrated to be very successful and a stronger incident shock could be generated with a lower decay rate in the similar initial conditions by comparing with the original JF10 shock tunnel.

(2) The test duration of the improved JF10 shock tunnel was extended by two times even though the new FDC driver is 40 % shorter than the original forward-running detonation driver. The 6 ms test duration is achieved, which makes aerodynamic force measurements much easier.

(3) The uniform core flow region of the improved tunnel is about 700 mm in length and 400 mm in diameter.

3 Detonation-Driven Hypervelocity Shock Tunnel JF12

Air-breathing hypersonic vehicles usually operate at relatively low altitudes in order to maintain the relatively high dynamic pressures required for maximum engine performance, and the scaling law is not valid due to the chemical reaction in its engine. Advanced hypersonic test facilities have been under development for more than 60 years [7], but there is still a lack of facilities for generating high-enthalpy flows with Mach numbers higher than M = 7 for hypersonic propulsion tests due to huge technological barriers in wind tunnel techniques, especially for the facility damage due to severe heat transfer problems for long test duration.

In order to meet the requirement of ground experiments of air-breathing hypersonic vehicles, several types of facilities were developed, including combustion-heated wind tunnels, arc-heated hypersonic wind tunnels, conventional air-heated wind tunnels and high-enthalpy shock tunnels. To clarify the real gas effect on hypersonic flows and the test-gas contamination effects on supersonic combustion, the shock tunnel appears to be the most promising test facility to provide hypersonic propulsion tests with the pure airflow of hypervelocity and the required total temperature. However, the test duration provided with shock tunnels is usually quite short and it is usually only a few milliseconds for most of the high-enthalpy shock tunnels. The short test-duration imposes severe technical barriers on the tests of supersonic combustion and aerodynamic force measurement. Moreover, it becomes even more difficult for hydrocarbon fuel combustion since the fuel injection, evaporation, mixing and ignition take time. Therefore, the development of the long test-duration shock tunnel is of significant importance for the study of air-breathing hypersonic vehicles.

A research project was launched in the State Key Laboratory of High Temperature Gas Dynamics (LHD), Institute of Mechanics, CAS in 2002 to investigate the techniques for long test-duration and hypervelocity shock tunnels. Much progresse has been achieved successively since then. Jiang et al. [21] proposed a forward detonation cavity (FDC) driver, and the test duration has been prolonged to three times than the original one. Li et al. [26] investigated the wave/interface interaction in shock tubes, and the tailored condition was proposed for detonation-driven shock tunnels. The shock/boundary interaction near the end of the driven section was examined numerically in detail, and a steel ring is proposed to be installed around the nozzle throat by Li et al. in order to delay driver-gas mixing with test gases [27]. Wave-reflection damping in the vacuum tank is solved with an E-shaped configuration of the vacuum system. With these techniques and the backward-running detonation driver, a large hypervelocity detonation-driven shock tunnel has been designed and calibrated successfully. This shock tunnel, as shown in Fig. 10, is about 265 m in total length, and the diameter of the nozzle exit is 2.5 m [27].

Fig. 10 Photo of long-test-duration hypervelocity detonation-driven shock tunnel

3.1 Descriptions of the JF12 Shock Tunnel

Under the support of National Major Project of Scientific Instrumentation R&D, the Long-test-duration Hypervelocity Detonation-driven Shock Tunnel (LHD^{ST}) was developed based on the backward-running detonation driver with several innovative techniques by the end of May 2012, as shown by the photo in Fig. 10. The shock tunnel, JF12, is named after shock tunnels under the serial number in LHD laboratory. Professor Sasoh, from Nagoya University, suggested an endearment name of the hypervelocity detonation-driven real gas shock tunnel (Hyper-Dragon for short) to this facility after his visit. Its performance tests demonstrated that the facility is capable of reproducing the pure airflow with Mach numbers from 5 to 9 at altitude of 25–50 km with 100 ms test duration, and will be a useful tool for testing engine/frame-integrated hypersonic vehicles and investigating into fundamental physical issues in hypersonic and high-temperature gas dynamics.

Figure 11 shows schematically the entire system of the JF12 shock tunnel. From its right to left, the first part is the vacuum system for damping wave-reflection during nozzle starting process, and its vacuum tank with a volume of 600 m^3 is 50 m in length. The second part is the test section that is 15 m in length and 3.5 m in diameter. The contoured nozzle is 15 m in length and 2.5 m in diameter, and another nozzle is smaller, having an exit of 1.5 m in diameter for Mach number of 5–7. Next to the nozzle, there is the driven section being 89 m in length and 720 mm in diameter. The detonation driver is 99 m in length and 400 mm in diameter. The driver operates in the backward-running detonation mode, that is, the

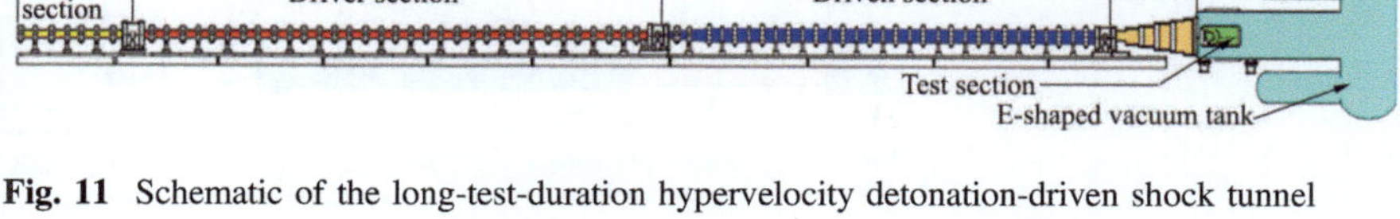

Fig. 11 Schematic of the long-test-duration hypervelocity detonation-driven shock tunnel

detonation is ignited at its right end. The detonation driver and the driven section are connected with the transient part by which the tube diameter is gradually reduced from 720 to 400 mm. Between the detonation driver and the transient part, there is the diaphragm rig that is used to produce the incident shock wave in the shock tunnel after the direct detonation initiation. The damping section is located at the far left end of the facility, and is 19 m in length and 400 mm in diameter.

For experimental measurements, the JF12 shock tunnel is equipped with a 384-channel digital data acquisition system that is able to acquire, amplify, digitize and store the data in real time. Immediately after each test run, the data are transferred to the computer in the control room for processing. Three 6-component force and moment balances are designed based on the 100 ms test duration and calibration experimental data showed that the balances work pretty well.

3.2 New Techniques Developed for the JF12 Shock Tunnel

A typical wave diagram for the detonation-driven shock tunnel operated in the backward-running detonation mode is shown in Fig. 12. The detonation is initiated between the detonation driver and the driven section, and its wave graph is schematically presented in Fig. 12. The nozzle starts once the incident shock is reflected from the reservoir end and the reservoir state is interfered when the reflected wave from the left end of the shock tunnel arrives. When building up a long test-duration and hypervelocity shock tunnel for Mach numbers of 5–9, there are several physical issues that affect the test time in different ways. These physical issues are briefly described as follows. The first issue is the powerful detonation driver, and the shock tunnel performance will benefit from the detonation driver with high pressure and required temperature. The second issue is the tailored interface operation condition and the wave reflection from the interface separating the driver gas from the test gas will ruin the test flow quality in an untailored operation condition. The third issue is the precursor wave reflection in the vacuum tank during the nozzle-working process, and the test duration will be terminated once reflected waves propagate into the test flow region. A large amount of laboratory space may

Fig. 12 Physical issues for developing long test-duration shock tunnel

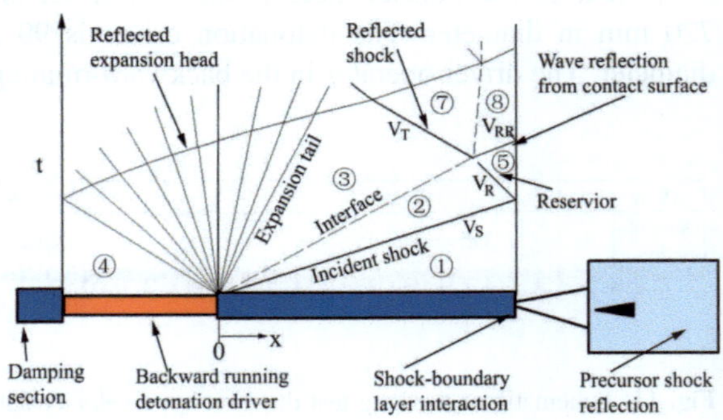

be occupied since the vacuum tank must be long enough to accommodate the wave propagation. Shock/boundary interaction is another key issue for shock tunnel test time. Immediately after the incident shock reflection, the driver gas moves within the boundary layer near the end of the driven section; leading to driver/driven gas mixing and result in test gas contamination. The last issue arises from the dumping section. The extremely high pressure from detonation front reflections must be carefully concerned in this section. These physical issues will be discussed in the following sections, and the corresponding techniques for constructing shock tunnels having long test-duration will also be discussed.

3.2.1 Backward-Running Detonation Driver

The backward-running detonation driver is well known to be good at producing a column of driver gas with uniform pressure regardless of the fact that its capability for incident shock generation is less powerful than the forward-running detonation driver. However, it is fully capable of producing hypersonic flows with Mach number from 5 to 10 because the required incident shock is not very strong. Furthermore, the total temperature of the test gas is about 1500–2300 K for the flows within a Mach number range of 5–7, and detonation drivers are usually too powerful to apply directly. Because the initial temperature of the test gas in front of the incident shock is the room temperature, the low temperature driver gas after detonations is required for the tailored interface operation that matches the sound speed of the driver gas with the test gas. Thus interface matching is extremely important to eliminate wave reflections from the contact face located between the driver gas and the test gas. On the other hand, the detonated driver gas must have high enough pressure to produce a strong incident shock, but not too high temperature for properly tailoring the interface condition. This is the key issue for developing a shock tunnel with long test time.

According to detonation physics, different detonable gas mixtures have different detonation temperature and pressure. The high detonation pressure driver gas with low temperature can be achieved by careful choice of the gas mixtures. Numerous gas mixtures were tested to examine the performance of a backward-running detonation driver by using the JF10 detonation-driven shock tunnel. Two results from the experimental data obtained are shown in Fig. 13, showing pressure variations during tunnel operation duration. Figure 13a shows P_5 of the hydrogen/oxygen gas mixture at the initial pressure of 3.5 MPa for the first case, and thus the generated incident shock has the Mach number of 2.5. Figure 13b presents the data of the acetylene/oxygen gas mixture at the initial pressure of 3.0 MPa for the second case, and the obtained incident shock Mach number is 2.47. It is observable that the pressure history shown in Fig. 13b reaches about 800 bars even though the P_1 is lower than the one in the second test case, and the P_5 value is 50 % higher than the one in Fig. 13a. Therefore, the backward-running detonation driver with the acetylene/oxygen gas mixture is suitable for generating hypersonic flows at the low Mach number region, that is, Mach number of 5–7, where the total temperature is not very high.

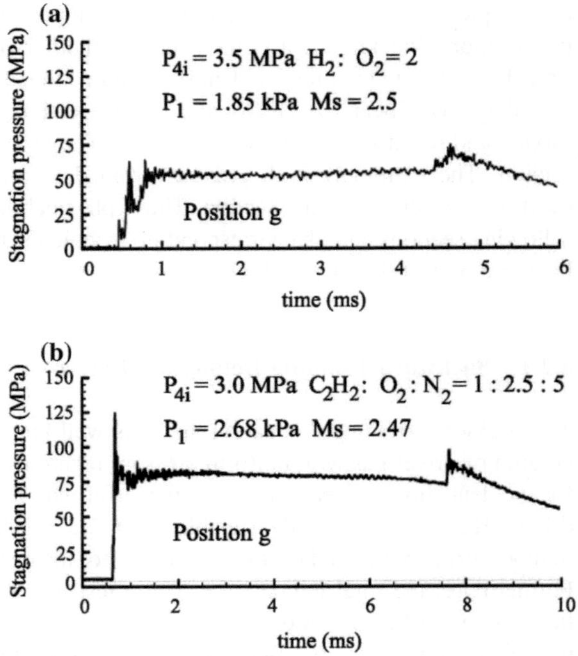

Fig. 13 Pressure variations obtained with a backward-running detonation driver at the same position for two test cases: **a** hydrogen/oxygen gas mixture; **b** acetylene/oxygen gas mixture

3.2.2 Tailored Interface Operation Condition

The second issue is the wave reflection from the contact surface that separates the driver gas from the driven gas. If the reflection takes place, the total pressure and temperature of the test gas at the end of the driven section will be changed during the nozzle operating, resulting in the early termination of shock tunnel test. In order to avoid the wave reflection, the sound speed of the test gas in front of the incident shock must meet a certain requirement for the sound speed of the detonation gas behind the contact interface. For the backward-running detonation drivers, the tailored interface condition is derived as follows:

$$\frac{a_4}{a_1} = \frac{2}{\gamma_1 + 1}\left(M_s - \frac{1}{M_s}\right)$$
$$\left\{\left[\frac{(\gamma_1 - 1)M_s^2 + 2}{2\gamma_1\left(M_s^2 - 1\right)}\right]\left[\gamma_4^2 + \frac{\gamma_1\gamma_4(\gamma_4 + 1)\left(M_s^2 - 1\right)}{(\gamma_1 + 1)M_s^2 + 2}\right]^{1/2} + \frac{\gamma_4 - 1}{2}\right\}$$

where a_4 is the sound speed of the detonated driver gas, γ_4 is its specific heat ratio, M_s is the incident shock Mach number, a_1 is the sound speed of the test gas behind the incident shock, and γ_1 is its specific heat ratio.

Based on the tailored-interface operation condition, calculated results are presented in Fig. 14, showing the relation between the sound speed ratio a_4/a_1 and the

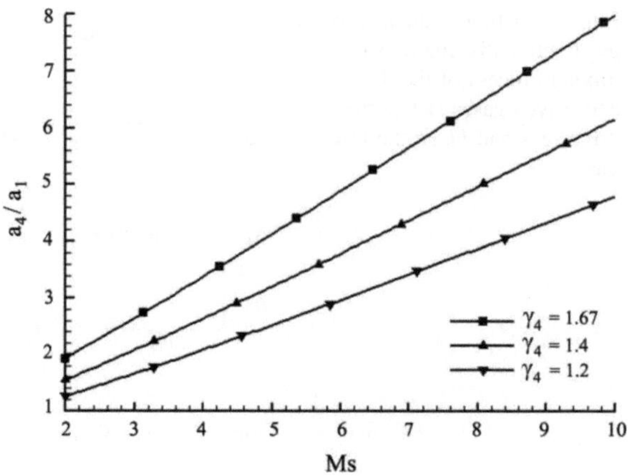

Fig. 14 Sound speed ratio a_4/a_1 varying with the incident shock March number M_S under the tailored-interface condition for a backward-running detonation driver with different detonable gas mixtures

incident March number M_S. For most of the experiments, air is used as a test gas at ambient temperature, and the sound speed ratio can be adjusted only by selecting different detonable gas mixtures. For the given incident shock, the larger the specific heat ratio is, the higher is the sound speed ratio. The acetylene/oxygen gas mixture is suitable for the low Mach number of hypersonic flows, and the hydrogen/oxygen gas mixture is for the high Mach number. There is another way to adjust the sound speed ratio by applying a diverging section between the detonation driver and the driven section, as mentioned by Laderman [29], and with the diverging section the detonation driver can be extended to the low Mach number region of hypersonic flows since the driver operation is limited by the direct detonation ignition.

3.2.3 Gas Contamination from the Shock/Boundary Interaction

The third issue is the shock/boundary interaction near the end of the driven section. The boundary layer will develop while the incident shock propagates in the driven section, and the shock/boundary interaction will take place after the incident shock reflects. The phenomenon can cause contamination of the test gas and greatly affect the test-duration of shock tunnels. Numerical simulations are carried out for examining the problem and the obtained results are presented in Fig. 15. Figure 15 shows a time sequence of gas fraction distributions in which the white region stands for the driver gas while black stands for the test gas. From the figure, it is observable that the driver gas goes along the driven section wall, enters the nozzle throat much earlier than what is expected, and leaves a relatively large percentage of the test gas behind. The test gas contamination is a severe problem for shock tunnels and the test duration could be reduced significantly. To prevent fast entering of the driver gas into the nozzle throat, a steel ring is suggested to be mounted around the nozzle throat. Numerical simulations demonstrated that the test duration is increased by 50 % due to the steel ring installation.

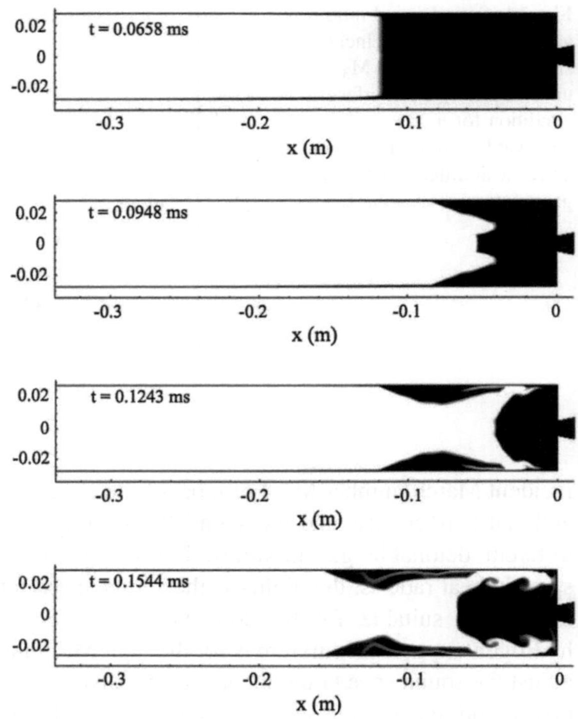

Fig. 15 A time sequence of gas fraction distributions from mixing process of the driver and driven gases (*white* the driver gas and *black* the test gas)

3.2.4 Wave-Reflection Damping in the Vacuum Tank

The last issue to be discussed here is the reflection-wave damping in the vacuum tank. For shock tunnels, its test duration is usually proportional to its length. This is the reason why the test time for shock tunnels is short, only a few milliseconds. If one wants to develop a shock tunnel having a hundred millisecond test duration, the total length of the shock tunnel will be extremely long because the wave speed is very high. Reduction of the shock tunnel length and the associated laboratory space is a meaningful work to do. A number of different configurations of the vacuum system were numerically simulated, and the results from two configurations are presented in Fig. 16. The upper half of the figure shows the pressure distribution in the E-shaped vacuum tank and the lower half depicts the T-shaped one. The overall volume and the length for the two test cases are set to be approximately same in the calculations. It is demonstrated that the E-shaped vacuum system has the better performance than the T-shaped configuration. Therefore, the JF12 shock tunnel is equipped with an E-shaped vacuum tank with about 600 m^3 in volume. The performance test shows that more than 200 ms test-duration was achieved for M = 7 test flow with the E-shaped vacuum tank being 50 m in length.

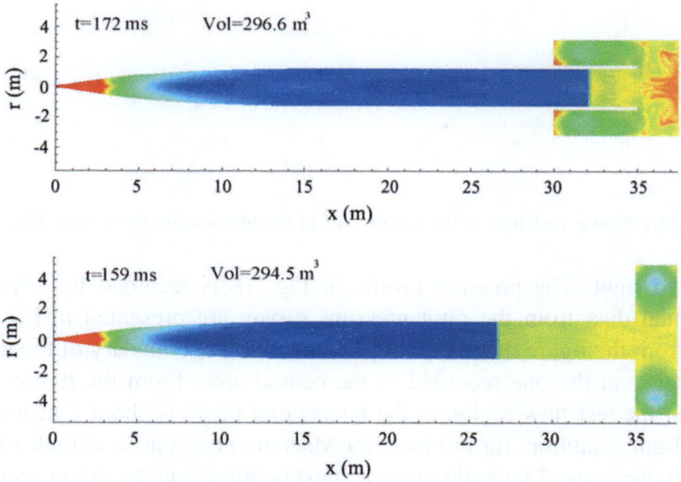

Fig. 16 Wave propagation and reflection in the E-shaped and T-shaped vacuum vacuums

3.3 The Results of Performance Tests

Using the above mentioned techniques, the project for the long-test-duration hypervelocity detonation-driven shock tunnel was initiated at the beginning of 2008, its installation was completed in March 2011, and the performance tests have been carried out since then. Some experimental data are presented in this chapter to show promising features of the JF12 shock tunnel.

Figure 17 presents reservoir pressure variations at the end of the driven section for a nozzle Mach number 7, and actually it is the P_5 pressure for the shock tunnel. The incident Mach number is measured to be 4.57. The total temperature of the test gas is about 2468 K, and the uniform pressure maintains as long as 130 ms. Figure 18 shows the pitot pressure variation at the nozzle exit in the test section for the same shot. The pitot pressure profile looks similar to the P_5 curve, and the test duration is confirmed.

The survey rake is made for the pitot pressure measurement, and is 2 m long in both vertical and horizontal directions. Forty pressure sensors are distributed along the rake in an equal interval. The sensors along the horizontal branch are numbered

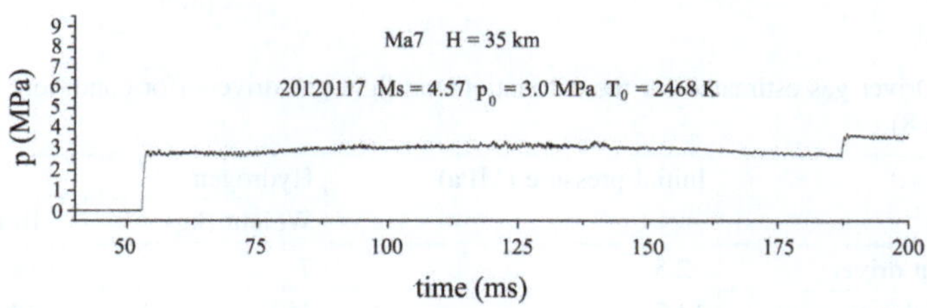

Fig. 17 Reservoir pressure (P_5) variation at the end of the driven section for nozzle Mach number 7

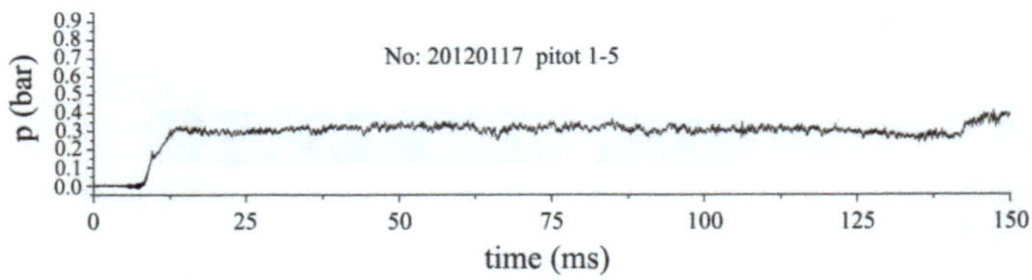

Fig. 18 Pitot pressure variation at the nozzle exit in the test section for nozzle Mach number 7

from left to right. The pressure profile in Fig. 18 is recorded by sensor No. 5. Experimental data from the pitot pressure survey are presented in Fig. 19. It is apparent from the figure that the pitot pressure near the boundary of the test flow is a bit higher than the one recorded in the central area. From the figure, it is concluded that the test flow region at the nozzle exit plane is about 2 m in diameter.

If the flight condition, for instance, for Mach number 8 at an altitude of 40 km is required to duplicate, 7 kg hydrogen gas must be filled into the driver section of the JF12 shock tunnel to reach the initial pressure of 2.5 MPa for the hydrogen-oxygen gas mixture. The hydrogen consumed for such a shot is about 18 bottles of compressed gas. If the light-gas driver is accepted for the JF12 shock tunnel, one shot will need about 380 bottles of hydrogen for the driver gas to reach an initial pressure of 14.2 MPa. The consumed driver gases estimated for the two drivers are given in Table 2. The heated-light gas driver is an important technique for shock tunnels, however, it may cause some problems for larger scale shock tunnels, not only the operation expanses become high, but heating, storage and transportation for a huge volume of light gases are very tough problems to deal with.

Fig. 19 Pitot pressure profile at the nozzle exit plane for Mach number 7

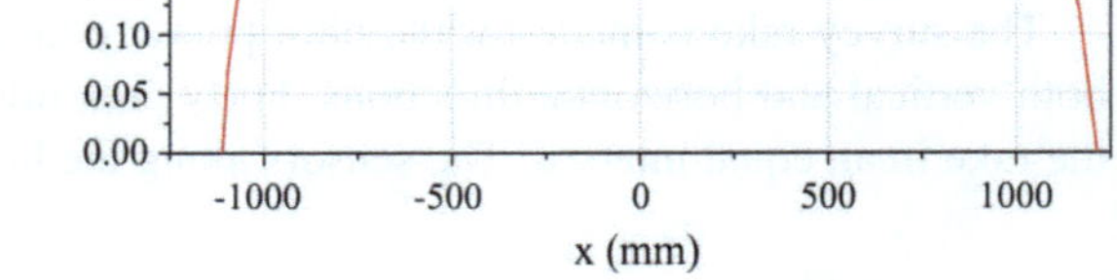

Table 2 Driver gas estimated for the detonation and light-gas drivers (for condition: H = 40 km and $M_\infty = 8$)

	Initial pressure (MPa)	Hydrogen	
		Weight (kg)	Bottles
Detonation driver	2.5	7	18
Light gas driver	14.2	152	380

Hydrogen bottle: $V = 0.04\ m^3$; $p = 12$ MPa; $m = 0.4$ kg

3.4 Concluding Remarks

The successful development of the long-test-duration hypervelocity detonation-driven shock tunnel was demonstrated with calibration tests. The shock tunnel performance was achieved due to the above discussed techniques, such as the detonation driver with the acetylene/oxygen mixture, the tailored-interface operation condition for detonation drivers, the shock/boundary interaction control at the end of the driven section, and the E-shaped configuration of the vacuum tank. The shock tunnel is capable of duplicating the flight conditions for Mach number 5–9 at the altitude of 25–50 km. Large scale test flows with more than 100 ms test duration can be generated at a reasonable operation expense. Hypersonic boundary layer development, scaled-model effects, hypersonic propulsion, vehicle-engine integrated models are possible to be tested with the new facility. It will be a useful tool for investigating into fundamental physics in hypersonic propulsion.

4 Detonation-Driven High-Enthalpy Expansion Tube JF16

Among the developed hypersonic test facilities, the high-enthalpy expansion tube appears to be a promising facility for generating near orbital speed flows for earth re-entry study. In the expansion tube, the unsteady flow expansion mechanism is applied to further speed up test gases to achieve high flow velocity, lower static temperature and less gas dissociation as compared with reflected shock tunnels. The concept of an expansion tube was proposed first by Resler and Bloxsom in 1952 [30]. Theoretical research was carried out later by Trimpi [31], but his experimental results showed that steady test flows were not achieved as was predicted. The Center for Hypersonics at the University of Queensland applied the free-piston driver to the expansion tube, and three expansion tubes, referred to as X1, X2, X3, were set up and a series of orbital speed experiments were carried out [32]. The JX-1 expansion tube with free-piston driver was built later by Sasoh at Tohoku University, Japan [33]. Based on the above introduced FDC driver [21, 23], the detonation-driven orbital-speed expansion tube, referred to as the JF16 expansion tube, was set up in 2006. The experimental data show that the velocity of the test gas reaches 8300 m/s. The test duration is around 50–100 μs and the total flow enthalpy is about 40 MJ/kg [25].

4.1 Descriptions of JF16 Expansion Tube

The JF16 expansion tube is shown schematically in Fig. 20. It consists of five main parts: a FDC driver, a shock tube, an expansion acceleration section, a test section

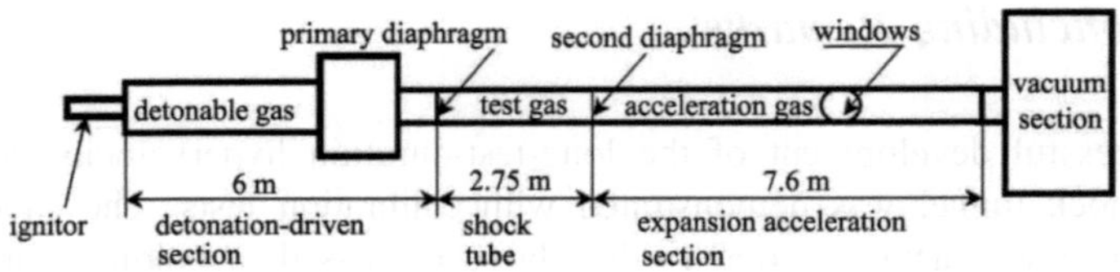

Fig. 20 Schematic of the JF16 expansion tube

and a vacuum section. This FDC driver is 6 m in length and 150 mm in inner diameter with an igniter installed at its left end to initiate a forward-running detonation. The driven shock tube is 2.75 m in length and 68 mm in inner diameter, and it can be replaced with a longer one of 4.75 m for different experimental requirements. The primary diaphragm is located between the FDC driver and the shock tube, separating the detonable gas mixture from the test gas. The expansion acceleration section is 7.6 m long with a square cross section of 60 mm × 60 mm. The second diaphragm is placed between the shock tube and the expansion acceleration section, separating the test gas from the acceleration gas. A set of test windows opposing each other is installed in the expansion acceleration section for flow visualization. A vacuum section is added to the right end of the facility to weaken wave reflections for minimizing their effects on the generated hypervelocity flows.

Surface pressure measurements were carried out along the shock tube and the expansion acceleration section. By setting the position of the second diaphragm as a starting point, the distances to each pressure transducer are set to be 1612, 2447, 3247 and 3847 mm, which are named as B, C, D, and E, respectively in the order from left to right. Ten surface ion probes were distributed along the shock tube and the expansion acceleration section with an equal interval to obtain the incident shock speed. The detonable gas applied is a hydrogen/oxygen mixture at 4:1 ratio. The igniter is a small diameter tube to provide a certain amount of hot gas to the FDC driver to ensure a direct initiation, and the gas mixture in the igniter is initiated with explosion wires.

4.2 Test Flow Evaluation

In the performance tests, the shock tube and the expansion acceleration section were filled with air at 20 and 0.15 mm Hg, respectively, as a basic test case. For special cases, the condition changes will be explained explicitly if there is any difference in the initial conditions. The FDC driver was filled with the H_2/O_2 mixture at an initial pressure of 1.0 and 1.5 MPa, respectively for the first two test cases. The wall pressure variations related to the test flow quality are presented in Figs. 21 and 22 for discussion.

Figure 21 shows the pressure histories measured at four measurement points of the first test case where the initial pressure in the FDC driver is set to be 1.0 MPa.

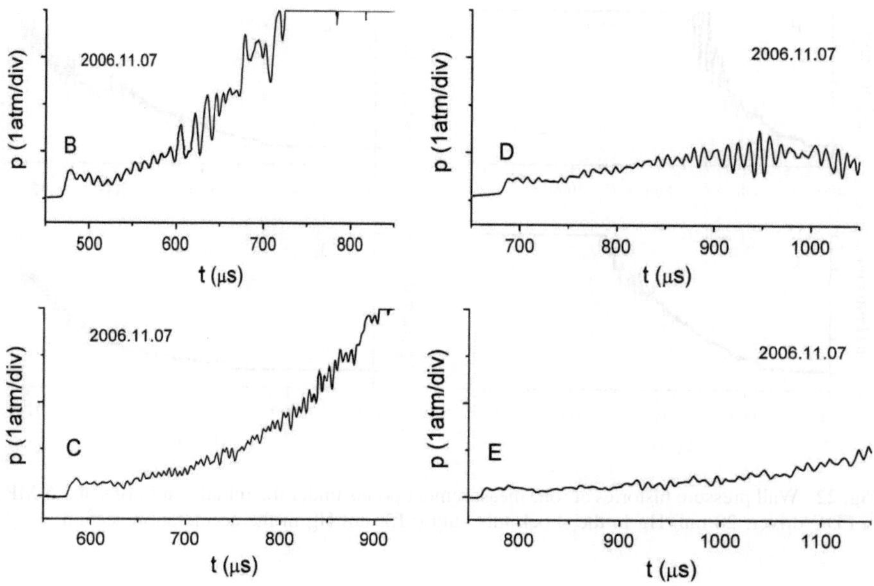

Fig. 21 Wall pressure histories at four measurement points under the initial conditions of 1.0 MPa in FDC driver, 20 mm Hg in the shock tube and 0.15 mm Hg in the acceleration section

It is observable that the pressure plateau appears behind the sharp jump induced by the incident shock at each measurement point. There is a pressure overshot at point B, but the overshot disappears at points C and D. This implies that uniform test flow is achieved from the position marked with C and D, and the test duration is measured to be about 50 μs.

Figure 22 shows pressure histories measured at four measurement points of the second test case where the initial pressure in the FDC driver is set to be 1.5 MPa, being higher than that used in the first case, but all other initial conditions are kept the same. It is observed that the pressure plateau appears at measurement point E. This implies that the uniform test flow is achieved from this position; the test duration is about 70 μs. Actually, it is not sufficient to determine the test time based only on the pressure plateau because there is no pressure change across the contact surface that separates the test gas from the driver gas. In the section on flow visualization, a series of conical shock photographs will be presented to support our test time determination because the temperature will vary across the contact surface.

By comparing Fig. 21 with Fig. 22, we find that the pressure plateau appears later in Case 2. This means that the stronger flow expansion leads to a longer development time of the test flow in the acceleration section. Therefore, high quality test flows can be obtained at different positions for different initial test conditions. According to the observed phenomena, the test section should be movable along the expansion section, so that experiments could be carried out at more uniform test flows of higher quality. Actually, the test flow needs space to

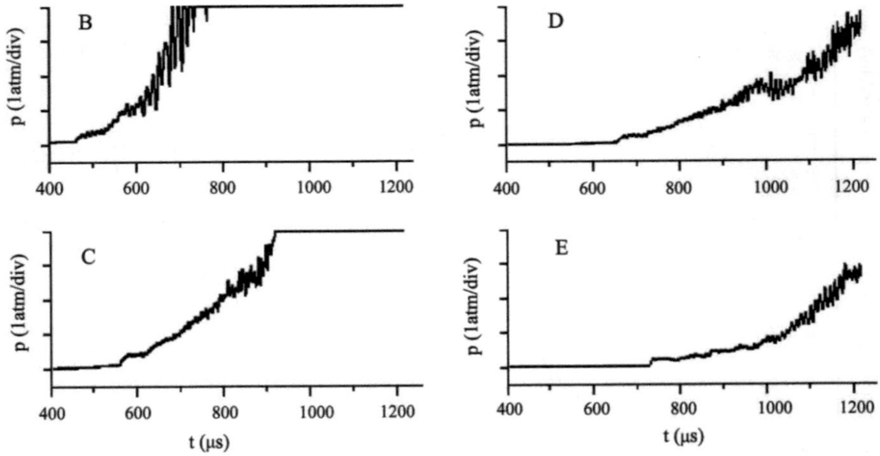

Fig. 22 Wall pressure histories at four measurement points under the initial conditions of 1.5 MPa in FDC driver, 20 mm Hg in the shock tube and 0.15 mm Hg in the acceleration section

develop, the higher the incident shock Mach number is, the longer is the flow expansion distance.

By using the measured incident shock speeds, some key parameters behind the incident shock during the flow development can be calculated with Mirels' theory. The results at 2 and 4 m positions starting from the second diaphragm along the expansion tube are presented in Table 3. In this table, u_s is the shock speed, u_c is the flow velocity, l_m and l are the maximum and measured distance between the incident shock and the interface behind the shock, respectively. From this table, it can be seen that the flow velocity at 2 m reaches 7190 m/s for Case 1, and 8336 m/s for Case 2. At the 4 m position, the flow velocity is a little lower, but still reached 6971 m/s for Case 1 and 8108 m/s for Case 2. This result implies that the incident shock will decay along the expansion tube due to viscous attenuation.

Table 3 Parameters during the flow development evaluated along the expansion tube at distance of 2 and 4 m from the second diaphragm along the expansion tube

	2 m		4 m	
	Case1	Case2	Case1	Case2
u_s (m/s)	7512	8472	7168	8163
l/l_m	0.51	0.83	0.81	0.99
l_m (mm)	89	25.3	90	25.3
l (mm)	45	21	73	25
u_c/u_s	0.957	0.964	0.973	0.973
u_c (m/s)	7190	8336	6971	8108

4.3 Other Parameters for Expansion Tubes

There are several important parameters in the expansion tube techniques that influence the performance of the JF16 expansion tube. These parameters were investigated, and some results are presented as follows:

The initial pressure in the shock tube was found to be a key parameter and plays an important role in flow stability. Setting the initial pressure in the FDC driver to be 1.0 MPa, several experiments were carried out by setting the shock tube pressure to 10 , 20 and 50 mm Hg. The pressure variations measured at the same position along the shock tube are presented in Fig. 23. It can be seen that the higher the shock tube pressure, the stronger is the flow pressure fluctuations. The pressure ratio across the incident shock wave changes as the shock tube initial pressure varies because of the constant pressure in the FDC driver. This means that the sound speed ratio (the driver gas to the test gas) will change across the incident shock wave, for example, the sound speed ratio is 0.826 in the case shown in Fig. 23a, 0.877 in Fig. 23b, and 1.086 in Fig. 23c, respectively. It is believed that the flow perturbations are generated

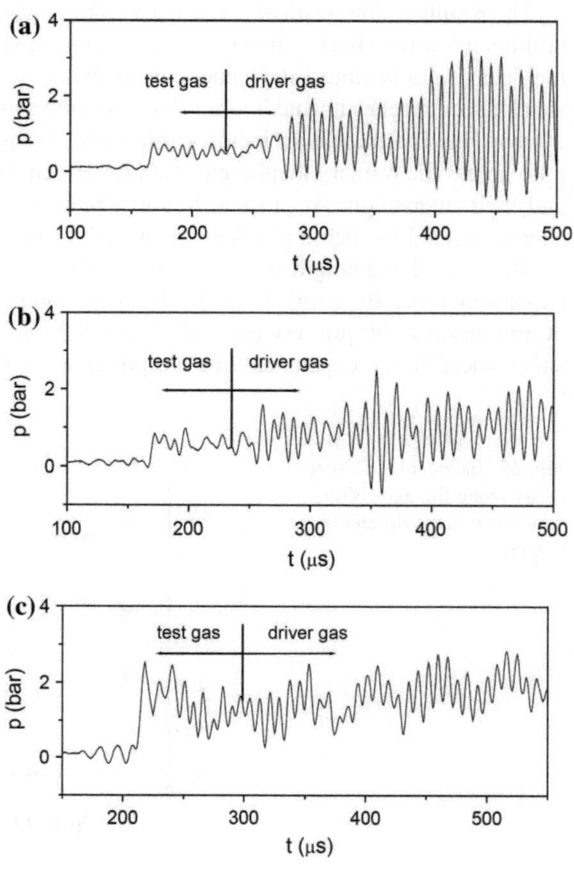

Fig. 23 Pressure history profiles measured in the shock tube at different shock tube pressures

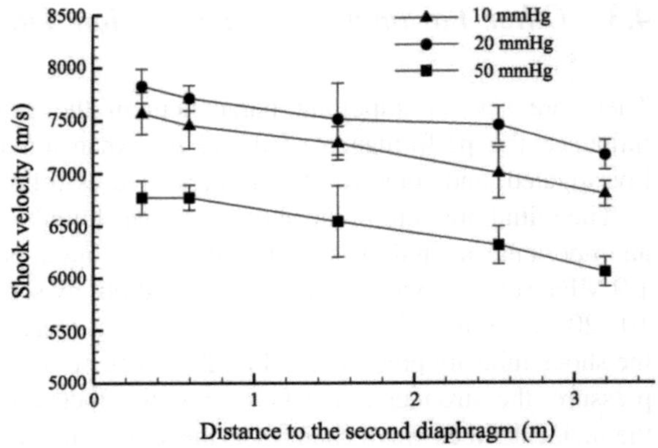

Fig. 24 Incident shock decay along the expansion acceleration tube at different pressures of the shock tube

mainly in the driver section and the higher sound speed ratio will lead to stronger perturbations in the driver gas. These perturbations could propagate rapidly into the test gas. Therefore, having the sound speed ratio being less than unity is a good choice for achieving high quality test flows.

The results of the incident shock decay along the expansion acceleration section at different initial shock tube pressures are presented in Fig. 24. It is observed that the decay ratio is almost the same, approximately 3 % per meter, and the incident shock speed reaches its maximum value when the shock tube pressure is taken to be 20 mm Hg. The maximum incident shock speed actually results from the pressure ratio across the primary diaphragm, the flow expansion after the second diaphragm, and their interaction. At present, this mechanism is not clear and more attention should be paid to such a problem when tuning an expansion tube.

The second diaphragm is found to be another important parameter for tuning an expansion tube. By using three kinds of diaphragms with different thicknesses: 90 μm terylene, 60 μm terylene and 25 μm cellophane. The results of the incident shock speed in the expansion tube are presented in Fig. 25. It is observed that the

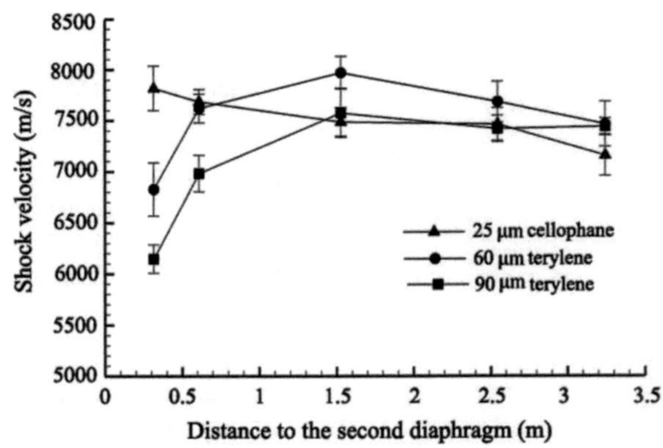

Fig. 25 Incident shock speed decay along the expansion tube when using different diaphragms

expansion tube performance benefits from the 25 μm cellophane. Thick diaphragms can lead to the shock speed decay upon the diaphragm rupture and flow re-acceleration later, as shown in Fig. 25.

4.4 Test Duration from Flow Visualization

In order to demonstrate the test flow stability and the effective test time, the flow visualization was carried out using a schlieren system and a high-speed camera. Photos of the conical shock in front of a sharp cone are presented in time order in Fig. 26.

In order to obtain the conical shock photographs, a steel 15° sharp cone made of steel is mounted in the test section. The test flow is generated at the condition being the same to Case 2, and the flow velocity is measured to be about 8100 m/s. A FASTCAM SA4 camera is used to for photographing from the test section, while exposure time of 1 μs is used, and the time interval between successive frames is 25 μs. Figure 26a (left hand top figure) shows the photograph taken 25 μs after the incident shock passes. By examining the conical shock angle from Fig. 26b–d, it can be seen that the shock configuration is the same. The flow field in Fig. 26a appears to be light blue, which shows the characteristics of the low-temperature gas filled initially into the expansion acceleration section. The flow field in Fig. 26e appears to be light yellow, which shows the characteristics of the driver gas. This fact indicates that the test flow thus generated is uniform since the shock angle keeps constant from Fig. 26b–d, and the test duration is not shorter than 50 μs, but not longer than 100 μs. Test flow quality is well demonstrated by these flow visualizations. Furthermore, the conical shock configuration is obtained at the flow velocity of 8100 m/s and the total flow enthalpy is about 39 MJ/kg according to the equilibrium theory. Therefore, such photographs are useful for CFD validation since there are no reports of such data at the near orbital speeds.

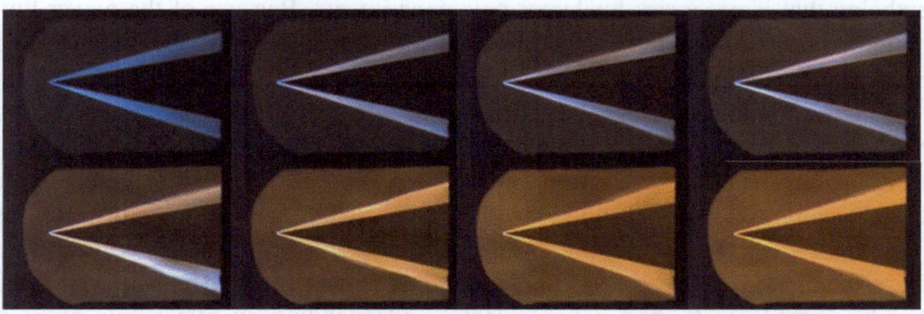

Fig. 26 Sequential photos of conical shocks in front of a 15° sharp cone placed in 8100 m/s hypersonic flow (the time interval is 25 μs between successive frames and the exposure time is 1 μs)

4.5 Concluding Remarks

The detonation-driven high-enthalpy expansion tube was developed successfully by combining up-to-date techniques of the detonation driver with an expansion acceleration section. The hypervelocity test flow of over 8 km/s and a total enthalpy up to 40 MJ/kg is generated in this facility for conducting experiments at the near orbital speed. From its performance tests, the hypervelocity flow was demonstrated to be relatively uniform and this indicates that the high quality test flow could be achieved by using the FDC driver. Experimental results also demonstrated that the ratio of the sound speeds across the driving and test gas interface has a critical impact on the magnitude of test flow perturbations, but the perturbations can be suppressed to a certain extent if the ratio does not exceed unity. The maximum velocity of the test flow can be obtained by adjusting the filling pressure of the shock tube while maintaining constant filling pressures in both the detonation driver and the expansion acceleration section. This implies that the optimization of test conditions is possible for the performance of the JF16 expansion tube in addition to the facility configuration optimization. Experiments also showed that the second diaphragm affects both the gas dynamics and the chemical kinetics of the test gas and the JF16 expansion tube performance benefits from light and thin diaphragms.

5 Summary

The first enclosed wind tunnel was invented in 1871 to measure the drag over a solid body by Wenham, a Council Member of the Aeronautical Society of Great Britain. Many wind tunnels for various purposes have been built in the world since then. Nowadays, to develop innovative flying vehicles, the wind tunnels become larger and larger in size, and require more and more high level of power for tunnel operation. For hypersonic flow generation, the power needed is so huge that the blow-down wind tunnels could not be employed to replicate the correct velocity or enthalpy that is necessary for hypersonic experiments. Because of the power limitation on continuous wind tunnels, shock tunnels were developed for ground-based hypersonic experiments. As the development of numerous advanced measurement techniques continues, the application of hypersonic shock tunnels is expected to become wider and wider. Detonation-driven high-enthalpy and hypersonic test facilities have been developed in the State Key Laboratory of High Temperature Gas Dynamics (LHD), Institute of Mechanics, Chinese Academy of Sciences (CAS), for years. The typical trajectories of various spacecrafts can be covered with these facilities and research on aero-thermo-dynamics and hypersonic propulsion can be carried out to gain a fundamental understanding on hypersonic physics. More especial is the long-test-duration hypervelocity detonation-driven shock tunnel (JF12) that is capable of duplicating the flight condition for Mach number 5–9 at an altitude of 25–50 km. The hypervelocity flow region of more than two meters in diameter can be generated for test duration more than 100 ms at reasonable operation cost.

The detonation is the violent combustion and its rapid energy conversion in explosives is the property which is very useful. A 20 m^2 detonation front operates at a power level equal to the total power the earth receives from the sun. The FDC driver and the backward-running detonation driver combining with a dump section make use of the detonation successfully and provide shock tunnels with the required power level at affordable expense. Techniques proposed for extending shock tunnel test duration are of significant importance because the long test duration cannot be achieved only by increasing shock tunnel length. Detonation-driven high-enthalpy and hypersonic test facilities so far developed and based on the present shock tunnel techniques demonstrated its several good properties, such as the stable reservoir pressure, the required enthalpy level, the long test duration and the cheap operation cost. These properties are very good for generating hypervelocity flows for ground tests. However, the heat transfer from detonated gases to facilities and the facility damage due to the harsh thermo-environment occur in the detonation-driven facilities, being even more severe. For advanced hypersonic test facilities, the detonation-driven shock tunnels are very promising, but their improvement both in its design and operation is necessary in future.

Acknowledgments The author would like to acknowledge the National Natural Science Foundation of China which provides funding for this research under the projects of 10632090 and 10621202. The authors would like to thank all the members from the group of shock wave and detonation physics for their kind assistance in experiments and facility development.

References

1. Bertin, J.J.: Hypersonic Aerothermodynamics. AIAA Education Series, Washington DC (1994)
2. Park, C.: Nonequilibrium Hypersonic Aerothermodynamics. Wiley, New York (1990)
3. Anderson, J.D.: Hypersonic and High Temperature Gas Dynamics. McGraw-Hill Book Company, New York (1989)
4. Bertin, J.J., Cummings, R.M.: Fifty years of hypersonics: Where we've been, where we're going. Prog. Aerosp. Sci. **39**, 511–536 (2003)
5. Hirschel, E.H.: Viscous effects. In: Space Course 1991, vol. 12, pp. 1–35, Aachen
6. Bertin, J.J., Cummings, R.M.: Critical hypersonic aerothermodynamic phenomena. Ann. Rev. Fluid Mech. **38**, 129–157 (2006)
7. Lu, F.K., Marren, D.E.: Advanced Hypersonic Test Facility (198). Progress in Astronautics and Aeronautics, Reston (2002)
8. Holden, M.S.: Design, development and calibration of the LENS facility (pp. 94–0161). AFOSR-TR (1994)
9. Holden, M.S., Wadhams, T.P., Candler, G.V.: A review of experimental studies in the LENS shock tunnel and expansion tunnel to examine real-gas effects in hypervelocity flows. In: 42th AIAA Aerospace Sciences Meeting and Exhibit Reno, AIAA Paper 2004-0916, NV, 5-8 Jan 2004

10. Holden, M.S., Chadwick, K.M., Kolly, J.M.: Hypersonic studies in the LENS facilities, AIAA 95-6040. In: 6th International Aerospace Planes and Hypersonic Technologies Conferences, Chattanooga, Tennessee, 3–7 April 1995
11. Stalker, R.J.: A study of the free-piston shock tunnel. AIAA J. **5**, 2160–2165 (1967)
12. Stalker, R.J., Morrison, W.R.B.: New generation of free piston shock tunnel facilities. In: Proceeding of the 17th International Symposium on Shock Tube and Waves, Bethlehem (1989)
13. Hornung, H.G.: The piston motion in a free-piston driver for shock tubes and tunnels, GALCIT Rep. FM88-1, 1988
14. Itoh, K., Ueda, S., Komuro, T., et al.: Improvement of a free piston driver for a high-enthalpy shock tunnel. Shock Waves **8**, 215–233 (1998)
15. Itoh, K., Ueda, S., Tanno, H., Komuro, T., Sato, K.: Hypersonic aerothermodynamic and scramjet research using high enthalpy shock tunnel. Shock Waves **12**, 93–98 (2002)
16. Bird, G.A.: A note on combustion driven tubes. Royal Aircraft Establishment, vol. 146, AGARD Report, May 1957
17. Yu, H.: Recent developments in detonation drivers. In: Proceedings of the Fifth National Symposium on Shock Waves and Shock tubes, Xi'an, China, Nov 1988
18. Yu, H., Esser, B., Lenartz, M., Grönig, H.: Gaseous detonation driver for a shock tunnel. Shock Waves **2**, 245–254 (1992)
19. Olivier, H., Jiang, Z., Yu, H., Lu, F.K.: Detonation driven shock tubes and tunnels Advanced Hypersonic Test Facilities, pp. 135–203. AIAA, USA (2002)
20. Jiang, Z., Yu, H., Takayama, K.: Investigation into converging gaseous detonation drivers. In: Ball, G.J., Hillier, R., Roberts, G.T. (eds.) Proceedings of ISSW22, Shock Waves, London, UK, 18–23 July 1999
21. Jiang, Z., Zhao, W., Wang, C., Takayama, K.: Forward-running detonation drivers for high-enthalpy shock tunnels. AIAA J. **40**, 2009–2016 (2002)
22. Zhao, W., Jiang, Z., Saito, T., Lin, J.M., Yu, H.R., Takayama, K.: Performance of a detonation driven shock tunnel. Shock waves **14**(1–2), 53–59 (2005)
23. Jiang, Z., Lin, J., Zhao, W.: Performance tests of JF-10 high-enthalpy shock tunnel with a FDC driver. J. Hypersonics **2**(1), 29–35 (2011)
24. Jiang, Z., Yu, H.: Experiments and development of long-test-duration hypervelocity detonation-driven shock tunnel (LHDst). AIAA paper 2014-1012 (2014)
25. Jiang, Z., Gao, Y., Zhao, W.: Performance study on detonation-driven expansion tube. AIAA paper 2009-7237 (2009)
26. Li, J.P., Feng, H., Jiang, Z., Yu, H.R.: Numerical computation on the tailored shock Mach numbers for a hydrogen/oxygen detonation shock tube. Acta Aerodynamica Sinica **26**, 291–296 (2008)
27. Li, J.P., Feng, H., Jiang, Z.: Investigation into test gas contamination of shock tunnels induced by shock/boundary interactions. Acta Mechanica Sinica **40**, 289–296 (2008)
28. Jiang, Z., Li, J.P., Zhao, W., Liu, Y.F., Yu, H.R.: Investigation into techniques for extending the test-duration of detonation-driven shock tunnel. Chinese Journal of Theoretical and Applied Mechanics **44**, 824–831 (2012)
29. Laderman, A.L.: Shock-tube performance with area divergence at the diaphragm section. AIAA Journal **10**, 1904–1906 (1967)
30. Resler, E.L., Bloxsom, D.E.: Very High Mach Number Flows by Unsteady Flows Principles. Cornell University, Graduate School of Aeronautical Engineering, New York (1952)
31. Trimpi, R.L.: A preliminary theoretical study of the expansion tube: A new device for producing high-enthalpy short duration hypersonic gas flows. NASA TR-133 (1962)
32. Neely, A.J., Stalker, R.J., Paull, A.: High enthalpy hypervelocity flows of air and argon in an expansion tube. Aeronaut. J. **95**(946), 175–186 (1991)
33. Sasoh, A., Morgan, R.G., et al.: High-enthalpy expansion tube experiments with gas injection. AIAA J. **38**, 2253–2259 (2000)

Author Biographies

Zonglin Jiang is well known for his studies in shock wave and detonation physics, cutting-edge technologies for the long-test-time shock tunnel and hypervelocity shock-expansion tube, and numerical algorithms for shock-capturing. He has authored/coauthored hundreds of journal papers by far. He received his PhD degree from Peking University in 1993, and has been the director of State Key Laboratory of High-temperature Gas Dynamics since 2000. He and his team set up a large scale shock tunnel with a test time up to 130 milliseconds in 2011. He is now Professor of Institute of Mechanics, Chinese Academy of Sciences, President of International Shock Wave Institute, and Vice President of Chinese Aerodynamics Research Society. His current research interests include detonation initiation and propagation, hypersonic gas dynamics and propulsion, as well as hypersonic and hypervelocity test technology.

Hongru Yu is a famous scientist in aerodynamics who has been concentrating on studies of the theory, experiment and application of shock tube/tunnel. After graduating from Dalian Institute of Technology in 1953, he joined the university as a lecturer. He moved to Institute of Mechanics, Chinese Academy of Sciences (IMCAS) in 1957. Prof. Yu initiated the study on shock tube/tunnel technology at IMCAS and developed the first shock tube and the first shock tunnel in China with his research group in 1960s. He also expedited detonation-driver technology for the development of high-enthalpy shock tunnels during 1980s. Since 1991, he has been an academician of Chinese Academy of Sciences.

Hypersonic Short-Duration Facilities for Aerodynamic Research at ITAM, Russia

Vasily M. Fomin, Anatoliy M. Kharitonov, Anatoly A. Maslov, Alexander N. Shiplyuk, Valentin V. Shumskii, Mikhail I. Yaroslavtsev and Valery I. Zvegintsev

1 Introduction

Beginning from the first wind tunnels and until 1950s, it was assumed that aerodynamic experiments can only be performed under steady conditions. Therefore, all wind tunnels were long-duration facilities where the test time was longer than the characteristic times of the examined phenomena by orders of magnitude. However, it became clear in the 1950s that the existing facilities cannot satisfy the requirements for modeling high-velocity flight conditions because of energy constraints. One way to overcome these constraints was reduction of the test time, proportionally leading to reduction of the energy necessary to perform the experiments [1]. The main advantage of test time reduction to 1 s and less is the possibility of using new design solutions, which allow record-breaking parameters of the test gas to be reached.

V.M. Fomin · A.M. Kharitonov · A.A. Maslov · A.N. Shiplyuk · V.V. Shumskii · M.I. Yaroslavtsev · V.I. Zvegintsev (✉)
Khristianovich Institute of Theoretical and Applied Mechanics, Siberian Branch, Russian Academy of Sciences (ITAM SB RAS), 630090 Novosibirsk, Russia
e-mail: zvegin@itam.nsc.ru

V.M. Fomin
e-mail: fomin@itam.nsc.ru

A.M. Kharitonov
e-mail: khar@itam.nsc.ru

A.A. Maslov
e-mail: maslov@itam.nsc.ru

A.N. Shiplyuk
e-mail: shiplyuk@itam.nsc.ru

V.V. Shumskii
e-mail: shumsky@itam.nsc.ru

M.I. Yaroslavtsev
e-mail: yaroslav@itam.nsc.ru

O. Igra and F. Seiler (eds.), *Experimental Methods of Shock Wave Research*, Shock Wave Science and Technology Reference Library 9,
DOI 10.1007/978-3-319-23745-9_10

Historically, the first facilities with the minimum test time were shock tubes, where a gas flow with an extremely hot temperature could be generated for a short time [2]. The working process in the shock tube is based on instantaneous breakdown of a partition (diaphragm) between the high-pressure and low-pressure sections of a cylindrical tube. After that, a shock wave passes over the low-pressure gas, and a moderate-sized region of the test gas with high temperature and pressure is formed behind the shock wave. The resultant test gas flows through a nozzle attached at the end of the shock tube and forms a flow with prescribed parameters. A typical test time in shock tubes is 1 to 10 ms. Such a short test time allows research aimed at modeling of physical and chemical processes that occur on the surface of re-entry vehicles flying with high velocities. Because of the low accuracy of the results of such experiments, they were mainly considered as physical (qualitative) tests rather than highly accurate aerodynamic experiments.

Despite the long period of existence of shock tubes, they are continuously improved and are still used for experimental research. The major differences in the design of various shock tubes are the methods of obtaining the driver gas. The gas parameters are increased by using electric heaters, detonation of an explosive gas mixture, or adiabatic compression by a moving piston.

In the 1950s, hotshot wind tunnels were developed [3]. Their operation was based on supplying energy in the form of an electric discharge to a motionless test gas in a settling chamber with a limited volume. The pressure and temperature in the chamber after the discharge reach maximum values of 200 MPa and of 4000 K, respectively. The working process in the form of quasi-steady exhaustion of the test gas from the settling chamber begins immediately after the discharge. In contrast to shock tubes, the working process here is not limited to the motion of shock waves, and almost all of the stored gas is used for flow generation. Therefore, the test time in hotshot wind tunnels increases to hundreds of milliseconds. Currently, there are approximately 20 wind tunnels of this kind in the world; these tunnels have nozzle exit diameters from 0.3 to 0. 5 m. All these tunnels are designed to obtain flows with high velocities (Mach number M = 8–20 or more).

The advantages of hotshot wind tunnels are the simple design and comparatively moderate sizes. Like other short-duration facilities, these wind tunnels are cost-efficient because of the low level of energy consumption. Hotshot wind tunnels allow various methods of test gas heating and cooling, and there are practically no problems of thermal protection of the wind tunnel walls and models, because the process lasts for a very short time. Despite some adverse effects, for example, flow contamination because of the high temperature generated by the discharge and a continuous decrease in flow parameters because of exhaustion from a constant volume, experiments performed in hotshot wind tunnels provided many new results which are important for aerodynamic practice.

Further evolution of short-duration facilities followed two parallel paths. On the one hand, the goal was to obtain extremely high parameters of the test gas, first of all, a high enthalpy of the flow. This path involved tremendous engineering and design problems, which required the test time to be reduced to 1 ms and less. At such test times, the flow cannot be considered as steady even in a rough approximation. Problems of measurement errors and interpretation of results appeared. Therefore, this path more and more deviated from the classical methods of experimental aerodynamics toward purely physical research and was rightfully criticized by the aerodynamic community.

Simultaneously, another version of short-duration facilities was developed where researchers tried to expand the range of gas flow parameters and retain a sufficiently long duration of the test time. This approach turned out to be successful and substantially expanded the range of applicability of such facilities in experimental aerodynamics. At the same time, the measurement and the computational equipment was vigorously developed, which made it possible to reach an identical accuracy of measurements in short-duration and blow-down wind tunnels.

The existing short-duration wind tunnels are mainly designed for modeling flight conditions at high hypersonic velocities ($M > 10–12$). However, experience showed that short-duration wind tunnels can also serve as fairly simple and effective tool for generating flows for aerodynamic tests in the range of Mach numbers from 1 to 8. The idea to expand the range of applicability of short-duration wind tunnels to $M \leq 8$ was put forward in [4, 5]. Despite a large number of blow-down wind tunnels, the use of short-duration tunnels in this range can substantially decrease the operation costs and, moreover, increase the Reynolds numbers, Re, owing to an increase in pressure and in the size of the flow domain. It can be easily demonstrated that implementation of this approach requires a significant increase in the test gas mass and, consequently, the volume of the settling chamber of the wind tunnel (up to several cubic meters at $M = 1$), which requires that numerous engineering problems must be solved, in particular, the problem of energy supply.

A separate group of short-duration tunnels are the facilities with adiabatic compression [6]. In these wind tunnels, the test gas is heated by means of adiabatic compression by moving a massive piston. The piston is accelerated by the driver gas having a comparatively low pressure (20–40 MPa). During acceleration, almost the entire energy of the driver gas transforms into the kinetic energy of the piston motion and then to the internal energy of a moderate mass of the test gas. The large ratio of the driver gas and test gas masses and the high efficiency of energy conversion ensure unique characteristics of wind tunnels with adiabatic compression.

Various methods of test gas heating are shown and compared in Fig. 1 [1] in the coordinates ρ_o/ρ_a versus T_0, where ρ_o and T_0 are the density and temperature of the test gas in the settling chamber after its heating, and ρ_a is the gas density under standard conditions. The arrows show the directions of the thermodynamic process during heating. In usual wind tunnels, heating occurs at constant pressure, resulting in a proportional decrease in the gas density with increasing T_0.

In shock tunnels, where high stagnation temperatures can be reached, nonisentropic heating of the test gas occurs while crossing shock waves, which also leads to significant losses of pressure and, correspondingly, density. This fact not only restricts the Reynolds numbers that can be reached in these tunnels, but also provides a nonequilibrium flow in the wind tunnel nozzle at stagnation temperatures higher than about 4000 K.

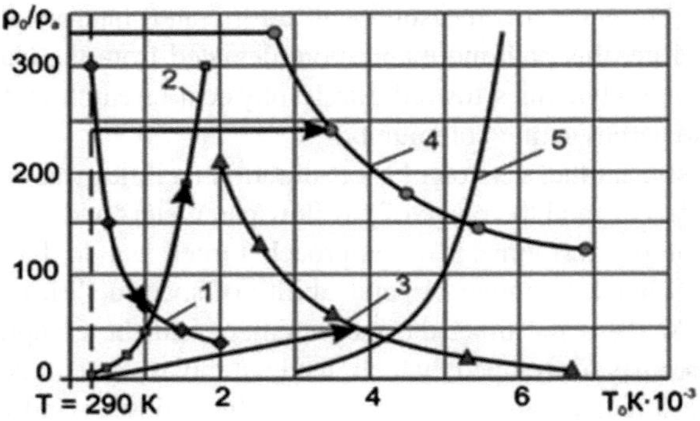

Fig. 1 Comparison of test gas heating methods in various wind tunnels: *1* usual wind tunnels (10 min and more)—heating at constant pressure; *2* wind tunnels with a heavy piston (0.1 s)—adiabatic heating; *3* shock tunnels (less than 0.01 s)—heating by shock waves; *4* short-duration wind tunnels (0.1 s)—heating at constant density; *5* boundary of the equilibrium flow in the nozzle

Isentropic heating by a heavy piston, which is beneficial from the viewpoint of obtaining high densities and high Reynolds numbers, does not allow high stagnation temperatures to be obtained.

It is only heating in a closed volume (at ρ_o = const), which is provided in short-duration (hotshot) wind tunnels, that allows obtaining high values of p_0, T_0, and ρ_o simultaneously, and, therefore, high values of Re_0 (as Re_0 is proportional to ρ_o).

Figure 2 shows the stagnation temperatures and pressures necessary to reproduce flight conditions as functions of the flight velocity and altitude.

As is seen from Fig. 2, for modeling flight conditions in the range of altitudes from 20 up to 50 km, the wind tunnel must ensure $P_0 \sim$ (10–2000) bar and $T_0 \sim$ (1000–4000) K because it is necessary not only to model the aerodynamic similarity criteria, such as M, Re, St, etc., but also to ensure real values of the stagnation pressure and temperature in the experiments. This is especially important

Fig. 2 Stagnation pressures and temperatures in the wind tunnel necessary to reproduce flight conditions; the gray area is the flight corridor

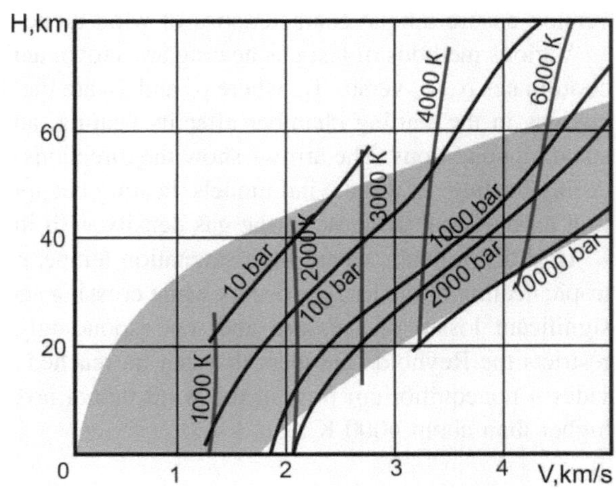

in studying flows with combustion, because the conditions of fuel inflammation, heat release in time and space, and chemical reaction rates depend to a large extent on pressure and temperature.

None of the existing ground-based facilities can satisfy all requirements necessary for duplication of flight conditions of high-velocity vehicles with air-breathing engines. However, in solving particular problems, experiments in high-enthalpy short-duration wind tunnels ensure the best approximation of real conditions in terms of both external and internal gas-thermodynamics. These are phenomena having characteristic times much shorter than the test time of the wind tunnel, namely, aerodynamic characteristics of the entire flying vehicle, including those with propulsion, compression of the test gas in vehicle's inlets of air-breathing engines, gas-thermodynamics of heat supply in the combustion chamber, mixing of fuels with the air flow, ignition and flame stabilization, provision of high combustion efficiency in high-velocity air-breathing engines (ramjets and scramjets), laminar-turbulent transition in boundary layers, formation and evolution of separation regions, determination of heat fluxes and local regions with anomalously high heat fluxes, and other typical phenomena observed in supersonic and hypersonic flows both in propulsion systems and on vehicle surfaces. The test time of the order of tenths of fractions of a second is fairly sufficient to study specific features of these phenomena and to optimize flow regimes and structural elements where these phenomena occur. Simultaneously, the short duration of the test time eliminates a severe problem of high-enthalpy ground-based wind tunnels; due to the short experimental time facility elements are not biased and destroyed by high stagnation parameters.

2 IT-301 Hotshot Wind Tunnel

The IT-301 hotshot wind tunnel was developed at the Institute of Theoretical and Applied Mechanics of the Siberian Branch of the Academy of Sciences of the USSR (now the Khristianovich Institute of Theoretical and Applied Mechanics of the Siberian Branch of the Russian Academy of Sciences, ITAM SB RAS) in the years 1968 to 70 on the initiative of V.V. Struminskii [4, 7, 8]. The wind tunnel followed the conventional structure of short-duration tunnels used in experimental aerodynamics since 1957. The wind tunnel ensured rapid heating of the test gas in the settling chamber owing to a discharge of a capacitor battery. It was equipped with a diaphragm that is broken during the test and with a vacuum tank for gas exhaustion.

In the photo and the schematic drawing of the wind tunnel shown in Fig. 3 one can observe the main elements of the IT-301 hotshot wind tunnel. They include the power system, a capacitor battery with stored energy of 0.5 MJ (1), settling chamber (4) with electrodes (2) and sensors for pressure monitoring (3), diaphragm (5), nozzle (6), test section (7), and vacuum tank for gas exhaustion (8).

Fig. 3 A photo and schematic drawing of the IT-301 hotshot wind tunnel

The wind tunnel operates as follows. First, the settling chamber is separated from the gas-dynamic duct of the wind tunnel by a diaphragm (5). The gas-dynamic duct, including the vacuum tank, is evacuated by vacuum pumps to a pressure of 10^{-2} to 10^{-3} mmHg. The settling chamber is filled with the test gas to an initial pressure of 20–150 bar, and the capacitor battery is charged. For starting the wind tunnel the capacitor battery is discharged resulting in temperature and pressure increase in the settling (discharge) chamber, and when a maximum or nearly maximum pressure is reached the diaphragm bursts and the gas is accelerated through the nozzle and flows into the test section, where the model is mounted. The test time in the IT-301 is 50–100 ms. In the course of exhaustion out of the closed volume of the settling chamber, the temperature and other flow parameters of the test gas continuously decrease, but the Mach number remains almost unchanged. The range of Mach numbers is M = 8–20. The wind tunnel power determined as the ratio of the energy input to the test time is $\sim$5 MW.

The IT-301 is equipped with a conical nozzle whose length can be varied within wide limits. The nozzle exit diameter can be varied from several centimeters to 0.5 m. Owing to such a structure, the wind tunnel serves as a convenient tool for conducting various aerophysical experiments, e.g., those aimed at creating powerful chemical and gas-dynamic laser sources, studying high-temperature equilibrium and nonequilibrium chemical reactions, etc., and for testing elements of promising hypersonic systems, including investigations of the processes of injection, ignition, and combustion of liquid and gaseous propellants in a supersonic air flow and in combustors of air-breathing engines in models with an "attached pipeline", where the combustion chamber is actually connected directly to the settling chamber of the hotshot wind tunnel, which offers a possibility of testing real-sized models at real flow parameters in some cases.

The IT-301 wind tunnel ensured stagnation temperatures up to 3000 K and stagnation pressures up to 1000 bar, which made it possible to perform tests at hypersonic velocities (M = 8–20 and more) with test times of 0.05–0.1 s. The first experimental studies showed that it is in this range of Mach numbers that hotshot wind tunnels have unique characteristics, which allow tests to be performed at supercritical Reynolds numbers with a turbulent boundary layer on the major part of the model, as happens in real flight conditions.

Since 1970, a large number of research and methodical experiments have been performed in the IT-301 [4, 5, 9–11]:

- Development of multichannel systems for pressure, heat flux, and aerodynamic force measurements.
- Development of systems of injection of liquid and gaseous fuels into models of air-breathing engines.
- Development of the system for synchronization of fuel injection, wind tunnel starting, and multichannel measurements.
- Verification and testing of operation modes of hotshot wind tunnels with the settling chamber consisting of two sections, which is equivalent to an increase in the settling chamber volume.
- The use of the N_2O decomposition reaction as an additional source of energy, which was later successfully applied in other hotshot wind tunnels.

3 IT-302 M Hotshot Wind Tunnel

The experience accumulated in IT-301 tests was used as a basis for the development and creation of a new hotshot wind tunnel IT-302 M [12, 13], which was commissioned in the year 1983 (Figs. 4 and 5). The IT-302 M wind tunnel has the same features as those of the IT-301 described previously. In contrast to all existing hotshot wind tunnels, the IT-302 M is equipped with a pressure multiplier, which ensures constant stagnation parameters during the test time.

The IT-302 M has the following technical characteristics: the energy stored in the capacitor battery is 1.75 MJ, the maximum volume of the settling chamber is 9.6 dm^3, the stagnation pressure is P_0 = (10–500) bar, and the stagnation temperature T_0 = (800–4000) K. The test section is made as an Eiffel chamber 1 × 1 × 1.5 m. The test time varies from 0.05 to 0.5 s.

Fig. 4 General view of the IT-302 M hotshot wind tunnel; left, with a nozzle 300 mm in diameter and right, with an attached combustor

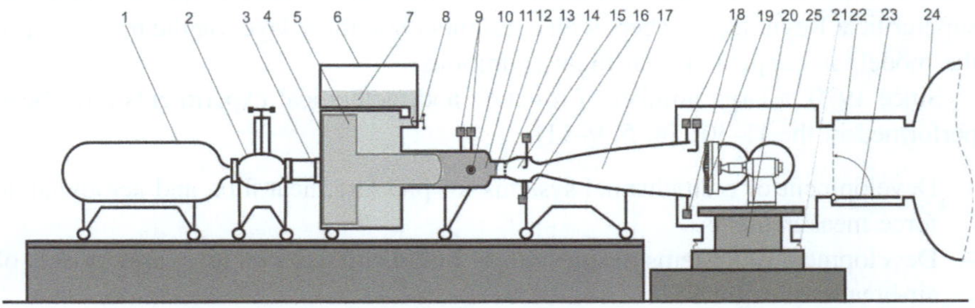

Fig. 5 Schematic description of the IT-302 M hotshot wind tunnel. *1* tank with the driver gas, *2* pneumatic valve D_y = 90 mm, *3* piston of the valve, *4* pressure multiplier casing, *5* two-stage piston, *6* tank for the damping liquid, *7* damping liquid, *8* gate, *9* pressure sensors, *10* electrodes, *11* settling chamber, *12* diaphragm, *13* throttling grid, *14, 16* pressure sensors, *15* throttling chamber, *17* nozzle, *18* flow regime control sensors, *19* model, *20* steel plate, *21* balance foundation, *22* vacuum gate casing, *23* shutter, *24* vacuum tank, *25* test section

Figure 6 shows the range of stagnation parameters ensured by the IT-302 M hotshot wind tunnel [indicated by (a)]. The lower boundary of this range is not a limiting case because the use of the double-section settling chamber in the gas-dynamic duct of the wind tunnel allows a several-fold decrease in the free-stream density. Figure 6 also shows the range of P_0 and T_0 (in zone b of Fig. 6) necessary for modeling supersonic combustion. Zone (b) is bounded from above and from below by the dynamic pressures q = 0 and 0.2 bar, respectively.

The IT-302 M wind tunnel is equipped with a contoured nozzle of exit diameters 300 and 400 mm, which ensure test regimes in the range of Mach numbers M = 4-8 with the root-mean-square deviation of M in the core flow within 1 %.

In addition to the flow regimes of the classical hotshot wind tunnel with decreasing parameters of the test gas (TG), i.e., gas exhaustion from a constant-volume chamber, IT-302 M ensures test regimes with constant TG parameters in the settling chamber and in the test section (stabilization of parameters). The most simple and reliable way to stabilize the TG parameters during the

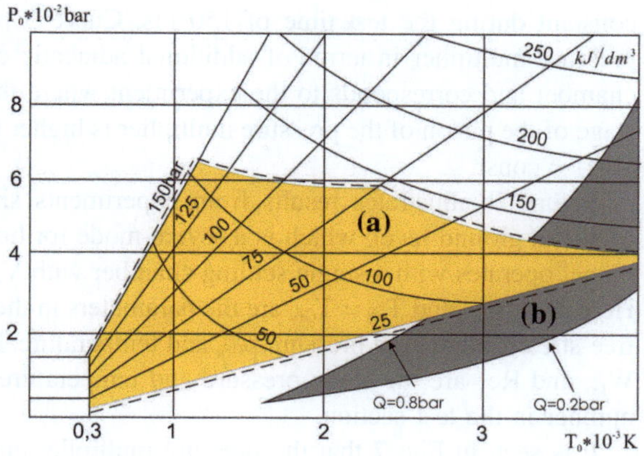

Fig. 6 Range of stagnation parameters of wind tunnels. **a** Range of stagnation parameters ensured by IT-302 M; **b** range of Po and To required for modeling combustors of air-breathing engines

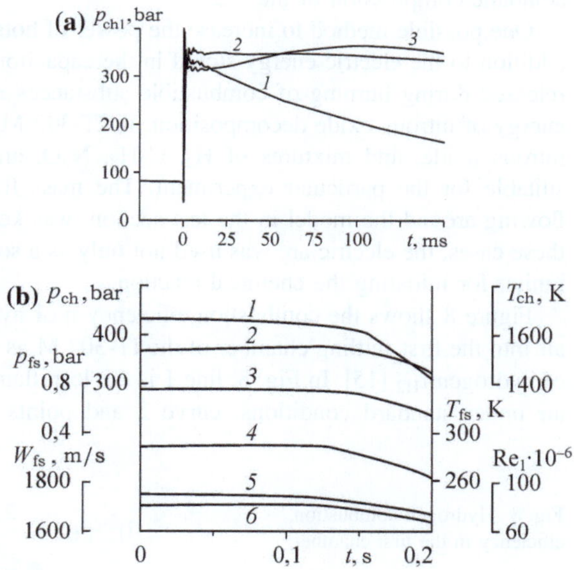

Fig. 7 TG Flow parameters versus the test time. **a** Recorded pressure in the first chamber without using a pressure multiplier (*1*) and with pressure multiplier (*2*, *3*); **b** TG parameters in the settling chamber (*1* p_{ch}, *2* T_{ch}) and at the nozzle exit (*3–6*) in experiment simulating Mach 5 flight conditions at ground level by using the pressure multiplier (*3* p_{fs}, *4* T_{fs}, *5* W_{fs}, *6* Re_1)

test time is to use a pressure multiplier [13]. Figure 7 illustrates the capabilities of the pressure multiplier in terms of stabilization of the TG parameters in the IT-302 M wind tunnel.

Figure 7a shows recorded pressure $p_{ch1}(t)$ in the first settling chamber (discharge chamber) in the case of wind tunnel operation with a double-section chamber. The volumes of the first and second chambers are $V_{ch1} = 9.6\ dm^3$ and $V_{ch2} = 12\ dm^3$, and the area of the throttling orifice between the chambers is $0.79\ cm^2$. Pressure decrease in the first chamber, in the case of wind tunnel operation without pressure multiplier is shown by Curve 1. Curve 2 illustrates the wind tunnel operation with stabilization of the TG parameters where the pressure multiplier parameters are chosen in such a way that the pressure in the first chamber is maintained almost constant during the test time of 150 ms. Curve 3 present the capabilities of the pressure multiplier in terms of additional adiabatic compression of the TG in the chamber and corresponds to the experiment where the pressure ahead of the large stage of the piston of the pressure multiplier is higher than that needed to maintain $p_{ch1}(t)$ = const.

Figure 7b illustrates results from experiments simulating flight conditions of M = 5 at ground level, which is a severe mode for hotshot wind tunnels. The wind tunnel operates with a single settling chamber with $V_{ch} = 10\ dm^3$ and $d_{th} = 10$ mm. Here $p_{ch} = p_{0fs}$ and $T_{ch} = T_{0fs}$ are the parameters in the settling chamber equal to the free-stream stagnation pressure p_{0fs} and temperature T_{0fs} in the test section; p_{fs}, T_{fs}, W_{fs}, and Re_1 are the static pressure and temperature, velocity, and unit Reynolds number in the test section.

It is seen in Fig. 7 that the pressure multiplier ensures reliable maintenance of constant TG parameters during the test time and ensures (if necessary) additional adiabatic compression of the TG.

One possible method to increase the power of hotshot wind tunnels is to use (in addition to the electric energy stored in the capacitor battery) the chemical energy released during burning of combustible substances in the settling chamber or the energy of nitrous oxide decomposition. In IT-302 M, we used hydrogen, propane, nitrous oxide, and mixtures of H_2, C_3H_8, N_2O, air, N_2, and O_2 in proportions suitable for the particular experiment. The mass fraction of oxygen in the TG, flowing around the model in the test section, was kept equal to 0.231 [14, 15]. In these cases, the electric arc was used not only as a source of energy, but also as an igniter for initiating the chemical reaction.

Figure 8 shows the combustion efficiency η of hydrogen pumped together with air into the first settling chamber of the IT-302 M as function of the molar fraction of hydrogen r_{H2} [15]. In Fig. 8, line 1 is the lean flammability limit of hydrogen in air under standard conditions; curve 2 and points 5 show experimental results

Fig. 8 Hydrogen combustion efficiency in the first chamber

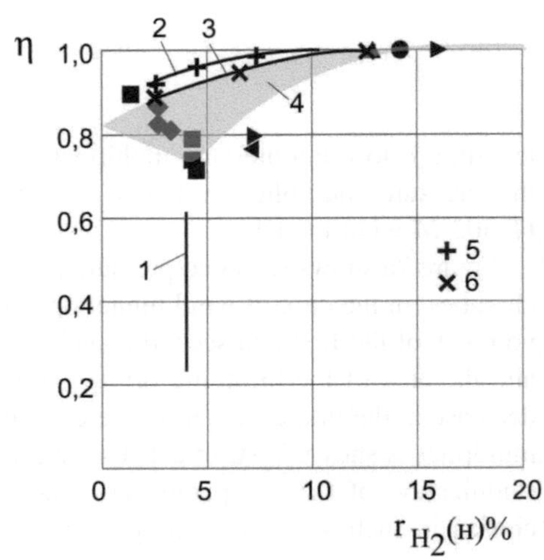

obtained while using a capacitor charging voltage of U = 3.0 kV and an electric energy input of 350 kJ; curve 3 and points 6 show results from experiments with U = 2.5 kV and electric energy input of 250 kJ. Domain 4 and the other points in Fig. 8 show the values of η calculated by gas-dynamic method [14] for U = 2.5 kV; points 5 and 6 were deduced from measurements in non-reacting hydrogen sample taken from the test section. It is seen that the fraction of unburned hydrogen is rather low (about 10 %) even in extremely lean hydrogen-air mixtures with hydrogen concentration lower than the lean flammability limit under standard conditions. The expansion of the flammability limit is caused by mixture preheating by electric discharge. The combustion efficiency rapidly approaches unity as the hydrogen concentration in the settling chamber and the electric energy input are increased.

It should be noted that the use of the chemical energy for increasing the wind tunnel power makes sense if the capabilities of heat supply by means of electric energy are exhausted, i.e., at the maximum value of U (for the IT-302 M, this maximum value is U = 5 kV) and at $r_{H2} > 5–7$ %. As seen in Fig. 8, this combustion efficiency is almost equal to unity in this case.

In most short-duration wind tunnels, the diaphragm bursts naturally, owing to its inability to carry the pressure increase in the chamber. If chemical energy is used for TG heating, a delay in the diaphragm breaking is expected due to the time needed for completing the reaction in the settling chamber. Moreover, the diaphragm breakdown delay can be used in the case of wind tunnel operation without chemical energy supply to increase the accuracy of results by means of eliminating transitional processes from the test time. The diaphragm in the IT-302 M can be broken in two different ways: Either due to its inability to withstand the pressure increase in the chamber or, by a forced breakdown induced by a signal from a control system providing a pre-arranged delay after the beginning of energy input into the settling chamber (controlled diaphragm).

The device used for a forced breakdown of the diaphragm is shown schematically in Fig. 9. The device is composed of pistons 2 and 4 with a liquid 3 located

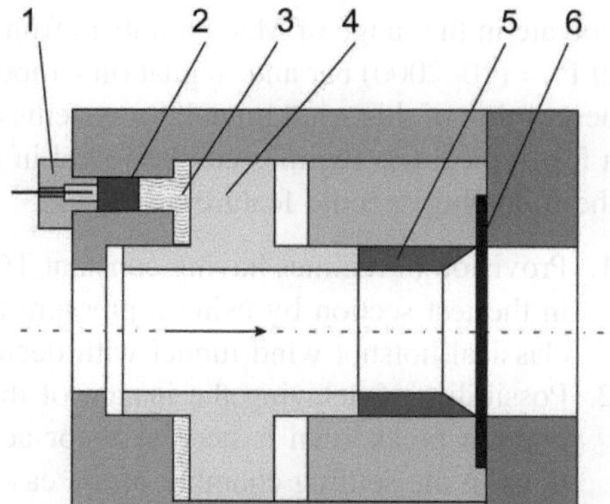

Fig. 9 Schematic drawing of the used diaphragm breaker

between them. When pyrocartridge 1 is initiated, the pressure between the pyrocartridge and piston 2 drastically increases and is transferred via piston 2 to liquid 3 between the pistons. As a result of the pressure increase in liquid 3, piston 4 together with knife 5 move to the right, and the knife cuts the diaphragm 6, separating the settling chamber from the gas-dynamic duct of the wind tunnel.

The system of controlled breakdown of the diaphragm ensures a delay from several milliseconds to several seconds after the electric arc discharge This controlling serves either for TG heating in the settling chamber or for initiation of the chemical reaction of H_2 or C_3H_8 burning or N_2O decomposition. The response time of the system, i.e. the time between the electric pulse transfer to the pyrocartridge 1 and diaphragm breakdown is less than 1 ms.

4 Project of Upgrading the IT-302 M Hotshot Wind Tunnel

The growing demand for experimental investigations of basic issues in hypersonic technologies for aeronautic and astronautic flows requires further improvement in the IT-302 M hotshot wind tunnel performance. Such improvement involves an increase in the settling chamber volume, i.e., an increase in the size of the generated flow and of the models to be tested. This modification will provide additional capabilities for studying processes whose modeling is determined by criteria including some linear size dependencies (e.g., Re) and obtaining a large amount of information owing to a large number of sensors mounted in the tested models. Another important task is to expand the range of wind tunnel operation to a Mach number range M = 4–6.

Figure 10 shows schematically the new high-enthalpy short-duration wind tunnel, which is a hypersonic wind tunnel; this is an improved version of IT-302 M. Functionally, the wind tunnel consists of two basic systems: gas-dynamic duct of the wind tunnel with a source of a high-enthalpy gas flow and a system of stabilization of the TG parameters during the test time. The wind tunnel is designed to operate in the range of Mach numbers from M = 4 to 20, stagnation pressure range of P_0 = (10–2000) bar and stagnation temperature of T_0 = (800–4000) K. A detailed description of this wind tunnel, its systems operation principles, and its operation in a few typical test regimes can be found in [16, 17]. The upgraded wind tunnel has the following specific features:

1. Provision of regimes having constant TG parameters in the settling chamber and in the test section by using a pressure multiplier, in addition to regimes of the classical hotshot wind tunnel with decreasing TG parameters.
2. Possibility of delaying the instant of diaphragm breakdown. The delay of diaphragm breakdown is necessary for activating and completing chemical reactions in the settling chamber in the case where chemical energy is used for TG heating. The delay also ensures a higher accuracy in determination of the gas

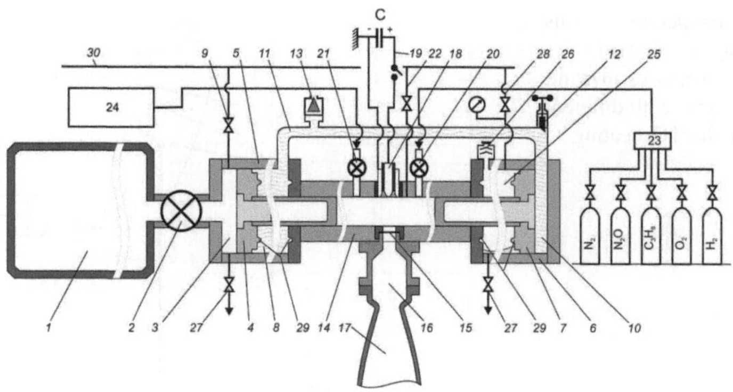

Fig. 10 Sketch of the short-duration wind tunnel with opposing pistons of the pressure multiplier: (*1*) receiver; (*2*) fast-response pneumatic valve; (*3*) pre-piston space of piston 4, (*4*) piston of the left part of the pressure multiplier, (*5*) left part of the pressure multiplier, (*6*) right part of the pressure multiplier, (*7*) piston of the right part of the pressure multiplier, (*8*) post-piston space of piston 4, (*9*) pipeline for technical works, (*10*) pre-piston space of piston 7, (*11*) hydraulic channel for the liquid, (*12*) flow-meter for the liquid, (*13*) check valve for evacuation of the liquid and exhaustion of gas bubbles, (*14*) settling chamber, (*15*) diaphragm unit, (*16*) second settling chamber, (*17*) nozzle, (*18*) coaxial electrodes, (*19*) capacitor battery, (*20*) pipeline for injecting gases participating in chemical reactions to the settling chamber, (*21*) pipeline for injecting high-temperature compressed air from an external heater 24 to the settling chamber, (*22*) pipeline for injecting compressed air to the settling chamber, (*23*) distributed system for injecting chemically reacting gases to the settling chamber, (*24*) external heater, (*25*) post-piston space of piston 7, (*26*) tank for exhaustion of air from space 25 and ejection of the damping liquid, (*27*) pipeline for release of excess pressure, (*28*) pipeline for returning the pistons to the initial positions, (*29*) labyrinth sealing, (*30*) air from gas holders

parameters by eliminating (from the test time) the period of transitional processes in the settling chamber and measurement channels due to shock loads induced by instantaneous heat input which can be, e.g., powerful electric discharge in the chamber or detonation burning of chemically reacting gases in the chamber.

3. Large volume of the settling chamber (V_{ch} up to 100 dm^3). The use of such a volume V_{ch} expands the capabilities of hotshot wind tunnels and increases the amount of information extracted from an experiment. In particular, it will be possible to generate a hypersonic flow in the test section by using nozzles with exit diameter $d_n = 1$ m with a test time up to 100 ms in combination with the second chamber. Otherwise, it will be possible to increase the test time up to 1 s by using moderate-size nozzles with $d_n \approx 0.3$ m.
4. Opposing arrangement of the pressure multiplier pistons for reduction of inertial loads on the wind tunnel (in the ideal case, to zero). There is a principal possibility of varying the chamber volume from one experiment to another by changing the initial distance between the pressure multiplier pistons.
5. Possibility of using various methods of TG heating: electric arc, chemical energy released during nitrous oxide decomposition or burning of combustible

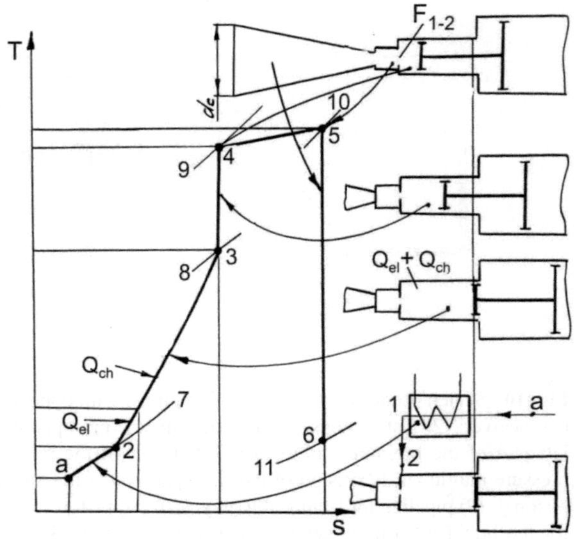

Fig. 11 A temperature versus entropy diagram showing operational processes in wind tunnel operating with different options for the TG heating

gases, adiabatic compression, heating in an external heat source with respect to the settling chamber, and combinations of these methods. All these methods of TG heating are separately used in various high-enthalpy short-duration wind tunnels. Some combinations of these methods of TG heating were tested in the IT-302 M and AT-303 hypersonic wind tunnels based at ITAM SB RAS.

Figure 11 shows the thermodynamic scheme where all methods of TG heating are used simultaneously and the possible paths are discussed below. Other combinations of individual methods of TG heating are also possible, which ensures the variety of test regimes in the proposed short-duration wind tunnel.

The T-S (temperature–specific entropy) diagram of Fig. 11 explains the wind tunnel operation that allows obtaining the maximum TG enthalpy by combining the above-mentioned heating methods of the TG with two settling chambers and stabilization of the TG parameters during the test time. A mixture of initial substances passes through an external heater 1, where it is heated from $T_a \approx 290$ K to T_{ch1}, and enters the first settling chamber of the wind tunnel (process a-2 shown in Fig. 11). During this step an amount of heat Q_{ext} is added while keeping the pressure constant along the isobar 7. When there is no external heating, then is $T_{ch1} = T_a$. In the isochoric process 2–3, the heat Q_{el} addition due to the electric discharge and the heat Q_{ch} due to exothermal chemical reactions are supplied to the initial mixture inside the first settling chamber. If the pressure in the first chamber, after the isochoric process at the intermediate point 3 (isobar 8), is smaller than the required value, then the TG is additionally compressed to the necessary pressure level (isobar 9) in the adiabatic process 3–4. When the necessary pressure level is reached (point 4), the diaphragm separating the first chamber from the second chamber bursts. The TG is throttled between the chambers through the throttling grid with the area of orifices $F_{1\text{-}2}$ (isenthalpic process 4–5) and exhausts through the nozzle (isentropic process 5–6), forming the nozzle exit flow having conditions appropriate to state *6* as shown in Fig. 11.

Point 4 in Fig. 11 characterizes the flow conditions in the first chamber during the entire test regime throughout the course of TG squeezing by the pressure multiplier until the piston system of the pressure multiplier reaches its extreme position (left in Fig. 11). Point 5 characterizes the flow conditions in the second chamber during the test regime after TG throttling from the first settling chamber while maintaining the enthalpy values $i_{ch1} = i_{ch2} = i_{0fs}$. Lines 7–11 are isobars at the corresponding points of the process. The total energy input to the TG in the first settling chamber is: $Q_{ext} + Q_{el} + Q_{ch}$ + work of adiabatic compression in process 3–4.

5 Transit-M Hypersonic Wind Tunnel

One of the promising directions in increasing the capabilities of short-duration wind tunnels in terms of their power is the use of a nondestructive controlled fast-response nozzle shutter instead of an expendable destroyed diaphragm with an uncontrolled instant of its breakdown. This makes it possible to use comparatively slow methods of TG heating and to increase the available stored energy by a factor of ten. In turn, the increase in the available energy allows the TG flow rate to be increased, thus provides a larger flow size and expanding the area of applicability of wind tunnels. In contrast to wind tunnels with fast energy input, which are usually called hotshot tunnels, these wind tunnels are more generally called short-duration facilities.

The Transit-M hypersonic wind tunnel with a controlled fast-response shutter was developed at ITAM SB RAS in the year 1985 [18–20]. This wind tunnel is designed for aerodynamic tests in the hypersonic range of velocities with elevated Reynolds numbers. In the test section of this wind tunnel a flow at Mach number from M = 4–8 and with Reynolds number, based on the nozzle diameter (300 mm), from 2×10^8 (at M = 4) to 10×10^6 (at M = 8) can be conducted. The test time is 0.3 – 0.6 s.

The general view of this wind tunnel is shown in a photo (Fig. 12) and in a drawing (Fig. 13). The basis of this facility is the settling chamber system, which is a source of the test gas and determines the characteristics of the test regime of the wind tunnel. This system includes the main and auxiliary settling chambers and also two additional tanks connected to the main settling chamber. The initial mass of the TG is accumulated before the experiment simultaneously in the main settling chamber and in the additional tanks, which yields a total volume of 110 dm^3 of compressed gas at a pressure up to 200 bar. The additional tanks accommodate four electric heaters having a power of 25 kW each, which are fed from an AC power source with frequency of 50 Hz and voltage of 380 V. The heaters heat the TG up to a temperature of 750 K. The heating process lasts for 30–60 s. To reduce heat losses, the additional tanks are equipped with thin-walled thermal protection screens inside.

Fig. 12 General view of the Transit-M wind tunnel

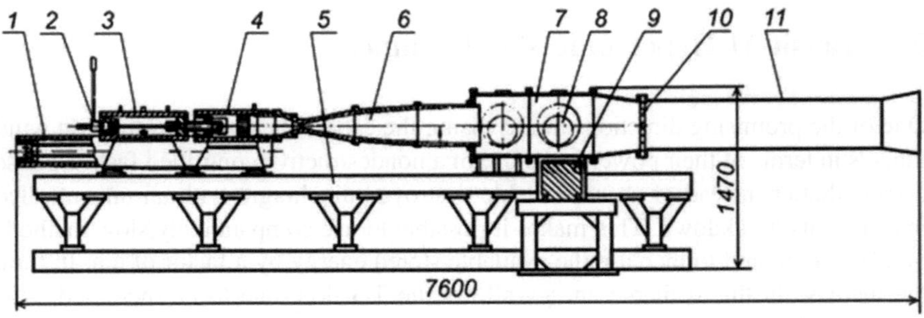

Fig. 13 Drawing of the Transit-M wind tunnel. *1* storage tank with an electric heater; *2* kickoff valve; *3* main settling chamber; *4* auxiliary settling chamber; *5* frame; *6* contoured nozzle; *7* test section; *8* optical windows; *9* platform insulated from vibrations; *10* grid; *11* exhaust diffuser. All dimensions are in mm

The main settling chamber contains a fast-response nondestructive shutter preventing gas exhaust into the auxiliary settling chamber and nozzle during gas accumulation for a test. The shutter is opened under the action of the TG pressure after actuation of the manually operated kickoff valve. The time of shutter opening is 5–8 ms.

After the shutter opening the compressed gas flows into the auxiliary chamber where the total pressure reduction (by throttling) and flow equalization ahead of the nozzle entrance occur. The structure of the Transit-M short-duration wind tunnel involves the use of replaceable axisymmetric contoured nozzles having a nozzle diameter of 300 mm. These nozzles generate a uniform gas flow at M = 4–8 around the examined model mounted inside the test section of the wind tunnel. The test section is made as an axisymmetric chamber and consists of two sections with optical windows for flow visualization. The lower part of one section contains a pocket for mounting a strain-gauge balance on the sting that holds the tested model; on a separate foundation insulated from vibrations.

From the test section, the gas flow passes through a diffuser made as a cylindrical tube 0.4 m in diameter and then either to the atmosphere or to a vacuum tank. The diffuser has a possibility of mounting a separating membrane 0.4 m in diameter, which makes it possible to provide a rarefaction wave in the test section and the nozzle before starting the wind tunnel flow. There is no need to use this membrane if a vacuum tank is used.

All elements of the wind tunnel are located on a common frame and can be moved along the axis with the use of special guides. Such motions are necessary for having access to the model between runs. The total length of the wind tunnel including the exhaust diffuser is 7600 mm; its width and height are 870 and 1470 mm, respectively.

Figure 14 shows typical pressure records taken in the settling chamber and in the test section of Transit-M during its testing in a Mach number range from M = 4–7: with exhaustion into the atmosphere, without using the vacuum tank. The experimental data agree well with calculated results obtained in the course of design, based on the above-described techniques.

In 2003, Transit-M was moved to the building of the AT-303 wind tunnel and was equipped with an exhaust vacuum tank having a volume of 6 m^3 (see Fig. 16). The vacuum tank was connected to the exhaust system of the AT-303 wind tunnel through a shutter 250 mm in diameter and a pipeline 200 mm in diameter and 10 m long.

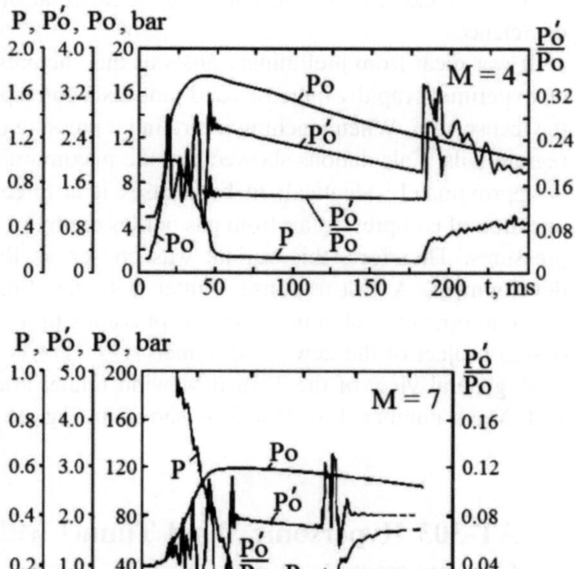

Fig. 14 Typical records of the operation regime of the Transit-M wind tunnel *Po* stagnation pressure; *Po'* Pitot pressure; *P* static pressure

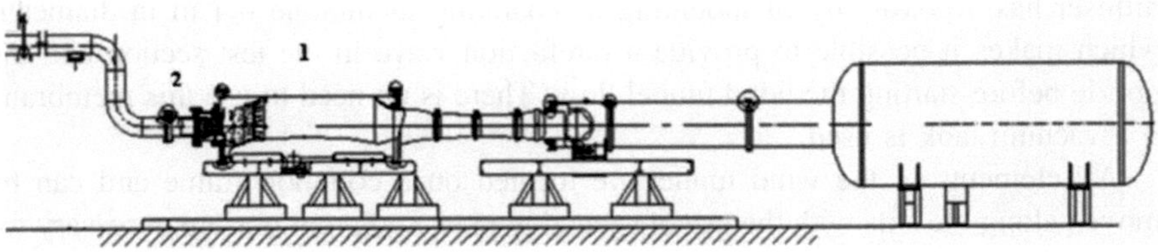

Fig. 15 Transit-M wind tunnel after its upgrading for operation at M = 2. *1* settling chamber 0.6 m^3; *2* fast-response shutter 160 mm

6 Upgrading of the Transit-M Wind Tunnel for Operation at M = 2

In order to perform experiments aimed at studying sonic boom problems, we considered the possibility of Transit-M operation with a large flow rate of the test gas at the Mach number M = 2 through a contoured nozzle with an exit diameter of 300 mm and throat diameter of 231 mm. Two variants of wind tunnel upgrading were considered: in the form of a vacuum tube with air input from the atmosphere, and in the form of a conventional wind tunnel with supply of compressed air from a system of gas holders of ITAM SB RAS, at the Khristianovich Institute of Theoretical and Applied Mechanics of the Siberian Branch of the Russian Academy of Sciences.

It was clear from preliminary analysis that the pressure in the vacuum tank after an experiment rapidly increases and induces a corresponding increase in pressure in the test section. When reaching a certain jet pressure ratio at the nozzle exit, the test regime fails. Calculations showed that the maximum test time in these two variants is approximately identical; in both cases it is three seconds or more. However, injection of compressed air from gas bottles ensures a much wider range of dynamic pressures. Therefore, this variant was chosen as the basic approach for further development. A fast-response shutter with the free cross-sectional diameter of 160 mm operating at compressed air pressures up to 20 bar was developed for the design project of the new wind tunnel.

A general view of the Transit-M wind tunnel after its upgrading for operation with Mach number flow M = 2, is shown in Fig. 15.

7 AT-303 Hypersonic Wind Tunnel with Adiabatic Compression

A new hypersonic wind tunnel with adiabatic compression (AT-303) was commissioned at ITAM SB RAS in the year 1999 [21, 22]. The general view of the AT-303 wind tunnel is shown in Fig. 16. The basis for this wind tunnel was the operational model of the working gas (TG) source (1:2.5), which was developed in

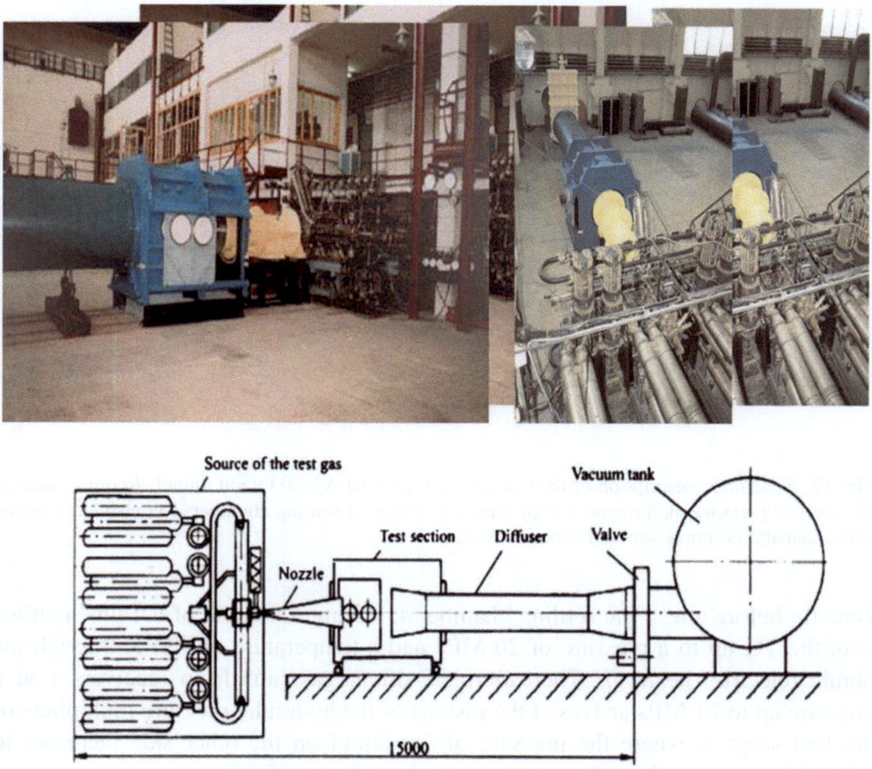

Fig. 16 General view and schematic drawing of the AT-303 wind tunnel

1983–1989 at the Siberian Branch of the Academy of Sciences of the USSR for the Central Aero-hydrodynamic Institute (TsAGI). The TG source ensures sufficiently high values of stagnation temperature (up to 2500 K) and stagnation pressure (up to 300 MPa) of the test gas (air) owing to comparatively slowly moving massive two-stage pistons (pressure multipliers) performing adiabatic compression of the gas. An optimal choice of parameters of the compression cycle ensures minimum energy expenditure and absence of TG contamination. An original system of controlling the velocity of the piston motion makes it possible to maintain a constant pressure in the course of gas exhaustion from the settling chamber. The TG source with adiabatic compression determines the basic operational characteristics of the wind tunnel. The AT-303 wind tunnel with an electric heater provides from 1 to 3.6 dm^3 of compressed air at a pressure from 100 to 300 MPa and temperature of 1300–2500 K. These values are sufficient for ensuring a test time from 50 to 200 ms with nozzles having exit diameters from 300 to 600 mm in the range of flow Mach numbers M = 8–20.

The TG source is shown schematically in Fig. 17. In the initial position, the valve in the settling chamber 4 is closed, and the pistons in the high-pressure section 3 are in extreme positions with the greatest possible distance between them.

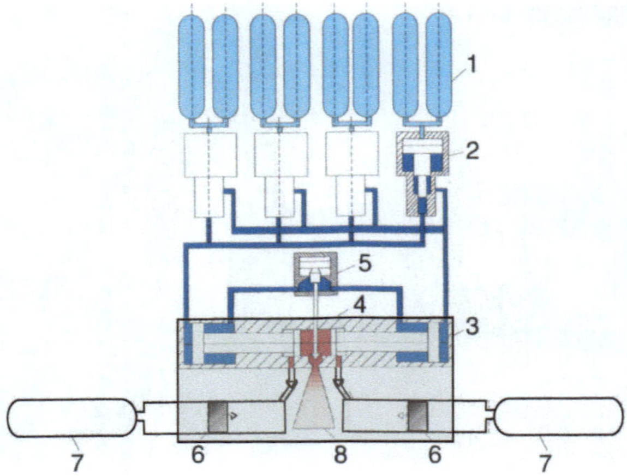

Fig. 17 Schematic description of the test gas source in the AT-303 wind tunnel. *1* compressed air receivers; *2* pressure multipliers; *3* high-pressure section; *4* settling chamber; *5* controller; *6* heater with adiabatic compression and *7* ohmic heater

Directly before a test, the settling chamber 4, having a volume of 6.4 dm^3 is filled with the TG up to a pressure of 20 MPa and a temperature of 950 K through the ohmic (electric) heater 7. Thereafter, the driver gas (air) from receivers 1 at a pressure up to 21 MPa arrives at the pistons of the hydraulic pressure multipliers of the first stage 2, where the pressure of the liquid on the other side increases to 141 MPa owing to the difference in areas by a factor of 6.7.

This liquid pressure is transferred to the pistons in the force cylinders 3, which are pressure multipliers of the second stage, and the pressure again increases there by a factor of 2.1. The pistons of the force cylinder start to move toward each other and adiabatically compress the TG in the settling chamber up to a pressure of 300 MPa. Thereby the temperature increases up to $\sim$2500 K. A particular value of the driver gas pressure is chosen to obtain a prescribed value of the maximum pressure in the settling chamber with allowance for the multiplication coefficient.

When compression is completed, a special shutter 4 is opened, and the TG escapes through the nozzle. In the course of TG exhaustion, the pistons continue to move, thus, maintaining almost constant stagnation parameters during the test time. The piston motion velocity is controlled by liquid overflow in controller 5. When the pistons stop, the test time is over. It should be noted that the piston's velocity does not exceed 5 m/s, and they move simultaneously and perpendicular to the wind tunnel axis; therefore, there is no problem with an axial impact owing to a kickback at the instant when the pistons stop their motion. This problem is typical for a wind tunnel having adiabatic compression by a heavy piston.

Typical records of the gas pressure in the settling chamber and the trajectories of motion of the pistons of the main force cylinder are shown in Fig. 18. Based on the measured results, the following conclusions can be drawn:

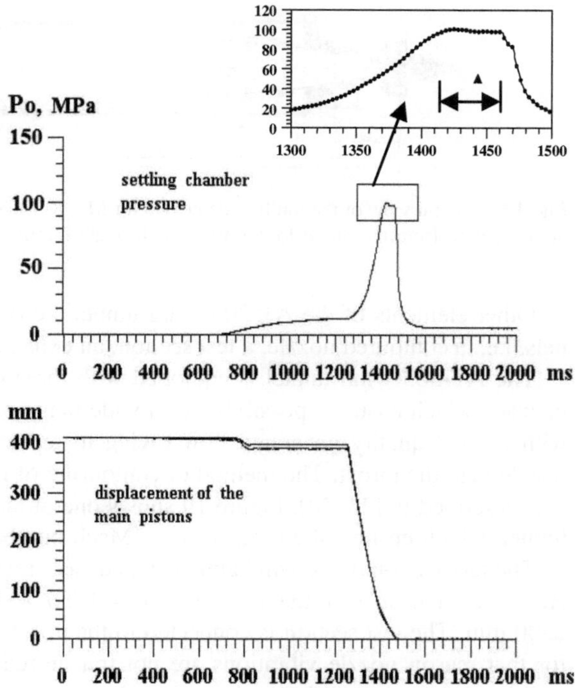

Fig. 18 Typical records of the basic parameters in the test regime of the AT-303 wind tunnel

1. Starting at t = 700 ms, the settling chamber is filled with the heated TG (air) up to initial pressure of P_o = 12 MPa.
2. Beginning from t = 125 ms, the pistons of the force cylinder start to move and compress the TG in the settling chamber up to 105 MPa.
3. At the time t = 1400 ms, the fast-response shutter in the settling chamber is opened, and the gas starts to escape from the nozzle into the test section
4. At the time t = 1470 ms, both pistons in the force cylinder stop their motion, and the operational cycle of the wind tunnel is finished.

The main fraction of energy for adiabatic compression of the TG is generated by the energy of the driver gas stored in receivers 1 of the wind tunnel. In each test, approximately 10–15 % of the stored driver gas or up to 4.9 MJ is spent on the TG compression owing to piston motion. The original structure of the TG source had a total volume of the receivers equal to 0.65 m^3 and a maximum pressure of 20 MPa, which corresponds to the maximum values of the stored energy equal to 39.5 MJ. In the year 2003, the driver gas system was upgraded: the receivers were replaced by twelve standard gas containers with a total volume of 0.96 m^3 designed for pressures up to 40 MPa, which corresponds to the maximum value of the stored energy equal to 117 MJ.

For filling the receiver with compressed air, AT-303 has its own compressor station including two VSh-2,3/400 compressors (produced by Ural Compressor Plant, Russia), which ensure air pressures up to 40 MPa, and 48 gas containers with a total volume of 3.8 m^3.

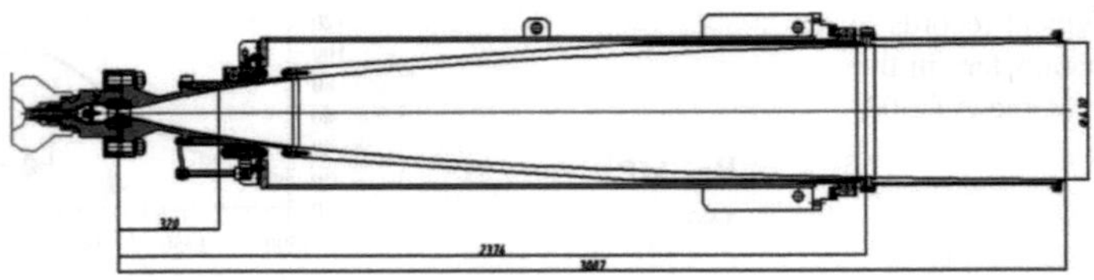

Fig. 19 General view of the multimode nozzle for M = 8–14. The nozzle diameter is 410 mm; the nozzle length from the throat to the nozzle exit is 2900 mm

Other elements of the AT-303 wind tunnel are typical to hypersonic wind tunnels, i.e., a contoured nozzle, a test section, an exhaust diffuser, and a vacuum tank.

The AT-303 wind tunnel is equipped with three original multimode contoured nozzles, which ensure a possibility of a wide range of test regimes from M = 6–20 with a good quality generated flow owing to replacement of a small part of the nozzle near the throat. The method of contouring of hypersonic multimode nozzles was described in [23, 24]. Figure 19 shows one of the nozzles of the AT-303 wind tunnel, which ensures the range of flow Mach numbers from M = 8–14.

The test section of the wind tunnel is made as a rectangular cabin (chamber). The cross-sectional size of the test section is 1750 × 1400 mm^2, and its length is 2500 mm. The test section is connected to the nozzle through a rubber gasket, and for that reason nozzle vibrations are not transferred to the test section. The side walls and the top wall have eight optical windows with 400 mm in diameter for optical measurements and flow visualization. Inside the test section, there is the so-called alpha-mechanism, which allows the angle of attack of the tested model and its position over the nozzle axis to be changed.

A movable replaceable diffuser is mounted at the end of the test section. The diffuser is placed inside a tube with 1200 mm in diameter, which connects the test section with the exhaust system of the wind tunnel.

The total volume of the exhaust vacuum system of the wind tunnel with allowance for the exhaust duct and test section is 260 m^3. Evacuation is performed by a system of vacuum pumps including two NV3-500 (produced by Nasosenergomash Sumy, Ukraine) backing vacuum pumps and two 2DVN-1500 pumps (produced by Nasosenergomash Sumy, Ukraine). For maintenance convenience, the test section is separated from the remaining exhaust system by a large-size vacuum gate with 1200 mm in diameter.

The TG parameters that can be obtained in the settling chamber of this wind tunnel are shown in Fig. 20 in the P_0–T_0 plane. The points corresponding to the initial stage of the TG before the compression process are located inside the rectangle in the left bottom corner of the figure. The upper boundary of this rectangle is the limiting initial temperature (∼900 K), which can be obtained by means of preliminary heating with the available electric heater. The line on the right shows the maximum initial pressure of the TG (20 MPa), which is limited by the strength of this heater. The relations between the initial and final parameters of the gas after the compression process are determined by isentropic relations owing to

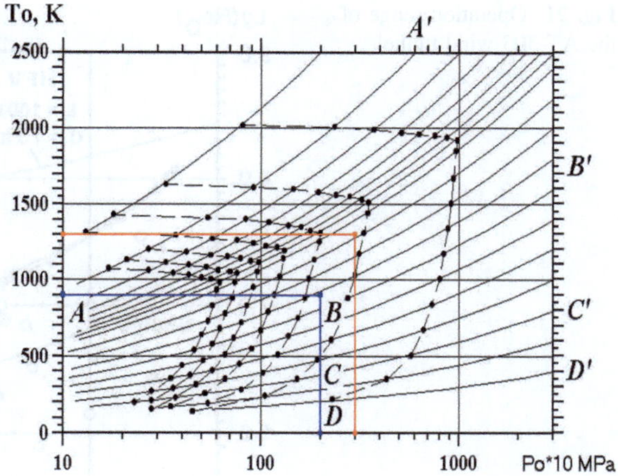

Fig. 20 Test gas parameters in the settling chamber of the AT-303

adiabaticity of the compression process. Here the inclined curves are isentropes calculated for real air. It should be noted that these curves would be straight lines for a perfect gas with a constant ratio of specific heats. The gas state in the course of adiabatic compression can be traced by moving along these lines. The parameters along the isentrope correspond to the final volume of the gas after its compression. The calculations show that the TG should be compressed up to a factor of 20 to obtain the maximum parameters (P_O = 300 MPa and T_O = 2500 K). Specifying required stagnation parameters at the end of the compression process, we can easily find from this graph which initial conditions should be provided in the settling chamber to obtain required conditions after gas compression.

The range of attainable Mach and Reynolds numbers for the AT-303 wind tunnel is shown in Fig. 21. The upper curve is the typical trajectory for a hypersonic flying vehicles of length L = 100 m with an air-breathing engine at a dynamic pressure of q = 75 kPa. The lower curve is the trajectory of reentry vehicles, such as Space Shuttle or Buran (L ≈ 45 m). The marked lines show the maximum values of the Reynolds number based on the nozzle diameter which can be obtained in AT-303 in nozzles with the exit diameters of 300 and 600 mm. The open points illustrate wind tunnel operation modes that have already been achieved. It is seen that AT-303 ensures adequate modeling of real flight conditions in terms of Mach and Reynolds numbers, including flight conditions of hypersonic airplanes with air-breathing engines.

Since the beginning of its regular operation from 2000 till 2012, more than 3000 runs were performed in AT-303 [25–29]. High TG parameters, large-size nozzles, sufficient duration of the test time, and high level of TG purity make this wind tunnel attractive for ground-based tests of promising hypersonic flying vehicles with air-breathing engines. The TG can be air, nitrogen, and other gases.

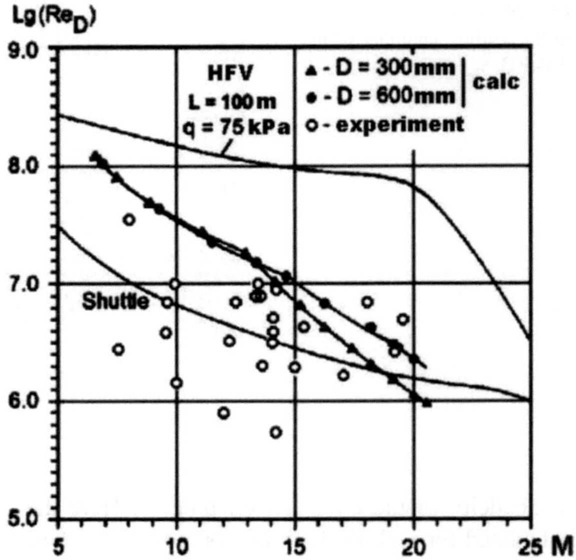

Fig. 21 Operation range of the AT-303 wind tunnel

8 The Upgrading Project of the aT-303 Wind Tunnel

The experience gained during the wind tunnel operation in 2000–2006 revealed some primary drawbacks of the TG source design. The main drawback is the short test time primarily caused by the small volume of the test section. The initial volume is approximately 6.5 dm^3, and the final volume after gas compression is smaller than 1 dm^3. The currently used TG source does not allow obtaining test regimes with Mach numbers smaller than M = 8, whereas the test regimes with M = 5–8 are the most important ones in testing modern hypersonic flying vehicles with air-breathing engines.

In 2006, design efforts were started for upgrading the TG source of the AT-303 wind tunnel [30]. Reconstruction of the TG source will make it possible to increase the test time in AT-303 by approximately a factor of five owing to a corresponding increase in the settling chamber volume with simultaneous alleviation of design and technological problems associated with a wind tunnel fabrication and subsequent operation.

The new TG source is schematically illustrated in Fig. 22. The source is designed for obtaining the test gas (air) with a pressure of 3000 bar and a temperature of 2500 K in a volume not smaller than 4 dm^3 after compression. Before the experiment, the TG is injected through the electric heater 1 and the back valve 2 to the adiabatic heater 3 and the force cylinder 6. In the course of filling these volumes, the TG pressure is 20–30 bar, and its temperature is 600 K. The adiabatic heater is a cylinder with thermally insulated walls 4 and a piston inside. When the heater is filled, the piston freely moves to the right and allows the heater volume to be filled by the TG. For adiabatic heating and compression of the TG, the piston is

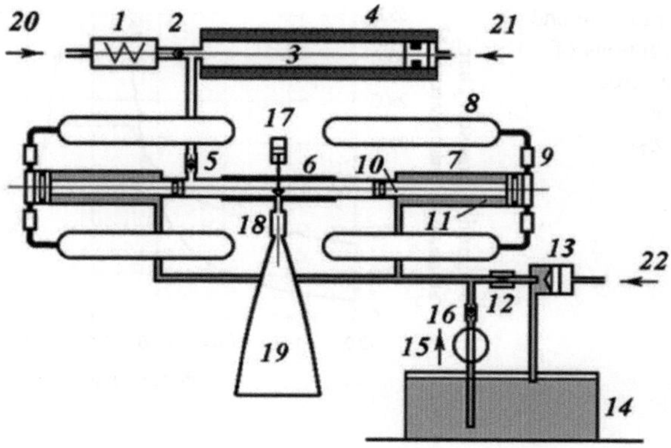

Fig. 22 Schematic description of the new TG source for reconstruction of the AT-303 wind tunnel. *1* electric heater; *2* back valve; *3* adiabatic heater; *4* thermal insulation; *5* back valve; *6* force cylinder; *7* driver cylinder; *8* driver gas holders; *9* pressure controller; *10* piston; *11* damping liquid; *12* throttle; *13* gate; *14* exhaust tank; *15* booster pump; *16* back valve; *17* nozzle shutter; *18* double-section settling chamber; *19* contoured nozzle; *20* TG injection; *21* injection of the gas for adiabatic heater; *22* injection of the gas for shutter control

set into motion by applying the driver gas 21 with a pressure up to 200 bar. The piston velocity is controlled by the flow rate of the driver gas 21. When the piston passes over the cylinder of the adiabatic heater, the entire TG is in the force cylinder 6 with a diameter of 100 mm and a volume of about 30 dm^3. The TG pressure increases to 200 bar, and the TG temperature increases to 1300–1400 K.

The main compression of the TG in the force cylinder 6 is performed adiabatically by two piston-multipliers 10. The pistons are set into motion by the compressed gas entering from the receivers 8, where it is stored at a pressure up to 350 bar. The diameter of the large stage of the driver piston is 400 mm, which allows the TG pressure in the force cylinder 6 with a diameter of 100 mm to be increased up to 3000 bar at the driver gas pressure of 200 bar. Pressure controllers 9 are used to stabilize the driver gas pressure at a level of 200 bar.

Before their motion, the pistons 10 are maintained in the initial position by the damping liquid 11, which is trapped in the cavity of the driver cylinder 7 by the gate 13 with a diameter of 180 mm.

To initiate the compression process in the force cylinder, the gate 13 is opened and the damping liquid starts to flow into the exhaust tank 14. The pistons 10 start to move and compress the TG in the force cylinder 6. The velocity of piston motion is constrained by a throttling device 12, which can reduce the damping liquid outflow velocity. In the course of piston motion, the TG pressure increases owing to adiabatic compression and is distributed over the length as is shown in Fig. 23.

After sevenfold compression of the TG in the force cylinder, the TG volume is approximately 4 dm^3, the TG pressure is 3000 bar, and the TG temperature is 2500 K. The forces on the two sides of the piston are equalized, and the pistons stop their motion.

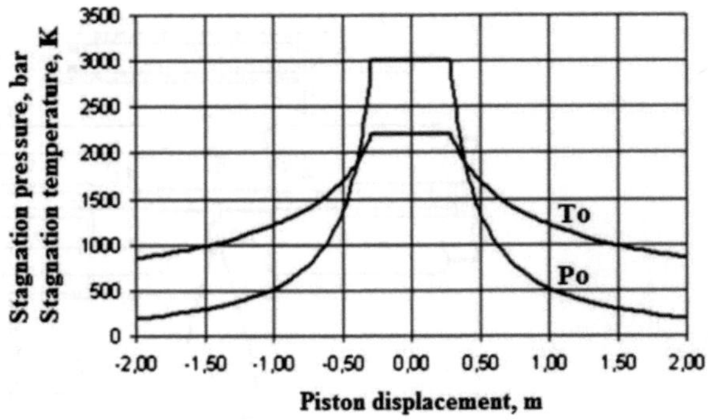

Fig. 23 Typical pressure and temperature distributions of the TG inside the force cylinder during the compression process

Immediately after the compression process, the nozzle shutter 17 is opened, and the TG passes through the settling chamber 18 and is accelerated in the contoured nozzle 19 to form a hypersonic flow with prescribed parameters, which is used to perform aerodynamic experiments. After the nozzle shutter 17 is opened, the pistons 10 start to move, compressing the exhausting gas in the force cylinder 6 and, thus, maintaining constant TG parameters in the exhaust process (during 0.1–0.2 s).

When the experiment is completed, the nozzle shutter 17 is closed, and pistons 10 return to their initial position with the help of the damping liquid that is supplied from exhaust tank 14 by the booster pump 15 through the back valve 16. Developing a pressure up to 350 bar, the booster pump 15 ensures not only returning of the pistons 10 to their initial positions, but also an increase in the driver gas pressure in the gas holders 8 to the initial level. After that, the wind tunnel is ready for the next experiment.

No more than 2 kg of test gas are spent in each run of the wind tunnel. The driver gas holders with a total volume of 960 dm^3 accommodate 400 kg of the gas at a pressure of 350 bar, which is pumped there only once and then is only replenished in case of leakage. To obtain high-pressure air, the UKS-400 compressor station is used. It is included in the AT-303 equipment and provides a supply of compressed air with a pressure of 350 bar and with a flow rate up to 100 kg/h.

Summary

1. Short-duration experimental facilities of ITAM SB RAS (at the Khristianovich Institute of Theoretical and Applied Mechanics of the Siberian Branch of the Russian Academy of Sciences, ITAM SB RAS) ensure a possibility of aerodynamic testing of models in the range of Mach numbers from M = 2 to 20. The following types of experiments are performed in these tunnels:
   - determination of aerodynamic characteristics and optimization of the shape of flying vehicles;
   - determination of the total aerodynamic characteristics of the airframe-engine system with an operating air-breathing engine;

- investigation of heat fluxes from the test gas to various structural elements, including local regions with anomalously high heat fluxes;
- investigation of the laminar-turbulent transition in boundary layers and flow evolution in near-wall regions both on external surfaces of the vehicle and in the engine duct;
- investigation of the internal flow of the test gas in structural elements of high-velocity air-breathing engines (inlets, combustion chambers, nozzles), including situations with an external flow around the flying vehicle, and mutual effect of structural elements in the engine system on each other;
- investigation of the processes in combustion chambers of air-breathing engines and the influence of the mixing efficiency, fuel type, injection conditions, ignition, and combustion efficiency both in an insulated combustor (with an attached pipeline) and in a combustor operating in a combination with the inlet;
- determination of the total thrust-economic characteristics of air-breathing engines.

2. The short time of the test regime in these facilities provides some favorable factors for the development of hardware for aerodynamic tests:

   - It is possible to avoid cooling the model despite significant heat fluxes in the combustion chamber. This possibility substantially simplifies the model design because thermal protection at hypersonic velocities can hardly be provided even in projects of real-size vehicles; this problem becomes next to unsolvable for models because of the smaller size, expensive materials, and complexity of fabrication.
   - Various sensors located inside the model are protected from thermal overloads;
   - It is possible to ensure large flow rates of the gas per second with a small total amount of the gas used. This fact resolves the problem of fire safety of tests and simplifies the fuel supply system structure;
   - As the fuel amount needed for the experiment is rather small, it can be located directly inside the model, which ensures modeling of the external flow around the operating model with minimum distortions of the flow and with minimum distortions of the balance measurements;
   - A small amount of the fuel used in the experiments requires a small amount of energy needed for heating the fuel for modeling the mixing process;
   - It is possible to use various gases (air, N_2, CO, CO_2, and others) or various types of the fuel for engine models since the cost of these substances and the issues of safety are not very important due to their small amounts;
   - The shock starting of the wind tunnel eliminates the necessity of using control of the inlet during this process and makes it possible to avoid complications in the model structure;

3. The accumulated experience from conducted tests shows that short-duration wind tunnels can involve the use of high-accuracy measurement tools and experimental methods, which was previously expected to be used only in blowdown tunnels:

   - measurement of free-stream parameters: total pressure and heat flux rakes, miniature multichannel sensors of flow deflection, dynamic measurements of pressure by moving sensors;
   - measurement of pressure distributions over the model surface: fast-response miniature sensors for measurement of variable pressures, pressure-sensitive paints for panoramic visualization;
   - measurement of aerodynamic forces and moments: fast-response multi-component strain-gauge and piezoelectric balance systems;
   - measurement of heat fluxes on the model surface: temperature sensors and heat flux sensors of various types, coaxial thermocouples, thin-film sensors, thin-wall method with a high spatial resolution, liquid-crystal coatings, temperature-sensitive films, infrared thermography;
   - measurement of shear stresses on the model surface: heated film sensors mounted into the model surface, interferometry of oil spots, oil-film visualization;
   - Optical measurements: schlieren pictures, shadowgraphy, interferometry, video filming with a time resolution of less than 1 μs, visualization with an electric discharge, thermal imaging systems;
   - fast-response and high-accuracy control systems, data acquisition systems, and data processing systems.

4. A large cycle of activities performed at ITAM SB RAS with the use of short-duration wind tunnels shows that such facilities offer many prospects in studying models of hypersonic flying vehicles with air-breathing engines, including modeling of real values of aerodynamic and thermal criteria of similarity and reproduction of real stagnation pressures and temperatures of the free stream incoming onto the model.
5. Based on the experience gained in using high-enthalpy short-duration wind tunnels at ITAM SB RAS, approaches for further upgrading of these facilities are developed to expand the capabilities of the test facilities of ITAM SB RAS.

References

1. Lukasiewicz, J.: Experimental Methods of Hypersonics, p. 258. Marcel Dekker Inc., New York (1973)
2. Bezmenov, V.Y., Ivashchenko, A.I., Sjirmanov, P.M.: Some aspects of the development and application of shock tubes, Obzory ONTI TsAGI, 1977, No. 517, 8 p
3. Rate, S.R., Eaves Jr., R.H.: Recent advances in the performance and testing of the AEDC-VKF tunnel F (hot-shot) hypersonic facility. AIAA Paper No. 84 (1974)

4. Korolev, A.S., Boshenyatov, B.V., Druker, I.G., Zatoloka, V.V.: Hotshot Wind Tunnels in Aerodynamic Research, p. 80. Nauka, Novosibirsk (1978)
5. Zatoloka, V.V.: Hotshot Wind Tunnels, p. 143. Nauka, Novosibirsk (1986)
6. Kislykh, V.V., Rakhmatulin, KhA: Double-section facility with adiabatic compression. Teplofiz. Vys. Temp. **10**(2), 400–404 (1972)
7. Antonov, A.S., Boshenyatov, B.V., Dmitriev, V.A., Zatoloka, V.V., Puzyrev, L.N., Yaroslavtsev, M.I.: IT-301 hotshot wind tunnel with hypersonic velocities based at ITAM SB USSR AS. In: Aerophysical Research ITAM SN USSR AS, Novosibirsk, pp. 20–24 (1972)
8. Boshenyatov, B.V., Voronov, G.M., Zatoloka, V.V.: On hotshot wind tunnels with supercritical Reynolds numbers from hypersonic to transonic Mach numbers. In: Gas Dynamics and Physical Kinetics, ITAM SB USSR AS, Novosibirsk, pp. 173–176 (1974)
9. Baev, V.K., Boshenyatov, B.V., Pronin, Y.A., Shumskii, V.V.: Experimental study of ignition of hydrogen injected into a supersonic hot air flow. In: Gas Dynamics of Combustion in a Supersonic Flow, Novosibirsk, pp. 53–63 (1979)
10. Baev, V.K., Boshenyatov, B.V., Pronin, YuA, Shumskii, V.V.: Injection of liquid into a supersonic flow of a high-enthalpy gas. Comb. Expl. Shock Waves **17**(3), 301–304 (1981)
11. Boshenyatov, B.V., Puzyrev, L.N., Yaroslavtsev, M.I.: Study of operation of the IT-301 hotshot wind tunnel with a double-section settling chamber. In: Issues of Gas Dynamics, ITAM SB USSR AS, Novosibirsk, pp. 291–294 (1975)
12. Dmitriev, V.A., Trushnikov, Y.F., Puzyrev, L.N., Yaroslavtsev, M.I.: Hotshot wind tunnel, Author's Certificate USSR No. 1156462, G01M9/00
13. Puzyrev, L.N., Yaroslavtsev, M.I.: Stabilization of gas parameters in the settling chamber of a hypersonic hotshot wind tunnel, Izv. SO AN SSSR, Ser. Tekhn. Nauk, No. 5, pp. 135–140 (1990)
14. Shumskii, V.V., Yaroslavtsev, M.I.: Chemical heating of the working gas in a short-duration high-enthalpy facility. Comb. Expl. Shock Waves **43**(5), 518–529 (2007)
15. Shumskii, V.V., Yaroslavtsev, M.I.: Composition of the test gas in the test section of a high-enthalpy wind tunnel. Comb. Expl. Shock Waves **48**(1), 24–32 (2012)
16. Yaroslavtsev, M.I., Fomin, V.M., Maslov, A.A., Puzyrev, L.N., Shumskii, V.V.: Wind tunnel, Patent 2436058 RF, MPK G 01 M 9/2, Application No. 2010129903/28, 16.07.2010; published 10.12.2011, Bul. No. 34
17. Maslov, A.A., Shumskii, V.V., Yaroslavtsev, M.I.: Short-duration wind tunnel with combined heating and stabilization of parameters. J. Appl. Mech. Tech. Phys. **53**(6), 805–811 (2012)
18. Zvegintsev, V.I.: Application of short-duration wind tunnels at $M < 8$, Izv. SO AN SSSR, Ser. Tekhn. Nauk, No. 5, pp. 129–134 (1990)
19. Zvegintsev, V.I.: Transit short-duration gas-dynamic facility. Abstracts of the All-Union School on the Methods of Aerophysical Research, p. 217. ITAM SB USSR AS, Novosibirsk (1986)
20. Zvegintsev, V.I.: Impulse wind tunnel with ohmic heater. In: Proceedings of the International Conference on the Methods of Aerophysical Research (ICMAR-92), Vol. I, pp. 171–173 (1992)
21. Kharitonov, A.M., Zvegintsev, V.I., Fomin, V.M., Topchyan, M.E., Mescherjakov, A.A., Pinakov, V.I.: New-generation hypersonic adiabatic compression facilities with pressure multiplier. In: Lu, F., Marren, D. (eds.) Progress in Astronautics and Aeronautics, Vol. 198, pp. 585–619 (2002)
22. Kharitonov, A.M., Zvegintsev, V.I., Vasenev, L.G., Kuraeva, A.D., Nalivaichenko, D.G., Novikov, A.V., Paikova, M.A., Chirkashenko, V.F., Shakhmatova, N.V., Shpak, S.I.: Characteristics of the AT-303 hypersonic wind tunnel. Part. 1. Velocity fields. Thermophys. Aeromech. **13**(1), 1–16 (2006)
23. Aulchenko, S.M., Galkin, V.M., Zvegintsev, V.I., Shiplyuk, A.N.: Design of multimode axisymmetric hypersonic nozzles by using optimization methods. Inzh.-Fiz. Zh. **82**(6), 1109–1112 (2009)

24. Aulchenko, S.M., Galkin, V.M., Zvegintsev, V.I., Shiplyuk, A.N.: Numerical design of multimodal axisymmetric hypersonic nozzles for wind tunnels. J. Appl. Mech. Tech. Phys. **51** (2), 218–225 (2010)
25. Zvegintsev, V.I., Kharitonov, A.M., Chirkashenko, V.F., Chibisov, S.V., Fletcher, D., Paris, S.: Hyperboloid flare tests in the AT-303 hypersonic wind tunnel. In: Proceedings of the International Conference on the Methods of Aerophysical Research, Novosibirsk, ITAM SB RAS, Novosibirsk, Vol. IV, pp.334–341, 28 June–2 July 2004
26. Goonko, YuP, Zvegintsev, V.I., Mazhul, I.I., Nalivaichenko, D.G., Turko, I.S., Kharitonov, A.M., Chirkashenko, V.F.: Wind-tunnel tests of a model scramjet under high Mach and Reynolds numbers. Thermophys. Aeromech. **10**(3), 315–338 (2003)
27. Kharitonov, A.M., Adamov, N.I., Zvegintsev, V.I., Brodetsky, M.D., Mazhul, I.I., Muylaert, J.M., Kordulla, W., Paulat, J.-C.: Experimental investigations of aerodynamics of re-entry vehicles at natural Reynolds number. In: Proceedings of EUCASS, Moscow, Russia, p. 8, 4–7 July 2005
28. Adamov, N.I., Vasenev, L.G., Zvegintsev, V.I., Mazhul, I.I., Nalivaichenko, D.G., Novikov, A.V., Kharitonov, A.M., Shpak, S.I.: Characteristics of the AT-303 hypersonic wind tunnel. Part. 2. Aerodynamics of the HB-2 reference model. Thermophys. Aeromech. **13**(2), 157–172 (2006)
29. Adamov, N.P., Brodetsky, M.D., Vasenev, L.G., Zvegintsev, V.I., Mazhul, I.I., Kharitonov, A.M., Paulat, J.-C., Muylaert, J., Kordulla, W.: Aerodynamics of reentry vehicles at natural Reynolds numbers. Thermophys. Aeromech. **13**(3), 317–326 (2006)
30. Zvegintsev, V.I., Shiplyuk, A.N., Shpak, S.I.: Mathematical model of the working gas source for hypersonic wind tunnel AT-304. In: Proceedings of International Conference on the Methods of Aerophysical Research (ICMAR-2010), Novosibirsk, Abstracts, Part I, pp. 261–262 (2010)

Author Biographies

Vasily M. Fomin An academician and he is a famous specialist in mathematical modeling in the field of continuum mechanics, the author and co-author of more than 250 scientific papers, including three monographs: “Supersonic two-phase flows under the condition of velocity nonequilibrium of particles” (1980), “Conjugate and unsteady problems of mechanics of reacting media” (1984), and “Shock wave processes in two-component and two-phase media” (1992). The main activity of V.M. Fomin’s scientific research is the construction of mathematical models of the shock wave processes of high-speed impact, the effect of detonation products on condensed media, and heterogeneous particle-laden flows, as applied to the problems of aerodynamics, detonation, and solid-propellant rocket engines.

Together with his colleagues, V.M. Fomin has constructed a theory of combined discontinuity and shock waves in particle-laden flows, developed a method of differential analyzer of shock waves on the basis of numerical data, proposed a method of decreasing the drag of objects in a supersonic flow by using a system of particles acting upon the bow shock wave, developed and justified a theory of pulse mechanism of bodies disintegration at high impact velocities, and developed a software complex that predicts the behavior of various constructions under extreme loading and put this method into practice in industrial research institutes and design bureaus of machine-building organizations.

Anatoliy M. Kharitonov is an internationally recognized Scientist in the field of high-speed fluid mechanics. His research focuses on experimental aerogasdynamics and measurement techniques. He has authored more than 130 papers, which have been published in prestigious Russians and International Journals. He is a member of several Scientific Boards in Russia, Supersonic Tunnel Association International, and European Hypersonic Association. He often made presentations on scientific conferences and was invited in different universities for reading lectures.

Professional Experience: Senior Scientist, Head of Lab of Supersonic Aerodynamics, ITAM SB RAS since 1968 up to 2001, now Professor, Chief Scientist

Honors

Honored active worker of Science Russian Federation
Laureate, prizewinner of Russian Council of Ministers
Laureate, prizewinner of Jukovsky

Anatoly A. Maslov is well-known for his studies in Fluid Mechanics at the Institute of Theoretical and Applied Mechanics of Novosibirsk in Russia. His research mainly deals with stability and transition of laminar super- and hypersonic boundary layers; gasdynamic of weakly ionized gases; application of active flow control to aerial vehicles; sensing and control of unsteady flows and separation; and microscale and nanoscale flows. He received his "Dipl.-Phys." degree in physics from the State University of Novosibirsk in Russia and his "Dr." and "Prof." degrees from the Institute of Theoretical and Applied Mechanics of Novosibirsk. Also the Novosibirsk Technical University has awarded him the rank "Professor" and he is a lecturer for Fluid Mechanics. His current interest is focused on the fundamental studies in the areas of experimental fluid dynamics and applied aerodynamics.

Alexander N. Shiplyuk is well-known for his studies in hypersonic boundary layers instability and controls at the Institute of Theoretical and Applied Mechanics of Novosibirsk in Russia. His research mainly deals with investigation of receptivity, stability, transition and control of high-speed boundary layers, development of controlled wave packet method for high-speed flows, hot-wire anemometer application for high-speed flows, development of novel hot-tube sensors based on nano-technology, aerodynamic forces and heat transfer measurements in high-speed flows. He received his "Dipl.-Eng." degree in aero-hydrodynamics from the Novosibirsk Electro-Technical Institute of Novosibirsk in Russia and his "Dr." and "Prof." degrees from the Institute of Theoretical and Applied Mechanics of Novosibirsk. He is Corresponding member of the Russian Academy of Sciences. His current interest is focused on the fundamental studies in the areas of experimental gasdynamics.

Valentin V. Shumskii Doctor of technical sciences, is a leading researcher at the Khristianovich Institute of Theoretical and Applied Mechanics of the Siberian Branch of the Russian Academy of Sciences. His scientific interests include various issues of power and propulsion of high-velocity engines with heat and mass supply and the use of short-duration wind tunnels for studying the performance and force characteristics of such engines.

Mikhail I. Yaroslavtsev Ph.D. in technical sciences, has been a researcher at the Khristianovich Institute of Theoretical and Applied Mechanics of the Siberian Branch of the Russian Academy of Sciences since 1968. His research interests are in the field of development and improvement of hypersonic technologies, processes of heat and mass transfer, propulsion, and supersonic combustion. He was the founder of a new type of a short-duration high-enthalpy wind tunnel, which is updated on a regular basis and is used for versatile experiments aimed at studying promising schemes of intakes and combustion chambers of air-breathing engines. A specific feature of this wind tunnel is the possibility of stabilization of flow parameters during the run time. The current research deals with solving various issues of design of large-size (nozzles 1-1.5 m in diameter) high-enthalpy wind tunnels. The results of these studies are protected by several patents.

Valery I. Zvegintsev is well-known specialist in the field of experimental aerodynamics. He is the designer of new type short-duration wind tunnels, which are of interest over the world. He has published more than 150 scientific papers and his monograph "Short-duration gas-dynamic facilities. P.1. Facilities for scientific research" is among them. The scope of his current interests concerns the experimental investigations of hypersonic vehicles with air-breathing propulsion with help of short-duration gasdynamic facilities.

Shock Tunnels and Hypersonic Research in IISc, Bangalore

Rengarajan Sriram and Gopalan Jagadeesh

1 Historical Overview

The growing national interest in hypersonic missions in the early 1970s demanded experimentation with high enthalpy flow field; such that the high temperature and real gas effects characterizing hypersonic flows are considered. While the pre-heating of the test gas in hypersonic wind tunnels is sufficient to avoid liquefaction in the expanded flow, the simulation of high total enthalpies associated with the hypersonic flow required impulse facilities like shock tunnel; where the test gas to be expanded to the required hypersonic Mach number is compressed and heated by means of shock wave propagation in a shock tube. The systematic experimental study of shock waves and shock dominated phenomena in IISc, Bangalore begun with the installation of its first shock tube in 1972. The laboratory was effectively instituted, as High Enthalpy Aerodynamics Laboratory (HEAL), with the commissioning of the country's first hypersonic shock tunnel HST-1 in the year 1973. A photograph of HST-1 is given in Fig. 1. The shock tunnel consisted of an aluminum shock tube, 8.4 m long and 50 mm in diameter. The end of the shock tube was connected to a conical nozzle, whose throat area can be varied so as to simulate Mach numbers ranging from 4 to 13 in the test section of 300 mm × 300 mm cross section ending into a dump tank [1].

HST-1 was operated in conventional fashion. Driver and driven sections of the shock tube were separated by a metal diaphragm. The material and thickness of the metal diaphragm dictates its rupture (driver) pressure; the metal diaphragm was ruptured by filling and pressurizing the driver section with the driver gas. The driver pressure, and the pressure maintained in the driven section containing the test gas

R. Sriram · G. Jagadeesh (✉)
Department of Aerospace Engineering, Indian Institute of Science, Bangalore 560012, India
e-mail: jaggie@aero.iisc.ernet.in

R. Sriram
e-mail: sriram@aero.iisc.ernet.in

O. Igra and F. Seiler (eds.), *Experimental Methods of Shock Wave Research*,
Shock Wave Science and Technology Reference Library 9,
DOI 10.1007/978-3-319-23745-9_11

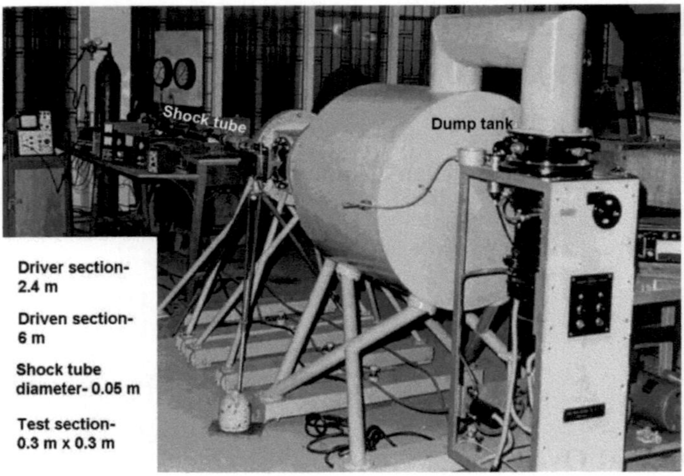

Fig. 1 A photograph of HST-1: India's first hypersonic shock tunnel

(by means of a vacuum pump) dictate the shock Mach number in the shock tube, which in turn dictates the flow enthalpy and pressure. The dump tank was maintained at much lower pressure than the driven section by means of a powerful vacuum pump; this was enabled by a paper diaphragm at the junction of the driven section and the nozzle. The test time of the tunnel ranged from 500 μs to 1 ms, and accordingly the tunnel was equipped with fast response diagnostics (especially for pressure, heat transfer rate and aerodynamic force measurements) and data acquisition systems.

A summary of the research investigations in HST-1 from 1975 to 1995 was presented by Reddy et al. [2]. The important experimental results include the measurement of aerodynamic forces on missile shaped bodies [3, 4], (qualitative) visualization of shock waves in the shock tunnel flow field over various geometries using electric discharge technique [5] and heat transfer rate measurement using platinum thin film sensors (painted on the thermally insulating Pyrex substrate) [6]. Apart from academic research, HST-1 also facilitated measurements useful for missions of national interest. This is exemplified by the measurement of heat transfer rates on (scaled model) of Indian Space Research Organization's (ISRO) Satellite Launch Vehicle SLV-3; the measurements were also performed in the SLV-3 model with the presence of boundary layer trip [7].

With the commissioning of more sophisticated shock tunnels with upgraded vacuum systems and better materials that can withstand higher pressures (in turn allowing simulation of higher enthalpies), HST-1 has become outdated and the research works in HST-1 have long been shelved. Nevertheless, the experience gained from the research in HST-1 has provided important guidance in developing the advance facilities and equipping them with advanced diagnostics, and illuminated the directions for the future course of hypersonic research in the laboratory.

The contemporary research in the laboratory may thus be demarcated as the post HST-1 research. The shock tunnel facilities, currently operational, includes—a conventional hypersonic shock tunnel HST-2 (1995–96), a free-piston tunnel HST-3 (2004–05), another conventional tunnel HST-4 (2005–06) with larger test section (enabling possible boundary layer transition-turbulence), and the tunnel HST-5 (2008–09) which can be driven by combustion in the driver section. Apart from the shock tunnels for hypersonic flow research, the laboratory has developed a number of shock tubes towards the fundamental and applied research on shock waves. This includes two shock tubes for the study of high temperature chemical kinetics CST-2 (2006–07) and CST-3 (2010–11), while another shock tube towards chemistry research CST-1 (2001) is currently not operational. A diaphragm-less shock tube (DST) has been under development; with certain automatic valves in the place of diaphragm giving promising results. A vertical shock tube (2014) for the study of shock propagation in liquids and solids, with such automatic valve in the place of diaphragm, has been commissioned and is being calibrated. Elaborate studies on shock tubes of small scales is underway, for which a shock tube of 1 mm diameter driven using solar power is developed, which shall as well be used for blast wave generation and research. With the growing horizons of research, especially in the direction of shock wave research and applications, the laboratory acquired its current name Laboratory for Hypersonic and Shock Wave Research (LHSR) in the year 2010, and has been actively collaborating with researchers from various domains including Chemistry, Material Science and Engineering, Nano research and Biological Sciences. A description of the important shock tunnel facilities in LHSR, the diagnostics and the prominent research activities that have been going on with the facilities shall be presented subsequently.

2 Hypersonic Shock Tunnels and Diagnostics

It was just mentioned that four hypersonic shock tunnel facilities (HST-2 to 5) with different capabilities are presently operational in LHSR. Expectedly they have very short test times ranging from $\sim$200 μs to a few milliseconds. Further, contemporary research in high speed flows require resolution of high frequency phenomena which are often expected (cavity tones, for example). Hence the instrumentation should have fast response, and the rate of data acquisition must be high. There is also a requirement of good spatial resolution of the data since drastic spatial variations may be expected of shock dominated flows. This requires miniaturized sensors for measurements. The tunnels are equipped with standard (commercial) fast response sensors for pressure measurements (such as miniature PCB and kulite sensors) and force measurements (PCB accelerometers). Further, there are sensors developed in-house with capabilities of the required spatial and temporal resolution, such as the platinum thin film sensors for heat transfer rate measurement and piezo-resistive Micro Electro-Mechanical System (MEMS) based sensors for closely spaced pressure measurements. The measured data is acquired digitally in systems capable of acquiring at high rates, from 1 to 10 mega samples per second (from National Instruments, Dewetron and Yokogawa). The flow fields in the tunnels are visualized by means of time resolved schlieren technique with the use of high speed cameras. The diagnostics (especially the in-house made sensors) and their measurements are illustrated better, when discussing the different flow fields where they are applied. Detailed description of the shock tunnels and the important flow field investigations pursued in them are subsequently presented.

2.1 HST-2

Hypersonic Shock Tunnel HST-2 is a conventional shock tunnel facility as mentioned before, similar to HST-1 but with upgraded vacuum systems and materials; it has a steel shock tube of 50 mm internal diameter (with 12.5 mm thick walls), with driver and driven sections of length 2 and 5.12 m respectively, separated by a metal (usually Aluminum) diaphragm. The end of the shock tube is connected to a conical nozzle whose throat area can be varied for simulating the required Mach number. The exit of the nozzle, 300 mm in diameter, leads to a rectangular test section of 300 mm × 300 mm cross section and length 450 mm. The test section ends in a dump tank of volume 1 m^3. A paper diaphragm separates the shock tube and the nozzle, so that the driven section and the dump tank are maintained at different vacuum levels. The dump tank is evacuated to about 10^{-5} mbar by means of vacuum pumps (a combination of rotary and diffusion pumps) to ensure a weaker starting shock as well as better establishment of the hypersonic flow. Figure 2 presents a schematic of HST-2. A photograph of the tunnel is given in Fig. 3.

Various test gases can be used by varying the driven gas. The driver gas is usually Helium (since a lighter gas is preferable). With air as test gas the facility can simulate Mach numbers ranging from around 6 to 12, based on the operation mode —'straight through' (without throat insert) or 'reflected' mode with suitable throat inserts (which decides the area ratio); the Mach numbers can however be higher with test gas like Argon for same nozzle area ratio. The tunnel can simulate a maximum of 3 MJ/kg, considering safety of the shock tube walls which may have to withstand high pressures for higher enthalpies. The Reynolds number ranges from a million to 5 million per meter with air as test gas. The end of the shock tube portion (driven section) is equipped with three PCB pressure sensors, which are used to measure the shock speed as well as the reservoir pressure for the tunnel at the nozzle entry after the shock reflects at the shock tube end. From the speed of the shock propagating towards the end of driven section and the speed of reflected shock (calculated from the instances the sensors show jump in pressure), the reservoir conditions (especially the total temperature/total enthalpy) for the hypersonic flow can be estimated.

The establishment of steady freestream in the test section can be ensured from the measured pitot pressure signal—the pressure measured by a PCB sensor at the stagnation point of a small hemispherical body in the freestream, thus sensing the

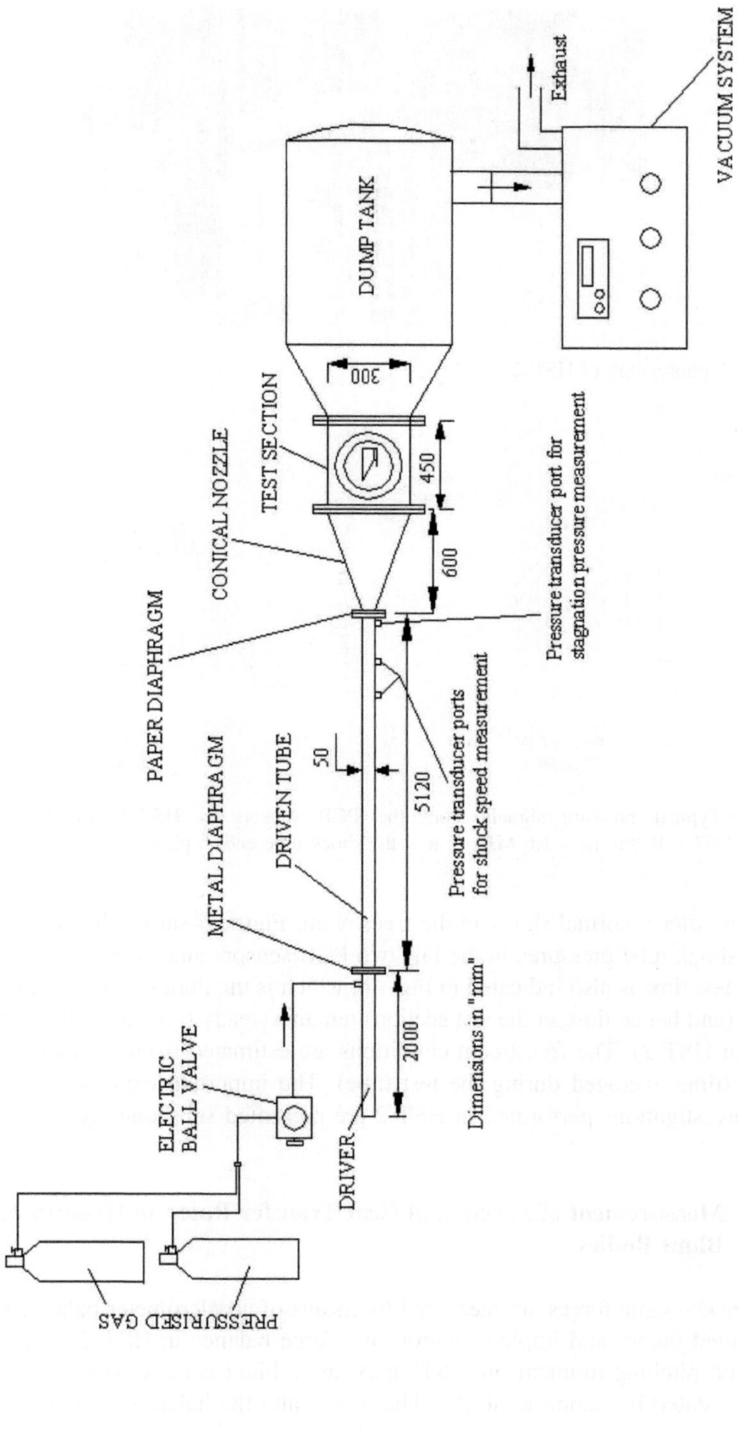

Fig. 2 Schematic diagram of HST-2

Fig. 3 A photograph of HST-2

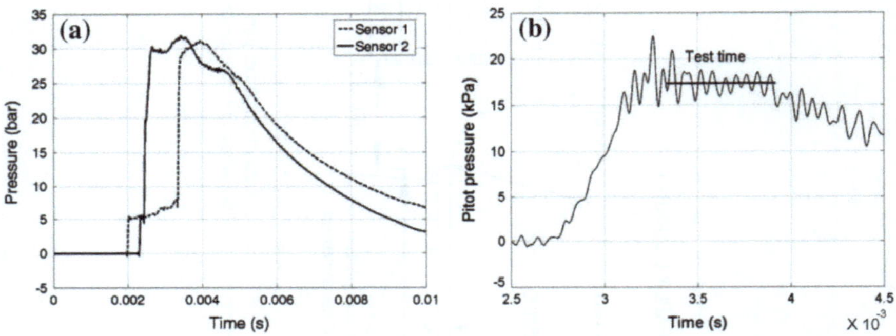

Fig. 4 Typical pressure signals from the PCB sensors in HST-2 (for $M_{\infty} = 8.67$, $Re_{\infty} = 1.67 \times 10^6$/m, $h_0 = 1.6$ MJ/kg), **a** at the shock tube end, **b** pitot

pressure after a normal shock in the freestream. Figure 4 shows the typical signals of (a) shock tube pressures in the last two PCB sensors and (b) pitot pressure. The tunnel test time is also indicated in Fig. 4b, which is the duration for which the pitot signal (and hence flow in the test section) remains steady (typically about 500 μs to 1 ms in HST-2). The freestream conditions are estimated from the measured pitot signal (time averaged during the test time). The important experiments and flow field investigations performed in HST-2 are presented subsequently.

2.1.1 Measurement of Forces and Heat Transfer Rates on Hypersonic Blunt Bodies

The aerodymanic forces are measured by means of accelerometer balance systems. A detailed theory and implementation of a force balance in HST-2 for measuring drag and pitching moment on a 60° apex angle blunt cone at Mach number 5.75 was presented by Sahoo et al. [8]. The model and the balance are held by rubber

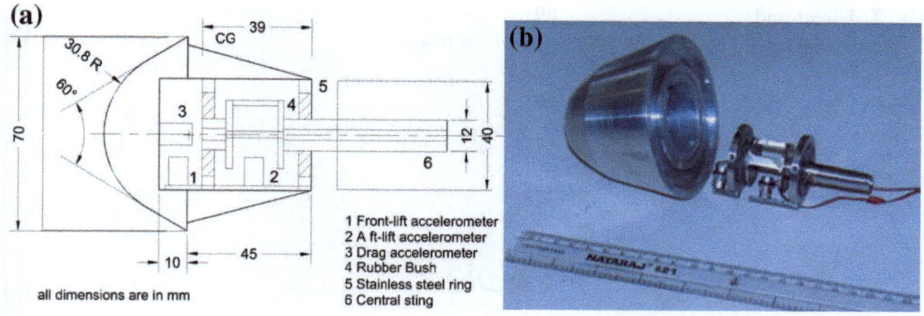

Fig. 5 Test model equipped with accelerometer force balance [8]: **a** schematic (of 60° blunt cone), **b** a photograph (of 60° blunt cone with 15° flare)

bushes which ensure unrestrained free floating of the system within the short test time. A schematic of the model equipped with force balance system and a photograph of the model are given in Fig. 5a and b respectively. Figure 6 shows a comparison of the accelerometer signal with the pitot signal (both in terms of volts) for the flow with test time of 450 μs.

It is seen that the accelerometer signal for the blunt body is qualitatively similar to the pitot signal; the rise time and the time for which the signals remain (statistically) steady are the same. From the sensitivity of the accelerometer in terms of volts/g, the acceleration can be obtained. From the acceleration (averaged during the test time) and the mass of the model, the force on the model can be calculated; with the assumption that within the test time the model if freely floating, since the displacements are small for the spring action of the rubber bushes to be significant. However for models of larger size or models made of materials with complicated properties, the structural vibrations significantly affect the accelerometer response. Hence, to be precise in measuring the aerodynamic forces, the model with the accelerometer balance is calibrated by applying a known impulse force (using an impulse hammer fitted with a sensor); the output (response) of the accelerometer to the known impulse force is measured, and the output is the convolution integral of a response function and the input [9]. From the impulse input and output calibration

Fig. 6 Comparison of accelerometer and pitot signals [8]

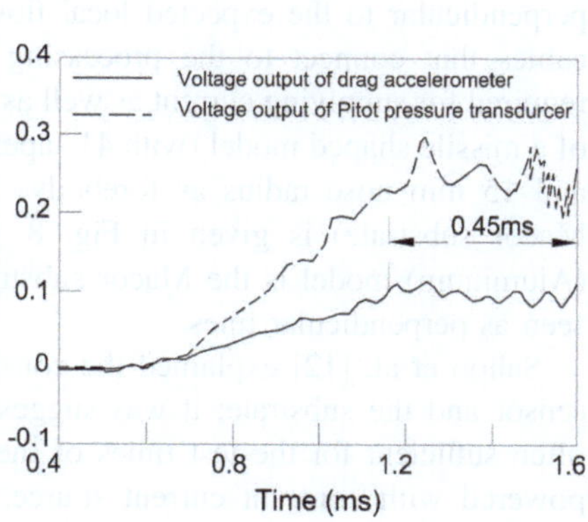

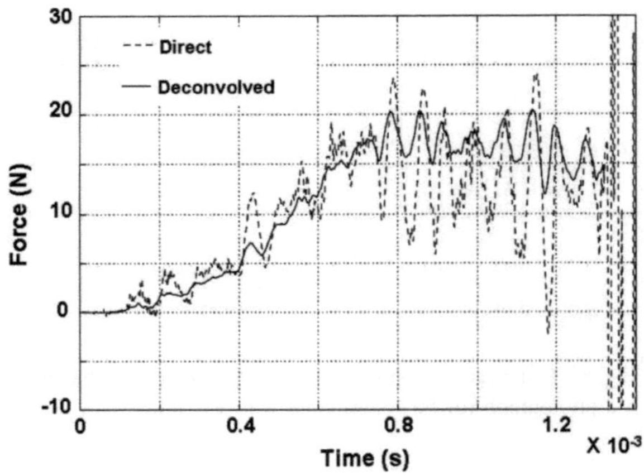

Fig. 7 Direct and de-convolved force signals [10]

signals, the force on the model during the flow event can be obtained by de-convolving the measured accelerometer signal. A comparison of the force (drag) signals (at Mach number 9.2) obtained by directly multiplying the accelerometer response with the mass and by de-convolution procedures, for a (120°) large angle blunt cone made of Hylem material, is shown in Fig. 7. The de-convolved drag signal not only seems to have noise of lesser amplitude, but also a considerably different (time averaged) value during the test time. The force measurements in the tunnels are thus obtained by means of the above mentioned calibration and de-convolution procedure.

Surface convective heat transfer rate measurement forms a very important part of hypersonic flow research, since the flows are dominated by high temperatures and heat fluxes; the measurement is also of practical importance in locating regions of peak heat transfer which is a very important concern in many hypersonic missions. LHSR employs in-house made platinum thin film sensors, sputtered on insulating Macor (a machinable ceramic) substrate for surface convective heat transfer rate measurements. The substrate with the sensors is flush mounted with the surface of model avoiding any protrusion on the surface that can affect the aerodynamic shape of the body. Typically the sensor appears as a thin line on the substrate oriented perpendicular to the expected local flow. The ends of the sensor are soldered to cables that connect to the processing and data acquisition systems which are required for supplying current as well as for measuring the voltage. The photograph of a missile shaped model (with 41° apex angle blunt cone of 51 mm base diameter and 15 mm nose radius as forebody) with flush mounted thin film sensors (on Macor substrate) is given in Fig. 8. The white strip in flush with the metal (Aluminum) model is the Macor substrate on which the thin film sensors can be seen as perpendicular lines.

Sahoo et al. [12] explained the considerations to be given while designing the sensor and the substrate; it was suggested that an about 1 mm thick substrate is often sufficient for the test times of the shock tunnels in LHSR. The sensors are powered with constant current source so that any change in resistance (due to

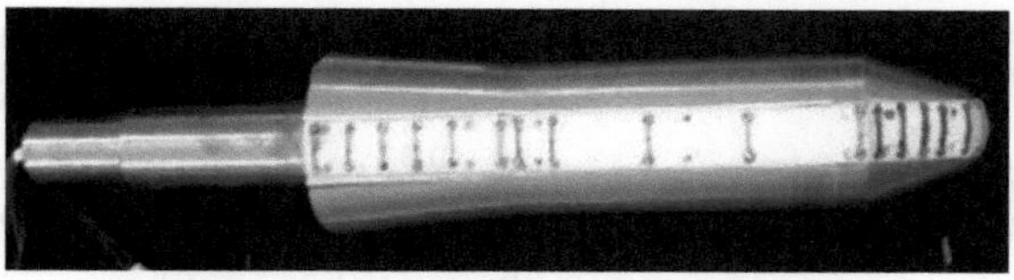

Fig. 8 Photograph of a model with flush mounted platinum thin film sensors [11]

change in temperature during the flow) can produce a corresponding change in voltage. The sensor output is connected to PC-based data-acquisition system. The change in voltage across the gauge with respect to time gives the temperature-time history at the gauge location on the model surface. Within the short test time, the heat conduction is sufficiently one dimensional (in the direction into the substrate), with the gradients in the other directions not significant in comparison. The unsteady 1-D heat conduction model can be solved (integrated) with the boundary conditions that at the surface of the substrate the temperature is the time varying temperature as measured by the sensor and at infinity (the other end of the substrate) there is no variation in temperature from the initial temperature. To solve the equation the temperature coefficient of resistance of the thin film sensor should be known, for which the sensor should be calibrated. Schultz and Jones [13] gave an expression for the wall heat flux in terms of the sensor temperature, which is evaluated numerically using the algorithm given by Cook and Felderman [14].

A measured temperature signal (in volts) at the stagnation point of the missile shaped model at Mach number 5.75 (total enthalpy of 2 MJ/kg) in HST-2 and the corresponding heat transfer rate deduced from the above mentioned procedure are given in Fig. 9 (a and b respectively). It can be seen that the temperature rise is parabolic; the parabolic temperature rise gives steady heat transfer rate on integration as may be seen in the figure. The variation of the convective heat transfer rates (normalized by the stagnation point heat transfer rate) along the missile shaped body is given in Fig. 10. Two different experimental data are plotted: one for the

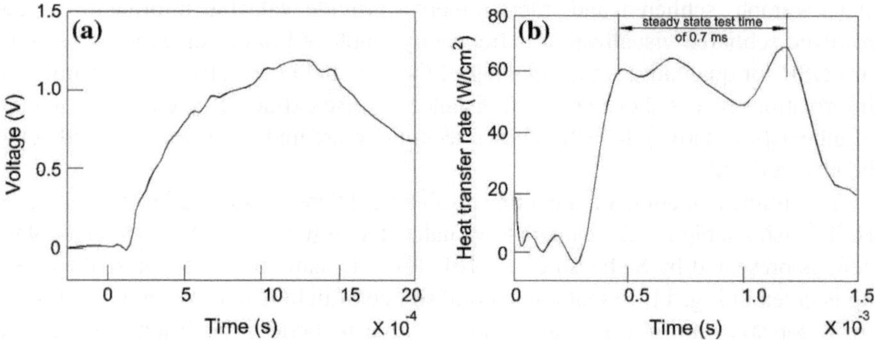

Fig. 9 Measurement using thin film sensor at stagnation point of missile shaped body [11]: **a** temperature signal **b** integrated heat transfer rate

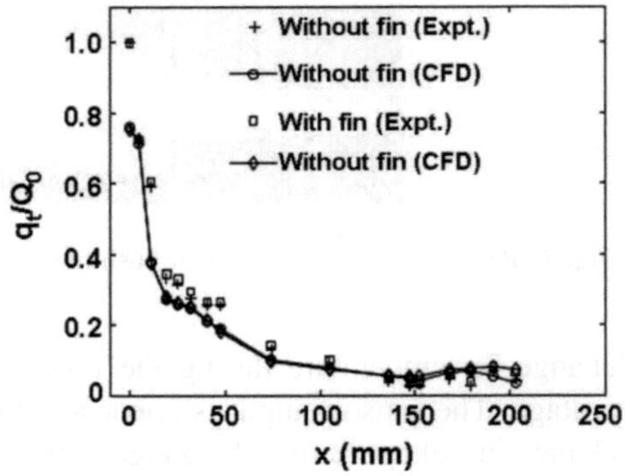

Fig. 10 Heat transfer distribution on the missile shaped model at Mach number 5.75 and total enthalpy 2 MJ/kg [11]

model without fin, other for the model with the addition of fin. A good comparison with the computations using a commercial CFD code is also demonstrated.

The heat transfer rates are not only measured on blunt bodies but also on many other shock tunnel flow fields such as the flat plate flow and flow field with separation over a backward facing step, to cite a few examples [15]. Concerning blunt bodies, the aerodynamic forces and surface convective heat transfer rates are often measured simultaneously, as exemplified by Sahoo et al. [12].

2.1.2 Schlieren Visualizations of Shock Tunnel Flow Fields

The visualization of the flow field is very important in illuminating various regions of the flow concretely, which can help in better interpretations of the measurements from the sensors (while there are also visualization techniques that provide quantitative information). Implementation of visualization techniques in shock tunnels are challenging since it requires the capturing of events in very short times. This is enabled by the modern high speed cameras. With the flow visualization by means of seeding (as in particle image velocimetry) being difficult and inaccurate in shock tunnels, the techniques based on light deflection due to density gradients (like shadowgraph, schlieren and interferometry) provide valuable information. Time resolved schlieren visualization is frequently employed in the shock tunnel studies in LHSR for qualitative understanding of the flow field (though certain quantitative information such as shock stand off distance are also extracted); the implementation of other (quantitative) deflection based techniques are under development and yet to be standardized.

The implementation of the time resolved schlieren visualization technique in HST-2 using a high speed camera to visualize the flow over a 120° apex angle blunt cone is presented by Satheesh et al. [16]. The schematic of the z-type schlieren set up is given in Fig. 11. Essentially a point source of light is rendered a parallel beam by a concave mirror (the plane mirrors serve to deflect the beams in required direction). The parallel beam passes through the (flow) field of interest, coming out with the shadow of the field. This is then focused by another concave mirror, and at

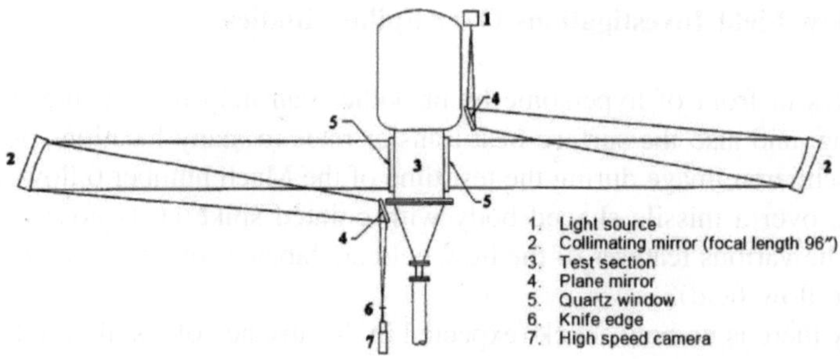

Fig. 11 Schematic of the z-type schlieren visualization arrangement for HST-2 [16]

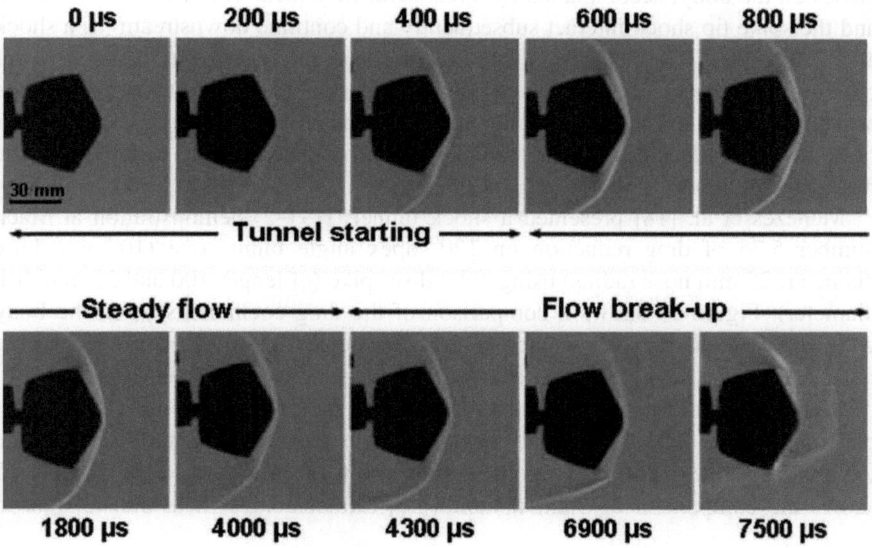

Fig. 12 Time resolved schlieren flow visualization over 120° apex angle blunt cone [16]

the focus the beam is cut by a knife edge. Depending on the density gradient in the field the beam is (locally) deflected towards or away from the knife edge [17]; thus enabling visualization of shocks and expansions in the high speed camera placed behind the knife edge. The time resolved flow field over the 120° apex angle blunt cone at Mach number 9.2 is shown in Fig. 12.

The events during the tunnel starting time, test time and the flow break up (as observed in pitot signal) are clearly indicated in the figure. From a known reference length, quantities such as shock stand off distances can be measured from these schlieren images. The schlieren images have complemented the measurements and many times illuminated many aspects of different flow fields in the shock tunnels; this shall be illustrated subsequently with the presentation of some of the more complex flow fields studied in HST-2 (and later, in other shock tunnels in LHSR).

2.1.3 Flow Field Investigations Over Spiked Bodies

Aero-spikes in front of hypersonic blunt bodies can help in reducing the aerodynamic drag, and also the surface heat transfer rates in many locations on the blunt body. A schlieren image during the test time of the Mach number 6 flow, simulated in HST-2, over a missile shaped body with pointed spike at its nose is shown in Fig. 13. The various features of the flow field are labeled (on one side of the visibly symmetric flow field).

Clearly there is no bow shock (expected in the absence of a spike) in front of the blunt body. The adverse pressure due to the expected bow shock induces the boundary layer over the spike to separate. The separation occurs at the spike tip, accompanied by a (separation or spike tip) shock. The separated shear layer reattaches on the body, accompanied by a reattachment shock. The reattachment shock and the spike tip shock interact subsequently and continue downstream (as a shock after reflection). The separation bubble (or 'dead air zone') has a much lower pressure than the pressure that the nose experiences behind the bow shock in the absence of a spike. This results in the reduction in aerodynamic drag. Heat transfer rates are also reduced in the separation bubble, but peak heating can occur in the reattachment zone.

Menezes et al. [18] presented a shock tunnel (HST-2) demonstration at Mach number 5.75 of drag reduction on 120° apex angle blunt cone (100 mm base diameter, 25 mm nose radius) using a flat disc spike (of length 100 and 25 mm disk diameter). Figure 14 shows a comparison of the drag coefficients c_d on the body with and without the disk spike, for various angles of attack. Significant drag reduction of up to 64.3 % was observed at lower angles of attacks (until 7°).

Significant heat transfer rate reduction with the spike was also observed for most part on the portions of the blunt cone inside the separation bubble, while in the vicinity of reattachment zone high heat transfer rates were observed [19]. Figure 15 shows the comparison of Stanton number S_t (dimensionless heat transfer rate in

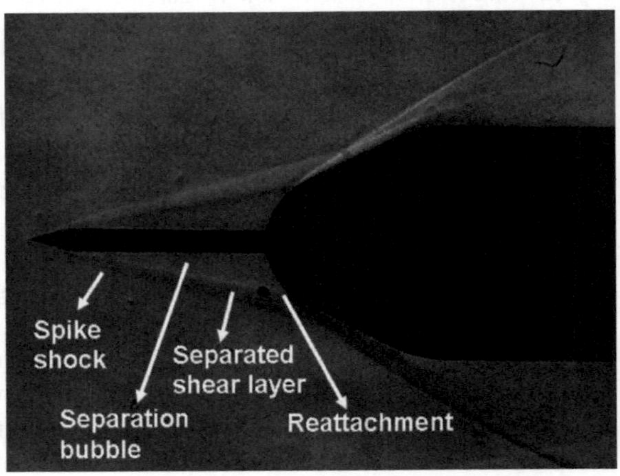

Fig. 13 A schlieren image in HST-2 of the Mach number 6 flow over a spiked missile shaped body

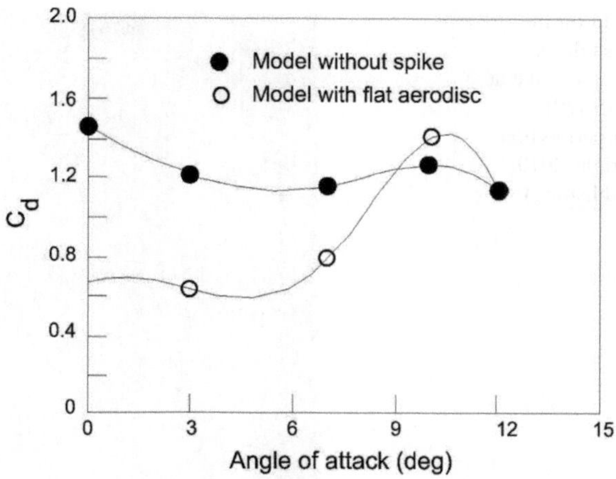

Fig. 14 Comparison of drag coefficient with and without spike at Mach number 5.75 (total enthalpy—1.2 MJ/kg) [18] (with kind permission of Springer Science+Business Media: Springer [18, p. 200])

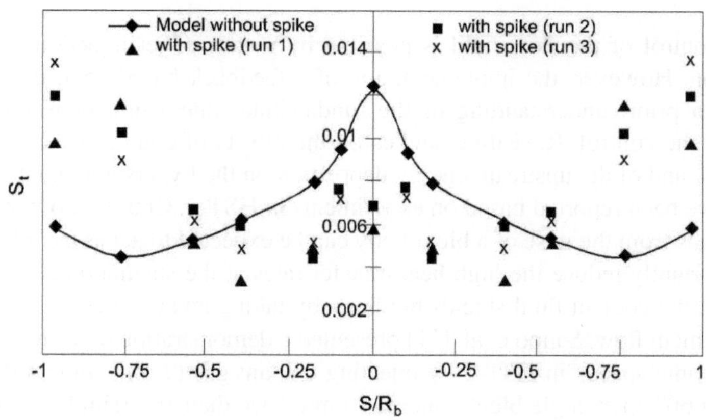

Fig. 15 Stanton number distributions at Mach number 5.75 on the 120° apex angle blunt cone with and without the flat disc spike [19]

relative terms to the thermal capacity of the fluid) distribution with and without the spike on the blunt cone; the Stanton number is plotted against the distance from the nose along the model surface S, normalized by the blunt nose radius R_b. After the reattachment it is seen that the heat transfer rate reaches such high values comparable to the stagnation point heat transfer rate value on the blunt body without spike. Thus despite significant reduction in drag, important concerns arise in the flow field with spike due to regions of high heat transfer rate. A better drag reduction at higher angles of attack (for the 120° apex angle blunt cone) was also reported by using a telescopic spike with multiple discs of varying diameter [20], the flow field over which is shown in the schlieren image in Fig. 16.

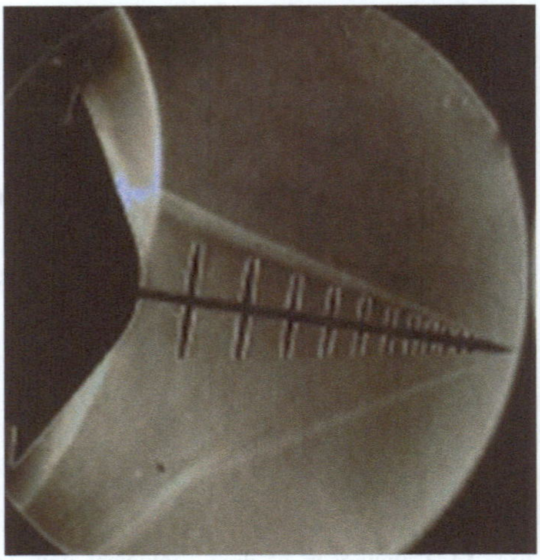

Fig. 16 Flow field on the blunt cone at 8° angle of attack with telescopic spike at Mach number 5.75 [20] (reproduced with permission from [20]. Copyright 2010, Multi-Science Publishing)

2.1.4 Aerodynamic Control by Counter Flow Jet Injection and Upstream Energy Deposition

Active control of the flow field is possible by means of jet injection and energy deposition. However, the implementation of a feedback based control mechanism requires a priori understanding of the fundamental interaction between the flow field and the control. Revealing studies on the effects of counter flow jet injection from nose and of the upstream energy deposition on the hypersonic blunt body flow fields have been reported based on experiments in HST-2. Counter flow injection of a (cold) gas from the nose of a blunt body can be expected to act as a coolant, which can importantly reduce the high heat transfer rates at the stagnation zone (nose) of the body; the coolant fluid shields the body by taking away the convective heating from the main flow. Sahoo et al. [21] presented a demonstration of such *film cooling* at hypersonic speeds in HST-2 by injecting coolant gas (2 mm sonic jet) from the nose of a 60° apex angle blunt cone of 70 mm base diameter (similar to the one in Fig. 5a) at zero angle of attack. Though a considerable increase in aerodynamic drag was observed (perhaps due to the reaction to jet momentum), significant reduction in heat transfer rates were observed along the model as shown in Fig. 17, showing percentage reduction in Stanton number S_t (or heat transfer rate) with injection (relative to the case without injection) plotted against the distance along the blunt cone from the nose s normalized by nose bluntness radius R_n; more reduction was observed at higher enthalpy (h_0 = 1.6), particularly in the nose portion (stagnation zone).

Higher mass injection of the coolant can improve the film cooling performance; however, the higher mass injection is accompanied by higher momentum flux of the injected jet. At higher momentum, the jet can significantly travel upstream and disturb the bow shock, such that as it reattaches on the model, local surface heat transfer peaks occur (although the coolant reduces the heat transfer downstream of

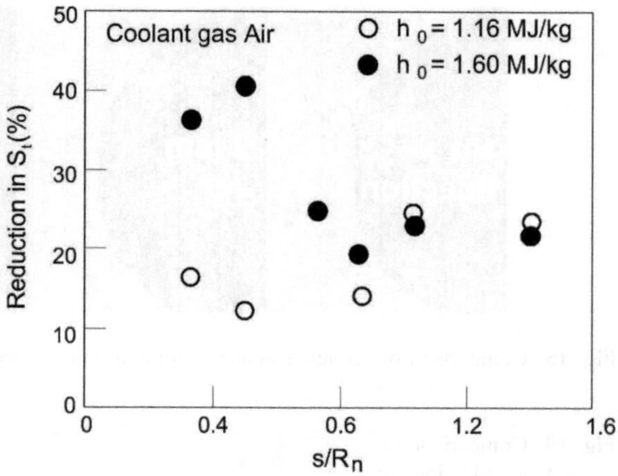

Fig. 17 Reduction in surface heat transfer rate with air injection at Mach number 5.75 [21]

reattachment). In order that this is optimized—that is the mass flow is increased for better film cooling, yet the momentum flux is reduced in order to avoid strong 'reattachment' of the coolant—Sriram and Jagadeesh [22] proposed the injection of the coolant through an array of closely spaced micro-jets. The demonstration in HST-2 was done on a nearly similar large angle blunt cone as before (but with base radius 80 mm). Rather than as a single jet, when injected (out of orifices from the settling chamber inside the model, where the coolant total pressure is also measured) as an array of closely spaced micro-jets of same effective exit area as the single jet, while the mass flow nearly remains the same, the coolant momentum is significantly reduced due to the viscous interaction between all the micro-jets in the array. Figure 18 compares the schlieren images without injection and with injection with jet total pressure to pitot pressure ratio of P = 1.45, as a single jet through 2 mm orifice and as array of micro-jets through 46 orifices of 300 microns placed within a space of 5 mm × 5 mm.

While the single jet can be seen to get almost close to and disturb the bow shock, the array of micro-jets seems to spread immediately on the body as it exits the orifices. This has lead to a better reduction in heat transfer rate with the array of micro-jets as shown in Fig. 19. In fact at the higher pressure ratio in the first sensor away from stagnation point, there is increase in heat transfer with single jet, while with the array there is nearly 40 % reduction at the same location. Expectedly the higher pressure ratio injection (due to corresponding higher mass flow) leads to better reduction with the array than the lower pressure ratio.

While the injection with low momentum is helpful in reducing the heat transfer rates over the blunt body, the injection of the gas at high momentum (high injection pressure ratios) can push the bow shock away—thus acting like a virtual spike—and thus reduce the aerodynamic drag, as demonstrated in HST-2 by Venukumar et al. [23] for the 60° apex angle blunt cone (of dimensions shown in Fig. 5) at Mach number 8. Figure 20a (schlieren image) depicts the various features: the high momentum (sonic) jet carries forward to push the bow shock significantly away

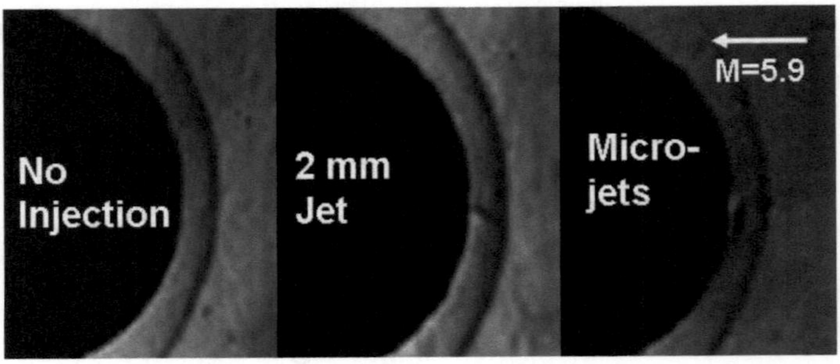

Fig. 18 Comparison of schlieren images with and without coolant injection [22]

Fig. 19 Comparison of percentage reductions in stanton number with coolant injection as single jet and array of micro-jets [22]

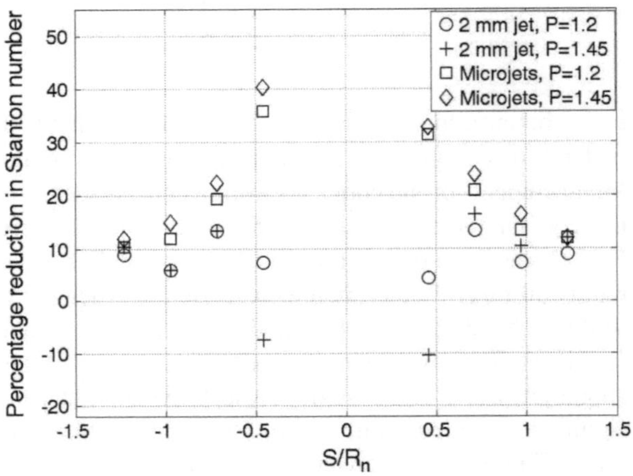

from the body; it is then turned back by the main (hypersonic) flow, reattaches on the body accompanied by a recompression (or reattachment) shock, leaving a region of dead air (or recirculation, similar to the separation bubble in spike flows). The low pressure in the dead air contributes to the reduction in drag. Higher the pressure ratio, more the bow shock is pushed and more spacious is the dead air region, which leads to better reduction as shown in Fig. 20b (showing the variation of percentage reduction in drag coefficient c_d with the ratio of jet total pressure P_{0j} to the freestream pitot pressure P_{0f}).

An interesting extension of the studies on drag reduction using counter flow jet injection was the enquiry on the effect of injecting a hot (plasma) jet from the nose of a blunt body. Over a hemispherical body of 60 mm diameter at Mach number 8.6, it was observed that at injection total pressure to pitot pressure ratio—referred simply as pressure ratio—of 72.5 (gas injection with total pressure of 6 bar), drag was found to be reduced by 28 % with a hot *argon* plasma jet (created by striking electric discharge in the injected jet prior to the exit) relative to a corresponding cold jet as shown in Fig. 21 [24]. For the specific case the value of the plasma current did not seem to have significant effect. The reduction was observed despite

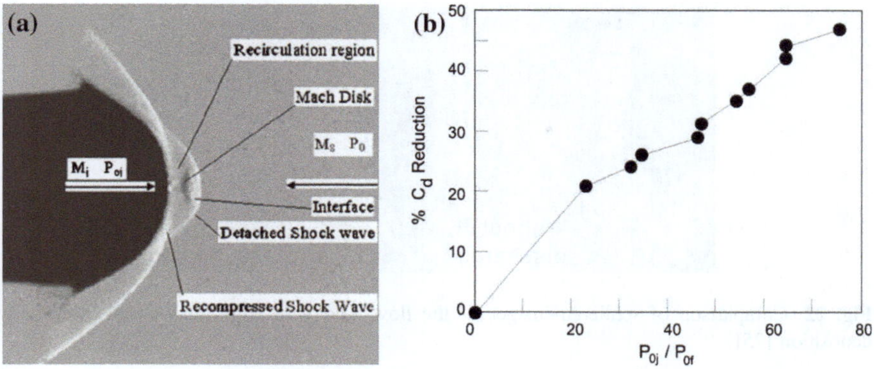

Fig. 20 Flow field over blunt cone with high momentum counter flow injection [23]: **a** a schlieren image, **b** variation of percentage reduction in drag with pressure ratio

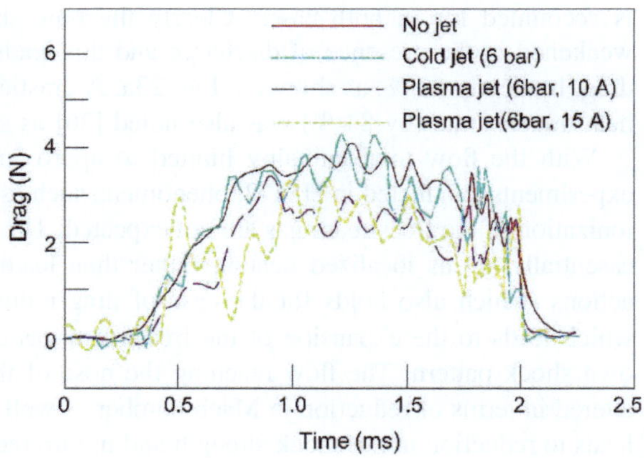

Fig. 21 Comparison of drag signals without fluid injection and with the injection of cold and plasma jets at pressure ratio of 72.5 [24]

the fact that at the same injection pressures hot fluid has lower momentum than cold fluid; the schlieren indicated that the shock stand off distances were nearly the same for both plasma and cold jet cases.

A related study on electric discharges for the control of blunt body flow field was the use of the discharge for energy deposition upstream of the blunt body. Satheesh and Jagadeesh [25] demonstrated reduction of drag on a 120° apex angle blunt cone of 60 mm base diameter at Mach number 9.1 with *argon* as the test gas. The discharge is created in the freestream during the run time by means of electrodes placed upstream of the nose (using a power supply of 10 kV and 10 A). It may be recalled that the drag reduction with counter flow plasma jet injection also had argon as the injected gas. Argon, being mono-atomic, takes all the supplied energy to the translational mode (before initiating electronic excitation and ionization) which determines the temperature. Hence for demonstrative purposes, argon is best suited. A comparison of schlieren images showing the flow field over the blunt

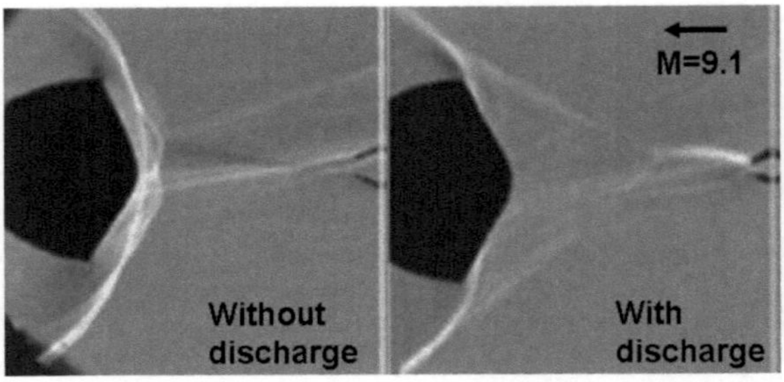

Fig. 22 Comparison of schlieren images of the flow field with and without upstream energy deposition [25]

cone, with and without upstream energy deposition (by discharge) is shown in Fig. 22; the electrodes can be seen in both images, indicating that the disturbance to the flow field due to the mere presence of the electrodes (without striking discharge) is accounted for in both cases. Clearly the bow shock in front of the body is weakened in the presence of discharge and this leads to a significant reduction in drag, by nearly 50 %, as shown in Fig. 23a. A drastic reduction in stagnation point heat transfer rate (by 84 %) was also noted [26] as shown in Fig. 23b.

With the flow total enthalpy limited to approximately 2 MJ/kg in the above experiments conducted in HST-2, phenomena such as those due to dissociation and ionization of the freestream gas are not expected. The local discharge is expected to essentially act as localized heating rather than leading to electro-magnetic interactions (which also holds for the case of drag reduction with argon plasma jet), which leads to the expansion of the freestream gas entering the zone, creating its own shock pattern. The flow reaching the nose of the blunt body is significantly altered in terms of reduction in Mach number as well as change in direction, which leads to reduction in the shock strength and in turn reduction in drag and stagnation point heating.

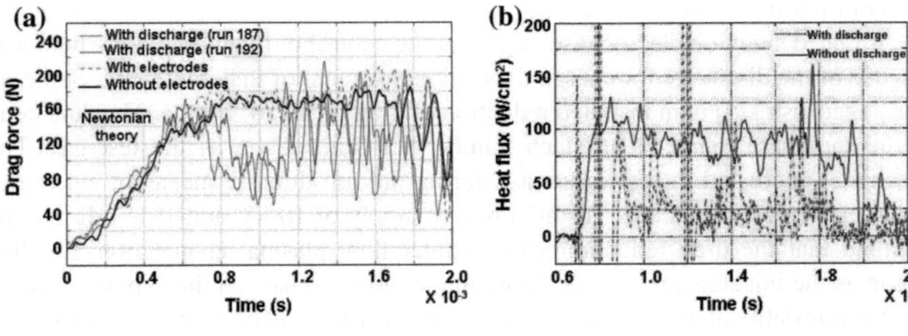

Fig. 23 Comparison of signals with and without upstream energy deposition: **a** drag [25], **b** stagnation point heat transfer rate [26]

2.1.5 Hypersonic Intakes and Shock-Boundary Layer Interactions

A very important research domain in hypersonic flows is the Scramjet and the hypersonic intakes. A number of investigations have been carried out on understanding and optimizing intake flow field, which is very essential for subsequent supersonic combustion. In the intake, the shock from the ramp impinges on the lip of the cowl plate (at design conditions). This provides the required shock compression for the flow entering the combustion chamber from the intake. Mahapatra and Jagadeesh [27] reported a study on shock interactions in generic scramjet intake at Mach number 8 in HST-2, for design and different off-design cases, by varying the position of the leading edge of the cowl relative to the ramp. A comparison of schlieren images of the intake flow fields for the design case and for an off-design case (when shock impinges on the cowl boundary layer) is shown in Fig. 24 (in Fig. 24, C1 is ramp shock, C2 cowl leading edge shock and C3 the separation shock).

At lower Mach number than the design, the shock misses the cowl plate causing spillage; at higher Mach number (equivalently when the cowl plate is long, with leading edge upstream of the ramp shock impingement location) the shock hits the cowl plate downstream of the leading edge, interacting with the cowl boundary layer (like the off-design case shown in Fig. 24). The interaction with the cowl boundary layer—especially as the boundary layer separates (as in Fig. 24 with separation S and reattachment R separated by a length L_{sep})—is particularly interesting, from the fundamental view point of shock boundary layer interaction (SBLI), but also from the practical implications due to its effects on the performance of intake and scramjet as well as the structures due to pressure and heating loads. This prompted a study of hypersonic impinging shock wave boundary layer interactions close to the leading edge of a flat plate, similar to the occurrences in off-design intakes; the study focused on strong interactions with *large* separation bubbles of length comparable to the distance of shock impingement from the leading edge [28]. The strong shock is generated by means of a wedge (shock generator) at angle 30.96° to the freestream, and the location of shock impingement relative to the leading edge is varied by moving the flat plate back and forth. A schlieren image of the flow field at Mach number 5.96 (enthalpy 1.3 MJ/kg), with

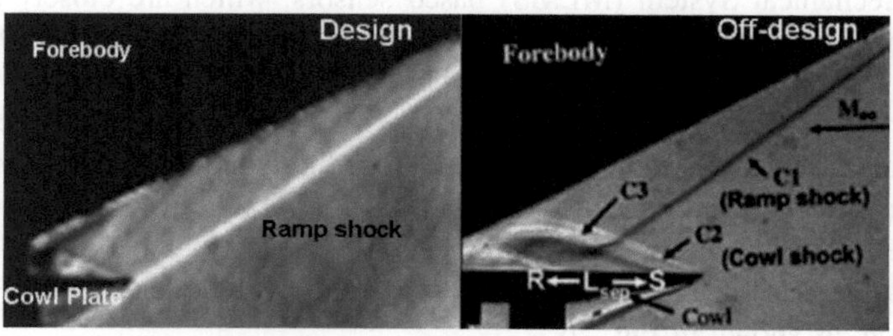

Fig. 24 Comparison of intake flow fields for design and off-design cases [27]

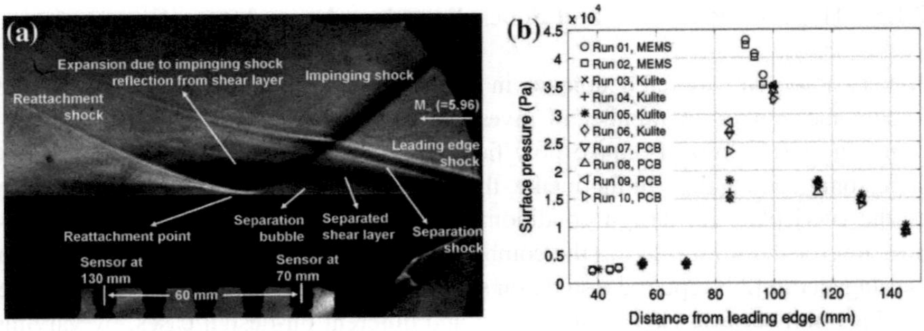

Fig. 25 Shock boundary layer interaction at Mach number 5.96, with shock impingement at 95 mm from the leading edge: **a** schlieren image, **b** surface pressure distribution [29]

(inviscid) shock impingement at 95 mm from the leading edge is shown in Fig. 25a. Corresponding surface pressure distribution is shown in Fig. 25b.

The various features of the interaction are shown in the schlieren image. The bubble due to the strong shock has a unique feature of being biased upstream. The separation shock can be seen very close to the leading edge shock, with the separation very close to the leading edge (nearly 15 mm from the leading edge for the above case) where the boundary layers are very thin. The separation length is about 75 mm, very much comparable to the distance of shock impingement from the leading edge. Relatively smaller separation bubbles, yet comparable in length to the distance of shock impingement to the leading edge, was observed at Mach number of around 8.5 at two different enthalpies of 1.6 and 2.4 MJ/kg. With the boundary layer thickness at the separation location negligible in comparison with the apparent separation length, the distance of shock impingement from the leading edge is clearly the scale for the separation length. The scaled separation length (for all experiments performed in HST-2 for enthalpies ranging from 1.3 to 2.4 MJ/kg) is found to follow an inviscid similarity law (independent of Reynolds number) [28]: it is directly proportional to the ratio of measured reattachment pressure to the freestream pressure and inversely proportional to the cube of freestream Mach number. In the regions such as separation and reattachment there are drastic variations of surface pressure, to resolve which closely spaced pressure measurements are required. This was achieved by means of in-house developed miniature Micro Electro-Mechanical System (MEMS) based sensors, which are closely spaced in arrays (each array/cluster containing 5 sensors in a space of 12 mm) [30].

Further, the control of the separation bubble at Mach number 5.96, with shock impingement at 95 mm from the leading edge, by means of boundary layer bleed at the separation location (which is close to the leading edge) was also demonstrated within the short test duration of the tunnel [29]. While the reattachment locations remain nearly the same, the separation locations with the boundary layer bleed can be seen farther (downstream) from the leading edge, with both (tangential) blowing and suction, as shown in Fig. 26.

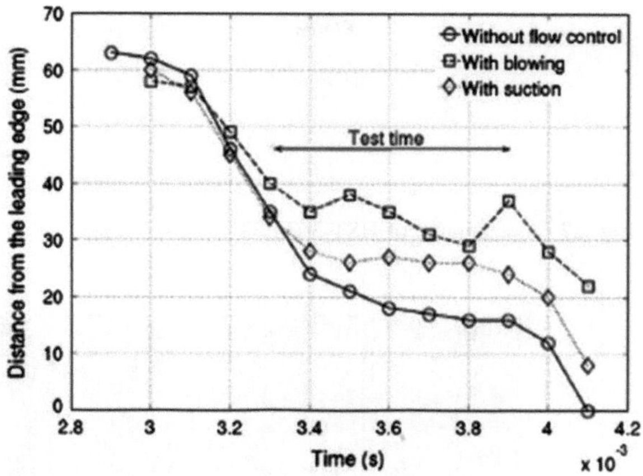

Fig. 26 Comparison of (time resolved) separation locations [29]

With blowing the average reduction in separation length is more (20 % reduction) than with suction (13.3 %). However, with blowing the separation (as well as reattachment) location seems to be unsteady; but with only one apparent cycle during the test time. Though the unsteadiness is apparent, the short test time of HST-2 does not allow the study of such low frequency phenomena. Thus in HST-2, (if not for high frequency phenomena) it is usually a static picture of the flow field that is studied at moderately high enthalpies, typically 1–2 MJ/kg. Experiments at higher enthalpies at which phenomena due to dissociations and reactions in the flowing gas becomes important, are performed in the free-piston tunnel HST-3, which shall be discussed subsequently.

2.2 *HST-3*

The study of the flow phenomena at higher total enthalpies at which dissociation (or even ionization) of the test gas must be considered, is possible by performing experiments in the free-piston shock tunnel HST-3 (also referred as FPST). In the free-piston tunnel the driver gas is compressed by means of a piston moving in the 'compression tube' (analogous to driver section in conventional tunnels); the piston compression (in very short time) also serves to increase the temperature of the driver gas adiabatically which also helps in achieving higher total enthalpy. The compressed driver gas ruptures the metal diaphragm, after which a shock propagates through the shock tube (driven section in conventional shock tubes); the subsequent operation is similar to conventional tunnels. A schematic (not to scale) of HST-3 is shown in Fig. 27. A photograph of the tunnel is shown in Fig. 28. The piston is moved by means of high pressure inside a reservoir of length 1.3 m and internal diameter of 400 mm (the red cylindrical portion in the photograph, with thick wall of 25 mm). A sudden exposure to the high pressure accelerates the piston inside the compression tube of length 10 m and internal diameter 166 mm (the long

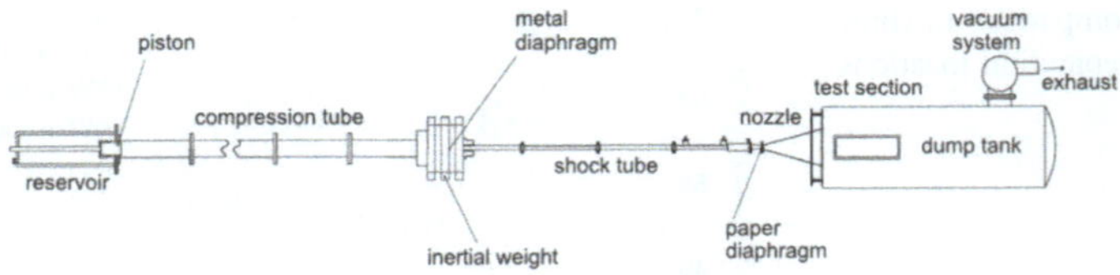

Fig. 27 A schematic of HST-3 (FPST)

Fig. 28 A photograph of HST-3 (FPST)

yellow tube in the photograph, wall thickness 27 mm). The driver gas in the compression tube is usually Helium, filled to appropriate pressure depending on the simulation requirements. The piston typically reaches a maximum speed of 100–120 m/s. The thicker walls of HST-3 allow higher pressures for the simulation of higher enthalpies; thicker (usually Aluminum) diaphragms that shall rupture at such high pressures as in the range of 150 bar can be placed separating the compression tube and the shock tube. The shock tube is 4.5 m long with internal diameter of 39 mm. The test section is 300 mm in diameter (fixed by nozzle exit diameter) inside the dump tank, evacuated by means of a high vacuum system.

With air as test gas the tunnel produces flows having Mach numbers ranging from 6 to 12, total enthalpy of up to 25 MJ/kg (though the maximum at which experiments have been carried out thus far is 6 MJ/kg) and Reynolds number in the range 0.1–2 million/m. As with HST-2, the PCB sensors at the end of the shock tube are used to measure the shock speed and reservoir pressure, which are used to obtain the reservoir conditions. From the reservoir conditions and the pitot pressure measured in the test section the freestream conditions are estimated. A typical pitot signal for the freestream with total enthalpy 6 MJ/kg is shown if Fig. 29; it can be seen that the test time is just around 250–300 μs. Such low test times are typical of high enthalpies, due to the higher speed of the waves catching up the compressed test gas. However for lower enthalpies simulated in FPST longer test times have been observed [31].

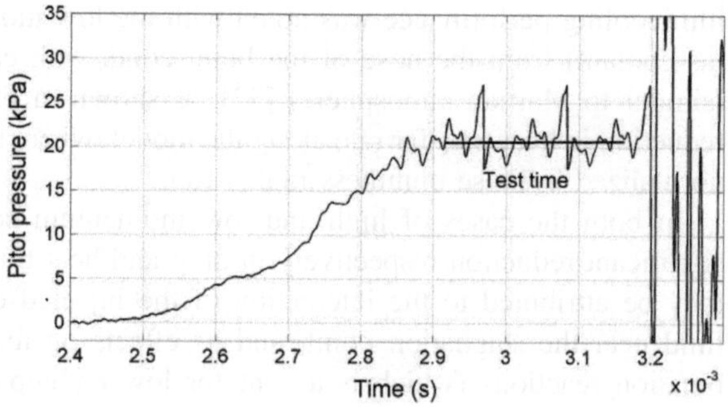

Fig. 29 A pitot signal in HST-3 (for $M_\infty = 8.61$, $Re_\infty = 0.3 \times 10^6$/m, $h_0 = 6$ MJ/kg)

In estimating the reservoir and freestream conditions the real gas effects must be accounted at high enthalpies. Based on simulation using the commercial software CHEMKIN for the conditions after the reflected shock in the shock tube, Kulkarni [31] observed that even at total enthalpy of about 4.5 MJ/kg about 13 % of air was dissociated (8.8 % NO, 4.1 % O and small traces of N); nearly 7 % dissociation was reported by Mohammed Ibrahim et al. [32] with carbon dioxide as test gas at 3.8 MJ/kg. Hence, the (equilibrium) flow conditions are estimated from the measured pressure signals (and shock speeds) by means of the code STN (Shock Tube and Nozzle calculations for equilibrium air) [33] using curve fits for thermodynamic properties [34], rather than calculating with perfect gas assumption. Thus in HST-3 the flow phenomena with the inclusion of real gas effects are studied. Some of the experimental studies in HST-2 are extended to include the role of real gas effects at higher enthalpies in HST-3.

For the case of high momentum counter flow jet injection from the 60° apex angle blunt cone, Kulkarni and Reddy [35] demonstrated enhanced drag reduction at higher enthalpies. The reduction in drag, at a nominal Mach number of 8, was almost twice for same injection to pitot pressure ratio at 5 MJ/kg simulated in HST-3, compared to that at 2 MJ/kg simulated in HST-2, as shown in Fig. 30a (c_d—drag coefficient, P_{0j}—jet total pressure, P_{0f}—pitot ptessure). Similarly, better

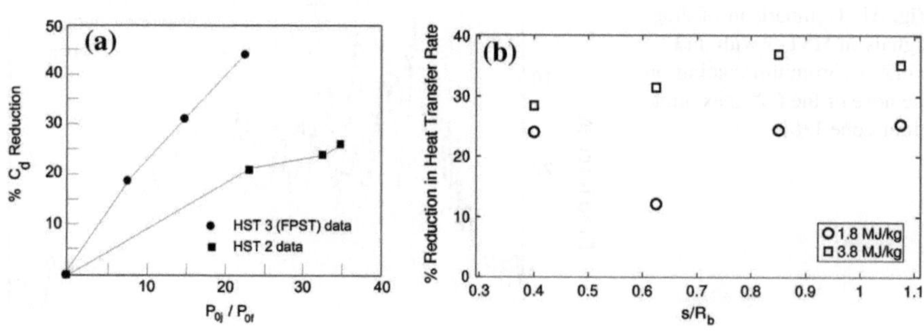

Fig. 30 Effect of counterflow injection on blunt body, a comparison between the HST-2 and the HST-3 flow fields: **a** percentage reduction in drag at various (high) injection pressure ratios [35] (reproduced with permission from [35]. Copyright 2008, AIP Publishing LLC). **b** Percentage reduction in heat transfer rate along the model with low momentum injection [32]

film cooling performance was noted with the low momentum injection of nitrogen (as coolant) from the nose of the blunt cone, with carbon dioxide as the test gas (similar to Martian atmosphere) [32], as shown in Fig. 30b (Showing percentage reduction in heat transfer rate along the model, where the distance s from the nose is normalized by nose bluntness radius R_b).

In both the cases of high and low momentum counter flow jet injection, the significant reduction respectively in drag and heat transfer rates at higher enthalpy may be attributed to the interaction of the injected cold gas with the dissociated fluid near the stagnation point, and its effects on molecular modes or on recombination reactions (which is absent for low enthalpy cases). Kulkarni et al. [36] further demonstrated a novel method for reducing drag on a blunt cone at high enthalpy (3.4 MJ/kg, at Mach number 8) by means of coating the nose with Chromium, which undergoes exothermic reaction with the stagnating fluid; the energy addition to the flow due to the exothermic reaction pushes the bow shock away from the body and reduces the drag, as shown in Fig. 31.

The densities are usually low when simulating higher enthalpies in HST-3, which affects the quality of schlieren visualization; further, the emissions in the flow field due to dissociated gas adversely affects the schlieren contrast/clarity. A schlieren image of impinging shock boundary layer interaction affected due to emissions from the flow field is shown in Fig. 32a. The issue was overcome with a powerful light source which enabled the visualization of the shocks as shown in Fig. 32b, however not with such good resolution as with low enthalpy flows simulated in HST-2.

For nearly same Mach number (of around 8.5) the separation bubble was much smaller (around 30–40 mm) for the HST-3 experiments at 6 MJ/kg, compared to the separation lengths in the experiments conducted in the HST-2 (about 60–70 mm) at 1.6 and 2.4 MJ/kg. The data at 6 MJ/kg was also not found to follow the similarity law (explained in Sect. 2.1.5) proposed for the *large* separation bubble from HST-2 experiments [28], as shown in Fig. 33 (L_{sep}—separation length, x_i—distance of

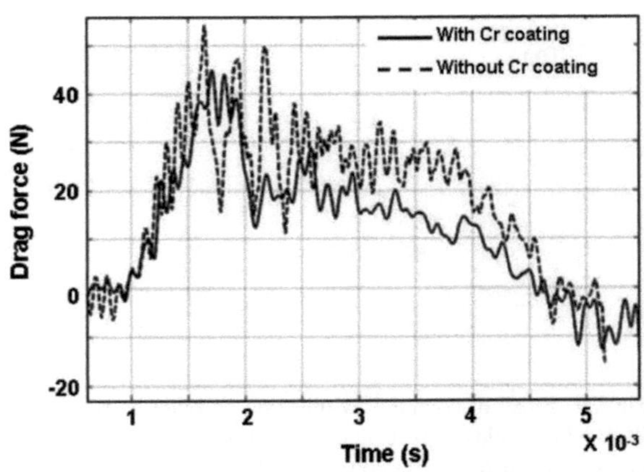

Fig. 31 Comparison of drag signals in HST-3 with and without chromium coating on the nose of the 60° apex angle blunt cone [36]

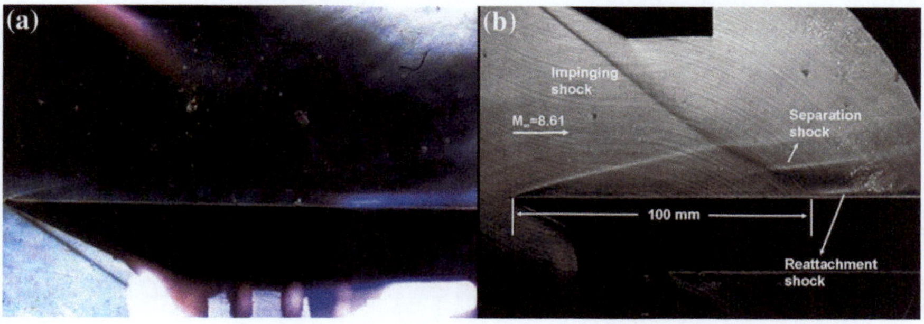

Fig. 32 Schlieren images of impinging shock boundary layer interaction in HST-3: **a** an image affected by emissions from the flow field, **b** image with better light source

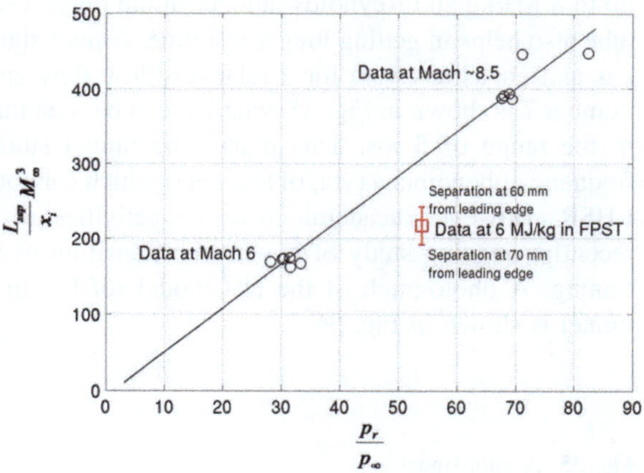

Fig. 33 Similarity law for the length of the impinging shock induced *large* separation bubble at hypersonic speed [28]

inviscid shock impingement location from the leading edge, M_∞—freestream Mach number, p_r—pressure at the reattachment, p_∞—freestream static pressure). This clearly is due to the prominence of real gas effects which are not considered in the proposed similarity law. It thus (HST-3) not only helps in complementing experiments in HST-2, but also resolves the phenomena arising uniquely out of high flow total enthalpy.

2.3 *HST-4*

HST-4 is another conventional shock tunnel, qualitatively similar to HST-2, but larger in scale; designed primarily for sponsored research purposes with large scale test models. It has a shock tube of internal diameter 165 mm (external diameter 200 mm), with 2 m driver and 10 m long driven sections. It has a test section of 1 m × 1 m. A photograph of HST-4 is shown in Fig. 34.

Fig. 34 A photograph of HST-4

The tunnel can simulate Mach numbers ranging from 6 to 15, total enthalpy of up to 8 MJ/kg and Reynolds number in the range 1–2 million/m. The longer shock tube also helps in getting longer test time. A pitot signal (with nitrogen as the driver gas and air as test gas) for a relatively low flow enthalpy of 0.7 MJ/kg at Mach number 7 is shown in Fig. 35, where it can be seen that the test time is much longer, in the range of 5 ms. This makes the tunnel suitable for studying certain low frequency phenomena (yet, of few kHz) which cannot be pursued in other tunnels in LHSR. However, academic research activities in the tunnel have begun only recently; ongoing study of hypersonic transition over a sharp cone is worth mentioning. A photograph of the test model (of 0.8 m axial length) mounted in the tunnel is shown in Fig. 36.

Fig. 35 A pitot signal in HST-4 (for $M_\infty = 7.1$, $Re_\infty = 2 \times 10^6$/m, $h_0 = 0.7$ MJ/kg)

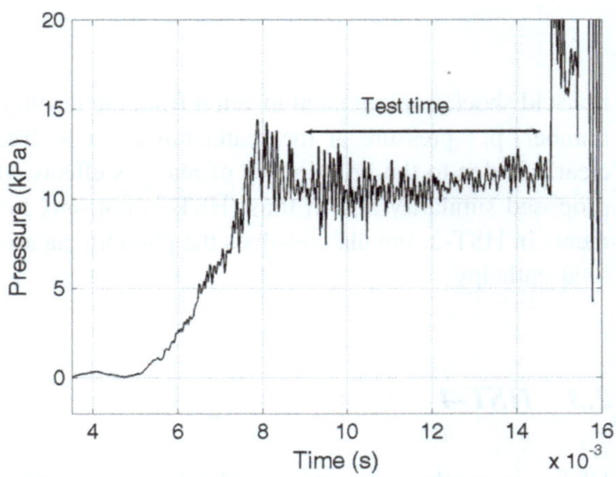

Fig. 36 A sharp cone model of 0.8 m length mounted in the test section of HST-4

2.4 *HST-5*

HST-5 is a combustion driven shock tunnel designed to simulate higher enthalpies, though it can also be operated like a conventional shock tunnel. In operating the tunnel as combustion driven, the driver section is filled with a mixture of helium, hydrogen, and oxygen in the ratio 10:2:1. The mixture is ignited by circumferentially mounted spark plugs in the driver section near the end where the diaphragm (separating the driven section) is mounted. The combustion increases the temperature and pressure in the driver section, which ruptures the metal diaphragm. The high temperature attained by the driver gas due to combustion helps in achieving higher enthalpies. A schematic of the tunnel is shown in Fig. 37. The shock tube is of internal diameter 102 mm (external diameter 130 mm), with 3 m driver section and 6.2 m driven section. The test section is 300 mm in diameter (fixed by nozzle exit diameter) inside the dump tank, similar to HST-3. A photograph of the tunnel is shown in Fig. 38.

While longer run times are observed in running the tunnel conventionally, the test time is much shorter when the tunnel is combustion driven. A pitot signal (at Mach number 6.2) for a combustion driven run is shown in Fig. 39; the useful test time is just approximately 300 microseconds (within which the signal is however oscillatory) even as the simulated total enthalpy is only 2.7 MJ/kg. For the flow conditions corresponding to the pitot signal, supersonic combustion by means of hydrogen injection in a (ramp) shock compressed flow between two parallel plates

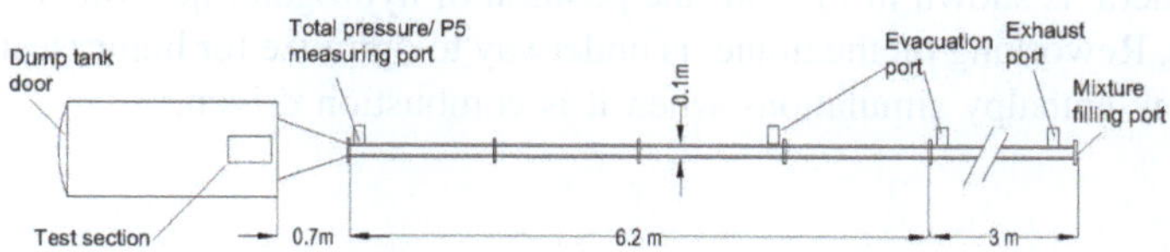

Fig. 37 Schematic of HST-5 [37]

Fig. 38 A photograph of HST-5

Fig. 39 A pitot signal in HST-5 (for $M_\infty = 6.2$, $Re_\infty = 2.3 \times 10^6$/m, $h_0 = 2.7$ MJ/kg) [37]

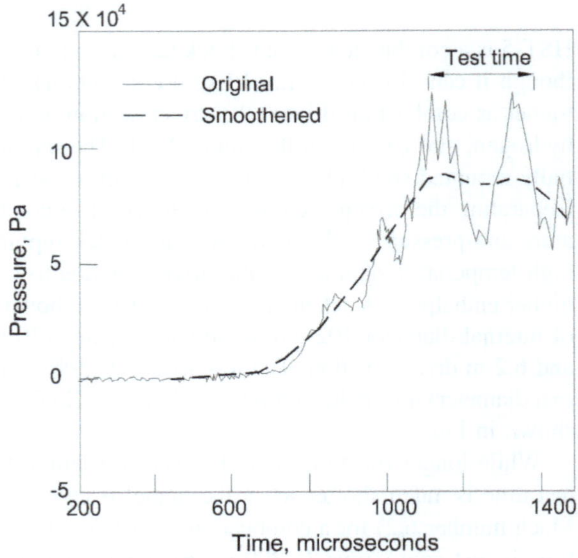

was demonstrated by Ratan and Jagadeesh [37] within the test duration; the emission due to supersonic combustion within the test time, captured using a high speed camera, is shown in Fig. 40 (the position of hydrogen injection is marked by an arrow). Reworking on the tunnel is underway to optimize for higher test time and higher flow enthalpy simulations when it is combustion driven.

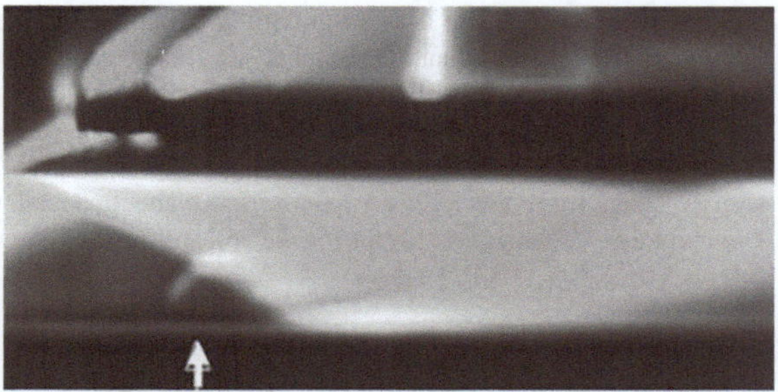

Fig. 40 Supersonic combustion (with hydrogen injection) of the ramp compressed oxygen rich flow between parallel plates [37]

3 Summary

The Laboratory for Hypersonic and Shock Wave Research is a distinguished experimental centre in the domain. It has four operational shock tunnels of varying capabilities: (1) A conventional tunnel HST-2 which has facilitated most of the academic research on hypersonic flow fields in the laboratory, (2) Free-piston tunnel HST-3 which simulates higher flow enthalpies to study the flow phenomena where real gas effects are prominent, (3) Another conventional tunnel HST-4 of larger scale with longer test duration which makes possible studying of low frequency unsteadiness and transition (due to the possibility of larger scale models), and (4) The combustion driven facility HST-5 which is also developed to simulate higher enthalpies; further development of HST-5 is underway to optimize its test time and enthalpy ranges. An overall view of the LHSR is given in Fig. 41. The important experiments performed in the different facilities were presented, exemplifying the capabilities of the tunnels, but also illuminating some of the important flow physics explored in the laboratory.

Fig. 41 A bird's view of LHSR

Acknowledgements The authors gratefully acknowledge the support of the members and collaborators. The authors thank G. Balakalyani for the schlieren image in Fig. 13, Tarandeep Singh for the model photograph Fig. 36, both of which are from ongoing works which are not yet published; and Ramesh Babu (Sr. Draughtsman, Department of Aerospace Engineering, IISc, Bangalore) for his help in improving the quality of some of the figures—mainly the plots and signals taken from the referred manuscripts. The financial support by Defense Research and Development Organization, India for the research in LHSR is gratefully acknowledged.

References

1. Reddy, N.M.: Development of a hypersonic shock tunnel. Proc. Indian Acad. Sci. Sect. C: Eng. Sci. **1**(1), 73–92 (1978)
2. Reddy, N.M., Nagashetty, K., Jagadeesh, G., Reddy, K.P.J.: Review of hypersonic research investigations in IISc shock tunnel (HST-1). Sadhana **21**(6), 741–773 (1996)
3. Joshi, M.V., Reddy, N.M.: Aerodynamic force measurements over missile configurations in IISc shock tunnel at $M_\infty = 5.5$. Exp. Fluids **4**(6), 338–340 (1986)
4. Raju, C., Reddy, N.M.: Aerodynamic force measurements over missile configurations in IISc shock tunnel at $M_\infty = 3.85$ and 9.15. Exp. Fluids **10**(2–3), 175–177 (1990)
5. Jagadeesh, G., Rao, B.K.S., Nagashetty, K., Reddy, N.M., Reddy, K.P.J.: Electrical discharge technique for three-dimensional flow field visualization in hypersonic shock tunnel. J Flow Vis. Image Process. **1**(1), 51–57 (1997)
6. Baskaran, V.:Heat transfer measurements over a flat plate in a shock tunnel at $M = 4$. Master of Engineering Dissertation, Department of Aerospace Engineering, Indian Institute of Science, Bangalore (1977)
7. Reddy, N.M., Viswanath, P.R.: SLV-3 nose cone heat flux measurements in the IISc shock tunnel, Report 77 FM 10. Department of Aerospace Engineering, Indian Institute of Science, Bangalore (1977)
8. Sahoo, N., Mahapatra, D.R., Jagadeesh, G., Gopalakrishnan, S., Reddy, K.P.J.: An accelerometer balance system for measurement of aerodynamic force coefficients over blunt bodies in a hypersonic shock tunnel. Meas. Sci. Technol. **14**, 260–272 (2003)
9. Mee, D.J.: Dynamic calibration of force balances for impulse hypersonic facilities. Shock Waves **12**, 443–455 (2003)
10. Satheesh, K.: The effect of energy deposition in hypersonic blunt body flow field. PhD thesis, Department of Aerospace Engineering, Indian Institute of Science, Bangalore (2007)
11. Saravanan, S., Jagadeesh, G., Reddy, K.P.J.: Convective heat transfer rate distribution over a missile shaped body flying at hypersonic speeds. Exp. Thermal Fluid Sci. **33**, 782–790 (2009)
12. Sahoo, N., Saravanan, S., Jagadeesh, G., Reddy, K.P.J.: Simultaneous measurement of aerodynamic and heat transfer data for large angle blunt cones in hypersonic shock tunnel. Sadhana **31**(5), 557–581 (2006)
13. Schultz, D.L., Jones, T.V.: Heat transfer measurements in short-duration hypersonic facilities. AGARDograph-AG-165 (1973)
14. Cook, W.J., Felderman, E.J.: Reduction of data from thin film heat transfer gauge; a concise numerical technique. AIAA J. **4**(3), 561–562 (1966)
15. Reddappa, P.: Shock tunnel investigations on hypersonic separated flows. PhD thesis, Department of Aerospace Engineering, Indian Institute of Science, Bangalore (2007)
16. Satheesh, K., Jagadeesh, G., Reddy, K.P.J.: High speed schlieren facility for visualization of flow fields in hypersonic shock tunnels. Curr. Sci. **92**(1), 56–60 (2007)
17. Settles, G.S.: Schlieren and Shadowgraph Techniques. Springer, Berlin (2001)
18. Menezes, V., Saravanan, S., Reddy, K.P.J.: Shock tunnel study of spiked aerodynamic bodies flying at hypersonic Mach numbers. Shock Waves **12**, 197–204 (2002)

19. Menezes, V., Saravanan, S., Jagadeesh, G., Reddy, K.P.J.: Aerodynamic drag reduction using aerospikes for large angle blunt cone flying at hypersonic Mach number. AIAA Paper No. 2002–2709 (2002)
20. Srinath, S., Reddy, K.P.J.: Experimental investigation of the effect of aerospike geometry on drag reduction and heat transfer rates at hypersonic Mach numbers. Int. J. Hypersonics **1**(2), 93–114 (2010)
21. Sahoo, N., Kulkarni, V., Saravanan, S., Jagadeesh, G., Reddy, K.P.J.: Film cooling effectiveness on a large angle blunt cone flying at hypersonic speed. Phys. Fluids **17**, 2005, pp. 036102: 1–11
22. Sriram, R., Jagadeesh, G.: Film cooling at hypersonic Mach numbers using forward facing array of micro-jets. Int. J. Heat Mass Transf. **52**, 3654–3664 (2009)
23. Venukumar, B., Jagadeesh, G., Reddy, K.P.J.: Counterflow drag reduction by supersonic jet for a blunt body in hypersonic flow. Phys. Fluids **18**, 118104: 1–4 (2006)
24. Mahapatra, D., Sriram, R., Jagadeesh, G.: Effect of counterflow argon plasma jet on aerodynamic drag of a blunt body at hypersonic Mach numbers. Aeronaut. J. **112**(1137), 683–687 (2008)
25. Satheesh, K., Jagadeesh, G.: Effect of concentrated energy deposition on the aerodynamic drag of a blunt body in hypersonic flow. Phys. Fluids **19**, 031701: 1–4 (2007)
26. Satheesh, K., Jagadeesh, G.: Experimental investigations on the effect of energy deposition in hypersonic blunt body flow field. Shock Waves **18**, 53–70 (2008)
27. Mahapatra, D., Jagadeesh, G.: Studies on unsteady shock interactions near a generic scramjet inlet. AIAA J. **47**(9), 2223–2231 (2009)
28. Sriram, R.: Shock tunnel investigations on hypersonic impinging shock wave boundary layer interactions. PhD thesis, Department of Aerospace Engineering, Indian Institute of Science, Bangalore (2013)
29. Sriram, R., Jagadeesh, G.: Shock tunnel experiments on control of shock induced large separation bubble using boundary layer bleed. Aerosp. Sci. Technol. **36**, 87–93 (2014)
30. Ram, S.N.: Measurement of static pressure over bodies in hypersonic shock tunnel using MEMS based pressure sensor array. MSc (Engg) thesis, Department of Aerospace Engineering, Indian Institute of Science, Bangalore (2011)
31. Kulkarni, V.: Investigation of flow modification techniques to reduce drag and heat transfer for large angle blunt cones in high enthalpy flows. PhD thesis, Department of Aerospace Engineering, Indian Institute of Science, Bangalore (2007)
32. Mohammed Ibrahim, S., Sriram, R., Reddy, K.P.J.: Experimental investigation of heat flux mitigation during Martian entry by coolant injection. J. Spacecraft Rockets **51**(4), 1363–1368 (2014)
33. Krek, R.M., Jacobs, P.A.: STN, Shock tube and nozzle calculations for equilibrium air. Research Report No. 2/93, The University of Queensland, Australia (1993)
34. Srinivasan, S., Tannehill, J.C., Weilmuenster, K.J.: Simplified curve fits for the thermodynamic properties of equilibrium air. NASA—R.P No. 1181 (1987)
35. Kulkarni, V., Reddy, K.P.J.: Enhancement in counterflow drag reduction by supersonic jet in high enthalpy flows. Phys. Fluids **20**, 016103: 1–4 (2008)
36. Kulkarni, V., Hegde, G.M., Jagadeesh, G., Arunan, E., Reddy, K.P.J.: Aerodynamic drag reduction by heat addition into the shock layer for a large angle blunt cone in hypersonic flow. Phys. Fluids **20**, 081703 (2008)
37. Ratan, J., Jagadeesh, G.: Investigations on supersonic combustion in a combustion driven shock tunnel. Proc. Inst. Mech. Eng. Part G: J. Aerosp. Eng. **226**, 1183–1191 (2012)

Author's Biography

Rengarajan Sriram is a research associate in the Department of Aerospace Engineering, Indian Institute of Science, Bangalore. His research primarily deals with experimental high speed fluid dynamics, with particular expertise on shock tunnel experiments and associated flow diagnostics. He received his 'PhD' degree in the Faculty of Engineering from Indian Institute of Science, Bangalore, in the year 2014, for his thesis on hypersonic impinging shock wave boundary layer interaction. His present research interests include effect of bluntness on shock boundary layer interactions near leading edge, pulsation in spiked cylinders, supersonic wall jets and Surface Dielectric Barrier Discharge (SDBD) induced flow field and flow control.

Gopalan Jagadeesh is a professor in the Department of Aerospace Engineering, Indian Institute of Science (IISc) Bangalore. He is also the Founder Chairman of Centre of Excellence in Hypersonics, IISc, Bangalore. His research areas include Hypersonic Aerothermodynamics and shock wave propagation in complex fluids. He has published extensively on various flow kinematic and energetic methods for drag reduction at hypersonic speeds. He is also well known for developing various industrial applications using shock waves including needless drug delivery. He has eight patents on various discoveries related to Shock Waves and hypersonic flow control. He is a Fellow of the Royal Aeronautical Society, (UK) and National Academy of Engineering (India) and Associate Fellow of American Institute for Aeronautics and Astronautics. He is on the Editorial Boards of Shock Waves, Current Science and International Journal of Aerospace Innovations.

Part III
Gun Tunnel

Gun Tunnels

David Buttsworth

1 Introduction

Gun tunnels are short duration wind tunnel facilities which have made important contributions to hypersonic flow research for more than half-century. Gun tunnels produce low enthalpy hypersonic flows with durations of several tens of milliseconds. For a capital investment similar to that of a comparably-sized, regular shock tunnel, gun tunnels offer at least an order of magnitude longer flow duration. This makes gun tunnels an attractive option for a wide range of hypersonic aerodynamics research, including the possibility of investigating unsteady aerodynamic effects.

An early review by Bray [1] illustrated the wide range of applications for gun tunnels with examples of aerodynamic experiments using force balances, and studies of separated flows, and hypersonic intake starting. The review by Bray [1] also mentioned the possibility of investigating unsteady hypersonic flows around oscillating models. Subsequent research using gun tunnel facilities over the years has certainly pursued these themes and others as well. Examples include separated flow studies [2, 3], inlet starting research [4–7], transitional and turbulent boundary layers [8, 9], film cooling and free shear layers [10, 11], and hypersonic aerodynamics using force balance and free flying models [12, 13].

Because of the longer flow duration produced by gun tunnels, the response time required of measurement systems is not as extreme as it is in the case of shock tunnels. For hypersonics research that does not require physical simulation of high temperature effects, the extended flow duration of gun tunnels relative to shock tunnels can lead to reductions in the cost of the experiments and improvements in the reliability of the data. Gun tunnels enable hypersonic aerodynamics research—in particular, research on separated and unsteady flows—which cannot be reliably performed in shock tunnels.

D. Buttsworth (✉)
University of Southern Queensland, Toowoomba 4350, QLD, Australia
e-mail: david.buttsworth@usq.edu.au

O. Igra and F. Seiler (eds.), *Experimental Methods of Shock Wave Research*,
Shock Wave Science and Technology Reference Library 9,
DOI 10.1007/978-3-319-23745-9_12

In this chapter, the arrangement and operation of gun tunnels will be reviewed briefly. The magnitude and uniformity of the temperatures that can be achieved with gun tunnels has a major impact on the quality of the hypersonic test flows they produce, but the measurement of flow temperature is significantly more difficult than flow pressure. Therefore this chapter also discusses temperature levels that can actually be achieved by gun tunnels. This was a focus of much of the early research on gun tunnels and quantifying the actual operating temperature remains important today because the precise definition of flow conditions is required for meaningful calibration of computational simulations.

2 Arrangement and Operation

The arrangement of a typical gun tunnel is illustrated in Fig. 1. The form of the gun tunnel is essentially the same as that of the conventional shock tunnel, except that a piston is used to separate the driver gas from the driven gas. During charging of the driver, the piston is isolated from the driver gas by one or two primary diaphragms. Using two primary diaphragms separated by a small volume which is charged to an intermediate pressure offers a convenient way to accurately control the rupture of the diaphragms. The barrel is filled with the test gas and by pre-heating the barrel, higher test gas stagnation temperatures can be achieved for a given compression ratio, and this is a technique that has been used in some experiments. The nozzle, test section, and dump volume would normally be evacuated to a low pressure (around 1 kPa is representative) which facilitates the rapid establishment of a steady hypersonic flow through the nozzle and free-jet test section. A light secondary diaphragm (like 'mylar') is used to retain the test gas within the barrel until the piston compression process occurs.

The gas dynamics of gun tunnel operation is described with reference to Fig. 1. Immediately after the rupture of the primary diaphragm(s), the piston accelerates and for typical gun tunnels, the piston achieves a relatively constant speed after travelling a short fraction of the overall barrel length. Light pistons are required for rapid attainment of the equilibrium piston speed and light pistons are also necessary in order to avoid large peak pressure during piston deceleration at the nozzle-end of the barrel. Peak pressures which exceed the equilibrium pressure during flow discharge through the nozzle can be eliminated if the piston mass can be chosen according to the equilibrium piston technique [14, 15] for the particular facility and operating condition.

For a given gun tunnel configuration, the ratio of initial pressure in the driver to that in the barrel p_4/p_1 affects the stagnation temperature which is achieved. Compression initially occurs through the action of a shock wave ('shock 1' in Fig. 1) which is driven ahead of the piston and reflects off the entrance to the hypersonic nozzle. Flow discharge through the nozzle begins with the rupture of the secondary diaphragm at the time of shock reflection. The reflected shock ('shock 2' in Fig. 1) imparts further entropy to the test gas and reflects off the piston (as 'shock

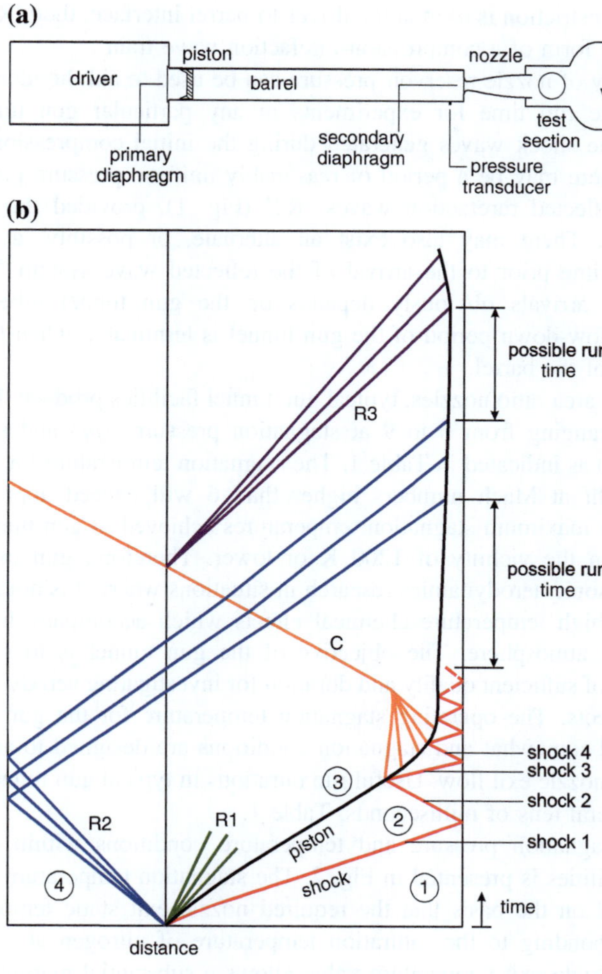

Fig. 1 Illustration of a representative gun tunnel: **a** facility arrangement; **b** distance versus time diagram for gun tunnel operation

3' in Fig. 1). Further shocks are reflected between the nozzle entrance and the piston, and a series of compression waves is established in the driver gas on the upstream side of the piston which coalesce into a single compression wave, designated 'C' in Fig. 1.

Rarefaction waves designated 'R1' and 'R2' which are centered on the primary diaphragm station will be propagate to the nozzle end of the barrel, and will be transmitted through the piston into the test gas in a modified form. The wave 'R3', which arises due to the transmission of the piston deceleration wave 'C' from the barrel into the larger diameter driver region, also has the potential to be registered in the test gas. If a restriction is used at the driver-to-barrel interface, then ‘R3’ would actually have the form of a compression-rarefaction wave train.

The uniformity of nozzle reservoir pressure can be used to aid the identification of an appropriate run time for experiments in any particular gun tunnel. For example, after the shock waves generated during the initial compression process have decayed, there may be a period of reasonably uniform pressure prior to the arrival of the reflected rarefaction waves ‘R2’ (Fig. 1), provided the driver is sufficiently long. There may also exist an alternate, or possibly, a preferred experiment run time prior to the arrival of the reflected wave system ‘R3’. The timing of wave arrivals obviously depends on the gun tunnel tube lengths. Ultimately the blow-down period of the gun tunnel is terminated when the piston reaches the end of the barrel.

With different area ratio nozzles, typical gun tunnel facilities produce flows with Mach numbers ranging from 6 to 9 at stagnation pressures (p_O) and stagnation temperatures (T_O) as indicated in Table 1. The stagnation temperature for terrestrial atmospheric flight at Mach numbers higher than 6 will exceed approximately 2000 K, whereas maximum stagnation temperatures achieved in gun tunnels have generally been in the vicinity of 1200 K or lower. Therefore, gun tunnels are applied to hypersonic aerodynamics research in situations where it is not necessary to simulate the high temperature chemical effects which accompany hypersonic flight within the atmosphere. The objective of the gun tunnel is to generate a hypersonic flow of sufficient quality and duration for investigating aerodynamic and heat transfer effects. The operating stagnation temperature for the gun tunnel is therefore relaxed somewhat and stagnation conditions are designed to avoid condensation in the nozzle exit flow. Useful run durations in typical gun tunnels would normally be several tens of milliseconds, Table 1.

A map of stagnation pressure and temperature conditions within cold-flow aerodynamic facilities is presented in Fig. 2. The stagnation temperature shown in Fig. 2 is defined on the basis that the required nozzle exit static temperature is 77.35 K, corresponding to the saturation temperature of nitrogen at 101.3 kPa. Targeting this nozzle exit temperature value allows a substantial margin for error because gun tunnel nozzle exit static pressures are generally much lower than 101.3 kPa. To simulate relevant aerodynamic effects under appropriate viscous conditions, the Reynolds number must be considered, and stagnation pressures required for different unit Reynolds numbers (Re_u) are presented as the solid lines in

Table 1 Indicative operating conditions of three gun tunnels in the UK

Quantity	Cranfield	Imperial	Oxford
M	8.2	9	6, 7, 8
Re_u (m^{-1})	9×10^6	$7–47 \times 10^6$	20×10^6
$p_{O,max}$ (MPa)	10	60	9
$T_{O,max}$ (K)	1290	1150	1000
t_{run} (ms)	80	20	30

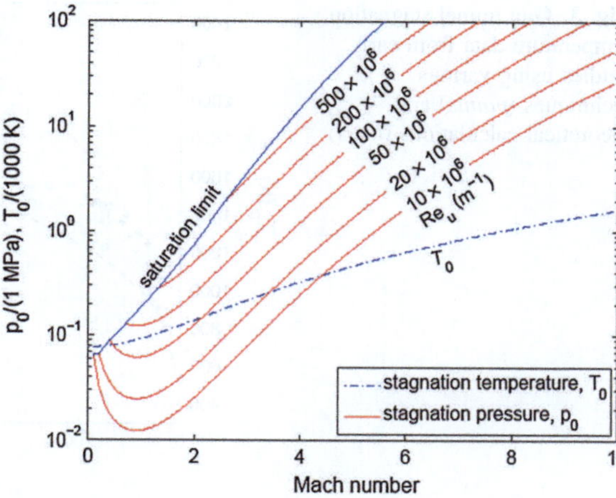

Fig. 2 Stagnation pressure and temperature values for different nozzle exit Re_u which avoid saturation conditions

Fig. 2. A useful range of unit Reynolds numbers can be achieved using gun tunnels at reasonable pressure levels. For example, at Mach 7, a unit Reynolds number of 50×10^6 m^{-1} can be achieve with a total pressure of about 20 MPa.

3 Early Stagnation Temperature Work

3.1 Calculation Methods

Stagnation temperature conditions generated by the shock compression process can be estimated from the pressure measurements that are normally obtained from a transducer located near the nozzle-end of the barrel. The strength of the incident and reflected shock waves can be accurately determined from the magnitude of the recorded pressure-jumps. The actual strength of the third shock, which is the shock wave that reflects from the piston, see Fig. 1, is more difficult to determine because expansion waves associated with the deceleration of the piston immediately follow the shock wave. The entropy rise associated with the third and subsequent shocks diminishes rapidly. Stagnation temperature calculation methods have generally considered the entropy rise across the first two or three shocks only, with isentropic compression (or expansion) thereafter to the final stagnation pressure measured during the specified run time [16, 17]. Results for the stagnation temperature based on a three-shock calculation from [17] are presented in Fig. 3 as the solid line showing the variation with the primary diaphragm pressure ratio p_4/p_1.

The calculation of stagnation temperature based on a two or three shock approach substantially over-estimates the stagnation temperature in the majority of gun tunnel experiments. Edney [17] estimated a stagnation temperature decrease relative to the idealized shock compression value using a shock tube boundary layer

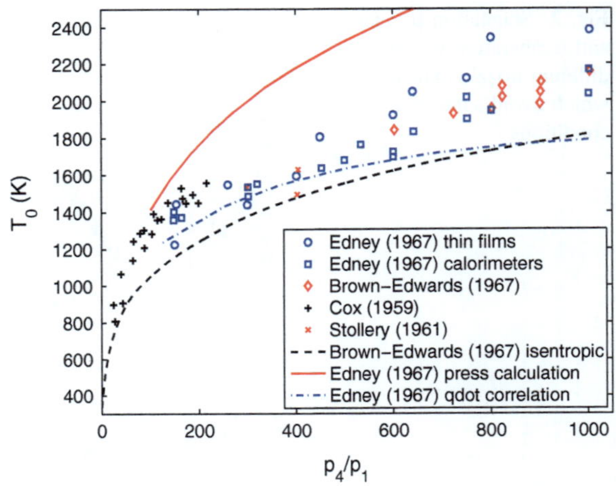

Fig. 3 Gun tunnel stagnation temperature data from early studies using various techniques (*points*) and theoretical calculations (*lines*)

heat transfer correlation and this is also presented in Fig. 3. A reduction in calculated stagnation temperature of around 35 % due to boundary layer cooling was determined for the highest pressure ratio case.

3.2 Measurement Techniques

Figure 3 presents a collection of early gun tunnel stagnation temperature data from various sources and shows that the calculation based on pressure measurements with the three-shock approach over-estimates the stagnation temperature in the majority of the cases. A wide variety of methods for gun tunnel stagnation temperature measurement have been developed including: the sodium line reversal technique [18, 19], calorimeter and thin film heat transfer gauges [17], disturbance velocity measurements [20–22], and thin wire thermocouple probe methods [23, 24].

The gun tunnel stagnation temperature increases with pressure ratio p_4/p_1, but values in excess of 400 were not used frequently, except by the FFA group in Sweden after an extensive period of piston development work [17, 19]. However, even for the lower pressure ratios $p_4/p_1 < 400$ which were more commonly used by other workers, the stagnation temperature data are significantly lower than the three-shock value. The gun tunnel stagnation temperature identified from the isentropic compression calculation happens to provide a lower estimate for the data in Fig. 3, although it certainly does not represent a theoretical lower limit. Heat transfer from the shock-compressed gas to the relatively cold barrel walls is an obvious mechanism by which stagnation temperatures can be reduced to values approaching, or even lower than, the isentropic compression value. For the pressure ratios below 400 that are normally used in gun tunnels, the standard deviation of the data in Fig. 3 from the calculation by Edney [17] which accommodates heat transfer effects is approximately 120 K.

The data in Fig. 3 do not provide a direct indication of possible stagnation temperature variations with time. The data should be regarded as representative values for gun tunnel stagnation temperatures, averaged over the run time. It is well established that gun tunnel stagnation temperatures are not constant with time; a significant rate of decrease in stagnation temperature is generally observed. Analysis by Lemcke [16] showed that stagnation temperature can decrease throughout the slug of test gas due to finite piston acceleration. Calculations by Edney [17] have correlated the time-averaged stagnation temperature data using a shock tube boundary layer heat transfer analysis, and a similar approach was adopted in [25] to deduce the time-dependent variation of gun tunnel stagnation temperature. Stagnation temperature decay rates from as high as 8000 K/s [22] down to about 1000 K/s [21] have been measured, and rates even lower than 1000 K/s [18] may also be possible.

Heat transfer from the test gas in the barrel is the primary cause of the observed rates of stagnation temperature decay and it depends on variables such as the gas and wall temperature difference, the barrel length and diameter, speed of compression, and the gas density. However, by taking a representative stagnation temperature decay rate of 4000 K/s and a flow duration of about 50 ms (Table 1), a change of stagnation temperature of around 200 K during the run time is suggested. If the mean stagnation temperature during the run time is 1100 K (Table 1), then the fractional change of flow temperature will be almost 20 % resulting in a fractional change of flow speed of about 10 % during the run time. Without careful measurements of flow stagnation temperature, it is difficult to confidently specify nozzle exit flow conditions that are required for the validation or calibration of computational simulations.

4 Thin Film Gauge Temperature Probes

Edney [17] was the first to use thin film gauges in a systematic study of gun tunnel stagnation temperatures. Hemispherical probes with platinum thin film gauges at the stagnation point of a Pyrex-like substrate were used in that work. More recently, Mallinson et al. [26] also used stagnation point platinum thin films, but on Macor hemispheres. In both works, the authors had reservations about the accuracy of the thin film data: Edney [17] was concerned mainly about the effect of gauge erosion and while this issue was also addressed in the work of Mallinson et al. [26], a more striking problem was the difference in the apparent stagnation temperatures deduced from the different probe diameters: the 80 mm diameter probe indicating a temperature 200 K higher than the 20 mm diameter probe.

In the conventional stagnation point thin film temperature probe arrangement [17, 26, 27], the voltage drop across the thin film gives the instantaneous wall temperature at the stagnation point T_w (through a prior temperature-resistance calibration), and the surface temperature history is analysed with an appropriate transient heat conduction theory for the substrate to yield the instantaneous heat flux at the surface q_w. To determine the flow stagnation temperature, the heat transfer coefficient at the stagnation point h is also required. The theory of Fay and Riddell [28] can be used to specify the stagnation point heat transfer coefficient, although an accurate calculation of the tangential velocity gradient external to the stagnation point boundary layer is required [29]. The stagnation point heat transfer coefficient is primarily a function of the probe geometry and the flow properties just upstream of the probe. The necessary flow properties can be determined with reasonable accuracy from stagnation pressure measurements at the end of the barrel and pitot pressure measurements at the temperature probe location. The convective heat transfer coefficient is a relatively weak function of flow stagnation temperature and the probe temperature, so the heat transfer law can be rearranged to give the instantaneous flow stagnation temperature directly,

$$T_0 = T_w + \frac{q_w}{h}. \tag{1}$$

With a conventional stagnation point temperature probe, errors in the measurement of heat transfer and the estimation of the heat transfer coefficient can have a major impact on the accuracy of the calculated value of stagnation temperature. Perturbations in measurement of the stagnation point surface temperature, and the calculation of the heat flux and convective heat transfer coefficient will affect the calculated flow stagnation temperature according to

$$\frac{\delta T_0}{T_0} = \frac{T_w}{T_0}\frac{\delta T_w}{T_w} + \left(1 - \frac{T_w}{T_0}\right)\frac{\delta q_w}{q_w} + \left(\frac{T_w}{T_0} - 1\right)\frac{\delta h}{h}. \tag{2}$$

Conventional thin film probes are operated with an initial temperature close to ambient so normally, $T_w/T_0 \ll 1$ throughout the gun tunnel run time. However, if thin film gauges are pre-heated in some way such that $T_w/T_0 \to 1$, the calculated stagnation temperature will be less sensitive to uncertainties in q_w and h, although it does become more sensitive to uncertainties in T_w. An accurate calibration for T_w is required in all cases and the calculation of heat flux q_w relies on this temperature calibration as well as other assumptions, so pre-heating the gauge has significant potential for reducing uncertainties in the flow stagnation temperature measurement.

Platinum thin film heat transfer gauges hand painted onto the rounded ends of fused quartz rods 3 mm in diameter have been used in the Oxford University Gun Tunnel [25, 30]. The thin films were located around the stagnation point and were pre-heated to different initial temperatures using an external resistance heating wire. Film temperatures and corresponding heat flux results for a particular operating condition in the Oxford University Gun Tunnel are presented in Fig. 4 for three different initial probe temperatures. By using multiple initial temperatures, the flow total temperature can be deduced without the need for an accurate stagnation point

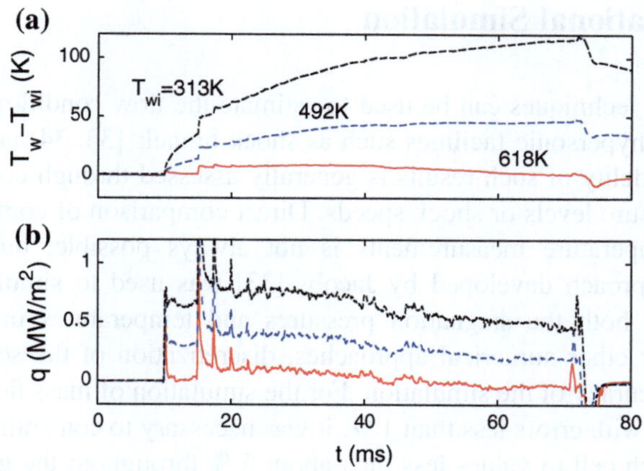

Fig. 4 Stagnation point thin film gauge results as a function of time in a gun tunnel flow [30]. **a** Thin film surface temperature T_w relative to the initial temperature T_{wi}. **b** Corresponding heat flux measurements

Fig. 5 Variation of heat flux with thin film surface temperature at a particular instant within a gun tunnel run [30]

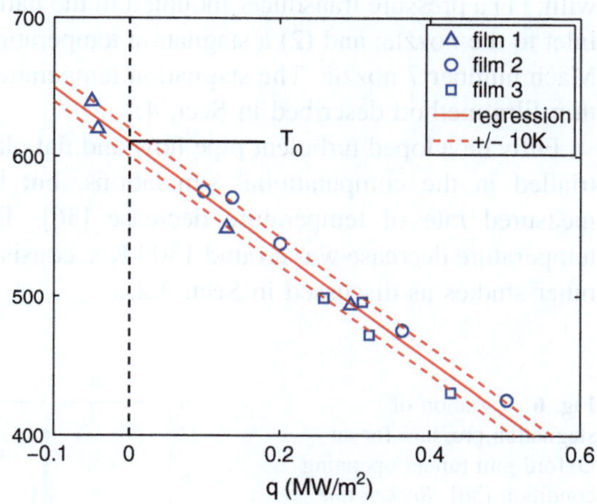

heat transfer correlation. In theory, only two different probe temperatures are required to determine the flow total temperature independent from a convective heat transfer coefficient, but in practice, uncertainties can be reduced if additional probe/temperatures are used.

Figure 5 illustrates the relationship between the thin film temperature and the convective heat flux, at a particular instant within a hypersonic gun tunnel flow. At each instant during the run time, the flow total temperature can be identified from the linear regression (as illustrated in Fig. 5) and is defined as the point on the regression line where the convective heat flux is zero. Additional details of the heated thin film probe method in various embodiments, and examples of its application in gun tunnels and other flows are provided elsewhere [25, 31, 32].

5 Computational Simulation

Computational techniques can be used to estimate the flow conditions produced in short duration hypersonic facilities such as shock tunnels [33, 34] and gun tunnels [35] and the fidelity of such results is generally assessed through comparison with measured pressure levels or shock speeds. Direct comparison of computations with stagnation temperature measurements is not always possible, but in [36], the Lagrangian approach developed by Jacobs [33] was used to simulate results for comparison to both the stagnation pressures and temperatures in a gun tunnel facility. As for other numerical approaches, discretization of the solution domain impacts the accuracy of the simulation. For the simulation of mass flow rates within the gun tunnel with errors less than 1 %, it was necessary to constrain the change of area across each cell to values less than about 5 % throughout the gun tunnel area changes [36].

Illustrative results from the computational simulation of the Oxford University Gun Tunnel are presented in Figs. 6 and 7. The details of the gun tunnel and its simulation are presented in [36]. Measurements in the gun tunnel were obtained with: (1) a pressure transducer mounted in the barrel about 150 mm upstream of the inlet to the nozzle; and (2) a stagnation temperature probe located at the exit of the Mach number 7 nozzle. The stagnation temperature probe was based on the heated thin film method described in Sect. 4.

Fully developed turbulent pipe flow and flat plate heat transfer correlations were trialled in the computational simulations, but both models underestimated the measured rate of temperature decrease [36]. The measured rate of stagnation temperature decrease was around 1800 K/s, consistent with the range of values from other studies as discussed in Sect. 3.2.

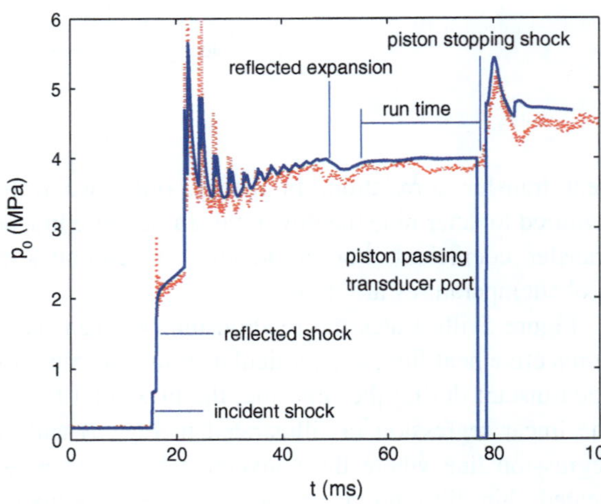

Fig. 6 Variation of stagnation pressure for an Oxford gun tunnel operating condition [36]. *Broken line* experimental data; *solid line* computational simulation

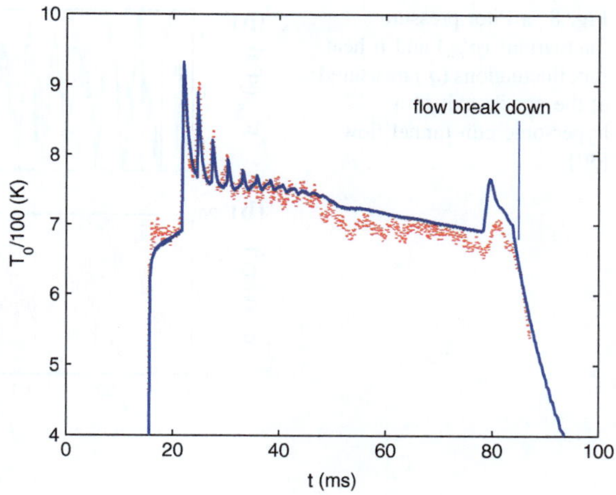

Fig. 7 Variation of stagnation temperature for an Oxford gun tunnel operating condition [36]. *Broken line* experimental data; *solid line* computational simulation

6 Fluctuations

Temporal variations in the flow properties have the potential to impact the fidelity of the physical simulation of hypersonic flight conditions achieved in wind tunnels. Quiet hypersonic wind tunnels, where the noise radiating from the boundary layers on the nozzle is greatly reduced, have been used in the study of boundary layer transition [37]. Even with careful management of nozzle boundary layers, gun tunnels cannot achieve such a low noise test environment.

The rapid piston acceleration in gun tunnels generates a shock wave which reflects between the end of the barrel and the piston. Although the strength of this wave rapidly diminishes, it can persist in the form of an isentropic disturbance well into the run time. Additional waves can also be generated through the interaction/reflection of the shock and/or isentropic waves with the barrel boundary layer or corner regions associated with the nozzle contraction and piston. Disturbances from such effects can propagate down the nozzle, into the test section.

Stagnation temperature fluctuations will arise due to pressure waves moving through the test gas in the barrel, and additional stagnation temperature variations will be present in the barrel due to turbulent heat transfer from the test gas to the barrel walls. Entropy 'spottiness' within the barrel will be convected through the nozzle and into the test section during the test flow drainage process. Again, it seems impossible to eliminate such temporal, and most probably, spatial variations in flow properties. However, it is possible to measure such fluctuations, and such work is important because it may be possible to demonstrate that the disturbances do not have an impact on the target flow regime of interest, or to provide quantitative information on the disturbances for specification of inflow boundary conditions to computational simulations.

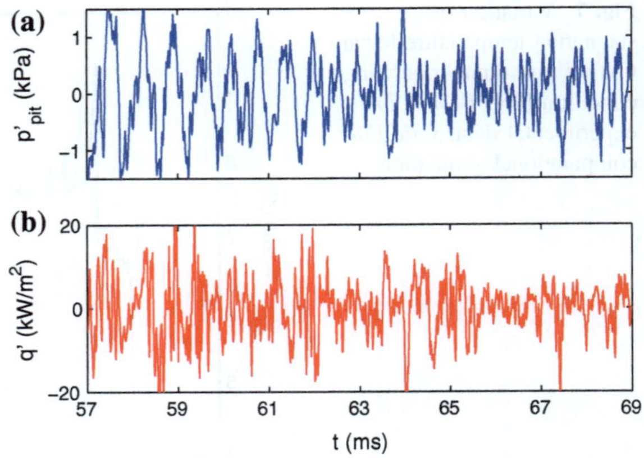

Fig. 8 **a** Pitot pressure fluctuations (p'_{pit}) and **b** heat flux fluctuations (q') measured at the nozzle exit of a hypersonic gun tunnel flow [30]

Hot wire anemometry has been used to characterise the nozzle exit flow in a hypersonic gun tunnel [38] and other instruments can also be used for this purpose. Examples of fluctuations in pitot pressure and stagnation point heat flux (from a platinum thin film probe) positioned at the exit of a hypersonic gun tunnel nozzle are presented in Fig. 8. In this case, the surface temperature of the thin film probe was very close to the flow stagnation temperature, so the fluctuations in heat flux arise primarily due to fluctuations in stagnation temperature, rather than fluctuations in the convective heat transfer coefficient. From this data, it was determined that rms fluctuations in stagnation temperature of 2.3 K were present, whereas isentropic temperature fluctuations associated with the pitot pressure fluctuations were around 0.6 K, so it was concluded that the measured stagnation temperature fluctuations are primarily attributable to fluctuations in entropy.

7 Related Facilities

Early gun tunnel work by Edney [17] and Brown-Edwards [19] using very careful piston design achieved stagnation temperatures up to about 2200 K, and shock compression with vitiation was used in a gun tunnel by Walton and Cain [39] to achieve elevated temperatures for scramjet combustor studies. However, gun tunnels are not regularly used to produce flow stagnation temperatures very much above 1200 K. Gun tunnels have generally been used to produce flows suitable for hypersonic aerodynamics research at low flow temperatures, while still avoiding condensation in the test gas. For aerodynamic testing at Mach numbers lower than about 8, it is not necessary to compress the test gas to stagnation temperatures higher than 1000 K, as illustrated in Fig. 2.

Hypersonic facility development work at the University of Southampton was initially focussed on gun tunnels [14, 40], but East [41] recognised that for the production of a Mach number 7 flow, sufficiently high stagnation temperatures (around 600 K) could be achieved using a relatively slow (shockless) free piston compression process. The facility developed by East [41] was operated on the principles established by Jones et al. [42] for light piston isentropic compression facilities, and offered an improvement on the Ludwieg tube with Isentropic Compression Heating (LICH) arrangement of Oldfield et al. [43] through the addition of a wave damper at the upstream end of the barrel. The barrel was also pre-heated and this was configured so that the initial temperature variations in the axial direction compensated for the non-uniform heat loss from the gas in the barrel during the compression and blow down period. High uniformity test flows with durations of up to 0.5 s were produced by the University of Southampton facility for dynamic stability studies [41].

For the production of hypersonic flows suitable for aerodynamic testing at Mach numbers up to about 7, gun tunnels can readily be converted to operate in the LICH mode as introduced by Oldfield et al. [43] by using a valve in place of the primary diaphram(s). The opening of this valve can be arranged so that the test gas is compressed in a shockless process which can improve test flow quality relative to shock compression used in gun tunnels. Such a facility is currently in operation at the University of Southern Queensland [44]. However, the rupture of the secondary diaphragm still generates expansion waves which can reflect off the piston and the upstream end of the barrel, thereby degrading the quality of the test flow. To attain very high levels of flow quality and long test flow durations, a wave damper and tailored barrel heating as introduced by East [41] should also be considered.

8 Conclusion

For more than half a century, gun tunnel facilities around the globe have been providing flows with sufficient duration and quality to make substantial contributions to knowledge in the area of hypersonic aerodynamics. Accessing high temperature effects is not normally the objective in gun tunnel experiments and operating stagnation temperatures are generally less than 1200 K. Knowledge of the flow temperature produced in the gun tunnel is required for definition of the flow state. Stagnation temperatures measured in gun tunnels are normally much lower than theoretical values calculated from shock pressure histories, and cooling rates well in excess of 1000 K/s are representative. Estimates of stagnation temperature based on semi-analytical or computational methods which accommodate barrel cooling effects through empirical boundary layer or pipe flow correlations offer reasonable accuracy with estimated uncertainties less than ±20 %. For higher precision in the definition of the flow state, direct measurements of flow stagnation temperature should be made and various techniques have proved suitable for this task. The use of thin film heat transfer gauges has provided reliable stagnation temperature data, and methods based on pre-heating such probes can enhance the accuracy and can also yield information on the fluctuations the flow. Challenges in the area of hypersonic aerodynamics remain great, so it is expected that gun tunnels around the globe will continue to make substantial contributions for the foreseeable future.

Acknowledgements I am indebted to Richard Morgan, Terry Jones, Terry Cain and Noel Morris for introducing me to the Oxford University Gun Tunnel in 1991.

References

1. Bray, K.N.C.: Evaluation of the hypersonic gun tunnel. In: Riddell, F.R. (ed.) Hypersonic Flow Research, Vol. 7 of Progress in Astronautics and Rocketry, pp. 547–579. American Rocket Society (1961)
2. Jackson, A.P., Hillier, R., Soltani, S.: Experimental and computational study of laminar cavity flows at hypersonic speeds. J. Fluid Mech. **427**, 329–358 (2001)
3. Nishio, M., Sezaki, S., Nakamura, H.: Measurements of capsule wake stabilization times in a hypersonic gun tunnel. AIAA J. **42**(1), 56–60 (2004)
4. Matthews, A.J., Jones, T.V., Cain, T.M.: Design and test of a hypersonic isentropic-spike intake with aligned cowl. J. Propul. Power **21**(5), 838–843 (2005)
5. Matthews, A.J., Jones, T.V.: Design and test of a modular waverider hypersonic intake. J. Propul. Power **22**(4), 913–920 (2006)
6. Lanson, F., Stollery, J.L.: Some hypersonic intake studies. Aeronaut. J. **110**(1105), 145–156 (2006)
7. Fan, X.-Q., Jia, D., Feng, D.-H., Li, H.: Experimental investigation on starting process of hypersonic inlet in gun tunnel. Tuijin Jishu/J. Propuls. Techol. **28**(1), 60–64 (2007)
8. LaGraff, J.E.: Observations of hypersonic boundary-layer transition using hot wire anemometry. AIAA J. **10**(6), 762–769 (1972)
9. Fiala, A., Hillier, R., Mallinson, S.G., Wijesinghe, H.S.: Heat transfer measurement of turbulent spots in a hypersonic blunt-body boundary layer. J. Fluid Mech. **555**, 81–111 (2006)
10. Richards, B.E., Stollery, J.L.: Laminar film cooling experiments in hypersonic flow. J. Aircraft **16**(3), 177–181 (1979)
11. Buttsworth, D.R., Morgan, R.G., Jones, T.V.: A gun tunnel investigation of hypersonic free shear layers in a planar duct. J. Fluid Mech. **299**, 133–152 (1995)
12. Kontis, K., Stollery, J.L.: Control effectiveness of a jet-slender body combination at hypersonic speeds. J. Spacecraft and Rockets **34**(6), 762–768 (1997)
13. Lewis, H., East, R.: Measurement of free-flight dynamic stability derivatives of cones in a hypersonic gun tunnel. In: 6th AIAA International Aerospace Planes and Hypersonics Technologies Conference, AIAA Paper 95-6082, Chattanooga, TN, 3–7 Apr 1995
14. East, R.A., Pennelegion, L.: The equilibrium piston technique for gun tunnel operation. Current Papers C.P. No. 607, Aeronautical Research Council (1962)
15. Davies, L., Regan, J.D., Dolman, K.A.: On the equilibrium piston technique in gun tunnels. Current Papers C.P. No. 982, Aeronautical Research Council (1968)
16. Lemcke, B.: Determination of final temperature in a gun tunnel. J. Aerospace Sci. **28**(10), 827–828 (1961)
17. Edney, B.E.: Temperature measurements in a hypersonic gun tunnel using heat-transfer methods. J. Fluid Mech. **27**(3), 503–512 (1967)
18. Stollery, J.L.: Stagnation temperature measurements in a hypersonic wind-tunnel using the sodium line reversal method. Nature **190**, 778–779 (1961)

19. Brown-Edwards, E.G.: Temperature measurements in a hypersonic gun tunnel using a modified line-reversal methods. J. Fluid Mech. **27**(3), 493–501 (1967)
20. Cox, R.N., Winter, D.F.T.: The light gas hypersonic gun tunnel and ARDE, Fort Halstead, Kent. AGARD Report No. 139 (1957)
21. Cox, R.N.: Hypersonic Flow. In: Collar, J. Tinkler, A.R., (eds.) Proceedings of the Eleventh Symposium of The Colston Research Society, University of Bristol, 6–9 Apr 1959
22. Merritt, G.E.: Velocity measurements in the University of Southampton hypersonic gun tunnel. Aero. and Astro. Report No. 172, University of Southampton (1961)
23. East R.A., Perry, J.H.: A short time response stagnation temperature probe. Current Papers C. P. No. 909, Aeronautical Research Council (1967)
24. Bartlett, R.P., Edwards, A.J., Hillier, R.: Development and calibration of a total temperature probe for the Imperial College Aeronautics Department gun tunnel. Aero. Report No. 97–02, Imperial College (1979)
25. Buttsworth, D.R., Jones, T.V.: A fast-response total temperature probe for unsteady compressible flows. J. Eng. Gas Turbines Power **120**(4), 694–702 (1998)
26. Mallinson, S.G., Hillier, R., Jackson, A.P., Kirk, D.C., Soltani, S., Zanchetta, M.: Gun tunnel flow calibration: defining input conditions for hypersonic flow computations. Shock Waves **10**, 313–322 (2000)
27. Matthews, A.J., Munro, S.I., Sweeney, W.R.: Stagnation temperature measurements for hypersonic intake testing in a gun tunnel flow. In: 14th AIAA/AHI International Space Planes and Hypersonic Systems and Technologies Conference, AIAA Paper 2006-8107, Canberra, Australia, 6–9 Nov 2006
28. Fay, J.A., Riddel, F.R.: Theory of stagnation point heat transfer in dissociated air. J. Aeronaut. Sci. **25**(2), 73–85 (1958)
29. Olivier, H.: Influence of the velocity gradient on the stagnation point heating in hypersonic flow. Shock Waves **5**(4), 205–216 (1995)
30. Buttsworth, D.R., Jones, T.V.: High bandwidth stagnation temperature measurements in a Mach 6 gun tunnel flow. Exp. Thermal Fluid Sci. **27**, 177–186 (2003)
31. Buttsworth, D.R., Jones, T.V., Chana, K.S.: Unsteady total temperature measurements downstream of a high pressure turbine. J. Turbomach. **120**(4), 760–767 (1998)
32. Buttsworth, D.R., Jones, T.V.: A fast-response high spatial resolution total temperature probe using a pulsed heating technique. J. Turbomach. **120**(3), 601–607 (1998)
33. Jacobs, P.A.: Quasi-one-dimensional modeling of a free-piston shock tunnel. AIAA J. **32**, 137–145 (1994)
34. Tani, K., Itoh, M., Takahashi, M., Tanno, T., Komuro, T., Miyajima, H.: Numerical study of free-piston shock tunnel performance. Shock Waves **3**, 313–319 (1994)
35. Groth, C.P.T., Gottlieb, J.J., Sullivan, P.A.: Numerical investigation of high-temperature effects in the UTIAS-RPI hypersonic impulse tunnel. Can. J. Phys. **69**, 897–918 (1991)
36. Buttsworth, D.R., Jacobs, P.A., Jones, T.V.: Simulation of Oxford University Gun Tunnel performance using a quasi-one-dimensional model. Shock Waves **11**, 377–383 (2002)
37. Schneider, S.P., Haven, C.E.: Quiet-flow Ludwieg tube for high-speed transition research. J. Aeronaut. Sci. **33**(4), 688–693 (1995)
38. Bergstrom, E.R., Raghunathan, S.: Nonstationarity in gun tunnel flows. AIAA J. **15**(9), 1362–1364 (1977)
39. Walton, C., Cain, T M.: Scramjet testing in a gun tunnel. In: Danesy, D. (ed.) Proceedings of the Fifth European Symposium on Aerothermodynamics for Space Vehicles, ESA SP-563, pp. 541–546, Cologne, Germany, 8–11 Nov 2004
40. Bray, K.N.C., Pennelegion, L., East. R.A.: A progress report on the University of Southampton hypersonic gun tunnel. Current Papers C.P. No. 457, Aeronautical Research Council (1959)
41. East, R.A., Qasrawi, A.M.S.: A long stroke isentropic free piston hypersonic wind tunnel. Reports and Memoranda R&M No. 3844, Aeronautical Research Council (1978)

42. Jones, T.V., Schultz, D.L., Hendley, A.D.: On the flow in an isentriopic light piston tunnel. Reports and Memoranda R&M No. 3731, Aeronautical Research Council (1973)
43. Oldfield, M.L.G., Jones, T.V., Schultz, D.L.: A Ludwieg tube with light piston isentropic compression heating. Paper No. 34 255, Aeronautical Research Council (1973)
44. Buttsworth, D.R.: Ludwieg tunnel facility with free piston compression heating for supersonic and hypersonic testing. In: Proceedings of the Australian Space Science Conference, pp. 153–162, Sydney, Australia, 28–30 Sep 2009

Author Biography

David Buttsworth works as an academic in the School of Mechanical and Electrical Engineering at the University of Southern Queensland (USQ) in Toowoomba, Australia. He also holds an honorary position at the University of Queensland where he collaborates with colleagues in the Centre for Hypersonics on a range of high- and low-temperature hypersonic research projects. David designed and commissioned 'TUSQ' which operates at USQ and provides hypersonic cold flows complementary to the flows generated in high enthalpy impulse facilities elsewhere in Australia

Part IV
Expansion Tubes

Expansion Tubes in Australia

David E. Gildfind, Richard G. Morgan and Peter A. Jacobs

1 Introduction

Hypersonics research at the University of Queensland (UQ) was set in motion by the arrival of Professor Ray Stalker in 1977. Stalker, the inventor of the free-piston driver [1], soon commenced work on a large free-piston driven reflected shock tunnel (RST), the T4 facility, funded by the Australian Research Council [2]. This facility was a larger scale development of the earlier T1, T2, and T3 machines built at the Australian National University. T4 theoretically had sufficient scale to provide the test times, stagnation pressures, and to accommodate the model sizes, required to conduct hypersonic combustion and propulsion studies, and the facility became operational in 1987 [3]. While T4 was destined to become UQ's workhorse for hypersonic flow experiments (T4 would fire its 10,000th shot in August 2008 [4]) around this same period, in the late 1980s, the new hypersonics group at UQ simultaneously began to investigate expansion tube operation. While RSTs dominated hypersonic ground testing in the 1980s and 1990s, it was always known that the stagnation of the test gas upstream of the supersonic nozzle limited them to sub-orbital flight speeds, and there remained the need for higher enthalpy ground testing capabilities.

The expansion tube concept was initially proposed by Resler and Bloxom in 1952 [5], and Trimpi [6, 7] derived the analytical framework to predict their

D.E. Gildfind (✉) · R.G. Morgan · P.A. Jacobs
Centre of Hypersonics, The University of Queensland, Brisbane 4072, QLD, Australia
e-mail: d.gildfind@uq.edu.au

R.G. Morgan
e-mail: r.morgan@uq.edu.au

P.A. Jacobs
e-mail: p.jacobs@uq.edu.au

O. Igra and F. Seiler (eds.), *Experimental Methods of Shock Wave Research*,
Shock Wave Science and Technology Reference Library 9,
DOI 10.1007/978-3-319-23745-9_13

performance. Expansion tubes could, in theory, produce a wide range of high enthalpy flow conditions [8]. As Miller and Jones notes [9] (p. 1):

> In theory, the expansion tube (so named by Trimpi) is capable of generating a wide range of hypersonic-hypervelocity flow conditions in air without being frustrated by the two principal difficulties with more conventional-type hypervelocity wind tunnels: namely, extremely high reservoir pressures and temperatures, and the problem of maintaining thermochemical equilibrium of the gas throughout the expansion in the nozzle.

Following several preliminary expansion tube studies across the US in the 1960s, including work with the 'Langley Pilot Model Expansion Tube' [9], NASA commissioned the Langley Expansion Tube/Tunnel, and in the 1970s Miller [10] conducted the first extensive experimental investigation into expansion tube operation. Miller found that for a given test gas, suitably steady test flows could only be obtained for a very narrow range of test conditions; "only a single flow condition, in terms of Mach number and Reynolds number…for a given test gas" [9] (p. 371). Many flow conditions had unacceptable levels of noise in the test flow [11], for reasons that were inexplicable at the time. Due to the relatively unsatisfactory findings of the Miller investigation, as well as "diminished programmatic needs" [9] (p. 371), the Langley Expansion Tube/Tunnel was deactivated in January 1983 [9], and the expansion tube concept lost traction in the years that followed this first phase of research activity.

Given its heritage with Professor Stalker and his free-piston driver invention, it was fitting that the UQ research group would be the first to use a free-piston driver to power an expansion tube [8]. In the late 1980s, NASA Langley contracted the UQ group to explore the possibility of driving an expansion tube with a free-piston compressor in order to produce high quality expansion tube flows. TQ, shown in Fig. 1, was a relatively small pilot facility, comprising a compression tube that was 2.3 m long with a 0.1 m diameter bore, used a 3.4 kg piston, and had a driven tube that was 5.26 m long with a 38.6 mm bore [8]. The driven tube was partitioned by a thin cellophane diaphragm into a 2.08 m long shock tube and 3.18 m acceleration tube [8]. It was thought that the 'versatility' of the free-piston driver might expand the window of useable test conditions [8].

The purposes of these initial experiments with TQ, which began in 1987 [12], were to replicate the Langley flow conditions in UQ's smaller facility, to establish if additional and acceptably steady test flows could be developed, and to investigate reasons for the poor test flows previously observed [8]. A systematic study with TQ eventually led to the discovery by Paull and Stalker [11, 13] of the cause of test

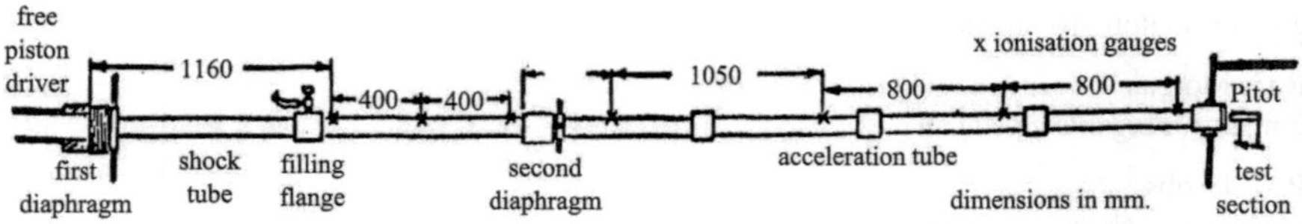

Fig. 1 Schematic of UQ's first expansion tube, TQ (adapted from [8])

flow unsteadiness (discussed later in this chapter) that had rendered so many flow conditions unusable in earlier studies. Not only did this finally explain the mechanism behind the observed test flow noise, but it also provided the means for rational flow condition design which would avoid the noise problem. Thus, faith in the potential of expansion tubes was gradually restored, and beginning with UQ, international research efforts with these facilities resumed.

This chapter provides an overview of UQ's experience with expansion tubes, which has involved four progressively larger facilities, TQ, X1 (a modification of TQ), X2, and X3. The chapter discusses the key developments at UQ, which begin with overcoming the problem of test flow noise, and finish at the present day with X3, the world's largest and highest performance free-piston driven expansion tube facility.

2 Principle of Operation

An expansion tube involves complex wave processes which are difficult to control and to measure. Referring to Fig. 2, the test gas is initially contained in a long steel tube (the shock tube), sealed at one end by a thick steel diaphragm (the primary diaphragm), and at the other end by a thin (typically) Mylar diaphragm (the secondary diaphragm). Upstream of the steel diaphragm is a larger diameter 'compression' tube, which contains a free-sliding massive piston and a light 'driver' gas such as helium; on the downstream side of the secondary diaphragm is another long

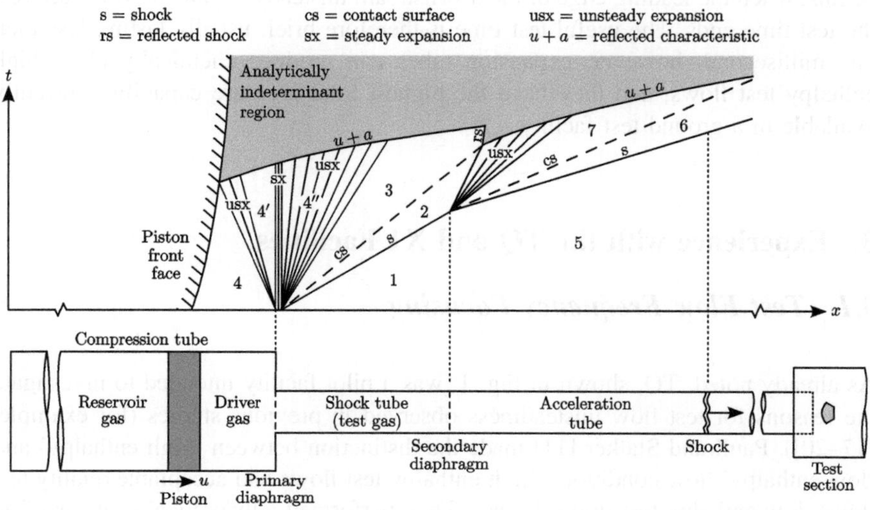

Fig. 2 Schematic diagram of a free-piston driven expansion tube (longitudinal scale greatly compressed)

tube (the acceleration tube) which extends to the test section, and initially contains air at very low pressure.

When the facility is operated, the heavy piston is fired towards the steel diaphragm, reaching maximum speeds of between 500 and 1000 km/hr. The piston compressively heats the helium driver gas in front of it, driving up the pressure and temperature until the gas explodes through the steel primary diaphragm. At the moment when the steel diaphragm ruptures, the driver gas pressure is tens of megapascals, and its temperature thousands of Kelvin. An extremely strong shock is driven into the test gas, compressing it and accelerating it down the tube. The test gas, now at high pressure and very hot, eventually arrives at the Mylar diaphragm, blasts through it, and suddenly encounters the low pressure (of order 10^0–10^2 pascals) of the downstream acceleration tube. This sudden change in boundary condition allows the test gas, already moving at several km/s, to expand forward towards the test section to speeds as high as 20 km/s.

The 'expansion tube' takes its name from this final unsteady expansion process. Total pressure and total temperature both increase when a supersonic flow undergoes an unsteady expansion. Unlike RST facilities, since the test gas is never stagnated, expansion tubes are not structurally limited by total temperature and pressure limits [14]. Furthermore, levels of dissociation and radiative losses can be minimised since the shock wave is no longer the only mechanism which adds energy to the test gas [15, 16].

The unsteady expansion process relies on the transfer of energy from the unexpanded upstream test gas to the expanded downstream test gas. Since only part of the test gas can be processed completely, test times are correspondingly reduced; "energy and total pressure are added to the flow at the expense of test time" [15] (p. 605). When the expanded test gas arrives at the test section the flow experiment begins; when the leading edge of the downstream unsteady expansion wave arrives, the test time ends. The useful test time is therefore brief, usually lasting less than one millisecond; however, expansion tubes can produce chemically clean high enthalpy test flows, and they have the highest total pressure capability currently available in a ground test facility.

3 Experience with the TQ and X1 Facilities

3.1 Test Flow Frequency Focusing

As already noted, TQ, shown in Fig. 1, was a pilot facility intended to investigate the reasons for test flow unsteadiness observed in previous studies (for example, [17–20]). Paull and Stalker [11] made the distinction between 'high enthalpy' and 'low enthalpy' flow conditions; high enthalpy test flows had acceptable quality test flows, low enthalpy test flows did not. They performed a theoretical analysis of an acoustic wave as it traverses an unsteady expansion [13], and proposed that the unsteady expansion at the secondary diaphragm, which produces a large drop in the sound speed of the test gas, has the effect of focusing all frequency components of noise present in the test gas into a narrow bandwidth of frequencies [11]. This is later characterised as strong disturbances in the test flow. It was also shown that this focusing effect only occurs for lateral acoustic waves (radial waves in an axisymmetric facility); the effect does not occur for longitudinal waves. Numerical simulations of the full Langley expansion tube [21] also showed radial disturbances from the tube area change at the primary diaphragm station focusing into a radial wave train that propagated into the expanded test gas.

This frequency focusing effect is an inherent feature of the unsteady expansion process, and occurs for both high and low enthalpy flow conditions. The reason that high enthalpy flow conditions have acceptable test flow quality is because for these conditions the test gas has very low levels of noise *prior* to the unsteady expansion, so that even after frequency focusing has occurred, noise levels remain acceptably low. In contrast, low enthalpy flows typically already have high levels of noise in the test gas prior to the unsteady expansion, and after frequency focusing these disturbances become unacceptably large. The characteristic difference between these two types of flows is the relative ratio between the sound speeds of the shock processed test gas, a_2, and the expanded driver gas, a_3 (refer to Fig. 2 for designation of gas regions).

Paull and Stalker determined that operating an expansion tube in a suitably over-tailored configuration (i.e. $a_2 > a_3$ in Fig. 2) can prevent acoustic disturbances present in the expanded driver gas from penetrating the test gas, the effect being likened to an 'acoustic buffer' [11]. The required ratio a_2/a_3 increases where higher frequency noise is to be suppressed, or where the driver gas sound speed is lower. Neely et al. [22] applied these principles to the TQ facility, successfully demonstrating steady air and argon test flows at flow velocities up to 9 km/s. In practice, Paull and Stalker [11] indicate that an acoustic buffer will be effective for $a_2/a_3 > 1.25$. This ratio is also supported by Morgan [15], which was published several years later following greater experience with the concept.

Paull and Stalker concluded that successful expansion tube operation would be limited to high enthalpy conditions (where the shock-processed test gas is very hot and therefore has a high sound speed) unless some means of reducing the noise in the driver gas could be devised. This seemed unlikely in a free-piston driven expansion tube; aside from the diaphragm rupture process itself, the expansion of driver gas through the area change has been shown to be a fundamental source of radial disturbances [23]. However, Morgan [15] would later propose that this sound speed increase could instead be achieved by using a shock-heated helium secondary driver, thus dramatically expanding the useful operating range of these facilities.

3.2 Operation with a Shock-Heated Secondary Driver

Henshall [24] first proposed the concept of a shock-heated secondary driver, and Stalker and Plumb [25] were the first to experimentally verify it (using a standard shock tube arrangement). Operation with a secondary driver involves placing a volume of, typically, helium between the primary diaphragm and the test gas. Referring to Fig. 3, the secondary driver is operated in the over-tailored mode such that $a_{sd2} > a_{sd3}$; i.e. the expanded primary driver gas (Region *sd*3) has a lower sound speed than the shock-processed secondary driver helium gas (Region *sd*2).

When a secondary driver is used, the shock-processed helium becomes the driver for the test gas. Observing Fig. 3, the pressure and velocity are constant across the *sd*2/*sd*3 interface. The extent of compressive shock heating which the Region *sd*2 gas undergoes, and therefore its temperature and sound speed, depends on its initial fill pressure. At sufficiently low fill pressures, the helium gas will be compressively heated so much that its temperature far exceeds the temperature of driver gas which could otherwise be achieved by free-piston compression alone. With a sufficiently high sound speed, the Region *sd*2 gas can potentially drive a faster shock through the test gas than the primary driver would be able to by itself.

The benefit of the secondary driver comes at the expense of duration, since the region *sd*2 gas expands more rapidly by virtue of its higher sound speed, although this may not be a concern for high enthalpy conditions. Furthermore, for a given total driven tube length, the secondary driver reduces the length of the remaining

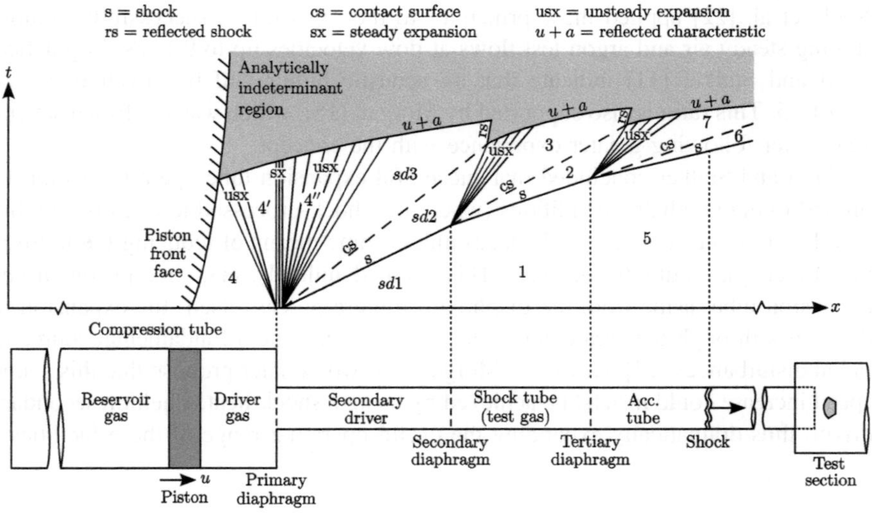

Fig. 3 Schematic diagram of wave processes occurring inside a free-piston driven expansion tube with shock-heated secondary driver (longitudinal scale greatly compressed). An *unsteady expansion* is assumed to form at the secondary diaphragm, which is typical for super-orbital operation of these facilities where the test gas is initially at low density

tubes, increases the complexity of internal flow processes, and adds to the amount of contaminants and debris in the flow (due to the addition of another diaphragm) [26].

Morgan and Stalker [27] first used a secondary driver with the X1 expansion tube in order to boost its super-orbital performance, and in the process achieved maximum shock speeds of 18.7 km/s. Neely and Morgan [28] further developed the "superorbital expansion tube concept", developing and characterizing a 13 km/s test flow in X1. This work also examined the influence of real gas effects and diaphragm rupture on expansion tube flow processes. Morgan [15] discusses the secondary driver concept in more detail, and details the operating regimes where this configuration can be beneficial.

As noted in the Sect. 3.1, a secondary driver can also be beneficial for lower enthalpy conditions. After Paull and Stalker [11] determined that under-tailored operation was the cause of noise transmission to the driver gas at lower enthalpy conditions, it was thought that adequate test flow quality could not be achieved at these conditions, since there was no acoustic buffer available to filter the driver noise. However, Morgan [15] later suggested that a secondary driver could be used as an alternative upstream means by which to achieve the required 'acoustic buffer', and this has subsequently opened up a much larger range of available test conditions with respect to the acoustic buffer requirement.

A couple of observations can be made about the use of a secondary driver as an acoustic buffer for low enthalpy conditions. Firstly, there is no acoustic buffer shielding the test gas from noise in the secondary driver gas, so disturbances introduced to the Region 3 gas (i.e. by Mylar diaphragm rupture, or during transit down the tube) would not be suppressed in the test gas (Region 2). Since the test gas itself lacks the intrinsic mechanism to prevent noise ingress, it would be expected that these conditions should be somewhat noisier than conditions where the test gas itself provides the acoustic buffer; without making any conclusive observations here, this does indeed reflect anecdotal experience with testing in X2 at high enthalpies with and without secondary drivers.

Secondly, while maximizing the ratio a_{sd2}/a_{sd3} may theoretically minimize noise transmission from the primary driver to secondary driver gases, this will also reduce the ratio a_2/a_3, which will increase noise transmission from the secondary driver gas to test gas. The optimum secondary driver configuration may be one which has sufficient sound speed to act as an acoustic buffer, but no more than this, although this question requires further investigation.

Finally, the critical wave processes associated with many low enthalpy conditions can be characteristically different to those associated with high enthalpy conditions. Observing Fig. 4, for sufficiently high initial test gas densities, the shock-processed secondary driver gas (Region *sd*2) will decelerate as it impacts the dense test gas (Region 1), and consequently will be compressively heated. This deceleration and compression occur due to a reflected shock (labeled 'rs' in Fig. 4) which arises at the secondary diaphragm (as opposed to the unsteady expansion shown in Fig. 3). The effect of this reflected shock on noise transmission may also require further investigation.

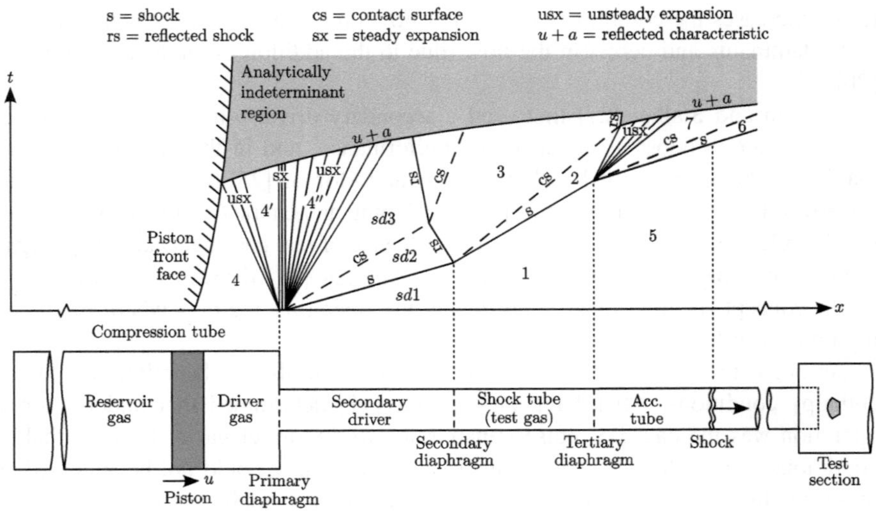

Fig. 4 Schematic diagram of wave processes occurring inside a free-piston driven expansion tube with shock-heated secondary driver (longitudinal scale greatly compressed). A *reflected shock* is assumed to form at the secondary diaphragm, which is typical for super-orbital operation of these facilities where the test gas is initially at high density

4 The X2 Facility

During the 1990s, Morgan [26] received Australian Research Council support to develop two larger expansion tube facilities, X2 and X3, which had total lengths of approximately 25 and 65 m respectively. Size is paramount for impulse facilities: a larger tube diameter permits testing of larger models; a longer tube provides a longer slug of test gas, and therefore more test time. The practical implications of increasing the facility scale become evident in Fig. 5, which compares UQ's three expansion tube facilities, beginning with TQ/X1. Despite the success of the pilot studies in TQ/X1, the driven tube bore was only Ø38 mm, and the test time was very short (for example, 15 μs of test time for a 13 km/s air condition [28]). Having established proof-of-concept with this pilot facility, the logical next step was to move to larger facilities.

X2 and X3 are each world-class facilities in their own right, however their development over the years has been strongly intertwined. X2 is a medium-sized machine which is manageable at a 'human' scale, and as such is a convenient and economical platform to trial new technologies and ideas. X3 is a much larger and heavier machine. Mechanical assistance is required to operate X3, design changes are more costly and difficult to incorporate, however this is the inevitable price to pay for X3's much higher performance. In this context, it is natural that the evolution of X2 has always been in advance of X3, and that X2 has been the initial platform for UQ's major developments with expansion tube facilities. This section focusses on X2, originally commissioned in 1995 [29], and discusses the major developments which have been accomplished with the facility.

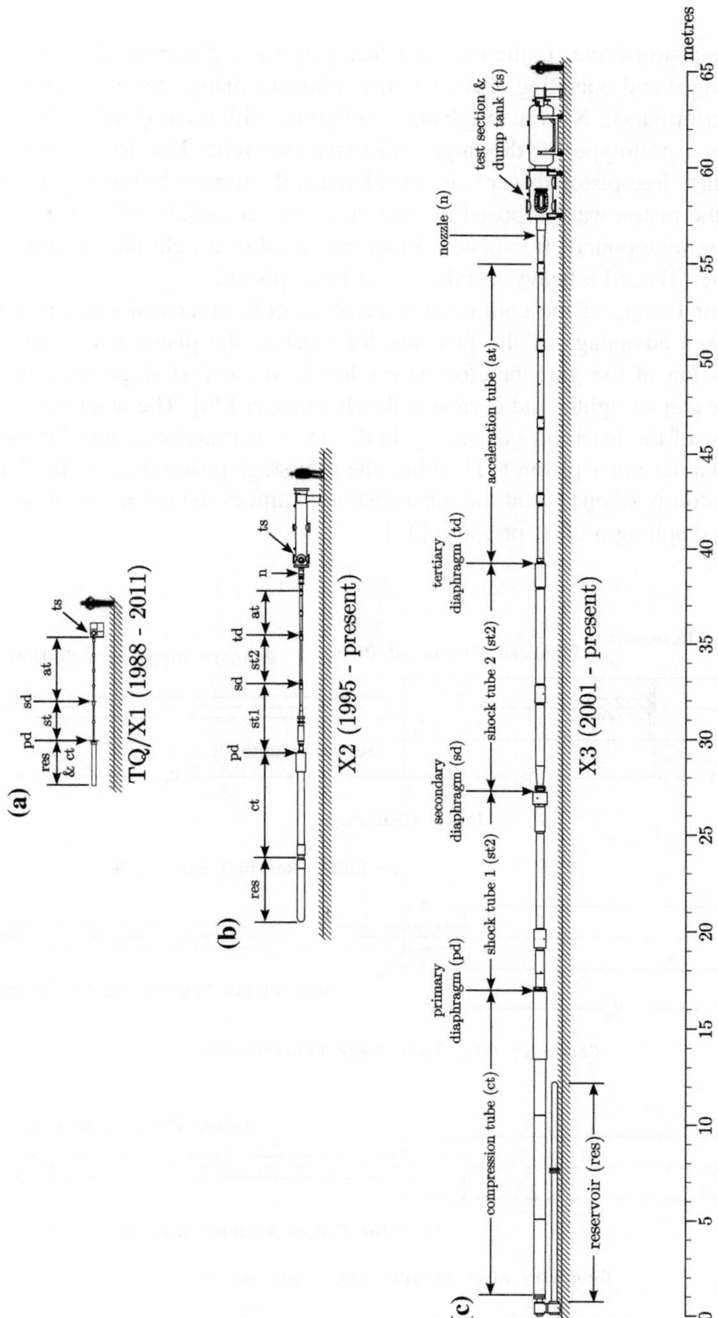

Fig. 5 The University of Queensland's X-series expansion tube facilities (to scale). Supporting carriage work, pressure manifold, and associated hardware, not shown

4.1 X2's Free-Piston Drivers

The free-piston driver is the engine which powers UQ's expansion tubes, and its performance and operating characteristics influence many aspects of the overall facility performance. X2 was originally configured with a compound driver in order to act as a prototype for the larger X3 expansion tube. Due to the size of X3, a single stage free-piston driver was considered at the time to be too expensive, so the compound piston was proposed in order to reduce costs [29, 30] (see Fig. 6). X2's two-stage (compound) free-piston driver consisted of a light aluminium outer piston, which carried a heavy stainless steel inner piston.

The first stage of the compression involves both inner and outer pistons. This stage takes advantage of the fact that for most of the piston cycle there is little compression of the gas, therefore stress levels in the first stage are low, and the structure can be lighter and correspondingly cheaper [30]. The outer piston is light, and most of the reservoir gas energy in this stage is transferred into kinetic energy in the heavier inner piston [31]. When the two stage piston reaches the buffer, the outer piston is stopped, and the inner piston continues the compression to the final primary diaphragm burst pressure [30].

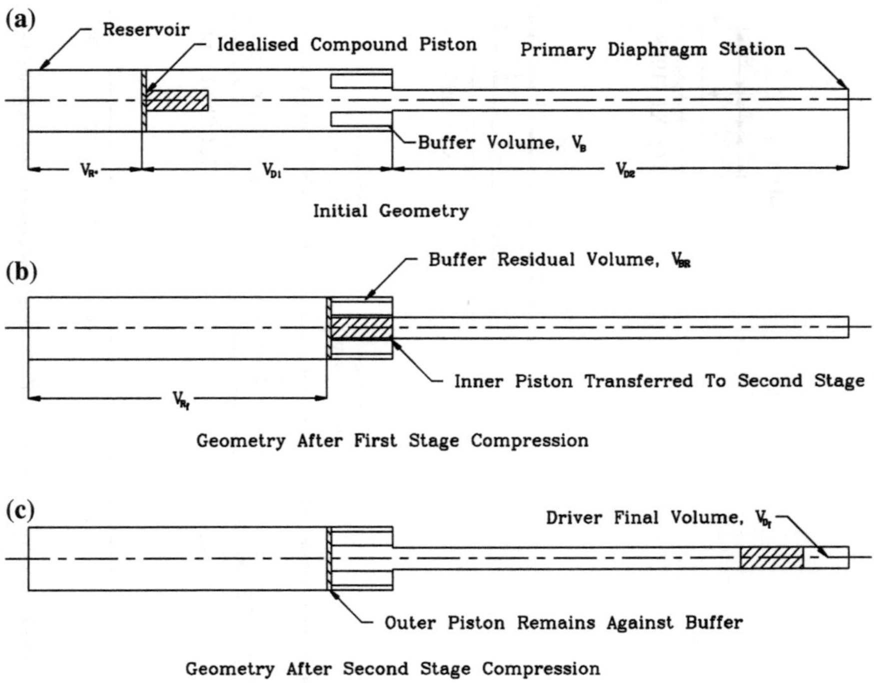

Fig. 6 Schematic of compound driver concept (reproduced from Doolan and Morgan [30])

The advantage of the compound driver is that it can achieve similar compression ratios to a single piston driver, but with a much shorter compression tube (for an inner/outer piston diameter ratio of 2, compression tube length can be reduced by approximately 75 %). Furthermore, the large diameter section of the compression tube can be made of thinner steel, since it does not have to contain the large pressures which arise at the primary diaphragm. Studies using X2's compound driver included holographic interferometry measurements of flow over cylinders [32] and flow over toroidal ballutes [33], flow tagging velocimetry [34], boundary layer development over flat plates [35], and hypersonic shock standoff on blunt bodies in ionizing flows [36].

The compound piston driver has three distinct disadvantages as compared to the single piston driver. Firstly, it is a more complicated device to operate [29]. Secondly, without the area change of the single piston, a longer slug of driver gas is required for adequate test time. Thirdly, there is a reduction in the maximum pressures and densities which can be achieved [29]. For these reasons, X2 was subsequently reconfigured with a single stage 35 kg piston, with an area ratio of 9 across the primary diaphragm. The installation of the new driver was completed in April 2004 [29].

The 35 kg piston was operated successfully for a range of flow conditions (studies included measurements of Mars entry radiation [37] and Titan entry radiation [38, 39], and Mach 10 scramjet combustion [40]). In 2009, attempts were made to simulate Mach 13 scramjet flight, but at much higher total pressures than the earlier Mach 10 scramjet combustion study [41]. It was found that the target test condition could not be achieved due to significant attenuation of the shock speed down the driven tube [41]. Figure 7 shows the primary shock speed along X2's driven tube. The dashed lines show the required shock speeds to achieve the Mach 13 condition; the data points show the experimentally measured shock speeds between adjacent pressure transducers located in the tube wall; the solid curve shows a later computation of the shock speed using the 1-D Lagrangian CFD code, L1d [42].

Referring to Fig. 7, strong rarefaction waves originating at the free-piston driver were responsible for the significant shock speed attenuation. Initial flow condition

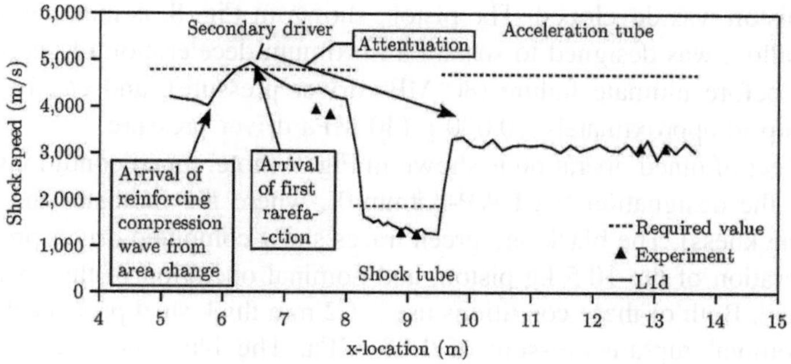

Fig. 7 Shock speed versus position for Mach 13 flow condition, using 35 kg piston with 100 % helium driver (reproduced from Gildfind et al. [41]). Primary diaphragm is located at x = 4.810 m. Maximum experimental uncertainty is ±3 %

calculations had used analytical methods which did not take into account transient wave processes in the driver; subsequent analysis with L1d, which included the full piston dynamics, indicated that the observed attenuation was due to rapid expansion of the driver gas following diaphragm rupture [41].

The 35 kg piston is relatively heavy compared with the length of X2's driver (4.5 m), and as a consequence, it is operated relatively slowly (speed is less than 100 m/s when the diaphragm ruptures). The results in Fig. 7 are for a 100 % helium driver gas operating at a compression ratio $\lambda = 42$. At this condition, the driver gas volume is small when rupture occurs, and the compressed helium is hot and therefore has a very high sound speed. The slow moving piston presents an approximately static boundary in comparison to the timescales associated with the unsteady expansion of the driver gas. This unsteady expansion soon reflects off the piston face, and is transmitted downstream as a $u + a$ wave. For higher enthalpy flow conditions, critical flow processes traverse the tube before this relaxation catches up. However, high total pressure scramjet flow conditions, which involve generating slower shock speeds (1–2 km/s) through much denser test gases, take longer to traverse the tube. At these lower enthalpy, higher total pressure conditions, the driver pressure relaxation must be delayed or else it will interfere with critical flow processes, as observed in Fig. 7 (see 'Arrival of first rarefaction').

One method to increase the useful driver gas supply time is to operate a 'tuned' free-piston driver [43]. This involves running the piston at sufficiently high speed, so that when the primary diaphragm ruptures, the piston displacement compensates for driver gas loss to the driven tube [41]. In practice the piston is run even faster than this—it is 'over-driven' [44]—such that even after the diaphragm ruptures, the driver pressure momentarily continues to increase. The piston, which is rapidly decelerating during this process, eventually slows to the point that the driver pressure drops. This process can, however, significantly extend the duration of time over which the driver gas is maintained at high levels.

The challenge with tuned operation in X2 was to be able to first accelerate the piston to high speeds (of order 200–250 m/s for X2), and then bring it to a soft landing at the other end of the tube, all within X2's 4.5 m compression tube length. Analysis indicated that the piston would have to be as light as possible, and a new 10.5 kg piston was developed. The piston, shown in Fig. 8, is manufactured from 7075-T6 alloy, was designed to sustain a maximum deceleration of approximately 40,000 g before ultimate failure (80 MPa driver pressure), and can be routinely operated up to approximately 20,000 g (40 MPa driver pressure).

The effect of tuned operation is shown in Fig. 9 (note: tuned conditions in Fig. 9 are given the designation 'x2-LWP-#.#mm-0', where #.# indicates the steel diaphragm thickness). The black and green traces show computed driver pressures for tuned operation of the 10.5 kg piston, and nominal operation of the 35 kg piston, respectively. Both of these conditions use a 1.2 mm thick steel primary diaphragm, with a nominal rupture pressure of 15.5 MPa. The blue and red curves show computed driver pressures for tuned operation with 2.0 and 2.5 mm thick diaphragms. It can be seen that tuned operation significantly changes the characteristics of the driver pressure trace.

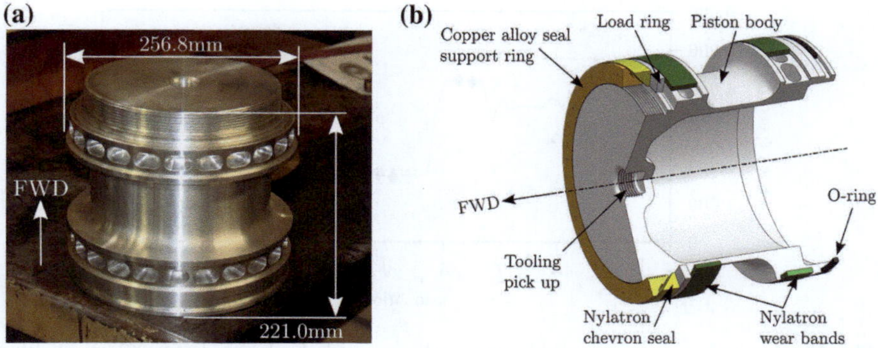

Fig. 8 X2's lightweight 10.5 kg piston. **a** Piston body, machined from 7075-T6. **b** Cutaway view of piston assembly. Adapted from Gildfind et al. [41]

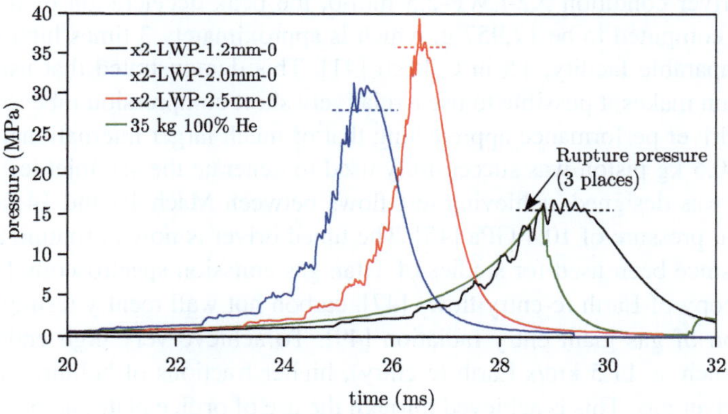

Fig. 9 L1d computed driver pressures. Driver pressure for 35 kg piston is shown for comparison (*green curve*); this pressure history has been time-referenced to align its rupture time with driver condition X2-LWP-1.2 mm-0 (both use the same 1.2 mm thick diaphragm, with nominal 15.5 MPa rupture pressure). Reproduced from Gildfind et al. [41]

Considering the black and green traces (for the 1.2 mm diaphragm), while it can be seen that tuned operation sustains the driver gas pressure for an order of magnitude longer duration, the tuned condition (black curve) has a lower compression ratio ($\lambda = 17.5$ compared to $\lambda = 42$), and uses a heavier driver gas (80 % He/20 % Ar compared to 100 % He), and therefore cannot drive the same *initial* shock speed as the 35 kg piston condition (green curve). However, operation of the tuned driver at higher diaphragm rupture pressures (the blue and red curves), and correspondingly higher compression ratios, achieves similar initial performance to the 35 kg piston condition, and furthermore, demonstrates no attenuation. The required, computed and experimental shock speeds for driver condition 'x2-LWP-2.5 mm-0' in Fig. 9 are shown in Fig. 10. It can be seen that experimental shock speeds are within 10 % of the required shock speeds, with no obvious attenuation.

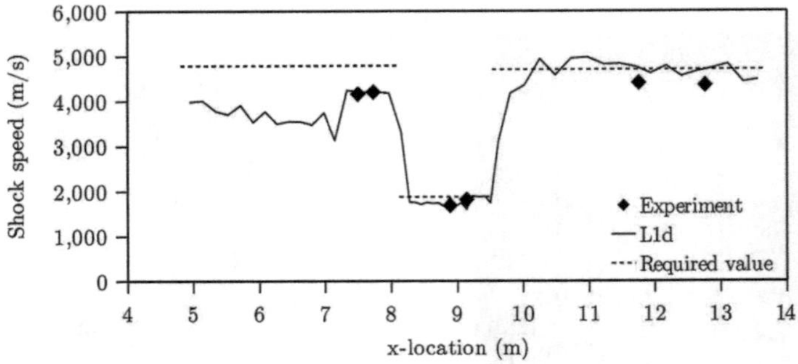

Fig. 10 Comparison between experimental and computed shock speeds for X2 tuned driver condition X2-LWP-2.5 mm-0 (reproduced from Gildfind et al. [41]). Maximum experimental uncertainty is ±2 %. No shock attenuation is observed

For driver condition X2-LWP-2.5 mm-0, the peak deceleration of the 10.5 kg piston is computed to be 17,957 g, which is approximately 3 times higher than the next comparable facility, T5 in Caltech [41]. This demonstrated that using a very light piston makes it possible to use a relatively short compression tube to achieve a level of driver performance approaching that of much larger international facilities.

The 10.5 kg piston was successfully used to generate the scramjet test flows for which it was designed, achieving test flows between Mach 10 and 14, at a maximum total pressure of 10.4 GPa [45]. The tuned driver is now in routine use in X2, and has since been used for studies of Titan gas emission spectroscopy [46], VUV spectroscopy of Earth re-entry flows [47], carbon hot wall reentry testing [48], and simulation of gas giant entry radiation [49]. To achieve very high enthalpy conditions (such as 11.5 km/s Earth re-entry), higher fractions of helium are required for the driver gas. This is achieved through the use of orifice plates at the driver area change [50], which reduce the choked diameter at the area change, and are sized to achieve the same volumetric flow rate as the original driver operating condition, thus ensuring that piston dynamics are preserved.

4.2 X2 Mach 10 Nozzle

Considering a facility such as X2, the basic core flow is constrained to be less than the diameter of the driven tube (Ø85 mm for X2), and significantly less when boundary layers are accounted for. The purpose of using a contoured nozzle is primarily to increase the test gas core flow size, although the test time may also increase slightly [51, 52]. Hypersonic nozzles are characterised by being purely diverging, with fully hypersonic flow throughout [53].

A steady expansion nozzle increases the size of model which can be tested, which has practical benefits in terms of model size and instrumentation. Increasing model size does not assist with meeting binary scaling (ρ-L) targets, since the

Fig. 11 The X2 Mach 10 nozzle (reproduced from Scott [54])

corresponding reduction in test flow density (approximately proportional to D^2) is greater than the increase in model size (proportional to D). However, due to the very high levels of total pressure achievable, the reduction of this ρ-L term due to nozzle expansion can often be compensated.

X2 has a contoured steel nozzle which is full capture and shock free [54]. The preliminary design of the nozzle used the method of characteristics, targeting an exit flow Mach number of 10 for an inflow Mach number of 7.3, and a 0° flow angle, assuming inviscid and irrotational flow [54]. This shape was then further optimised using a Nelder-Mead technique matched with the compressible flow solver SM_3D+ [55]. The nozzle, shown in Fig. 11, was tested by Scott [54] for three conditions (two air and one Titan atmosphere) and found to produce acceptable results.

The difficulty with contoured nozzles is that they are optimised for a single nozzle inlet Mach profile. Whilst the contoured nozzle may produce a uniform exit flow at the design Mach number, it is more susceptible to flow non-uniformities at off-design conditions [53]. Furthermore, these nozzles are susceptible to shock generation by the wall contour, and the high Mach numbers through the nozzle encourage growth of boundary layers, which may already be thick at the acceleration tube exit/nozzle inlet [53].

Scott's [54] nozzle design process accounted for the presence of boundary layers with its second stage SM_3D+ analysis, however it is noted that this applies to a steady inflow, whereas actual expansion tube inflows are transient. This optimisation process also required an inflow Reynolds number to be selected, therefore deviation from this Reynolds number during operation also constitutes an off-design condition.

For idealised supersonic flow in a diverging nozzle, the Mach number depends only on cross-sectional area, and is calculated by solving Eq. (1), where the subscripts i and o indicate nozzle *inflow* and *outflow* respectively. If Eq. (1) is used with the *geometric* area ratio, A_o/A_i, it will often over-predict the flow expansion in comparison to experiment, since thick and varying boundary layers will affect the *actual* area change of the test gas *core flow*. Instead, an *effective* area ratio is used, and experience with X2 indicates that this can only be established from a combination of experimentation and sophisticated CFD analysis (refer Sect. 4.3.2).

$$\frac{M_i}{M_o}\left[\frac{1+\frac{\gamma-1}{2}M_o^2}{1+\frac{\gamma-1}{2}M_i^2}\right]^{\frac{(\gamma+1)}{2(\gamma-1)}}=\frac{A_o}{A_i} \tag{1}$$

4.3 Test Flow Characterisation

Test flow characterisation involves quantifying relevant test flow properties. For an expansion tube these include:

- Static pressure, temperature, and velocity;
- Core flow diameter (the portion of steady flow across the nozzle exit plane which is not excessively affected by boundary layer effects). Pitot pressure measurements are typically used to establish the size of this region;
- The steadiness or unsteadiness of these properties as the test time progresses;
- Shock speed is measured since it provides critical information about primary wave processes before test flow arrival at the test section;
- Chemical and thermal composition of the test flow: levels of vibrational excitation and thermal nonequilibrium become increasingly important at the upper envelope of planetary entry flows and scramjet combustion.

In an expansion tube, it is exceedingly difficult to measure these flow properties directly, especially with intrusive instrumentation. For example, while it is possible to directly measure the static pressure along the tube wall, it is not possible to directly measure this property away from the wall without first interfering with the flow itself. Furthermore, the expansion tube presents an extremely harsh instrumentation environment. A probe placed in the path of an expansion tube flow is subject to the extreme heating of hypervelocity flow, and is impacted by debris which trail behind the test gas. Referring to Fig. 12, the Mylar film, initially used to separate gases in the driven tube, later becomes entrained in the flow as particulates; these particulates arrive after the test gas and act as hypersonic projectiles, potentially impacting and damaging sensitive hardware in the test section. To protect expensive pressure transducers they are shielded from the flow, but this significantly complicates their aerodynamic response, and can reduce their sensitivity and degrade their measurement.

The experimenter is therefore presented with only a limited range of direct and indirect measurements. CFD models of expansion tube flow processes are then used to build a more complete picture of the flow (i.e. to fully characterise it). CFD has progressed to the point that it is now routinely used to perform 'virtual' experiments for comparison with 'actual' experiments. These simulations fully characterise the flow, however the utility of these calculations depends on how well they correlate with experimental measurements. The codes continually evolve to include more detailed physical models, and are calculated with greater resolution. The critical limitation however is that the uncertainty in CFD calculations is directly dependent

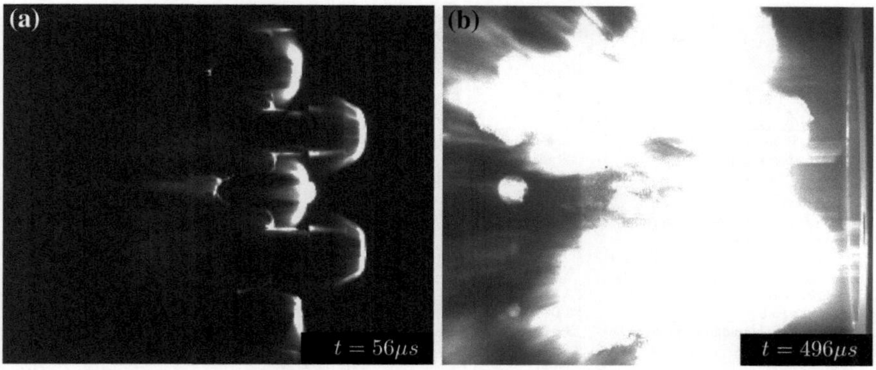

Fig. 12 High speed camera footage of flow in X2 (adapted from Gildfind [56]). **a** Test gas flowing from *right* to *left* over Pitot probes. **b** Later arrival of driver gas and entrained particulates

on the uncertainty in experimental measurements. Improving characterisation of expansion tube test flows therefore relies on the simultaneous improvement of both experimental and numerical techniques.

4.3.1 Test Flow Measurements

Experimental pressure measurements are used for routine characterisation of the test flow. Figure 13 shows the layout of piezoelectric PCB static pressure transducers along the length of X2, as well as other leading dimensions. These transducers are positioned in the tunnel wall, flush with the internal surface of the tube, and measure the static pressure of the flow as it passes. The microsecond response time of the gauges permits accurate identification of the time of shock arrival, and this data is used to calculate the average time of flight of the shock between adjacent transducer pairs. These measurements therefore provide experimental data about the static pressure of the flow, as well as the shock speed, and are an important diagnostic tool for reconstruction of the test flow properties.

The core flow is determined by sampling Pitot pressure across the tube exit. As discussed in Sect. 4.3 (and Fig. 12) the expansion tube presents an extremely harsh instrumentation environment, particularly for direct measurements of Pitot pressure. When pressure transducers are used, protective caps are utilised to prevent line-of-sight to the transducer sensing surface. Figure 14 shows various techniques used to protect transducers during Pitot pressure measurements. Figure 14a shows a standard Pitot cap, which has an internal brass diffuser to block line-of-sight of the flow to the internal transducer. Figure 14b and c show two variations of swirl cap which have four vortically aligned holes feeding the cavity containing the PCB sensing surface; these holes are designed to induce swirl in the flow, which is intended to dampen Helmholtz Resonance inside the Pitot cavity [57]. Even with

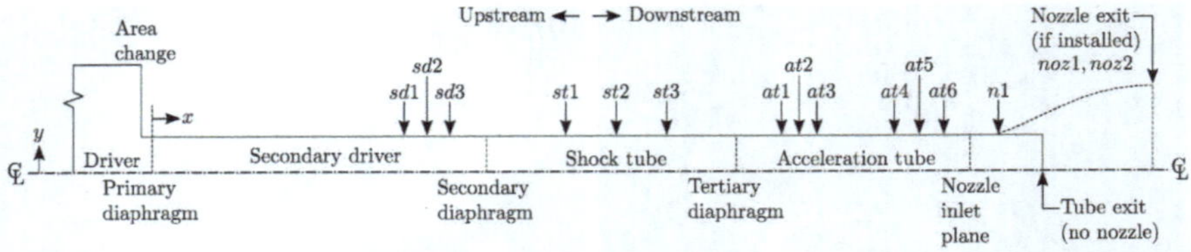

Description	x (mm)	y (mm)
Piston front face prior to launch	−4,589	-
Driver area change	−110	-
Primary diaphragm downstream face	0	-
Transducer *sd1*	2,577	42.5
Transducer *sd2*	2,810	42.5
Transducer *sd3*	3,043	42.5
Secondary diaphragm downstream face	3,418	-
Transducer *st1*	4,231	42.5
Transducer *st2*	4,746	42.5
Transducer *st3*	5,260	42.5
Tertiary diaphragm downstream face	5,976	-

Description	x (mm)	y (mm)
Transducer *at1*	6,437	42.5
Transducer *at2*	6,615	42.5
Transducer *at3*	6,796	42.5
Transducer *at4*	7,590	42.5
Transducer *at5*	7,846	42.5
Transducer *at6*	8,096	42.5
Nozzle inlet plane	8,343	-
Transducer *n1*	8,652	42.5
Tube exit (no nozzle)	9,105	42.5
Nozzle exit (if installed), *noz1*, *noz2*	10,731	100.8

Fig. 13 Schematic diagram of X2's experimental setup. Horizontal scale has been compressed for clarity. Free-piston driver is not shown. Location of the exit plane for configurations with and without the Mach 10 nozzle are shown

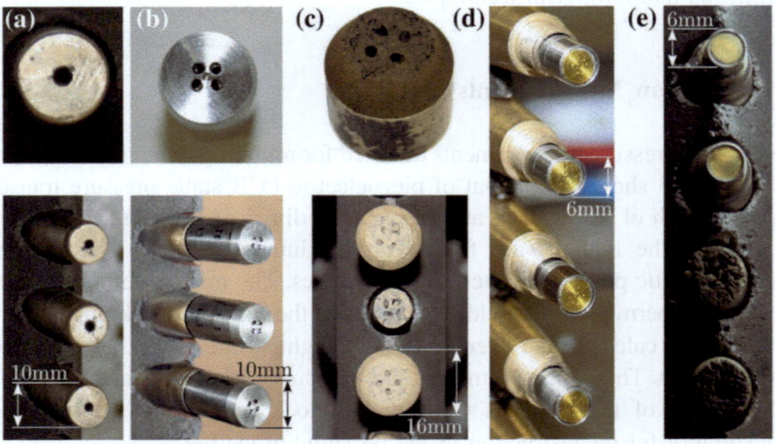

Fig. 14 Various Pitot probe configurations. **a** Standard brass single hole cap (used with an internal brass six hole impact plate to prevent line-of-sight between flow and the PCB transducer face). **b** Stainless steel 4 hole swirl cap. **c** Brass T4 RST 4 hole swirl cap (vibration isolated with rubber). **d** Example of cellophane protection across PCB transducer face. **e** Example of 0.05 mm thick brass shim protection across PCB transducer face

these shielding techniques, the internal cavity can still get extremely hot; Fig. 14d and e respectively show examples of cellophane and brass shim being applied to the PCB sensor surface to protect the transducer from the high transient heat loads.

A recent study in X2 [45] to develop scramjet flow conditions, with total pressures in the gigapascal range, found that normal Pitot measurement techniques, such as those detailed in Fig. 14, could not produce meaningful pressure measurements. These flow conditions induce high heat loads on the Pitot probe caps, and have large amounts of entrained Mylar since these diaphragms must be relatively thick to contain the high initial fill pressure (typically 0.1–0.2 mm Mylar thickness). A combination of brass shim and Pitot swirl cap was required to prevent transducer damage, after which the response became inadequate. While this combination has been used with RST facilities, the test times in X2 are much shorter (of order 100 μs, compared to order 1000 μs), and the increase in response time becomes prohibitive.

In order to reduce the Pitot transducer heat loads, a conical probe was designed based on a 15° half angle, using eight holes to fill the sensor cavity, as shown in Fig. 15. Whereas a Pitot probe measures the stagnation pressure behind a normal shock, the cone probe processes the flow with a conical shock. The static pressure at the cone surface still provides information about the flow density and velocity, however, the flow over the cone probe is significantly less severe, and each cone's pointed tip presents a smaller obstacle to oncoming debris entrained in the flow. Initial experimentation with these probes has provided promising initial results [45], however UQ is still establishing their full performance characteristics.

Finally, UQ has also measured Pitot pressure using bar gauges [58] (Fig. 16). A strain-gauged bar is placed into the flow, and the strain gauge response over time is measured. The flow is processed by a normal shock which forms over the disc at the exposed end of the bar; the average pressure in the bar cross-section can be

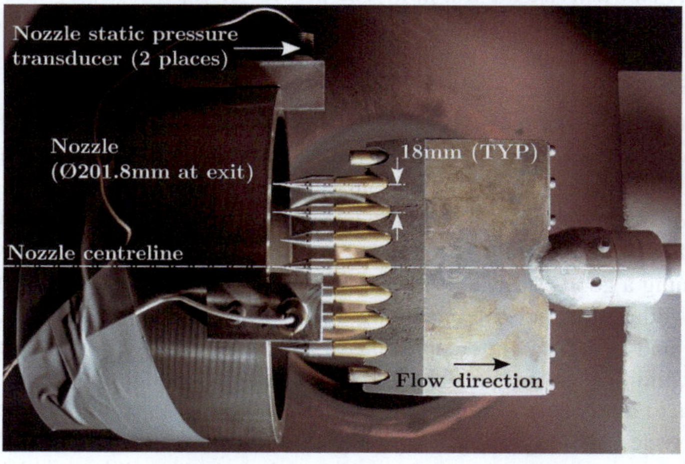

Fig. 15 Conical probe arrangement in test section, X2 with nozzle

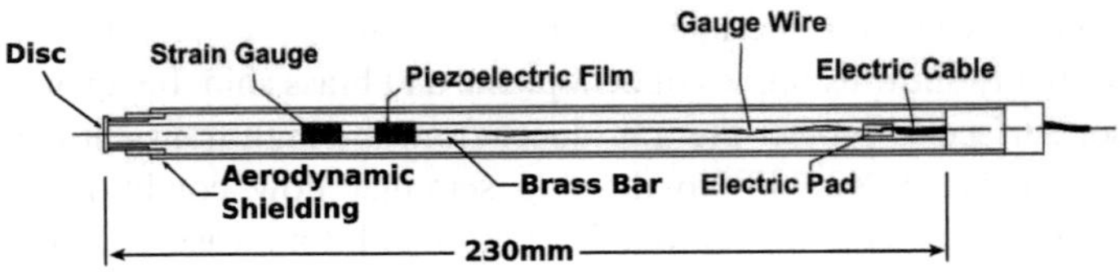

Fig. 16 Schematic of stress wave bar gauge (adapted from [58, 59])

closely correlated to the aerodynamic pressure acting on the disc at the stagnation streamline (i.e. the Pitot pressure). When the flow impacts the disc, stress waves are propagated down the bar. The bar is long and thin, so the stress waves are approximately planar. An axial strain measurement on the *surface* of the bar therefore provides a measurement of the axial strain *through* the bar, which can be used to calculate the axial force. Unless de-convolution techniques are used, this approach can provide a force measurement up until the stress waves reflect from the other end of the bar and return to the strain gauge. Chiu and Mee [58] report a rise time of approximately 5 μs, an uncertainty of ±7 %, and test times of 100 μs for a 230 mm long bar. This level of performance comes at the cost of complexity, durability, and test time however, and as a result these gauges have not been used for routine measurements in UQ.

4.3.2 Facility Numerical Simulation

UQ, with its extensive and long term experience with expansion tubes, has been involved in the majority of axisymmetric simulation work on these machines. These simulations have been performed with several purposes in mind:

1. *To fully characterise the test flow.* The process of experimentally characterising expansion tube test flows is very challenging, primarily due to short test times, harsh environmental conditions, space restrictions, and a lack of available non-invasive diagnostic techniques to make direct measurements. The advanced measurement techniques which *are* available, are correspondingly expensive, however even the most advanced equipment can only reveal a small portion of the flow properties, and typically only at a few spatial locations.
2. *Development of new flow conditions.* Axisymmetric CFD is the most accurate of the *practical* tools available to predict facility response for new flow conditions. Obtaining a good estimate of the facility response permits many potential issues to be addressed prior to conducting an experimental campaign (which can be costly and is usually time constrained). Once benchmarked against an existing flow condition experiment, the code can then be used for parametric design studies to improve or modify that flow condition. However, despite the utility of these codes, the experiment itself remains essential in order to establish the *actual* flow. The further away from the benchmarked conditions which the code

departs, the less reliable must the computed solution be assumed to be, and correspondingly greater care must be applied to its predictions.

3. *Validation of numerical codes.* Numerical codes need to be validated against experimental results, particularly in the hypersonic flight regime [60–65]. Validation data may comprise, for example, pressure or heat transfer measurements on the surface of a model in the test section, Pitot pressure measurements from a probe located in the path of the test flow, spectroradiometric measurements of the flow through an observation window into the test section, and so forth. A similar 'numerical experiment' is then performed, and the computed flow properties are compared to the experimental measurements. In order to accurately repeat the 'numerical experiment', it is typically necessary to first simulate part or all of the facility flow processes upstream of the test section, so that the complex transient test flow is accurately reproduced. To otherwise assume that the test flow is 1-D and uniform may be over-simplistic, and may constitute an additional source of discrepancies between computation and experiment which then compromises the validation process.

Current expansion tube numerical modelling at UQ adopts a hybrid solution approach (see Fig. 17), whereby only the low pressure acceleration tube and dumptank are modelled two-dimensionally. A radially uniform transient inflow is then defined at the start of the acceleration tube. The inflow is calculated using the 1-D code L1d [42], which models the entire facility including the piston dynamics, and captures the dominant longitudinal wave processes, including their complex interactions. The axisymmetric calculation is performed using the code Eilmer3, developed at UQ by Jacobs et al. [66–68]. Eilmer3 is "an integrated collection of programs for the simulation of transient, compressible flow in two and three spatial dimensions" [67]. The code, which has its origins in the early 1990s under the name 'cns4u', was originally developed for the simulation of reflected shock tunnel and expansion tube impulse facilities. Several aspects of the code make it particularly well-suited to this purpose [67]:

1. It solves the compressible Navier-Stokers equations using an upwinding approach, which can be very effective at capturing the strong shocks associated with these facilities.
2. It has multiple-block capability, which permits a reasonable solution time, using parallel processing, for models which typically are computationally very expensive.
3. It has thermochemistry and finite-rate chemistry capabilities, which are necessary for accurate predictions about the extreme flow processes which occur in these impulse facilities, particularly for superorbital flow conditions (6–15 km/s).

Wheatley et al. [59] argued that the hybrid approach is usually acceptable for two principle reasons. Firstly, the shock tube flow typically has a relatively thin boundary layer on account of its relatively low velocity and high density. Secondly, when the diaphragm separating the test and acceleration tube gases ruptures, the

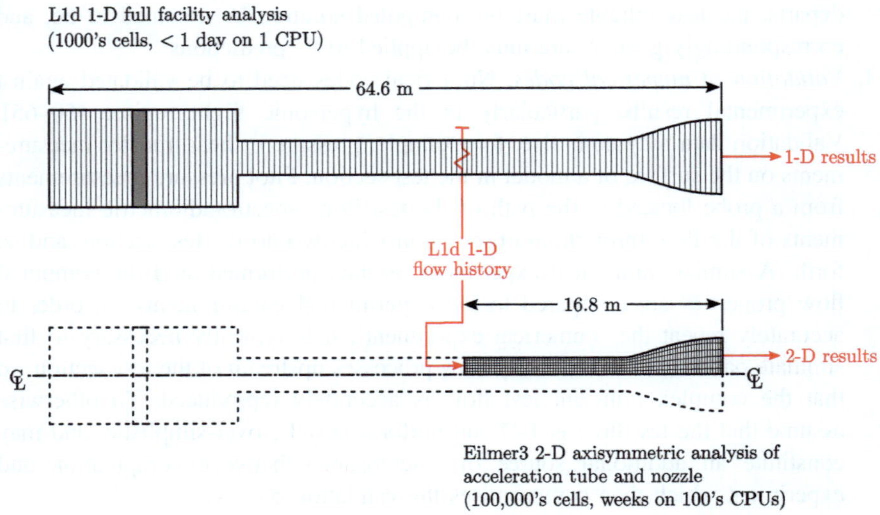

Fig. 17 Schematic diagram of X3 L1d/Eilmer3 CFD hybrid analysis model. Not to scale

downstream portion of the shock-processed test gas, which will eventually be the test flow, is located immediately behind the primary shock, where the boundary layer has only just begun to develop. It is often reasonable therefore to treat the shock-processed test gas as 1-D, and restrict the 2-D axisymmetric calculation to the low pressure acceleration tube and dumptank [59]. This is the approach which has typically been adopted at UQ, for example [54, 69, 70, 71].

Wheatley et al. [59] studied rarefied superorbital flows in UQ's X1 expansion tube (which was decommissioned in 2011). Experimentally measured shock speeds were used to calculate inflow conditions to the acceleration tube. Flow in the acceleration tube and dumptank was then computed using UQ's axisymmetric solver mb_cns (the original precursor to UQ's Eilmer3 code). An earlier study by Wendt et al. [72] on the same facility used a similar approach with mb_cns.

The next phase of studies used the L1d code to calculate the inflow to the shock tube, however these L1d calculations did not include piston dynamics. Instead, the volume between the piston face and the primary diaphragm was modelled as fixed [71]. The driver pressure at primary diaphragm rupture is known; the position of the piston downstream face, and the temperature of the driver gas, are then both adjusted until shock speeds, pressure histories, peak pressures, and pulse duration, satisfactorily match between experiment and the 1-D calculation [71]. This approach extends the reasoning of Wheatley et al. [59] to include a time history of flow properties in the shock tube. It has been used by Jacobs et al. for X3 [71], by Scott [54], McGilvray et al. [73], and Potter et al. [69] for X2, by McGilvray et al. [73] for the Hypervelocity Expansion Tube at the University of Illinois (using the UQ codes), and by Stewart et al. [74] for the RHYFL-X expansion tube concept (also using UQ codes).

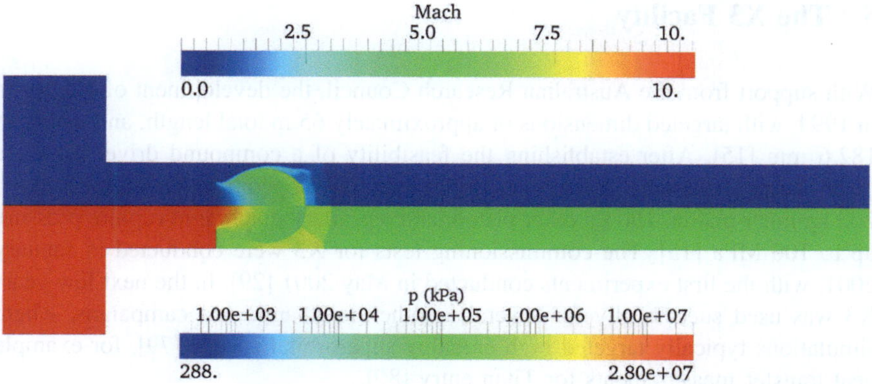

Fig. 18 Axisymmetric CFD simulation of primary diaphragm rupture in UQ's X2 expansion tube [56]

Gollan et al. [75] used a similar approach to simulate flow in X2 operated as a non-reflected shock tube. However, in this instance a more sophisticated L1d model was used, which modelled the full piston dynamics and primary diaphragm rupture. This approach is currently the routine methodology for hybrid simulation of expansion tubes, and has been adopted in recent studies such as Gildfind et al. [45, 76]. However, in order to adequately capture the full transient and spatial characteristics of expansion tube test flows, higher fidelity is required in the modelling of flow processes upstream of the acceleration tube. At the time of writing, piston dynamics is being coded into Eilmer3, and driver gas flow through the rupturing primary diaphragm has already been shown to introduce complex lateral wave disturbances to the driven gas (Fig. 18). Sophisticated high temperature gas models have also been developed for Eilmer3, which include finite rate chemistry [77] and radiation effects [78], which become particularly important for the higher enthalpy conditions (for example, Fig. 19).

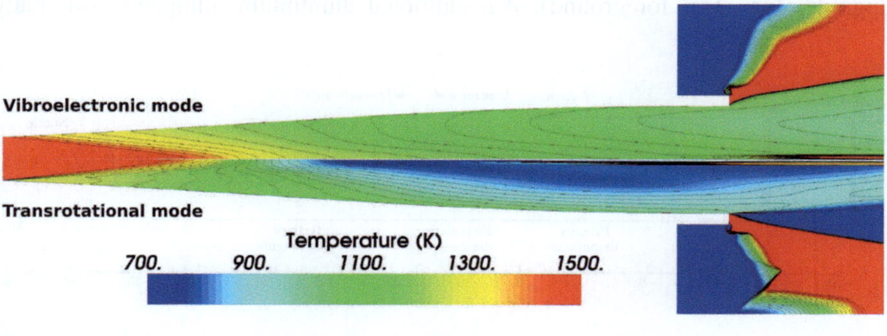

Fig. 19 Temperature contours for nonequilibrium X2 nozzle expansion calculation with Eilmer3 (adapted from [69])

5 The X3 Facility

With support from the Australian Research Council, the development of X3 began in 1994, with targeted dimensions of approximately 65 m total length, and a bore of 182.6 mm [15]. After establishing the feasibility of a compound driver in X2, a much larger two-stage free-piston driver was developed for X3, comprising a 200 kg inner piston, 100 kg outer piston, and a primary diaphragm rupture pressure up to 100 MPa [15]. The commissioning tests for X3 were conducted in January 2001, with the first experiments conducted in May 2001 [29]. In the next few years X3 was used successfully for several aerothermodynamic test campaigns, where simulations typically targeted high enthalpy superorbital speeds [79], for example heat transfer measurements for Titan entry [80].

Several difficulties with the original configuration of X3 did, however, lead to a series of major upgrades to the facility. Principle amongst these was the development of a new single-stage free-piston driver. The original dual-stage arrangement proved complicated to operate, and the requirement to decelerate the outer piston with a buffer arrangement set upper limits on the piston velocity, which in turn limited performance in the tube [79]. The use of a single piston also introduces an area change at the primary diaphragm, and thus provides the corresponding performance benefits this entails. The current experimental configuration of X3 is shown in Fig. 20; various photographs of the upgraded hardware are shown in Fig. 21; the new driver is shown in Fig. 21a.

X3 has been modified to incorporate a contoured Mach 10 nozzle, manufactured from fiberglass using fiber-winding around a mandrel [81]. UQ first applied this technique for the manufacture of contoured Mach 8 and Mach 10 nozzles for the T4 reflected shock tunnel [82]; this technique is also discussed in [51] and [83]. The new nozzle, which has an exit diameter of Ø440 mm, is shown in Fig. 21f; an unexpected consequence of fiberglass construction is light emission through the nozzle during facility operation, characterised as a bright flash during the passing of the hot accelerator gas.

X3's original dumptank was moved to X2 to enable scramjet testing [79], and a new and much larger stainless steel test section was procured for X3 as part of its upgrade (Fig. 21c; foreground). An additional aluminium dumptank, originally

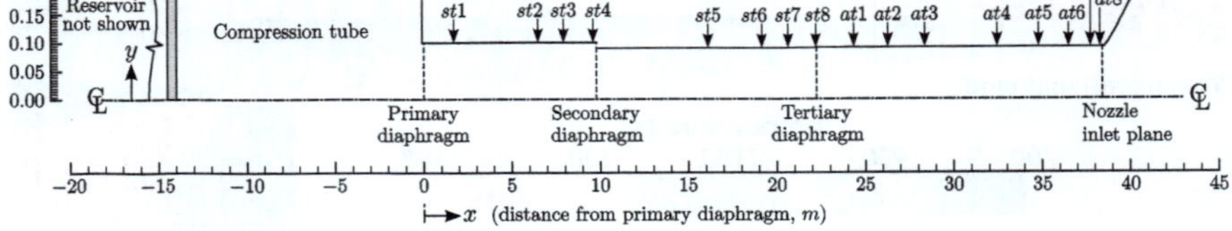

Fig. 20 Geometric layout of X3 expansion tube facility. Longitudinal scale compressed for clarity

Fig. 21 Photographs of X3's upgraded hardware. **a** Ø500 mm bore single stage driver compression tube (protective shielding removed to show capstan with belt drive). **b** Ø500 mm 100.8 kg lightweight piston. **c** Nozzle and test section with 50 mm thick perspex windows installed. **d** Pitot rake instrumented with 15° partial impact cone probes and a Pinckney static pressure probe [76, 85]. **e** Instrumented scramjet in test section (side covers removed). **f** Nozzle flash during facility operation due to flow luminosity [86]

used on the T3 reflected shock tunnel at the Australian National University in Canberra, was also donated to UQ for use with X3; this is attached to the rear of the test section, and approximately doubles the volume of the vacuum chamber (Fig. 21c; background). The new test section provides optical access to experiments, and is fully equipped for scramjet engine experiments. Wide field of view

50 mm thick perspex windows (Fig. 21c) are available for non-specialist optical applications including high speed imaging at up to megahertz frame rates; adaptors permit flexible placement of optical glass for more specialist optics studies such as Schlieren imaging, spectroscopy, holographic interferometry, and so forth.

The new single-stage driver was completed in 2011. Initial commissioning of the driver used a new 200 kg piston. Tuned operation with the piston was achieved with the use of an orifice plate at the driver in combination with helium/argon driver gas mixtures [76]. Flow condition development efforts have initially been directed towards high Mach number scramjet combustion studies. Scramjet testing has been conducted for Mach 10 flight, with 1–2 kPa static pressure, Ø200 mm core flow, and a test time of approximately 1 ms; preliminary characterisation of this test flow is detailed in [76].

While tuned operation of the 200 kg piston has successfully powered the Mach 10 scramjet test flows, higher enthalpy conditions will require higher driver gas sound speeds. As well, tuned operation of the driver at these conditions will require higher piston speeds, which in turn require a lower piston mass. A new lightweight piston has therefore been designed and manufactured (Fig. 21b). The piston is machined from 6061-T6 (the strongest aluminium alloy readily available in Ø500 mm billet), and has been sized for routine operation up to a maximum driver pressure of 40 MPa [84].

An initial commissioning study with the lightweight piston, involving blanked-off testing, was conducted in April 2013. A blanked-off test involves replacing the rupturing primary diagram with thick steel plate; when the piston is launched, instead of the diaphragm rupturing, the piston bounces back and forth. These tests were used to demonstrate the structural integrity of the piston, and to validate 1-D numerical models of X3's driver and piston dynamics [84]. It was found that X3's existing reservoir, at its maximum rated pressure of 14 MPa, has insufficient volume to propel the lightweight piston to the 250–300 m/s maximum speeds required for tuned operation at high driver gas sound speeds. An extension to X3's reservoir has subsequently been commissioned which will increase its volume by 50 %; the reservoir extension is scheduled for delivery and installation in 2015.

In 2014 the facility was used to conduct scramjet engine studies. Initial testing was focused on a 2-D scramjet engine at Mach 10, at a similar dynamic pressure to tests which were previously conducted on the same engine in the T4 RST [87], which approached the limit of T4's capability. Conducting these experiments at a flow condition common to both facilities provides an independent validation of the engine performance as a starting point for high Mach number testing, and in future will be followed by full free stream scramjet testing at Mach 12 and beyond. A new Mach 12 contoured nozzle, with exit Ø647 mm, is currently (2015) under construction and will be used for the Mach 12 scramjet experiments. These scramjet experiments will also require completion of the lightweight piston commissioning process with the upgraded reservoir. In addition to scramjet testing, the new driver and Mach 12 nozzle will also be used for radiation studies of planetary entry flows, using significantly larger models than has previously been possiblewith the UQ facilities.

In addition to the above, instrumentation and data acquisition systems have been improved, and a scramjet hydrogen fuelling system has been developed. Significant design work has also been directed towards improving the reliability and ease of operation of the facility, in order to prepare it for the higher usage rates which will be required into the future.

6 Conclusion

The University of Queensland first began experimenting with free-piston driven expansion tube facilities in the 1980s, and currently has two operational facilities, X2 and X3. Over 2500 shots have been conducted with X2, which has provided a reliable and high performance ground testing capability in support of a broad range of hypersonics studies. X2 has also been the platform for facility development studies, which have included the development of improved free-piston drivers, contoured nozzles, instrumentation, and specialised simulation codes. X3 is a much larger facility than X2, and provides greater capabilities in terms of test time and model size. The improvements achieved with X2 have now been incorporated into X3, which will offer an important ground testing platform for the Australian hypersonics community well into the future.

Acknowledgments The authors wish to thank: Mr F. De Beurs, Mr B. Loughrey, and the rest of the UQ Mechanical Engineering Workshop, for technical support; The Australian Research Council for support and funding; The Queensland Smart State Research Facilities Fund 2005 for support and funding; The Australian Space Research Program and UQ for their funding in support of the "Scramjet-based Access-to-Space Systems" (SCRAMSPACE) project. Many of the advances that have been made, both experimentally and computationally, have come through the PhD and Masters studies of the Group's many students.

References

1. Stalker, R.J.: Isentropic compression of shock tube driver gas. ARS J. **30**, 564 (1960)
2. Paull, A., Stalker, R.J.: Scramjet testing in the T3 and T4 hypersonic impulse facilities. In: Curran, E.T., Murthy, S.N.B. (eds.) Scramjet Propulsion, Progress in Astronautics and Aeronautics, pp. 1–46. AIAA, Reston, Virginia (2000)
3. Stalker, R.J., Morgan, R.G.: The University of Queensland free piston shock tunnel T4: initial operation and preliminary calibration. In: Proceedings of the 4th National Space Engineering Symposium, IEAust, Adelaide (1988)
4. http://hypersonics.mechmining.uq.edu.au/t4. Downloaded 19 Aug 2013
5. Resler, E.L., Bloxsom, D.E.: Very High Mach Number Flows by Unsteady Flow Principles. Cornell University Graduate School of Aeronautical Engineering, Limited Distribution Monograph, Jan 1952
6. Trimpi, R.L.: A preliminary theoretical study of the expansion tube, a new device for producing high-enthalpy short-duration hypersonic gas flows. NASA Technical Report R-133 (1962)

7. Trimpi, R.L.: A theoretical Investigation of Simulation in Expansion Tubes and Tunnels. NASA TR R-243, NASA Langley Research Centre, Langley Station, Hampton, VA (1966)
8. Paull, A., Stalker, R.J., Stringer, I.: Experiments on an expansion tube with a free piston driver. In: Proceedings of the AIAA 15th Aerodynamic Testing Conference, AIAA paper 88-2018, San Diego, CA, 18–20 May 1988
9. Miller, C.G., Jones, J.J.: Development and performance of the NASA Langley Research Center expansion tube/tunnel, a hypersonic-hypervelocity real-gas facility. In: Archer, R.D., Milton, B.E. (eds.) Proceedings of the 14th International Symposium on Shock Waves, pp. 363–373. Sydney, Australia, Aug 1983
10. Miller, C.G.: Operational Experience in the Langley Expansion Tube With Various Test Gases. NASA Technical Memorandum 78637, NASA Langley Research Center, Hampton, VA, Dec 1977
11. Paull, A., Stalker, R.J.: Test flow disturbances in an expansion tube. J. Fluid Mech. **245**, 493–521 (1992)
12. Stalker, R.J.: Recent developments with free piston drivers. In: Kim, Y.W. (ed.) Current topics in shock waves 17th International Symposium on Shock Waves and Shock Tubes, pp 96–105. Bethlehem, PA (1990)
13. Paull, A., Stalker, R.J.: The effect of an acoustic wave as it traverses an unsteady expansion. Phys. Fluids **3**(4), 717–719 (1991)
14. Bakos, R.J., Erdos, J.I.: Options for enhancement of the performance of shock-expansion tubes and tunnels. In: Proceedings of the 33rd Aerospace Sciences Meeting and Exhibit, AIAA 95-0799, Reno, NV, 9–12 Jan 1995
15. Morgan, R.G.: Free-piston driven expansion tubes. In: Ben-Dor, G., Igra, O., Elperin, T. (eds.) Handbook of Shock Waves, vol. 1, pp. 603–622. Academic Press, New York (2001)
16. Bakos, R.J., Morgan, R.G., Tamagno, J.: Effects of oxygen dissociation on hypervelocity combustion experiments. In: Proceedings of the AIAA 17th Aerospace Ground Testing Conference, AIAA-92-3964, Nashville, TN, 6–8 July 1992
17. Jones, J.J.: Some performance characteristics of the LRC 3-3/4-inch pilot expansion tube using an unheated hydrogen driver. In: Proceedings of the Fourth Hypervelocity Techniques Symposium, pp 7–26. Arnold Engineering Development Center, Arnold Air Force Station, Tennessee, 15–16 Nov 1965
18. Spurk, J.H.: Design, operation, and preliminary results of the BLR expansion tube. In: Proceedings of the Fourth Hypervelocity Techniques Symposium, pp. 111–144. Arnold Engineering Development Center, Arnold Air Force Station, Tennessee, 15–16 Nov 1965
19. Norfleet, G.D., Lacey, J.J., Whitfield, J.D.: Results of an experimental investigation of the performance characteristics of an expansion tube. In: Proceedings of the Fourth Hypervelocity Techniques Symposium, pp. 49–110. Arnold Engineering Development Center, Arnold Air Force Station, Tennessee, 15–16 Nov 1965
20. Miller, C.G.: Operational experience in the Langley expansion tube with various test gases. NASA Technical Memorandum 78637 (1977)
21. Jacobs, P.A.: Numerical simulation of transient hypervelocity flow in an expansion tube. ICASE Interim Report 20/NASA Contractor Report 189601, Langley Research Center, Hampton, Virginia, Jan 1992
22. Neely, A.J., Stalker, R.J., Paull, A.: High enthalpy, hypervelocity flows of air and argon in an expansion tube. Aeronaut. J. **95**(946), 175–186 (1991)
23. Jacobs, P.A.: Numerical simulation of transient hypervelocity flow in an expansion tube. Comput. Fluids **23**(1), 77–101 (1994)
24. Henshall, B.D.: The use of multiple diaphragms in shock tubes. A.R.C. Technical Report C. P. No. 291, Ministry of Supply, Aeronautical Research Council (1956)
25. Stalker, R.J., Plumb, D.L.: Diaphragm-type shock tube for high shock speeds. Nature **218** (5143), 789–790 (1968)
26. Morgan, R.G.: A review of the use of expansion tubes for creating superorbital flows. In: Proceedings of the 35th Aerospace Sciences Meeting and Exhibit, AIAA 97-0279, Reno, NV, 6–10 Jan 1997

27. Morgan, R.G., Stalker, R.J.: Double diaphragm driven free piston expansion tube. In: Proceedings of the 18th International Symposium on Shock Waves, Sendai, Japan, 21–26 July 1991
28. Neely, A.J., Morgan, R.G.: The superorbital expansion tube concept, experiment and analysis. Aeronaut. J. **98**, 97–105 (1994)
29. Scott, M.P., Morgan, R.G., Jacobs, P.A.: A new single stage driver for the X2 expansion tube. In: Proceedings of the 43rd AIAA Aerospace Sciences Meeting and Exhibition, AIAA-2005-697, Reno, Nevada, 10–13 Jan 2005
30. Doolan, C.J., Morgan, R.G.: A two-stage free-piston driver. Shock Waves **9**(4), 239–249 (1999)
31. Doolan, C., Morgan, R.G.: Test-time optimization of a large scale hypervelocity expansion tunnel. In: Proceedings of the Eighth National Space Engineering Symposium, The Institution of Engineers, Australia, Brisbane, 20–21 Sept 1993
32. Eichmann, T.N., McIntyre, T.J., Bishop, A.I., Vakata, S., Rubinsztein-Dunlop, H.: Three-dimensional effects on line-of-sight visualization measurements of supersonic and hypersonic flow over cylinders. Shock Waves **16**, 299–307 (2007)
33. McIntyre, T.J., Lourel, I., Eichmann, T.N., Morgan, R.G., Jacobs, P.A., Bishop, A.I.: Experimental expansion tube study of the flow over a toroidal ballute. J. Spacecraft Rockets **41** (5), 716–725 (2004)
34. Littleton, B.N., Bishop, A.I., McIntyre, T.J., Barker, P.F., Rubinsztein-Dunlop, H.: Flow tagging velocimetry in a superorbital expansion tube. Shock Waves **10**, 225–228 (2000)
35. Hayne, M.J., Mee, D.J., Gai, S.L., McIntyre, T.J.: Boundary layers on a flat plate at sub- and superorbital speeds. J. Thermophys. Heat Transfer **21**(4), 772–779 (2007)
36. Higgins, C., Inger, G.R., Morgan, R., Eichmann, T., McIntyre, T.: Shock standoff on hypersonic blunt bodies in multi-temperature ionizing nonequilibrium gas flows. In: Proceedings of the 8th AIAA/ASME Joint Thermophysics and Heat Transfer Conference, St Louis, Missouri, 24–26 June 2002
37. Eichmann, T.N.: Radiation measurements in a simulated Mars atmosphere. PhD Thesis, School of Mathematics and Physics, The University of Queensland, Brisbane, Australia (2012)
38. Jacobs, C.M.: Radiation in low density hypervelocity flows. PhD Thesis, School of Mechanical and Mining Engineering, The University of Queensland, Brisbane, Australia (2011)
39. Brandis, A.M., Morgan, R.G., McIntyre, T.J.: Analysis of nonequilibrium CN radiation encountered during Titan atmospheric entry. AIAA J. Thermophys. Heat Transfer **25**(4), 493–499 (2007)
40. McGilvray, M., Morgan, R., Jacobs, P.: Scramjet experiments in an expansion tunnel: evaluated using a quasi-steady analysis technique. AIAA J. **48**(8), 1635–1646 (2010)
41. Gildfind, D.E., Morgan, R.G., McGilvray, M., Jacobs, P.A., Stalker, R.J., Eichmann., T.N.: Free-piston driver optimisation for simulation of high Mach number scramjet flow conditions. Shock Waves **21**(6), 559–572 (2011)
42. Jacobs, P.A.: Quasi-one-dimensional modelling of a free-piston shock tunnel. AIAA J. **32**(1), 137–145 (1994)
43. Stalker, R.J.: A study of the free-piston shock tunnel. AIAA J. **5**(12), 2160–2165 (1967)
44. Itoh, K., Ueda, S., Komuro, T., Sato, K., Takahashi, M., Myajima, H., Tanno, H., Muramoto, H.: Improvement of a free piston driver for a high-enthalpy shock tunnel. Shock Waves **8**(4), 215–233 (1998)
45. Gildfind, D.E., Morgan, R.G., Jacobs, P.A., McGilvray, M.: Production of High-Mach-Number scramjet flow conditions in an expansion tube. AIAA J. **52**(1), 162–177 (2014)
46. Porat, H., Zander, F., Morgan, R.G., McIntyre, T.J.: Emission spectroscopy of a mach disk at Titan atmospheric entry conditions. In: Proceedings of the 29th International Symposium on Shock Waves, University of Wisconsin, Madison, 14–19 July 2013

47. Sheikh, U.A., Jacobs, C., Laux, C.O., Morgan, R.G., McIntyre, T.J.: Measurements of radiating flow fields in the vacuum ultraviolet. In: Proceedings of the 29th International Symposium on Shock Waves, University of Wisconsin, Madison, 14–19 July 2013
48. Zander, F., Morgan, R.G., Sheikh, U., Buttsworth, D.R., Teakle, P.R.: Hot-wall reentry testing in hypersonic impulse facilities. AIAA J. **51**(2), 476–484 (2013)
49. James, C.M., Gildfind, D.E., Morgan, R.G., McIntyre, T.J.: Working towards simulating gas giant entry radiation in an expansion tube. In: Proceedings of the 29th International Symposium on Shock Waves, University of Wisconsin, Madison, 14–19 July 2013
50. Gildfind, D.E., James, C.M., Morgan, R.G.: Free-piston driver performance characterisation using experimental shock speeds through helium. Shock Waves, **25**(1), 169–176 (2015)
51. Bakos, R.J., Calleja, J.F., Erdos, J.I., Auslender, A.H., Sussman, M.A., Wilson, G.J.: Design, calibration and analysis of a tunnel mode of operation for the hypulse facility. In: Proceedings of the 19th AIAA Advanced Measurement and Ground Testing Technology Conference, AIAA 96-2194, New Orleans, LA, 17–20 June 1996
52. Stewart, B.S., Jacobs, P.A., Morgan, R.G.: The starting process of an expansion tube nozzle. In: 23rd International Symposium on Shock Waves, The University of Texas at Arlington, USA, 23–27 July 2001
53. Chue, R.S.M., Bakos, R.J., Tsai, C.-Y., Betti, A.: Design of a shock-free expansion tunnel nozzle in hypulse. Shock Waves **13**(4), 261–270 (2003)
54. Scott, M.P.: Development and modelling of expansion tubes. PhD Thesis, Centre for Hypersonics, Department of Mechanical Engineering, University of Queensland, Brisbane, Australia (2006)
55. Craddock, C.S.: Computational optimization of scramjets and shock tunnel nozzles. PhD Thesis, Department of Mechanical Engineering, The University of Queensland, Brisbane, Australia (1999)
56. Gildfind, D.: Development of high total pressure scramjet flow conditions using the X2 expansion tube. PhD Thesis, Division of Mechanical Engineering, School of Engineering, The University of Queensland, Brisbane, Australia (2012)
57. McGilvray, M., Jacobs, P.A., Morgan, R.G., Gollan, R.J., Jacobs, C.M.: Helmholtz resonance of pitot pressure measurements in impulsive hypersonic test facilities. AIAA J. **47**(10), 2430–2439 (2009)
58. Chiu, H.S., Mee, D.J.: Modified bar gauges. Research Report Number 2003/22, Division of Mechanical Engineering, The University of Queensland (2003)
59. Wheatley, V., Chiu, H.S., Jacobs, P.A., Macrossan, M.N., Mee, D.J., Morgan, R.G.: Rarefied, superorbital flows in an expansion tube. Int. J. Numer. Meth. Heat Fluid Flow **14**(4), 512–537 (2004)
60. Hillier, R., Boyce, R.R., Creighton, S.A., Fiala, A., Jackson, A.P., Mallinson, S.G., Sheikh, A. H., Soltani, S., Williams, S.: Development of some hypersonic benchmark flows using CFD and experiment. Shock Waves **12**(5), 375–384 (2003)
61. Roy, C.J., Oberkampf, W.L., McWherter-Payne, M.A.: Verification and validation for laminar hypersonic flow fields, Part 2: validation. AIAA J. **41**(10), 1944–1954 (2003)
62. Anderson, G., Kumar, A., Erdos, J.: Progress in hypersonic combustion technology with computation and experiment. In: AIAA Second International Aerospace Planes Conference, AIAA-90-5254, Orlando, FL, 29–31 Oct 1990
63. Anderson, G.Y., McClinton, C.R., Weidner, J.P.: Scramjet performance. In: Curran, E.T., Murthy, S.N.B. (eds.) Scramjet Propulsion, Progress in Astronautics and Aeronautics, pp. 369–446. AIAA, Reston, Virginia (2000)
64. Candler, G.V., Nompelis, I.: CFD validation for hypersonic flight: real gas flows. In: Proceedings of the 40th AIAA Aerospace Sciences Meeting and Exhibit, RTO-TR-AVT-007-V3/AIAA 2002-0434, Reno, NV, 14–17 Jan 2002
65. Marvin, J.G.: A CFD Validation Roadmap for Hypersonic Flows. NASA Technical Memorandum 103935, Ames Research Center, Moffett Field, California (1992)

66. Jacobs, P.A., Gollan, R.J.: The Eilmer3 Code: User Guide and Example Book. Mechanical Engineering Report 2008/07, Department of Mechanical and Mining Engineering, The University of Queensland (2009)
67. Jacobs, P.A., Gollan, R.J., Denman, A.J., O'Flaherty, B.T., Potter, D.F., Petrie-Repar, P.J., Johnston, I.A.: Eilmer's Theory Book: Basic Models for Gas Dynamics and Thermochemistry. Mechanical Engineering Report 2010/09, Department of Mechanical and Mining Engineering, The University of Queensland (2010)
68. Gollan, R.J., Jacobs, P.A.: About the formulation, verification and validation of the hypersonic flow solver Eilmer. Int. J. Numer. Meth. Fluids **73**, 19–57 (2013)
69. Potter, D.F., Gollan, R.J., Eichmann, T., McIntyre, T.J., Morgan, R.G., Jacobs, P.A.: Simulation of CO2-N2 expansion tunnel flows for the study of radiating shock layers. In: Proceedings of the 46th AIAA Aerospace Sciences Meeting and Exhibit, AIAA 2008-1280, Reno, Nevada, 7–10 Jan 2008
70. Morgan, R.G., McIntyre, T.J., Jacobs, P.A., Buttsworth, D.R., Macrossan, M.N., Gollan, R.J., Capra, B.R., Brandis, A.M., Potter, P., Eichmann, T., Jacobs, C.M., McGilvray, M., van Diem, D., Scott, M.P.: Impulse facility simulation of hypervelocity radiating flows. In: Proceedings of the 2nd International Workshop on Radiation of High Temperature Gases in Atmospheric Entry, 6–8 Sept, Rome, Italy. ESA Publication, Noordwijk, The Netherlands, Nov 2006
71. Jacobs, P.A., Silvester, T.B., Morgan, R.G., Scott, M.P., Gollan, R.G., McIntyre, T.J.: Superorbital expansion tube operation: estimates of flow conditions via numerical simulation. In: Proceedings of the 43rd AIAA Aerospace Sciences Meeting, AIAA-2005-0694, Reno, NV, 10–13 Jan 2005
72. Wendt, M., Macrossan, M., Jacobs, P., Mee, D.: Pilot study for a rarefied hypervelocity test facility. In: Proceedings of the 13th Australasian Fluid Mechanics Conference, Monash University, Melbourne, Australia, 13–18 Dec 1998
73. McGilvray, M., Austin, J.M., Sharma, M., Jacobs, P.A., Morgan, R.G.: Diagnostic modelling of an expansion tube operating condition. Shock Waves **19**(1), 59–66 (2009)
74. Stewart, B., Morgan, R.G., Jacobs, P.A., Hayne, M.: Flow establishment in large-scale high-performance expansion tubes. In: Proceedings of the AIAA/AAAF 11th International Space Planes and Hypersonic Systems and Technologies Conference, Paper No. 5239, Orleans, France, 29 Sept–4 Oct 2002
75. Gollan, R.J., Jacobs, C.M., Jacobs, P.A., Morgan, R.G., McIntyre, T.J., Macrossan, M.N., Buttsworth, D.R., Eichmann, T.N., Potter, D.F.: A simulation technique for radiating shock tube flows. In: Klauss, H., Seiler, F. (eds.) Proceedings of the 26th International Symposium on Shock Waves, pp. 465–470. Göttingen, Germany, 15–20 July 2007
76. Gildfind, D.E., Sancho, J., Morgan, R.G.: High Mach number scramjet test flows in the X3 expansion tube. In: Proceedings of the 29th International Symposium on Shock Waves, University of Wisconsin-Madison, Memorial Union, 14–19 July 2013
77. Gollan, R.J.: The computational modelling of high-temperature gas effects with application to hypersonic flows. PhD Thesis, Division of Mechanical Engineering, School of Engineering, The University of Queensland, Brisbane, Australia (2008)
78. Potter, D.: Modelling of radiating shock layers for atmospheric entry at Earth and Mars. PhD Thesis, Division of Mechanical Engineering, School of Engineering, The University of Queensland, Brisbane, Australia (2011)
79. McGilvray, M.: The use of expansion tube facilities for scramjet testing. PhD Thesis, Centre for Hypersonics, Department of Mechanical Engineering, The University of Queensland, Brisbane, Australia (2008)
80. Capra, B.R., Morgan, R.G.: Radiative and total heat transfer measurements to a Titan explorer model. J. Spacecraft Rockets **49**(1) (2012)
81. Davey, M.G.: A hypersonic nozzle for the X3 expansion tube. Bachelor of Engineering Thesis, Department of Mechanical Engineering, The University of Queensland, Brisbane, Australia (2006)
82. Jacobs, P.A., Stalker, R.J.: Design of axisymmetric nozzles for reflected shock tunnels. Report 1/89, Department of Mechanical Engineering, University of Queensland (1989)

83. Hass, N.E., Shih, A.T., Rogers, R.C.: Mach 12 and 15 scramjet test capabilities of the hypulse shock-expansion tunnel. In: Proceedings of the 43rd AIAA Aerospace Sciences Meeting and Exhibit, AIAA 2005-690, Reno, Nevada, 10–13 Jan 2005
84. Gildfind, D.E., Morgan, R.G., Sancho, J.: Design and commissioning of a new lightweight piston for the X3 expansion tube. In: Proceedings of the 29th International Symposium on Shock Waves, University of Wisconsin-Madison, Memorial Union, 14–19 July 2013
85. Pinckney, S.Z.: A short static-pressure probe design for supersonic flow. NASA TN D-7978 (1975)
86. Zander, F.: Personal communication (2012)
87. McGilvray, M., Kirchhartz, R., Jazra, T.: Comparison of Mach 10 scramjet measurements from different impulse facilities. AIAA J. **48**(8), 1647–1651 (2010)

Author Biographies

David E. Gildfind Centre for Hypersonics, School of Mechanical and Mining Engineering, The University of Queensland, Brisbane, Australia; David Gildfind's research is concerned with the development of hypersonic impulse facilities. His main efforts have been directed towards optimising free-piston driver operation, expansion tube flow condition development, and test flow characterisation. David graduated as an aerospace engineer from RMIT University in 2001, and worked for a number of years on various aircraft platforms in Australia and overseas. He later undertook his PhD and several years of post-doctoral work in hypersonics at the University of Queensland (UQ), and is now a lecturer within UQ's School of Mechanical and Mining Engineering.

Richard G. Morgan Centre for Hypersonics, School of Mechanical and Mining Engineering, The University of Queensland, Brisbane, Australia; Richard Morgan has been the Director of the Centre for Hypersonics at The University of Queensland since its inception in 1997, and lectures in the School of Mechanical and Mining Engineering. Richard developed the superorbital expansion tube concept, which enables flows at speeds in excess of 50,000 km/h to be generated in the laboratory. He is an expert in modelling and testing the influences of radiation from the hot flows that occur in the bow waves of vehicles entering the atmospheres of the earth, other planets and moons. He has collaborative research programs with NASA, ESA, Oxford University, Ecole Centrale (Paris) and AOARD in radiating flows, as well as continuing Australian Research Council (ARC) support in this area since 1990. He was involved as a flight team member in the 2010 airborne observation of the Japanese 'Hayabusa' asteroid sample return mission, for which he was a co-recipient of the NASA Ames 'honour' award for 2010. He received his undergraduate and postgraduate degrees from The University of Oxford.

Peter A. Jacobs Centre for Hypersonics, School of Mechanical and Mining Engineering, The University of Queensland, Brisbane, Australia; Peter Jacobs has worked on shock tube and expansion tube simulations for more than 25 years. His work on expansion tubes started at NASA Langley Research Center, following discussions with Charles Miller, Judy Shinn and Bob Trimpi. Since then, much of what he has learned about the gas dynamic processes in these machines has been encoded into a set of flow simulation programs that are maintained by the Compressible Flow CFD group at The University of Queensland. (See http://cfcfd.mechmining.uq.edu.au) He is a graduate of The University of Queensland and spends most of his days teaching or thinking about computation and gas dynamics.

Hypersonic Ludwieg Tube

Rolf Radespiel, Malte Estorf, Dirk Heitmann, Federico Muñoz and Torsten Wolf

1 Introduction

The aerothermal design of vehicles operating at hypersonic speeds calls for a good knowledge of aerodynamic and thermal loads on the vehicle components. While significant progress has been achieved in the development of accurate and efficient numerical flow simulation methods there still exist a number of hypersonic flow phenomena where flow experiments are needed to achieve acceptable uncertainty levels in vehicle design. Among these are laminar-turbulent transition of boundary layers and shock-boundary layer interactions. Hypersonic flow research over several decades has resulted in significant theories and a large amount of measured data on these subjects. Nevertheless, the flow physics is extremely complex, and numerical flow simulations cannot resolve the complete range of temporal and spatial flow scales involved. Hence there is the need for fundamental flow experiments in these fields.

Hypersonic flight usually takes place at high flow velocities and high flow enthalpies. This makes representative flow experiments in wind tunnels very difficult as the high enthalpies introduce extreme heat loads on the facility and on the investigated wind tunnel models, not to speak of the extreme kinetic energy rate needed to establish a high-enthalpy hypersonic air stream. Hence, realistic flight enthalpy levels can only be simulated in a few, expensive facilities worldwide and only during a few milliseconds of intermittent runtime. The experimentation opportunities offered by these facilities are therefore too limited in fulfilling the

Contribution to “Experimental Methods of Shock Wave Research”.

R. Radespiel (✉) · M. Estorf · D. Heitmann · F. Muñoz · T. Wolf
Technische Universität Braunschweig, 38106 Braunschweig, Germany
e-mail: r.radespiel@tu-bs.de

O. Igra and F. Seiler (eds.), *Experimental Methods of Shock Wave Research*,
Shock Wave Science and Technology Reference Library 9,
DOI 10.1007/978-3-319-23745-9_14

needs of research institutions interested in the fundamentals of hypersonic flow physics. This is why many fundamental problems in hypersonics are investigated in flow facilities where the Mach and Reynolds numbers are properly simulated but at lower than in a real flight flow enthalpy.

The typical research institutes involved in these fundamental research works can only afford a limited investment into a facility and relatively low operation cost. Typical values for these cost constraints are one Million US dollars for the facility and a few US dollars for the consumables needed per run. In addition, a suitable facility should be operated by the flow researcher alone, without assistance of a specially trained technician.

Top-level requirements for a hypersonic research facility call for a Mach number high enough that typical hypersonic flow behaviors are achieved for blunt and slender configurations. Furthermore, in order to enable testing of laminar and turbulent flows, the facility should provide a Reynolds-number of 15×10^6, based on the model length. The test time should be in the order of 100 ms in order to allow for a range of useful flow measurement techniques.

Continuous progress in wind tunnel engineering and in instrumentation available for flow measurements in hypersonics over the last decades has resulted in a specific wind tunnel configuration that meets these requirements. This is the hypersonic Ludwieg tube. Several of these gas dynamics facilities are in operation worldwide. The present contribution describes one of the successful designs of this kind, along with data that characterize the quality and capabilities of this facility.

2 Ludwieg Tube Design

The working principle of the Ludwieg tube goes back to the 1950s [1]. Hubert Ludwieg analyzed the cost effectiveness of high-speed wind tunnels, then in operation. He pointed out that the relative cost advantage of the blow-down wind tunnel relative to the in draft tunnel increases with tunnel Mach number. If the needed run time for testing is only in the order of one second or less, one can achieve significant cost reductions by choosing a long storage tube to function as a high-pressure reservoir in the following way: If a valve at the tube end opens the flow path into a Laval nozzle, an almost constant flow rate results, as long as the resulting expansion wave has not returned from its way through the storage tube. This flow phenomenon is used to avoid costly control mechanisms for the flow total pressure and the total temperature, during the tunnel run. The principle of using a blow-down storage tube can be combined with a vacuum tank at the low pressure tunnel end, in order to increase the possible range of test section pressures. The general layout is displayed in Fig. 1. This invention was the basis for the well-known Göttingen Ludwieg tube facility, RWG, that employs three different storage tubes and six different nozzles in order to cover a large Mach number range, from 4 to 12 [2]. This facility has a run time of 0.3 s.

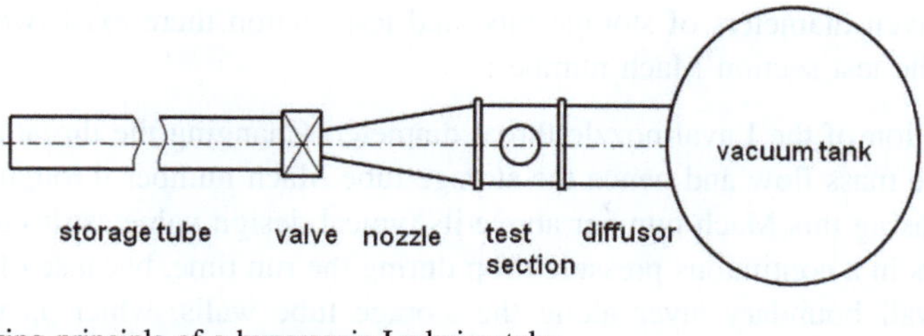

Fig. 1 Working principle of a hypersonic Ludwieg tube

Fig. 2 Schematic drawing of the fast acting valve in closed end position

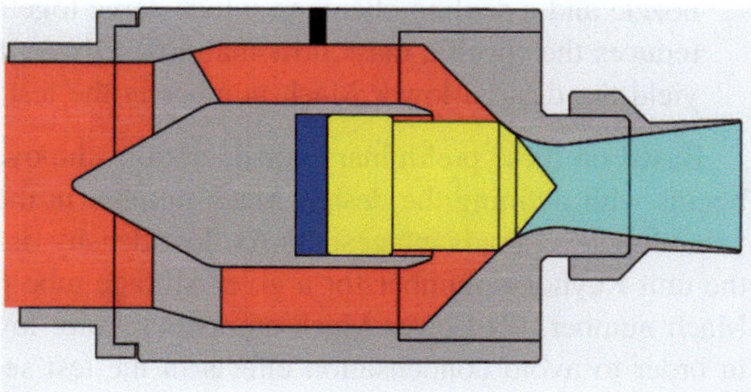

Over the years progress in flow measurement instrumentation further reduced the run-time requirement. Furthermore, it was recognized that the valve design ahead of the nozzle was critical for further reducing the facility cost. Georg Koppenwallner introduced a new fast acting valve, which was now centered within the flow path towards the nozzle [3]. Here a central piston is pneumatically operated by controlling its back pressure and this feature results in opening times of about 10 ms (Fig. 2).

The gas expansion in the hypersonic nozzle significantly reduces its static temperature. That causes the gas to fall below its saturation curve, if not heated previously. One option is to heat only the part of the storage tube that contains the air which is pushed out during one run. Hence, the storage tube length for a given tunnel run time can be kept as short as possible. Note, that the partially heated tube implies a step in the tube temperature which has to be compensated by a step in the tube diameter, in order to avoid disturbances from the step to be reflected upstream when the expansion wave passes. A reasonable compensation is achieved if the Mach number of the flow behind the expansion wave does not change across the location where the tube temperature changes. Hence, the product of tube area A and speed of sound, i.e. the temperature T, must be constant [3]:

$$A_1\sqrt{T_1} = A_2\sqrt{T_2}$$

For given diameters of storage tube and test section there exist two options to change the test section Mach number:

- Variation of the Laval nozzle throat diameter: Changing the throat changes the nozzle mass flow and hence the storage tube Mach number through the tunnel. Increasing this Mach number above its typical design value as discussed below results in a continuous pressure drop during the run time, because of increase in the wall boundary layer along the storage tube walls, which is undesirable. Hence, the test section Mach number can be merely increased by inserting smaller throats or even complete nozzles.
- Use of a tandem nozzle: Here the original Laval nozzle is replaced by a steep nozzle and a settling chamber, which serve together as a throttling device. This reduces the specific mass flow that now fills the throat of the second nozzle, to yield the desired lower Mach number in the test section [3].

Based on these preliminary considerations the overall layout of a Ludwieg tube begins with defining the design Mach number in the test section. Choosing a low value has several advantages: (1) As flow density drops rapidly with Mach number the unit Reynolds number for a given storage tube pressure also drops with rising Mach number. (2) Lower Mach numbers require lower storage tube temperatures, in order to avoid condensation effects in the test section. This creates less severe problems with natural convection in the storage tube, and entropy fluctuations in the test section flow will be smaller. (3) Lower storage tube temperatures allow using oil particles for seeding the wind tunnel flow. This is important for application of Particle Image Velocimetry. As a consequence the test section Mach number is chosen as low as possible to achieve typical hypersonic flow behaviors in inviscid gas dynamics and boundary layer flows. The general belief is that these behaviors are reasonably well obtained with a test section Mach number of 6.

The test section size is determined by technical and operational reasoning. While a given Reynolds number depends linearly on the storage tube pressure and the model size, practical model manufacture and instrumentation considerations call for choosing the test section as large as possible. For example, measurements of boundary layers in transition research make it very difficult to resolve the higher frequencies and smaller spatial scales of flow instabilities observed on smaller models, at the same Reynolds number. Also, it can be shown that the relative disturbance level due to nozzle boundary layer noise is larger for smaller nozzles [4]. Hence, smaller test sections generally exhibit a larger unit Reynolds number effect on boundary layer transition. Moreover, significant benefits result from the possibility to share wind tunnel models between existing wind tunnels. As many existing hypersonic wind tunnels of leading research establishments have a test section diameter of around 500 mm, this size is advantageous in this respect. It can be concluded that a test section diameter of 500 mm represents an excellent choice for a facility dedicated to fundamental flow research.

An important choice concerns the placement of the tunnel valve. Locating the valve just at the end of the storage tube, as in Fig. 1, will usually create significant disturbances of the flow that enters the nozzle. The result will be a turbulent nozzle boundary layer and hence noise levels observed in the test section will be at conventional, noisy levels. An alternative is to place the valve behind the test section. Hence, the entrance to the wind tunnel nozzle can be shaped in a completely smooth way, thereby allowing for laminar boundary layers [5]. A low-cost mechanical solution is to place a diaphragm within the tunnel diffuser. However, this location has two disadvantages: (1) Wind tunnel nozzle and test section are now part of the high-pressure wind tunnel components and need a corresponding construction and safety rating. Similarly, the wind tunnel models experience much higher mechanical loads during the tunnel start-up process. (2) The start-up process of the steady tunnel flow is much more affected by model blockage. This limits the possible model size and the Reynolds number. As a thumb rule, the allowed model size of a blunt geometry is twice as large for a tunnel valve located upstream of the nozzle, compared with the downstream position.

Next, the storage tube diameter needs to be determined. As the tunnel throat diameter is given by the test section Mach number and its diameter, the storage tube diameter directly determines the storage tube Mach number. A low storage tube Mach number generally leads to better flow uniformity in the test section. Moreover, due to the growing boundary layer in the storage tube, static pressure slightly drops during tunnel run and this effect becomes larger for increasing storage tube Mach number. It turns out that this effect is rather small for storage tube Mach numbers of around 0.05 while an undesired pressure drop would result at significant larger values [3, 6]. Finally, the desired test time determines the storage tube length. For low values of the storage tube Mach number the test time is approximately determined by

$$t = \frac{2L_1}{a_1} + \frac{2L_2}{a_2}.$$

Here the two terms take into account, that the storage tube may be composed of two parts with different lengths, L, and with two different temperatures and correspondingly different sound speeds, a. A suited test time is assumed to be 100 ms.

3 Tunnel Components

These considerations of hypersonic wind tunnel layout lead to the actual design of the Hypersonic Ludwieg Tube Braunschweig (HLB) [7]. The overall design drawing is displayed in Fig. 3.

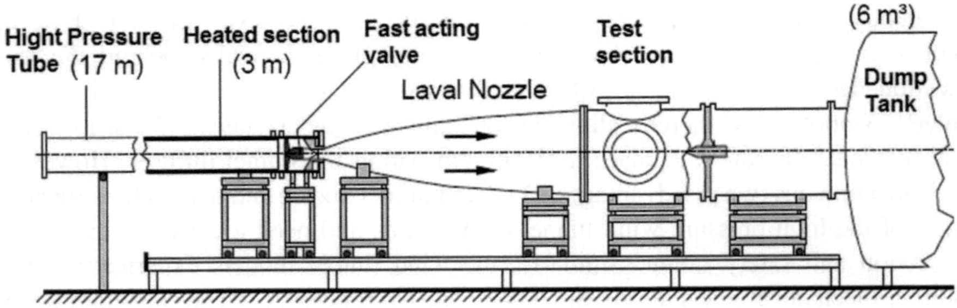

Fig. 3 Overall design drawing of the hypersonic Ludwieg tube Braunschweig

3.1 Storage Tube and Valve

The high pressure section consisting of a 17 m long storage tube with a 3 m long heated section and a valve housing that can be pressurized up to 30 bar. The chosen Mach number of the storage tube placed upstream of the nozzle throat, $M_1 = 0.059$ determines the throat area ratio, $A^*/A_1 = 0.101$. The static temperature T_1 of the flow behind the expansion wave depends on the stagnation temperature T_0 upstream of the wave and the adiabatic coefficient κ as,

$$\frac{T_0}{T_1} = \left(1 + \frac{\kappa - 1}{2} M_1\right)^2$$

Our experience gained by systematic variations of the storage tube temperature shows that heating the tube up to a temperature around 450 K is sufficient to avoid condensation in the test section. As mentioned above, natural convection causes significant stratification of the heated air in the storage tube, if the cold part is not isolated from the heated part. To avoid this unwanted convection an isolated shutter is installed at the end of the heated section [8]. It is opened only shortly before a run. The cross-sectional area blocked by the opened shutter is smaller than the area difference between heated and non-heated section. Therefore the shutter should not impose any additional disturbances on the passing expansion wave. With this shutter the temperature difference due to stratification is reduced to about 20 K. It should be noted that future constructions of similar Ludwieg tubes should employ a longer, completely heated and fully isolated storage tube, to avoid stratification in the storage tube as much as possible.

The fast opening valve used to control the wind tunnel run is sketched in Fig. 2. The valve housing also holds the flow probes used to define the flow state in the tunnel. These are a static pressure gauge and thermocouples located at three circumferential positions in the annular clearance around the valve. The thermocouples have a rise time of about 4 ms. The data from these probes is also used for monitoring homogeneity of the tunnel static temperature during the tunnel run.

3.2 Nozzle

The low pressure section of the HLB consists of the nozzle, the test section, the diffusor and the 6 m^3 dump tank. The geometrical definition of the hypersonic nozzle is seen in Fig. 4.

It is comprised of a convergent, conical entry, a subsonic circular contour towards the throat, a transition section to the divergent conical section with an opening angle θ_B, and a polynomial that determines the flow straightening to the final divergence angle θ_A. These geometry parameters were determined by using results of numerical flow simulations as a guideline [9], see Fig. 5.

These simulations assumed fully turbulent boundary layer flow. The analysis of a reference configuration that employed a short transition section with a constant radius of curvature and a simplified polynomial for the flow straightener revealed two sources of compression waves in the nozzle: (1) the overexpansion of the flow in the transition section behind the throat and (2) the curvature distribution in the flow straightener section. These compression waves are observed in the nozzle flow field as depicted in Fig. 5. Note that in all further design calculations the final

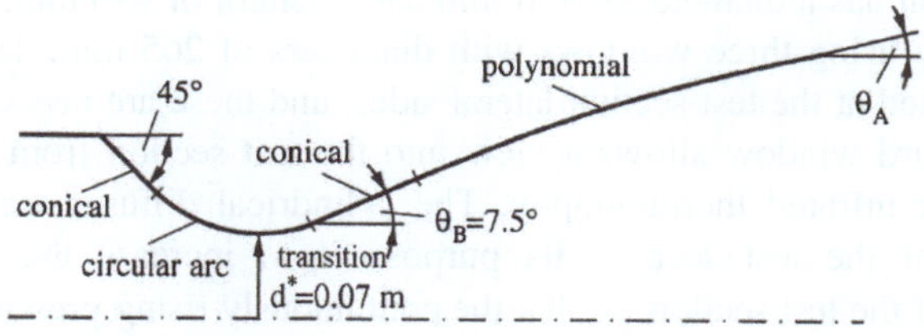

Fig. 4 Nozzle geometry

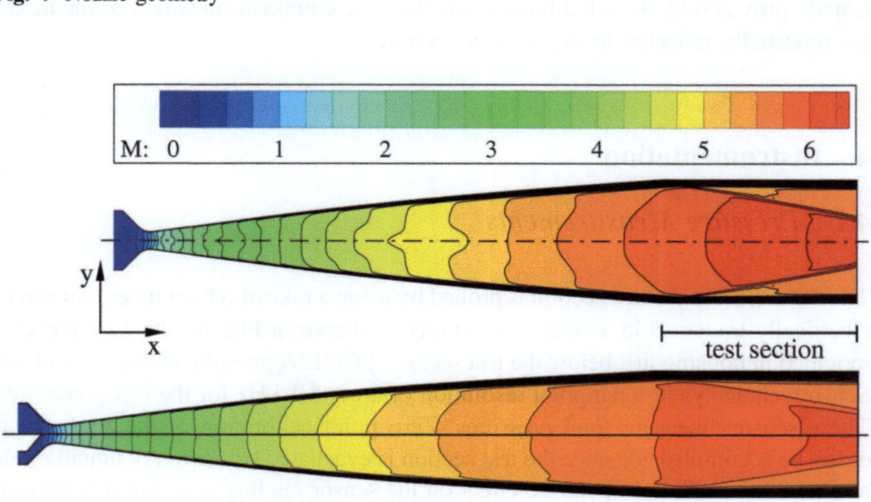

Fig. 5 Mach number contours of the reference nozzle (*top*) and the optimized nozzle (*bottom*), where the effect of simulating the valve geometry is also displayed [9]

divergence angle, $\theta_{A,}$ was fixed at a value of 3°. This value was chosen based on previous observations, that hypersonic nozzles are sensitive to small changes in flow conditions, and that a fully expanded nozzle would represent an unnecessary risk. The compression wave caused by overexpansion of the supersonic flow into the divergent nozzle section was successfully diminished by using a suitable super-ellipse segment instead of the circular arc. This allowed control of the curvature distribution. The compression wave created by a discontinuous curvature in the initial polynomial of the flow straightener can be avoided as well, by using an improved polynomial with a continuous curvature distribution along the nozzle axis. This yielded a rather smooth flow in the final nozzle geometry. Also seen in Fig. 5 is the influence of the centrally mounted valve on the supersonic nozzle flow. Due to losses in the valve boundary layer, the flow near the axis exhibits a wake-type velocity distribution. This local effect is very limited, i.e. the related deviations in test section Mach number are in the order of 0.02.

3.3 Test Section and Diffusor

The test section has a diameter of 496 mm and a length of 940 mm. Optical access is provided by using three windows with diameters of 265 mm. Two of the windows are located at the test section lateral sides, and these are needed for schlieren optics. The third window allows a view into the test section from above. This is often used for infrared thermography. The cylindrical diffuser section is located downstream of the test section. Its purpose is to increase the static pressure downstream of the test section, so that the continuously rising pressure in the dump tank does not lead to early flow break down. Pressure recovery in the diffusor is usually provided by the wind tunnel model, as the compression waves of the model are repeatedly reflected in the diffuser section.

4 Instrumentation

4.1 Pressure Measurements

The mean flow in the test section is probed by using a rake of 6 Pitot tubes that can be electrically traversed in x- and z-directions as shown in Fig. 6. The xz-traverse is mounted in housing just below the test section. SCC15A pressure transducers of Co. Sensortechnics yield a temporal resolution of around 1 kHz for the Pitot pressures. These sensors measure total pressures. Zero point calibration of the sensors can easily be accomplished, since the test section is evacuated before every tunnel run to around 2 mbar. The temperature effect on the sensor reading was also determined.

High-resolution pressure fluctuation measurements are performed with M131A31 sensors (Co. PCB Piezotronics, Inc.). In the past years, these sensors are

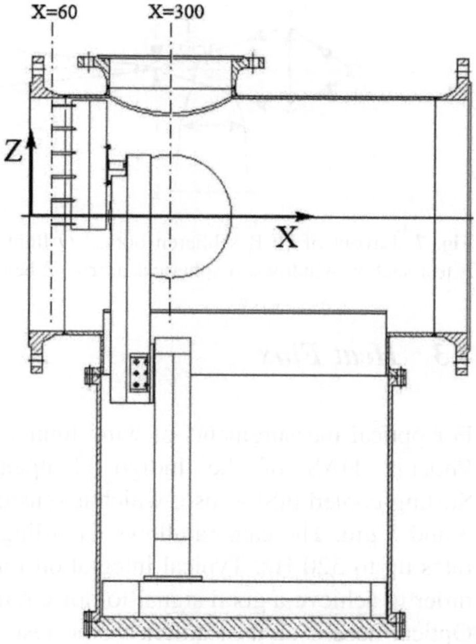

Fig. 6 Drawing of test section with Pitot rake and xz-traverse, from [7]

widely used for measurements of hypersonic boundary layer instabilities (see, e.g., among many others, Estorf et al. [10] or Heitmann et al. [11]). According to the manufacturer's specification the resonance frequency of the sensor is larger than 1 MHz and the output signal is high-pass filtered at 10 kHz. The nominal diameter of the active sensor face is 3.18 mm, however, it is known that the active sensor area is much smaller, i.e. $0.76 \times 0.76\ \mathrm{mm}^2$ [12]. Hence the sensor can resolve very small flow structures. Power is supplied by two PCB instruments (M482A22 and M483A), which, at the same time, perform signal conditioning. The sensors have sensitivities in the range of 0.02 mV/Pa and a nominal resolution of 7 Pa. The pressure data is sampled with a Spectrum M2i.4652 transient recorder, having a sampling frequency of 3 MHz. Time traces of 265 ms length are usually recorded, consisting of about 180 ms before/after and 80 ms of data during the tunnel run.

4.2 Schlieren

The schlieren set up used in the HLB is sketched in Fig. 7. A virtual image of the light source Q is generated at the knife edge (a disc with a central hole) K, so that only part of the light beam can pass through. The part of the beam that passed through K is focused at the spherical mirror S that has a curvature radius of 6 m. As mentioned the schlieren knife-edge, K provides selective shading of the light passing through the test section. It is redirected towards the camera by beam splitter T.

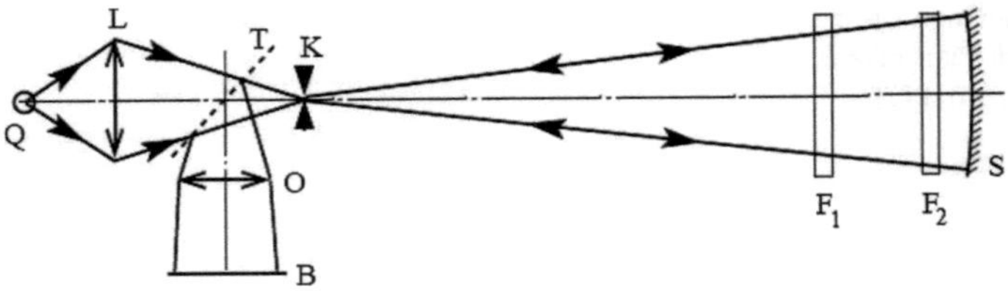

Fig. 7 Layout of HLB schlieren optics: *Q* light source, *L* lense, *K* knife edge, schlieren edge, *F* test section windows, *S* spherical mirror, *T* beam diverter, *O* camera lense, *B* schlieren image

4.3 Heat Flux

For optical measurements of wind tunnel model temperatures an infrared camera, Phoenix DAS of the Indigo Company is available. The camera uses a Stirling-cooled InSb-sensor which is sensitive in the middle infrared range, between 3 and 5 μm. The camera allows recording images with 320 × 256 pixels at frame rates up to 320 Hz. Typical integration time at temperature of 300 K are 3 ms, in order to achieve a good signal-to-noise ratio. This results in frame rates of 170 Hz. Optical access of the camera to the test section is obtained by using a sapphire window.

For temperature calibrations a black radiator is installed in place of the model before measurements. The temperature calibration curve is computed using a quadratic fit of the calibration points. The non-uniformity of the pixel-sensitivities is eliminated by a two-point correction, assuming that the pixel sensitivities differ only in the linear terms.

Wind tunnel models suited for infrared measurements are made from Plexiglas (black 811-Rhm-GS) and coated with Nextel Velvet Coating 811-21 at thickness of approximately 50 μm. This choice is the result of extensive studies in model materials [13], where the surface emissivity and material absorption were carefully measured. Note that the effects of temperature dependent thermophysical properties of the model material were also investigated [14, 15].

Image coordinates are mapped to space coordinates by using a calibration grid which is applied to the surface of the model before measurements. The currently implemented calibration software detects the image coordinates of the grid nodes and determines numerical functions for mapping curved surface coordinates into a Cartesian system [13]. This spatial calibration method is well suited for the images of axisymmetric model geometries.

1D nonlinear heat conduction perpendicular to the surface is assumed for calculating the surface heat fluxes from the transient temperature data. For this purpose the conduction into the Plexiglas is modeled by a finite difference method. The unknown heat flux is found by iterative regularization method: The surface temperature calculated from an iteratively improved heat flux is compared with the measured value in a least squares formulation. The gradient for iterative correction of the heat flux is found by solving the adjoin problem. Iteration end is adjusted to the noise level of the measured temperature data. Details of heat flux calculation with an extensive discussion of measurement errors can be found in [13]. There the overall uncertainty of the measured heat flux is estimated to be about 4 %.

High-frequency time resolved heat flux measurements are performed with a fast response heat flux sensor (ALTP, Atomic Layer Thermopile) [16]. The measuring method is based on the transverse Seebeck effect which allows measuring directly the heat flux density. The sensitive area has a size of 2 mm × 1 mm. As the ALTP sensors are powered with a low-noise amplifier from Cosytech, the output signals are amplified with an adjustable gain from 100 to 800 for the DC output and fixed amplified with a rate of 5000 for the AC output. The thermal insulation of the gauge is achieved via a ceramic housing and has a diameter of 8 mm.

5 Wind Tunnel Flow

5.1 Flow Start

The low pressure section of the HLB comprising of the nozzle, the test section, the diffuser and the 6 m^3 dump tank are evacuated to about 3 mbar before a tunnel run. The storage tube is pressurized up to 30 bar. Figure 8 displays an example of the measured transient distributions of static pressure recorded by the sensor in the annular duct around the valve and the test section Pitot pressure, upon the valve opening. After the initial transient, virtually constant pressures are observed. After 90 ms run time, pressures drop due to the travelling expansion wave in the storage tube that reaches the nozzle. The rising pressure in the dump tank causes break down of the test section flow at 110 ms, which is seen in the sudden rise in the recorded Pitot pressure. Shortly thereafter, the valve closes. The oscillations of the storage tube pressure at the beginning of the tunnel run could be traced back to the first mode of transversal oscillations in the tube, according to their frequency [15]. A measurable effect of these oscillations in the test section has not been observed.

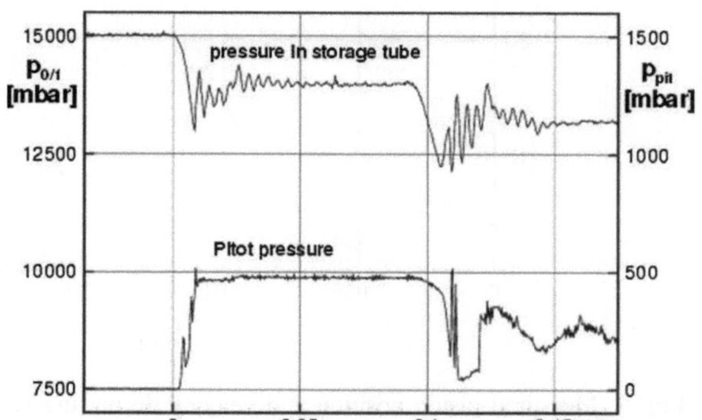

Fig. 8 Example of recorded pressure distribution in the storage tube and in the test section

The initial transient of the tunnel flow initiation was further investigated by using time-accurate, numerical flow simulations [15]. Because of limited computational resources, the assumption of axially symmetric flow within the storage tube, valve and nozzle was invoked. Numerical resolution of shock waves and contact discontinuities was accomplished by employing a hybrid flux vector scheme, AUSMPW+, for discretizing the inviscid fluxes. Turbulent wall boundary layers were modeled by employing the Menter-SST eddy viscosity model. The computations with the DLR TAU flow solver [17] used a deforming mesh in order to account for the piston motion during the valve opening. Therefore, the piston motion was measured as shown in Fig. 9. The piston velocity exhibits an oscillatory behavior during the opening phase which results from the dynamics of the pneumatic system involved.

The numerical results are used for understanding the complex flow physics during the valve opening and the initial transient flow into the test section. Figure 10 displays computed density gradients at several time instances. The initial flow shortly after the piston motion started displays the primary shock and the

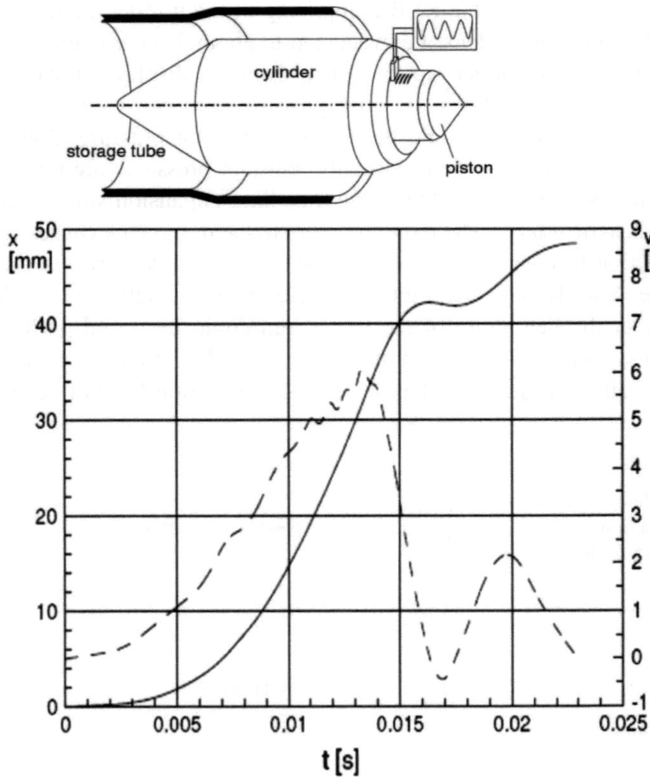

Fig. 9 Measured piston position and velocity during the valve opening. With kind permission from Springer Science+Business Media: [15], Fig. 2

contact discontinuity which run into the nozzle. Later, the initial expansion wave that propagates into the storage tube is seen. The initial flow structure in the nozzle is characterized by a very small annular throat between the piston and the converging nozzle contour. This configuration leads to strong expansion-flow rates and a number of shocks, which generate significant flow separation in the divergent part of the nozzle.

As the piston moves backwards, the increasing annular throat delivers a larger mass flow into the nozzle and the flow separation disappears (t = 5 ms). At this point the critical flow state is still located between the piston and the converging nozzle contour and hence, a strong oblique shock wave is generated with multiple reflections along the nozzle. This situation changes when the critical flow state moves to the geometrical nozzle throat at about 9.25 ms. From this time on a rather smooth nozzle flow develops which is increasingly less distorted by the piston.

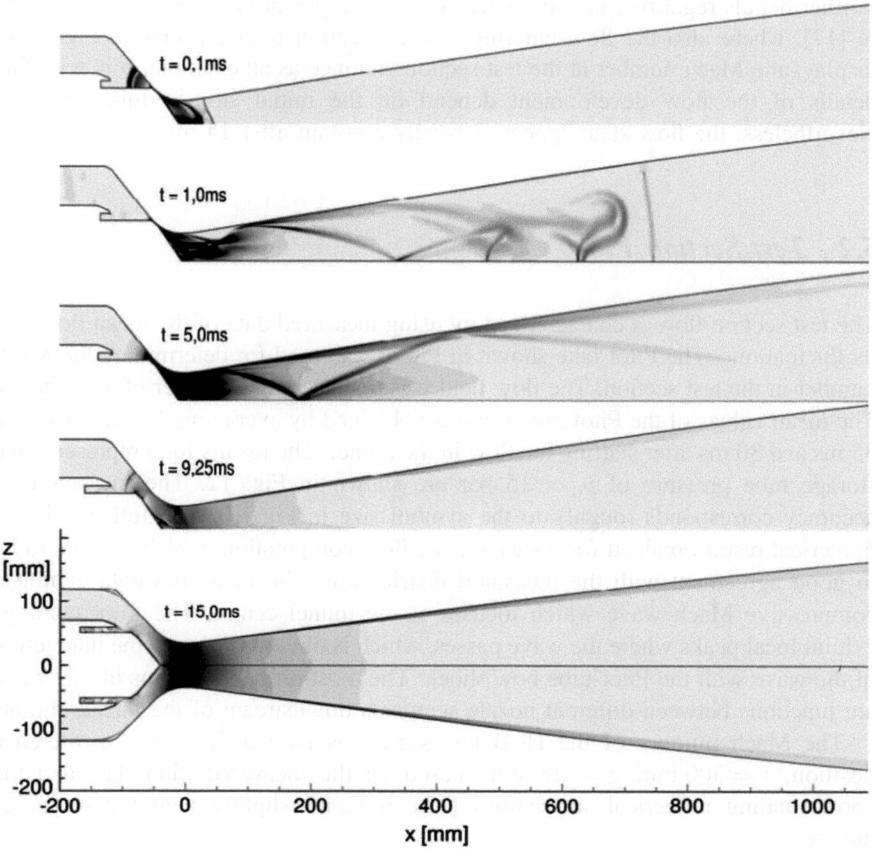

Fig. 10 Temporal flow development in HLB nozzle [17]. Shown are density gradient contours for an initial storage tube pressure of 3.25 bar. The initial pressure in the nozzle is 2 mbar. With kind permission from Springer Science+Business Media: [15], Fig. 7

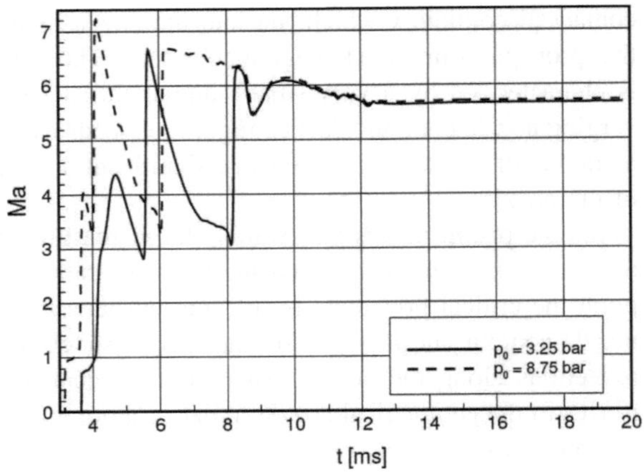

Fig. 11 Time history of computed Mach number at test section entrance for two storage tube pressures. With kind permission from Springer Science+Business Media: [15], Fig. 9

Further details regarding the computed flow development are available in [15] and in [17], where also the flow entering the test section is characterized. Figure 11 displays the Mach number at the test section entrance as an example. It is seen that details of the flow development depend on the initial storage tube pressure. Nevertheless, the flow state appears virtually constant after 14 ms.

5.2 Test Section Flow

The test section flow is characterized by using measured data of the mean flow and its fluctuations. The Pitot rake shown in Fig. 6 was used for determining the Mach number in the test section. The flow field was traversed at a number of x-positions. The mean values of the Pitot pressure were obtained by averaging the data between 35 ms and 80 ms after starting the flow in the tunnel. The results for a representative storage tube pressure of $p_0 = 15$ bar are shown in Fig. 12. The measurement accuracy corresponds roughly to the symbol size in Fig. 12. The full line is the numerical result obtained for a steady-state flow computation, which appears to be in good agreement with the measured distribution. The measured data exhibit a compressive Mach wave which focuses at the tunnel centre. The Pitot readings exhibit local peaks where the wave passes, which is obviously due to the interaction of the wave with the Pitot-tube bow shock. The most probable origins of this wave are junctions between different nozzle segments downstream of the nozzle throat.

The Mach number of the HLB test section is then defined at the reference position, $x = 300$ mm, $z = 100$ mm, based on the measured Pitot data and the corresponding numerical simulations [13]. It varies slightly with the Reynolds number:

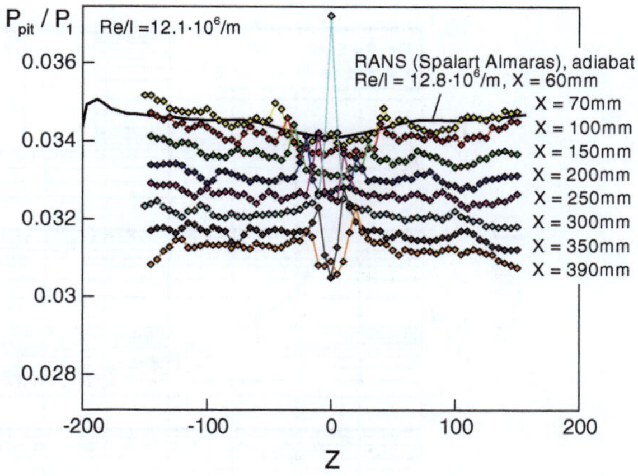

Fig. 12 Distribution of Pitot pressures in the test section for p_0 = 15 bar

$$M_\infty = \begin{matrix} 5.91 \pm 0.03 & \textit{for} & p_0 = 15.0\text{ bar} \\ 5.85 \pm 0.03 & \textit{for} & p_0 = 3.25\text{ bar} \end{matrix}$$

The assumption of isentropic nozzle flow and the reference Mach number are used for determining the reference free-stream pressure:

$$p_\infty = (0.9327 \pm 0.0007)\left(1 + \frac{\gamma - 1}{2} M_\infty^2\right)^{\frac{\gamma}{1-\gamma}} p_0,$$

where p_0 is the storage tube pressure measured before flow initiation. Readings of the thermocouples in the annual clearance around the valve are used for obtaining the total temperature of the tunnel flow. Further measurements of test section temperatures available in [18] indicate that the measured total temperature is a reasonable representation for the reference test section position with an uncertainty of ±5 K. Hence

$$T_\infty = \left(1 + \frac{\gamma - 1}{2} M_\infty^2\right)^{-1} T_{1t} = (0.1256 \pm 0.0011) T_{1t} \pm 0.6\text{ K}$$

where T_{1t} denotes the storage tube total temperature as measured at the valve housing. The Reynolds number per unit length in the test section can be computed from these values and by assuming a suitable viscosity law [13]. Similarly, a reference heat flux density is computed in order to determine the non-dimensional Stanton numbers from measured heat fluxes. The given uncertainties for freestream Mach number, pressure, and temperature can be combined to obtain uncertainties for the reference values of the Reynolds number and the heat flux as detailed in [13].

Pressure and heat flux fluctuations were determined by using stagnation probes. Figure 13 shows amplitude spectra of normalized Pitot pressure fluctuations, at a

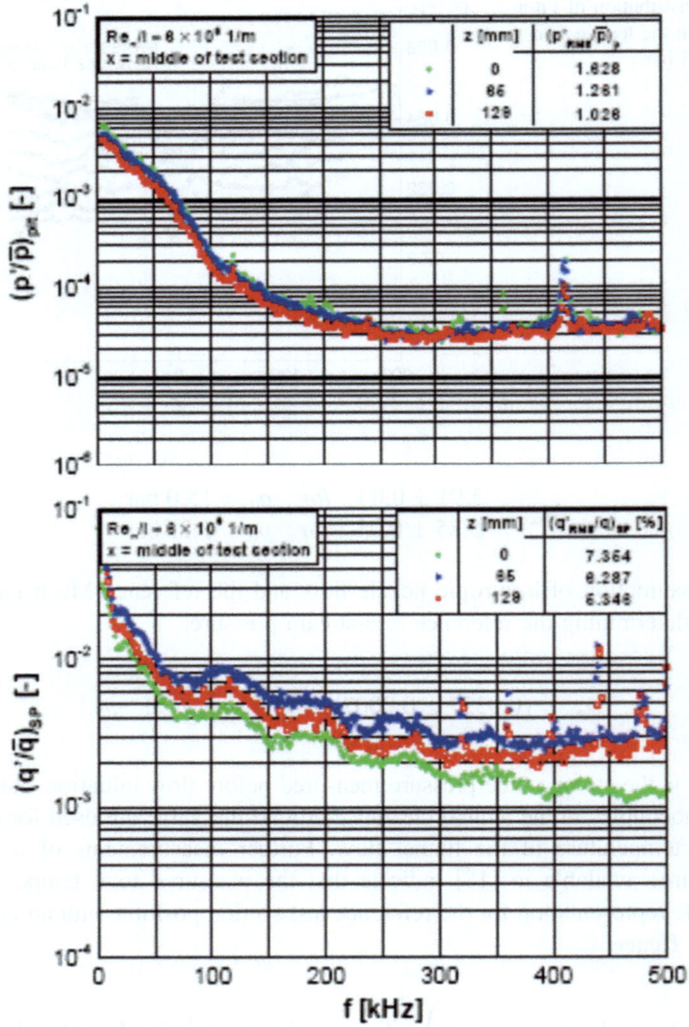

Fig. 13 Spectra of normalized Pitot pressure fluctuations (*left*) and normalized heat flux fluctuations (*right*) at a unit Reynolds number of 6×10^6/m and $T_{1t} = 480$ K [19]

unit Reynolds number of 6.0×10^6/m at three positions in the vertical symmetry plane of the test section. All three spectra show a noticeable peak at about 410 kHz, representing the natural frequency of the used Entran EPIH-113-1B/Z1 sensor. A typical decay of the fluctuation level versus frequency without considerable dominant peaks is visible in this figure. The rms-values of the normalized Pitot pressure fluctuations are between 1–1.6 % and increase towards the centerline. Figure 13 also shows the amplitude spectrum of the heat flux fluctuation in the stagnation point of the blunt body normalized by the corresponding mean heat flux density obtained simultaneously. The rms-values of the normalized fluctuations are between 5.3–7.3 % and increase towards the centerline, like the pressure fluctuations.

5.3 Reference Model Testing

Testing of two reference model geometries yields further insight into the quality of the used instrumentations and the behavior of the test section flow. The first model geometry is chosen for evaluating the infrared based heat transfer measurements. The geometry is displayed in Fig. 14. It is an axially-symmetric representation of the HERMES re-entry vehicle with deflected control flap [20]. One wind tunnel model was manufactured from black Plexiglas (PMMA) and painted with Nextel Velvet Coating 811-21. An alternate wind tunnel model was provided by the DLR in Göttingen. This is a thin-wall model which consists of a 0.35 mm nickel layer, to which Nichrome wires are soldered to provide thermocouple contact [21]. The measured temperature change with time is proportional to the heat flux, as detailed in [13]. The flow over the model develops a thin, laminar boundary layer along the

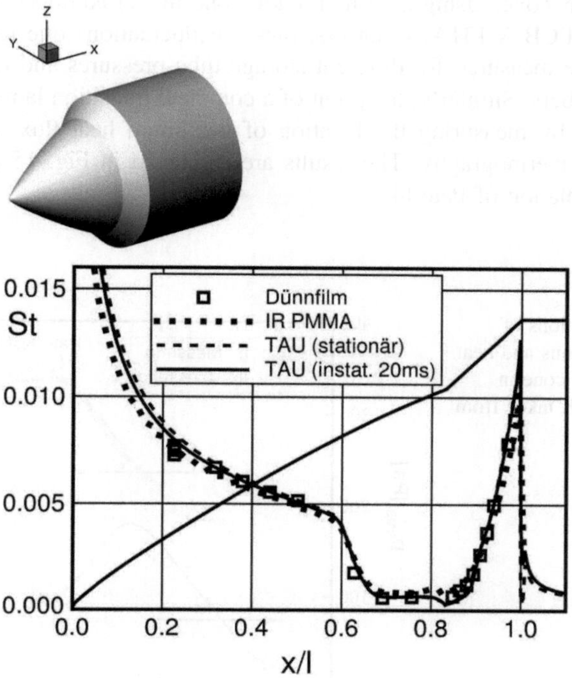

Fig. 14 Stanton number distributions for laminar flow over the hyperboloid and flare configuration, $M_\infty = 5.85$, $p_0 = 3.25\,\text{bar}$, $T_{1t} = 485$ K, taken from [17]

forebody. The interaction of the boundary layer and the oblique flare shock wave leads to a separation bubble with locally low heat transfer. This is followed by flow re-attachment with a strong overheating. The infrared-based heating result was obtained by inverse solution of the one-dimensional unsteady heat conduction equation at a tunnel run time of 75 ms. The resulting heat flux values were averaged over a model segment of 20° in circumferential direction. Figure 14 displays very good agreement between the two measurements and the two corresponding numerical flow simulations. One of these flow simulations, with the DLR-TAU Code, used the unsteady start-up of the test section flow shown in Fig. 10 as a non-stationary onset flow. The other numerical simulation used the flow conditions at the test section reference point as described above, and assumed a parallel free stream flow. Both simulations resulted in very similar results for the Stanton number, which are in line with the two experimental realizations.

This good agreement confirms the experimental approach of defining suited reference flow conditions for the HLB test section flow. Further analysis of uncertainties for the infrared measurements is available in [13], it revealed that the remaining uncertainties due to coating transparency and due to errors of the wind tunnel reference heat flux are similarly large. Both errors yield uncertainties in the Stanton number of 3–4 % of the measured value.

The level of the free stream disturbances in HLB is indirectly assessed by investigating laminar boundary layer instabilities and boundary layer transition on a pointed circular cone. Using a 7° half-angle cone model equipped with a streamwise array of PCB M131A31 sensors, pressure fluctuations due to second-mode instability were measured for different storage tube pressures and hence, different Reynolds numbers. Similarly, the point of a complete transition laminar-turbulence was identified by measuring the location of maximum heat flux along the cone using infrared thermography. The results are displayed in Fig. 15 along with the transition correlation of Pate [4].

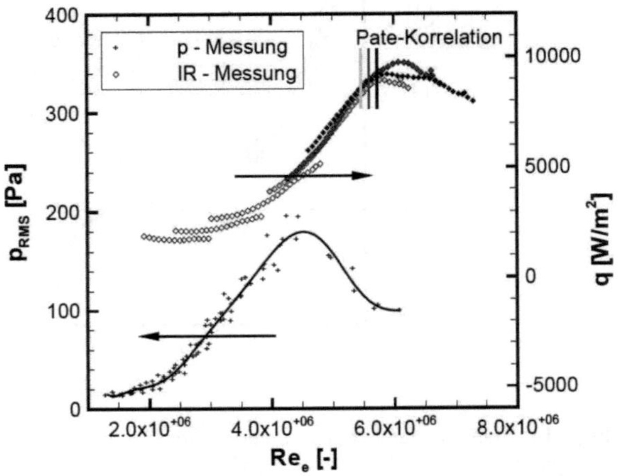

Fig. 15 Distributions of pressure fluctuations and heat flux along the 7° cone in HLB, T_{t1} = 480 K taken from [22]

Here, computed flow quantities at the boundary layer edge of the cone flow are used to plot the measured boundary layer data. The Pate correlation estimates the effect of wind tunnel noise from the turbulent nozzle boundary layer on hypersonic transition for cones. It is seen that transition in HLB appears somewhat downstream of the Pate correlation, presumably because fluctuations in Ludwieg tubes are relatively low compared to other blow-down facilities. However, the expected effect of unit Reynolds number is represented in the HLB test section.

6 Experiments in Transitional Hypersonic Boundary Layers

Recent boundary layer research in the Hypersonic Ludwieg Tube Braunschweig aims at identifying and characterizing the instabilities of the laminar boundary layer that lead to laminar-turbulent transition. So far the work focused on harmonic modes with exponential growth. Progress in the development of suited wind tunnel models and model instrumentation has resulted in the ability to determine the frequency and amplification of harmonic modes in the boundary layer, the orientation of wave fronts and the size of typical wave packets.

The experimental work used the circular 7° sharp cone as a generic model of slender hypersonic vehicles. Several wind tunnel models were manufactured of coated Plexiglas and instrumented with PCB M131A31 sensors. These sensors exhibit a very high sensitivity and their temporal resolution is sufficient for resolving high-frequency modes observed in hypersonic boundary layers [10]. Infrared analysis of the surface temperature allows monitoring the flow break down to turbulence and can be used to detect stationary instability modes as well.

A recent experimental approach assumes that instabilities may be characterized by using space-time cross correlations. Variation of the spatial distance in these correlations can be accomplished with sensors placed on rotatable discs. Figure 16 displays such a wind tunnel model.

For model accuracy, the discs are mounted on a single model axis with high precision bushings. 10 PCB sensors are installed on three rotatable discs with their sensing head mounted flush to the surface. The first disc has a stainless steel tip and carries sensors 0, 1, 2 used to measure initial growth of the boundary layer instabilities. Sensors 3, 4, 8 can be rotated, relative to sensor 2. They are installed closely behind sensor 2. This arrangement is used for measuring space-time correlations.

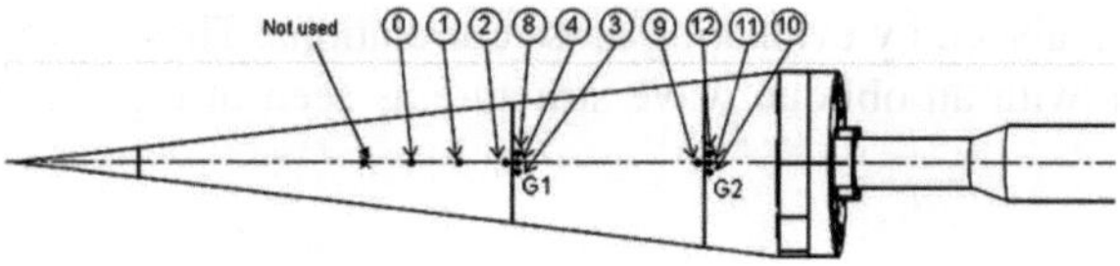

Fig. 16 Sharp cone model with three rotatable discs

A similar group is formed by sensors 9, 10, 11, 12. With these two groups the unit Reynolds number can be varied during experiments. The final model disc accomplishes the model attachment to the model support sting, at the predefined angle of attack.

The dominating instability of streamwise hypersonic boundary layer flow is an acoustic instability. It is called 2nd mode, according to Mack [23]. The amplification rate of the mode in hypersonic flow is much larger than the Tollmien-Schlichting instability (1st mode). Signals from sensors mounted on the cone, installed at zero angle of attack show that the 2nd modes appear as wave packages that grow while travelling downstream. First modes are not detected by the PCB sensors. Figure 17 displays measured data of the 2nd mode behavior. The spectral analysis of the measured fluctuations is also presented. It shows the growth of the modes. Note that the initial growth rates are in good agreement with the result of linear stability analysis (not shown here) [10, 22].

The temporal distributions of two sensor readings can be correlated in order to extract more structural data of the wave packages. Using the measured phase velocity one can interpret the temporal correlation signal as the spatial distribution of wave fronts in a statistically averaged wave package. This is illustrated in Fig. 18 which shows the wave fronts measured at two unit Reynolds numbers. The wave fronts are normal to the streamwise direction, as expected. The measured wave length is close to two boundary layer thicknesses. The lateral extension of the average wave package is 4–5 boundary layer thicknesses.

The flow over the sharp cone at angles of attack develops a more complex set of boundary layer instabilities. These are streamwise 2nd modes as well as instabilities due to the boundary-layer cross flow. The observed infrared-based surface heat flux shown in Fig. 19 underlines the complexity of the 3-dimensional boundary layer flow. Plotted are colored heating contours as a function of the cone x-coordinate and the cone azimuthal angle, θ. The value $\theta = 0°$ denotes the windward side of the cone, whereas $\theta = 180°$ is at the leeside. The break down to turbulence is indicated by the rapid color change from green to red.

At the windward side the transition front displays a smooth behavior and only 2nd mode instabilities have been measured in this area so far. At the cone shoulder a zigzag behavior of the transition front is observed and longitudinal flow structures appear in the heat flux distribution. This is typical for transition dominated by stationary cross flow vortices. However, the signals from the PCB sensors at locations between 70° and 110° reveal that other instabilities are also present. Figure 20 presents a spectrum obtained with the sensors located at $\theta = 70°$.

One observes the 2nd mode at a high frequency of 350 kHz. Moreover, there is a distinct peak in the signal of the sensor group 2, at 30 kHz. The time traces were then similarly analyzed by evaluating cross correlations. These revealed rather large wave packages with an oblique wave structure as seen in Fig. 21. The wave front

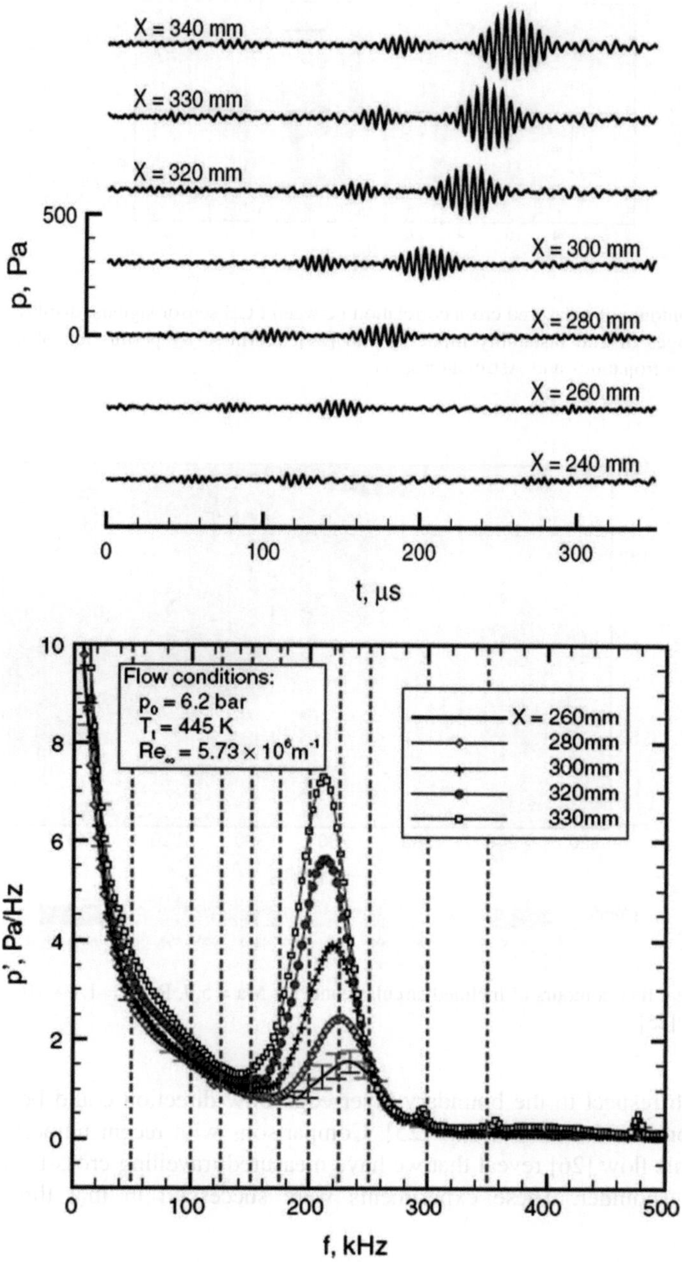

Fig. 17 Time traces and disturbance spectra from PCB M131A31 sensors on a 7° cone, $Re/m = 5.7 \times 10^6/m$. From [24]; reprinted by permission of the American Institute of Aeronautics and Astronautics, Inc.

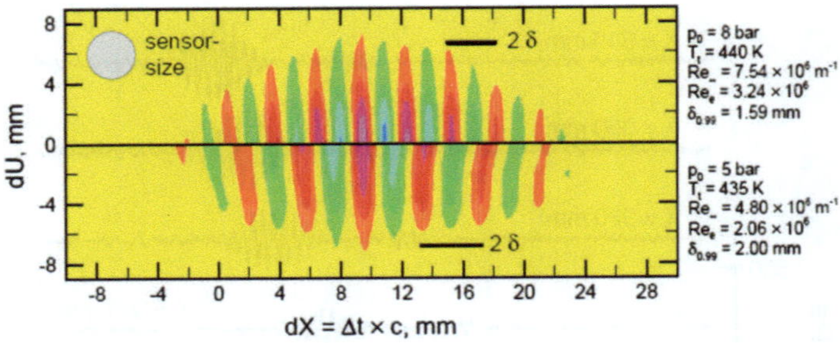

Fig. 18 Contours of measured cross correlation between PCB sensor signals, displaying averaged wave packages of 2nd instability modes. From [24]; reprinted by permission of the American Institute of Aeronautics and Astronautics, Inc.

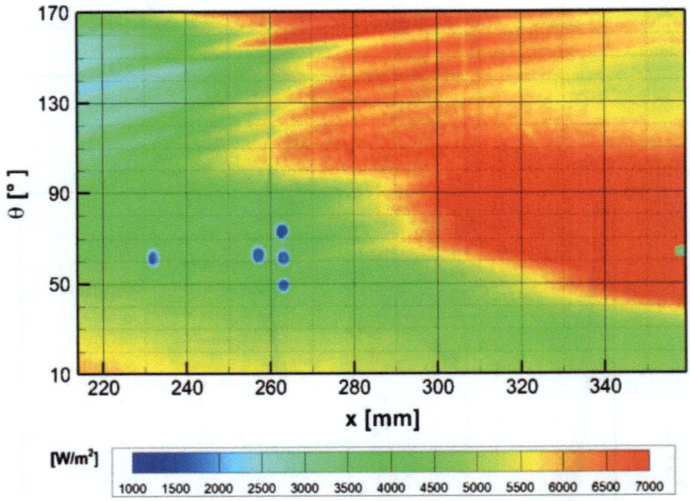

Fig. 19 Heat flux contours of inclined circular cone, for Ma = 5.9, Re/m = $1.1 \times 10^7\ \mathrm{m}^{-1}$, $\alpha = 6°$, taken from [25]

angle with respect to the boundary layer edge flow direction could be also determined from the correlation data [25]. Comparisons with recent numerical simulations of this flow [26] reveal that we have measured travelling cross-flow modes at the cone shoulder. These experiments were successful in that three different

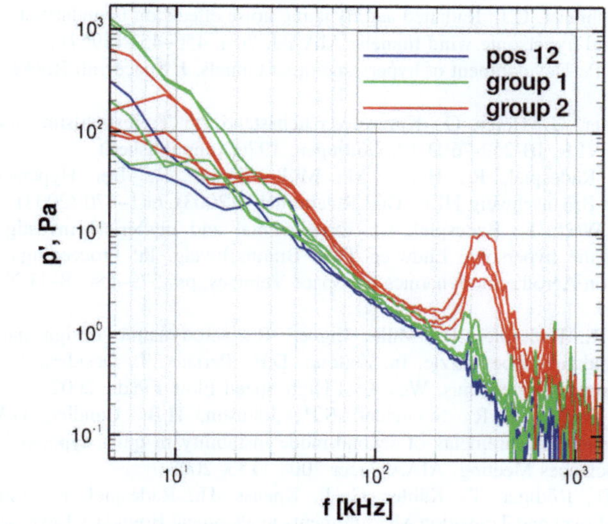

Fig. 20 Spectra of PCB sensors on the circular cone of Fig. 16 at Re/m = 3.64×10^6, $\alpha = 6°$ and $\theta = 70°$, taken from [25]

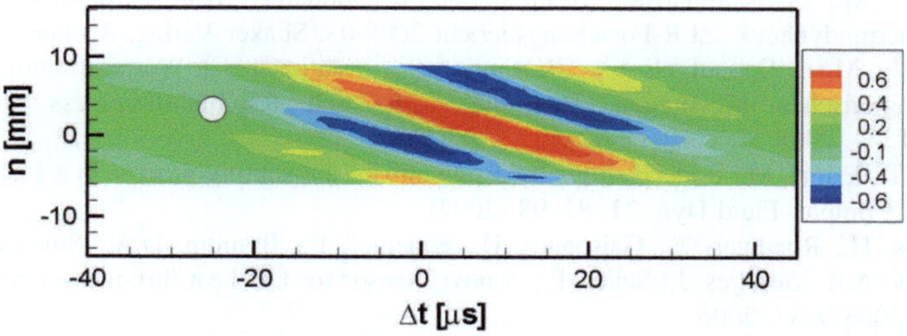

Fig. 21 Space-time correlation from band-pass filtered PCB signals of the circular cone at x = 360 mm and $\theta = 70°$ and Re/m = 3.64×10^6, $\alpha = 6°$, taken from [25]

boundary layer instabilities were simultaneously measured for the first time. As the flow is complex, more investigations are planned to understand better the sensitivities of cross flow dominated transition in hypersonic boundary layers.

References

1. Ludwieg, H.: Der Rohrwindkanal. Z. f. Flugwiss. **3**, 206–216 (1055)
2. Ludwieg, H., Hottner, T., Grauer-Carstensen, H.: Der Rohrwindkanal der Aerodynamischen Versuchsanstalt Göttingen, Jahrbuch 1969 der DGLR, 1969, pp. 52–58
3. Koppenwallner, G., Müller-Eigner, R., Friehmelt, H.: HHK Hochschul Hyperschall-Kanal: Ein "Low Cost" Windkanal für Forschung und Ausbildung, DGLR Jahrbuch **2** (1993)

4. Pate, S.R., Schueler, C.J.: Radiated aerodynamic noise effects on boundary-layer transition in supersonic and hypersonic wind tunnels. AIAA J. **7**(3), 450–457 (1969)
5. Schneider, S.P.: Development of hypersonic quiet tunnels. J. Spacecraft Rockets **45**, 641–664 (2008)
6. Koppenwallner, G., Hefer, G.: Kurzzeitversuchsstand zur Treibstrahlsimulation bei hohen Drücken. DFVLR, IB 252-76 H 12, Göttingen, 1976, (unpublished)
7. Estorf, M., Radespiel, R., Heine, M., Müller-Eigner, R.: Der Hyperschallwindkanal Ludwiegrohr Braunschweig HLB. DGLR-Jahrbuch **1**(2003), 661–670 (2003)
8. Estorf, M., Wolf, T., Radespiel, R.: Experimental and numerical investigations on the operation of the hypersonic Ludwieg Tube Braunschweig. In: Proceedings 5th European Symposium on Aerothermodynamics for Space Vehicles, pp. 579–586. 8–11 Nov 2004, ESA SP-563, 2005
9. Kozulovic, D., Radespiel, R., Müller-Eigner, R.: aerodynamic design parameters of a hypersonic ludwieg tube nozzle. In: Zeitoun, D.E., Periaux, J., Desideri, J.A., Martin, M. (eds.) Conference Proceedings: West East High Speed Flow Fields, 2002
10. Estorf, M., Radespiel, R., Schneider, S.P., Johnson, H.B., Candler, G.V., Hein, S.: Surface-pressure measurements of second-mode instability in quiet hypersonic flow. AIAA Aerospace Sciences Meeting, AIAA Paper 2008-1153, 2008
11. Heitmann, D., Rödiger, T., Kähler, C. J., Knauss, H., Radespiel, R., and Krämer, E., Disturbance-Level and Transition Measurements in a Conical Boundary Layer at Mach 6, 26th AIAA Aerodynamic Measurement Technology and Ground Testing Conference, AIAA Paper 2008-3951, 2008
12. Rufer, J., Berridge, D.C.: Experimental study of second-mode instabilities on a 7-Degree Cone at Mach 6, AIAA Paper 2011-3877, 2011
13. Estorf, M.: Ortsaufgelöste Bestimmung instationärer Wärmestromdichten in der Aerothermodynamik. ZLR-Forschungsbericht 2008-03, Shaker Verlag, Aachen, 2008
14. Estorf, M.: Ortsaufgelöste Bestimmung instationärer Wärmestromdichten aus thermografischen Messungen, Deutscher Luft- und Raumfahrtkongress 2008, Paper No. 81344, 2008
15. Wolf, T., Estorf, M., Radespiel, R.: Investigations of the starting process in a Ludwieg Tube. Theor. Comput. Fluid Dyn. **21**, 81–98 (2007)
16. Knauss, H., Roediger, T., Gaisbauer, U., Kraemer, E., Bountin, D.A., Smorodsky, B.V., Maslov, A.A., Scruljes, J., Seiler, F.: A novel sensor for fast heat flux measurements, AIAA Paper 2006-3637, 2006
17. Wolf, T.: Strömungsanalyse und Wärmeübergang beim Startvorgang eines ventilgesteuerten Ludwieg-Rohres. ZLR-Forschungsbericht 2007-06, Shaker Verlag, 2007
18. Estorf, M., Wolf, T., Radespiel, R.: Experimental and numerical investigations on the operation of the Hypersonic Ludwieg Tube Braunschweig. In: 5th European Symposium on Aerothermodynamics for Space Vehicles, pp. 579–586. ESA SP 563, 2005
19. Heitmann, D., Kähler, C.J., Radespiel, R., Rödiger, T., Knauss, H., Krämer, E.: Disturbance-level and transition measurements in a conical boundary layer at Mach 6, AIAA Paper 2008-3951, 2008
20. Schwane, R.: Description of the testcase: MSTP Workshop 1996 "Reentry Aerothermodynamics and Ground to Flight Extrapolation". Technical Report YPA/1889/RS, ESTEC, Noordwijk, 1996
21. Krogmann, P.: Hyperboloid/flare experiments at Mach 6.8 in RWG, Technischer Bericht DLR-IB 223-94 C 44, Institut für Strömungsmechanik, DLR Göttingen, 1994
22. Heitmann, D.: Transitionsuntersuchungen in hypersonischen Grenzschichten mit laserinduzierten Störungen. CFF Forschungsbericht 2011-09, Shaker Verlag, 2011
23. Mack, L.M.: Linear stability theory and the problem of supersonic boundary-layer transition. AIAA J. **13**, 278–289 (1975)
24. Heitmann, D., Radespiel, R., Knauss, H.: Experimental study of boundary-layer response to laser-generated disturbances at Mach 6. J. Spacecraft Rocket **50**(2), 294–304 (2013)

25. Munoz, F., Heitmann, D., Radespiel, R.: Instability modes in boundary layers of an inclined cone at Mach 6. AIAA-Paper 2012-2823, 2012
26. Perez, E., Reed, H.L., Kuehl, J.J.: Instabilities on a hypersonic yawed straight cone. AIAA Paper 2013-2879, 2013

Author Biographies

Rolf Radespiel is known for his boundary layer research in high-speed flows at Technische Universität Braunschweig, Germany. He has been extensively using the Braunschweig Ludwieg Tube for this research and his research group presented a number of advanced measuring techniques. He received his "Dipl.-Ing." degree in engineering science from Technische Universität Braunschweig and his "Dr.-Ing." degree from the same University. He presently holds the Chair of Fluid Mechanics in Braunschweig. His current research interests include models of transition and turbulence, further high-lift augmentation and drag reduction by means of flow control.

Malte Estorf worked in the aerothermodynamics group at the Institute of Fluid Mechanics in Braunschweig between 2002 and 2008. He put the Braunschweig Ludwieg Tube into operation and established advanced measurement techniques. He received his "Dipl.-Ing." degree in engineering science from Technische Universität Braunschweig and his "Dr.-Ing." degree from the same University. He presently works for the Volkswagen AG.

Dirk Heitmann received his "Dipl.-Ing." degree in engineering science from Technische Universität Braunschweig in 2005 and his "Dr.-Ing." degree from the same University in 2011. His research focused on boundary layer transition in high speed flows. This work was mainly conducted in the Braunschweig Ludwieg tube. His research interests include boundary layer transition, high speed flows and measurement techniques. He presently works at the DLR Braunschweig in the aeroacoustics department.

Federico Muñoz graduated in 2009 from National University of La Plata, Argentina as Aeronautical Engineer. He has since conducted doctoral studies at TU Braunschweig, by characterizing second modes and cross flow instabilities in hypersonic boundary layers.

Torsten Wolf received his „Dipl.-Ing." degree in engineering science and his "Dr.-Ing." from Technische Universität Braunschweig. His thesis covers the numerical rebuilding of the starting process of the Braunschweig Ludwieg Tube. He currently works in the Turbine Aerodynamics department at Rolls-Royce Deutschland as aerodynamics engineer and project lead of research projects.

Appendix

In the following, a list of active shock tube and tunnel laboratories that are not described in the present Volume due to lack of space and/or information availability is given. This is not a complete list and the order in which the different laboratories appear is fully arbitrary; it has no implication on their quality ranking. In this listing only research laboratories that are not described in details in the present volume are listed.
For a few research laboratories with shock tube facilities and gun tunnels only a laconic description was available; it appears after the list with the corresponding laboratory addresses. Interested readers should feel free to approach any of the laboratories cited for a more detailed information.

1 List of Active Shock Tube Laboratories not Described in the Present Volume

1. The University of New South Wales, School of Aerospace, Civil and Mechanical Engineering, Australian Defence Force Academy, Canberra, Australia
2. China Aerodynamics Research and Development Center, Mianyang, China
3. University of Science and Technology of China, Department of Mechanics and Mechanical Engineering, Hefei, China
4. University Duisburg-Essen, Institute for Combustion and Gasdynamics, Germany
5. Ehine University, Department of Chemistry, Matsuyama, Japan
6. Yokohama University, Graduate School of Engineering, Yokohama, Japan
7. Instituto de Estudos Avançados (IEAv), São José dos Campos, SP, Brasil
8. Semenov Institute of Chemical Physics, Heterogeneous Combustion Laboratory, Moscow, Russia

O. Igra and F. Seiler (eds.), *Experimental Methods of Shock Wave Research*,
Shock Wave Science and Technology Reference Library 9,
DOI 10.1007/978-3-319-23745-9

9. Central Aerohydrodynamic Institute (TSAGI), Moscow, Russia
10. Central Research Institute of Machine Building Russian Space Agency (TsNIIMash), Moscow, Russia
11. General Applied Science Laboratory (ATK GASL *Advanced Systems*), Ronkonkoma, NY, USA
12. Calspan-University of Buffalo Research Center (CUBRC), Buffalo, NY, USA
13. University of Texas, Aerodynamic Research Center, Austin, TX, USA
14. Arnold Engineering Development Complex (AEDC), Tullahoma, TN, USA
15. University of Manchester, Aerospace Research Institute, Manchester, Great Britain
16. NASA Ames Research Center, Ames Technology Capabilities and Facilities, Moffett Field, CA, USA
17. Purdue University, School of Aeronautics and Astronautics, W. Lafayette, IN, USA
18. University of Texas at Arlington, UT Arlington Aerodynamics Research Center, Arlington, Texas, USA
19. Instituto Superior Técnico, Instituto de Plasmas et Fusão Nuclear, Lisbon, Portugal.

2 Description of Shock Tube Facilities

2.1 Los Alamos National Laboratory, Los Alamos, NM, USA

Extreme Fluids Team in Physics Division
Kathy Prestridge, Adam A. Martinez, Greg Orlicz
Ricardo Mejia-Alvarez, Brandon Wilson
P.O. Box 1663
Los Alamos, NM 87545, USA
kpp@lanl.gov
http://www.lanl.gov/projects/shocktube

(a) **Vertical Shock Tube Facility**
Membraneless, pneumatic driver.
Total length: 7 m
Shock Mach Number: 1–5
Test Section: 250 mm × 250 mm × 1 m

(b) **Horizontal Shock Tube Facility**
Membraneless, pneumatic driver.
Total length: 7.5 m
Shock Mach Number: 1–5
Test Section: 75 mm × 75 mm × 0.75 m

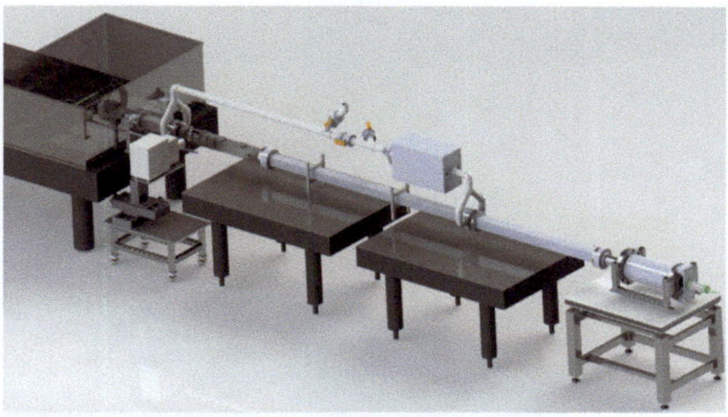

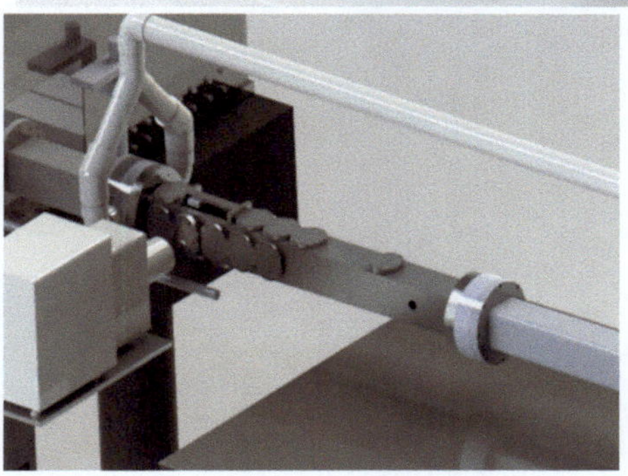

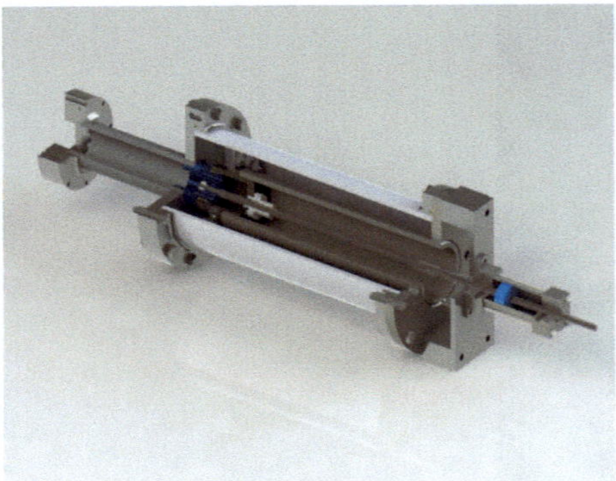

Pneumatic Driver

2.2 University of Wisconsin, Madison, WI, USA

Department of Engineering Physics
Riccardo Bonazza, David Rothamer, Jason Oakley
1500 University Avenue
Madison, WI 53706, USA
bonazza@engr.wisc.edu
http://www.engr.wisc.edu/ep/faculty/bonazza_riccardo.html
Vertical shock tube, total length: ~9.2 m,
Shock Mach number: 1–5,
test section size: 250 mm × 250 mm^2 × 1.5 m.

2.3 University of Illinois at Urbana-Champaign, Urbana, IL, USA

Mechanical Science and Engineering Department
Nick Glumac, Herman Krier
Mechanical Engineering Building, 1206 W. Green St.
Urbana, IL 61801, USA
glumac@illinois.edu
http://illinois.edu/

Horizontal tube: total length: ~11.4 m
Shock Mach number: 1–5,
Test section size: various; typically 3.9 cm ID, 0.5 m long

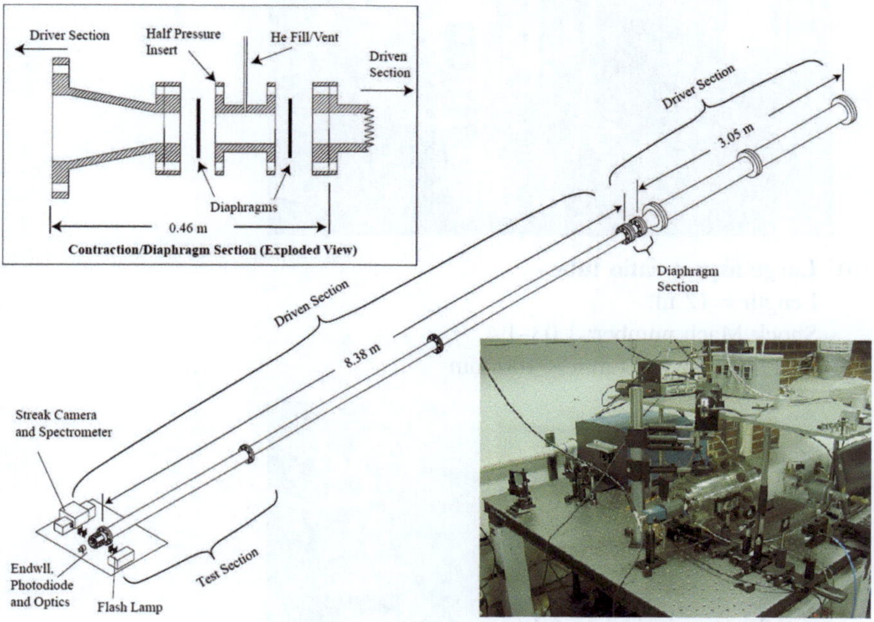

2.4 University of the Witwatersrand, Johannesburg, South Africa

School of Mechanical, Industrial and Aeronautical Engineering
Beric Skews
1 Jan Smuts Avenue
Braamfontein 2000

Johannesburg, South Africa
beric.skews@wits.ac.za
http://www.wits.ac.za

(a) **Automated shock tube**
Known as the Seitz tube, length = 12 m
Shock Mach number: 1.03–3
Test section: rectangular, 180 mm × 76 mm

(b) **Large aspect ratio tube**
Length = 12 m
Shock Mach number: 1.03–1.4
Test section: 1100 mm × 100 mm

(c) **450 mm tube**
Length = 12 m
Shock Mach number: 1.1–1.6
Test section: uses large aspect ratio test section

(d) **Bifurcated tube**
Length: 11 m
Shock Mach number: 1.2–1.6
Test section: 300 mm × 300 mm

(e) **Diffraction tube**
Length: 6 m
Shock Mach number: 1.2–5
Test section: 300 mm × 300 mm

(f) **Transonic shock tube**
Length: 66 m
Shock Mach number: 1.2–2.3
Test section: rectangular, 300 mm × 150 mm

Driver Driven

(g) **Expansion tube**
A shock tube with test section in the driver
Shock Mach number: 1.1–1.4
Test section: rectangular, 160 mm × 160 mm

(h) **Shock profile tube**
Test section added to 450 mm tube
Shock Mach number: variable
Test section: variable shape

(i) **General purpose tube (hypersonic)**
Length: 18 m
Shock Mach number: 1.2–1.6 as shock tube
Mach numbers = 4, 5, 6 with fixed nozzle blocks

(j) **Vertical (Conical) shock tube**
Length: 4 m
Shock Mach number: 1.2–1.6 at inlet
Test section: open-ended

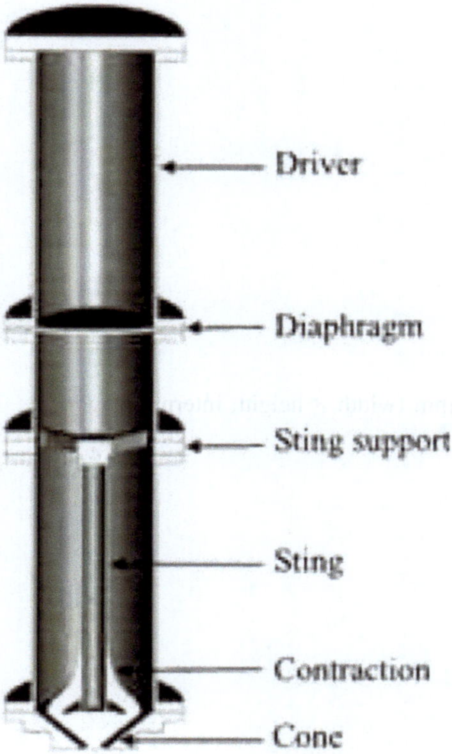

2.5 Korea Advanced Institute of Science and Technology (KAIST), Daejeon, Republic of Korea

Department of Aerospace Engineering
Gisu Park—gisu82@kaist.ac.kr
Keun Shik Chang—kschang@kaist.ac.kr
Chul Park—cpark216@kaist.ac.kr
373-1 Guseong-dong, Yuseong-gu
Daejeon 305-701, Republic of Korea
http://ae.kaist.ac.kr/common/People/Professors.php

(a) **Shock tube**
Total length: about 5 m
Shock Mach number: 2–5
Test section size: 47.5 mm (internal diameter, axially symmetric)

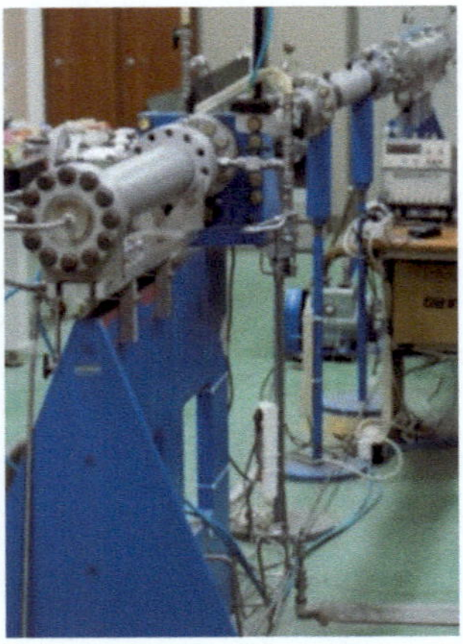

(b) **Shock tube**
Total length: about 6 m
Shock Mach number: 1–3
Test section size: 65 mm × 65 mm (width × height, internal length)

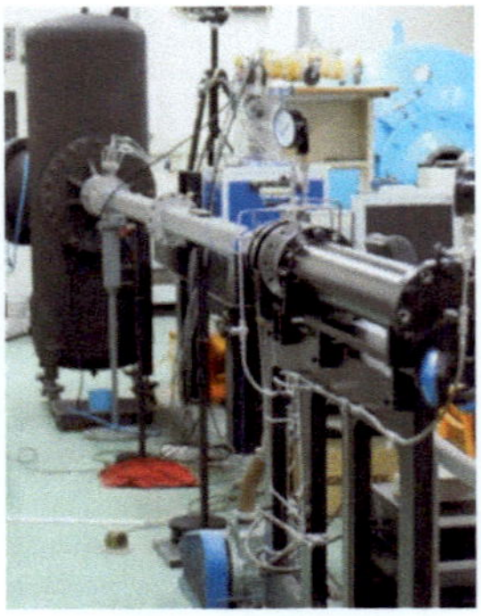

(c) **Shock tunnel**
Total length: 12.5 m (driver tube—helium gas driven)
Nozzle-exit Mach number: 4–5
Test section size: 60 mm × 150 mm (width × height, internal length)

2.6 Seoul National University, Seoul, Republic of Korea

Department of Aerospace Engineering
In-Seuck Jeung
1 Gwanak-Ro, Gwanak-Gu
Seoul 151-744, Republic of Korea
enjis@snu.ac.kr
http://apcl.snu.ac.kr

(a) **Shock Tunnel**
Total length: ~10,000 mm,
Driven tube diameter: 6.7 cm
Nozzle: (1) Mach number 7 conical nozzle with 15 cm exit diameter
(2) Mach number 7 contoured nozzle with 20 cm exit diameter
Test section size: 300 mm × 300 mm (width × height)

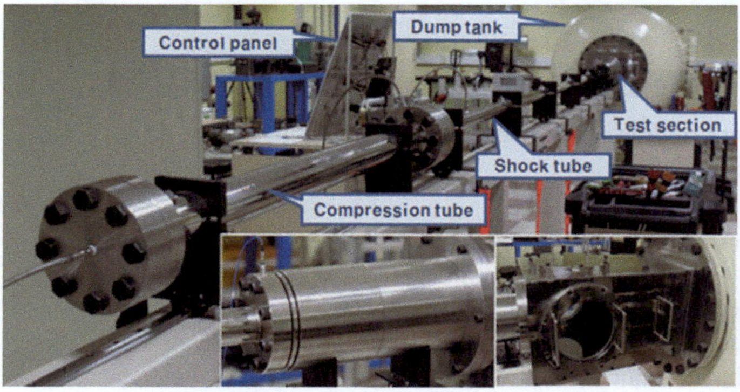

(b) **Shock Tunnel**
Total length: ~16,000 mm
Driven tube diameter: 10 cm
Nozzle: Mach number 7 conical nozzle with 34 cm exit diameter
Test section size: 1000 mm × 3000 mm (diameter × length)

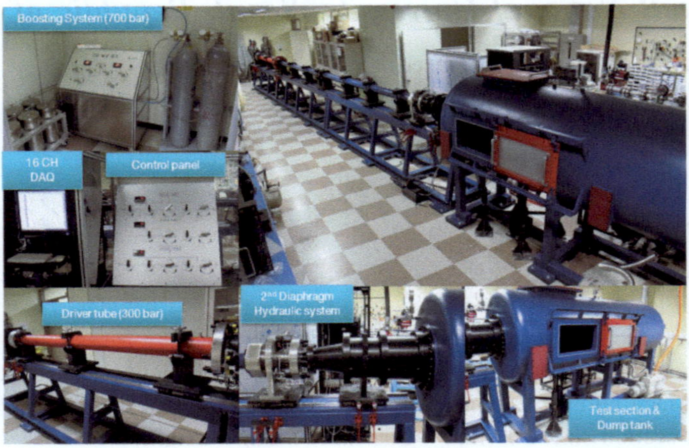

2.7 University of Arizona, Tucson, AZ, USA

Department of Aerospace and Mechanical Engineering
Jeffrey Jacobs
1130 N. Mountain Ave.
Tucson, AZ 85721, USA
jwjacobs@email.arizona.edu
http://ame.engr.arizona.edu/jeffrey-w-jacobs
Vertical shock tube, total length: 5–10 m
Shock Mach number: 1.1–1.3,
Test section size: 89 mm × 89 mm (square) × 0.9 to 2.0 m.

2.8 Petersen Research Group, Texas A&M University, College Station, TX, USA

Department of Mechanical Engineering
Eric L. Petersen
3123 TAMU
College Station, TX 77845, USA
epetersen@tamu.edu
http://petersengroup.tamu.edu/
High Pressure Shock Tube

Length: ~2.5 m driver, 4.7 m driven
Maximum Pressure: 100 atm
Driven section diameter: 15.24 cm
Uses: chemical kinetics; spectroscopy

Low Pressure Shock Tube
Length: ~7.1 m
Maximum Pressure: 10 atm
Driven section: 10.8 cm × 10.8 cm
Uses: dust-layer tests; heated capability; chemical kinetics; aerosol shock tube

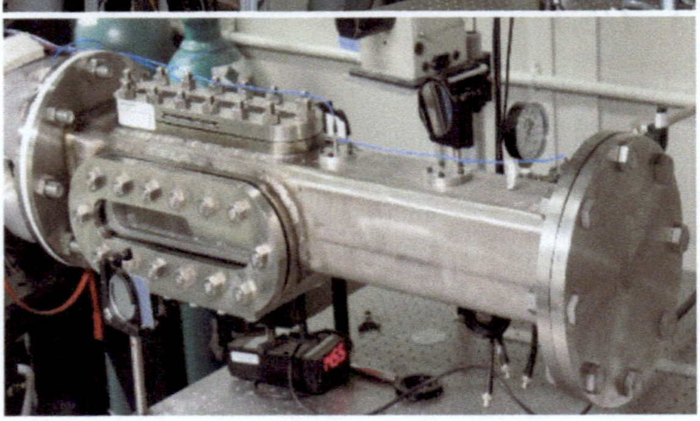

Aerospace Shock Tube
Length: ~10.8 m
Maximum Pressure: 30 atm
Driven section diameter: 16.2 cm
Uses: chemical kinetics; spectroscopy

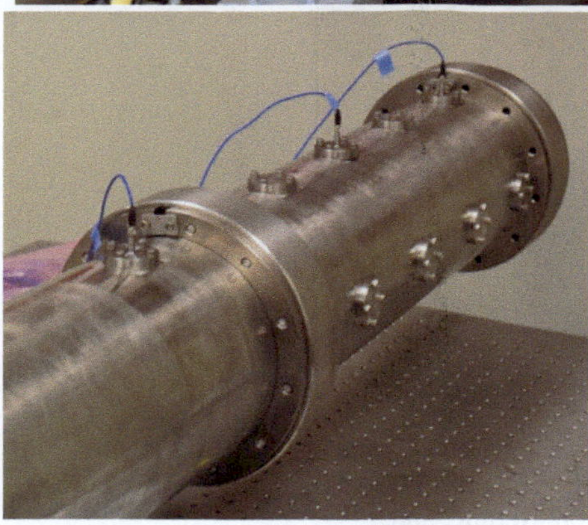

2.9 Georgia Institute of Technology, GA, USA

George W. Woodruff School of Mechanical Engineering
Devesh Ranjan
771 Ferst Drive
Atlanta, GA 30332, USA
devesh.ranjan@me.gatech.edu
http://me.gatech.edu/faculty/ranjan

Variable Inclination Shock Tube
Inclination angle: 0–90°.

Total Length: ~9 m
Shock Mach number: 1–3 in atmospheric air
Test section size: 114 mm × 114 mm (square)

The shock-tube facility at Georgia Tech is used to study turbulent mixing of gaseous interfaces with various initial conditions for reacting and non-reacting systems.

Facility, Diagnostics, Test Section Module, and Broken Diaphragm

Inclined shock tube (under construction phase) showing platforms used for changing diaphragm

3 Description of Hypersonic Gun Tunnel

3.1 Department of Aeronautics, Imperial College, London, UK

Richard Hillier, Paul Bruce
Exhibition Road
London, SW7 2AZ, UK
n.maccarthy@imperial.ac.uk
http://www.imperial.ac.uk/engineering/departments/aeronautics/

Hypersonic Gun Tunnel Data:

Test section size: 0.6 m × 1 m
Maximum model size: Slender 0.8 m length
Working gas: Nitrogen

Max speed: Made number 9
Reynolds Number: 7 – 47 × 10^6/m
Total pressure: 600 bar
Maximum temp: 1150 K
Run time: 20 ms Recharge time: 1 h.

MIX
Papier aus verantwortungsvollen Quellen
Paper from responsible sources
FSC® C105338

FSC
www.fsc.org

If you have any concerns about our products,
you can contact us on
ProductSafety@springernature.com

In case Publisher is established outside the EU,
the EU authorized representative is:
Springer Nature Customer Service Center GmbH
Europaplatz 3, 69115 Heidelberg, Germany

Printed by Libri Plureos GmbH
in Hamburg, Germany